T0338280

Transparent Oxide Electronics

Transparent Oxide Electronics

From Materials to Devices

Pedro Barquinha, Rodrigo Martins, Luis Pereira
and Elvira Fortunato

CENIMAT-I3N, Departamento de Ciência dos Materiais and CEMOP-UNINOVA, Faculdade de Ciências e Tecnologia, FCT, Universidade Nova de Lisboa

A John Wiley & Sons, Ltd., Publication

This edition first published 2012

Registered Office
John Wiley & Sons, Ltd, The Atrium, Southern Gate, Chichester, West Sussex, PO19 8SQ, United Kingdom

For details of our global editorial offices, for customer services and for information about how to apply for permission to reuse the copyright material in this book please see our website at www.wiley.com.

Library of Congress Cataloging-in-Publication Data

Transparent oxide electronics : from materials to devices / Pedro Barquinha ... [et al.].
p. cm.
Includes bibliographical references and index.
ISBN 978-0-470-68373-6 (cloth) – ISBN 978-1-119-96700-2 (ePDF) – ISBN 978-1-119-96699-9 (oBook) – ISBN 978-1-119-96774-3 (ePub) – ISBN 978-1-119-96775-0 (Mobi) 1. Semiconductor films. 2. Oxides–Electric properties. 3. Thin film transistors. 4. Transparent semiconductors. I. Barquinha, Pedro.
TK7871.15.F5T73 2012
621.3815′2–dc23

2011042396

A catalogue record for this book is available from the British Library.

HB ISBN: 9780470683736

Set in 10/12pt Times by SPi Publisher Services, Pondicherry, India
Printed and bound in Singapore by Markono Print Media Pte Ltd

Elvira Fortunato and Rodrigo Martins would like to dedicate this book to Catarina,
her lovely daughter, for her inestimable stimulus and understanding of our
scientific work, fully supported by the love that unites us!

Pedro Barquinha would like to dedicate this book to Martinho and Maria,
his parents, and Ana, his wife, for their love,
support and comprehension.

Luis Pereira would like to dedicate this book to his parents,
his wife and his little princess Inês.

Contents

The Impact of Oxides …

"The rapid development of semiconductor techniques is requiring an ever greater application of materials of varying physical and physic-chemical properties. This demand cannot be fully met by a handful of elements having semiconducting properties, nor by the few relevant chemical compounds, which are already understood. Therefore increasing attention is being paid to studies on less known chemical compounds able to act as semiconductors.

Among them, compounds of metals or semiconducting elements with oxygen may be considered as most promising."

Extract from the Introduction of the book: "Oxide Semiconductors", by Z.M. Jarzebski, Pergamon Press, 1973.

… and the physics behind field effect …

"The particular device … consists of a very thin layer of semiconductor placed on an insulating support. This layer of semiconductor constitutes one plate of a parallel-plate capacitor, the other being a metal plate in close proximity to it. If this capacitor is charged, with the metal plate positive, then the additional charge on the semiconductor will be represented by the increased number of electrons. … Consequently, the added electrons should be free to move and should contribute to the conductivity of the semiconductor. In this way, the conductivity in the semiconductor can be modulated, electronically, by a voltage put on the capacitor plate. Since this input signal requires no power if the dielectric is perfect, power gain will result."

Extract from the book: "Electrons and Holes in Semiconductors", by W. Shockley Van Nostrand Reinhold Company Inc., 1950.

Preface

As far as novel materials and devices are concerned, the scientific challenge today is to develop new green materials and technologies that allow both for new product concepts and applications and innovation in terms of physical performance. These materials should be possible to process with low cost and using non-fab conventional manufacturing technologies, away from the dominant field of silicon, the most important material used nowadays to produce electronic devices. These requirements are met by the novel multicomponent amorphous oxides, which can be used in a wide range of applications, as they offer exceptional electronic performance as active semiconductor components or can be tuned for applications where high transparency and electrical conductivity are demanded. Here, one of the most interesting features is related to transparency and the ability to deposit these oxides at low temperatures, giving rise to a plethora of new application opportunities in sectors where the discreetness of the devices as much as their optoelectronic functionality are critical. This is the case, for instance, with thin-film transistors (TFTs) that are fully based on oxides, to which the present authors have been contributing over recent years; or the production of paper electronics, a new area where the authors are world pioneers and which emerge as a key field for research application in the years to come, mainly for the so-called 100 % low cost green disposable electronics.

This book provides an overview of the world of transparent electronics, chiefly the processing of oxide semiconductors and their application to transparent TFTs, being essentially focused on the work developed over recent years in our laboratory. The book is organized into seven chapters as follows. Chapter 1 is a short introduction to transparent electronics and related (semi)conducting materials. Chapter 2 provides some fundamental background regarding the properties and applications of n-type oxide semiconductors (both binary and multicomponent). Special emphasis is then given to the effect of composition and (post-) processing parameters on the structural, electrical and optical properties of the materials. Even though most of the work presented is related to sputtered oxides, recent results obtained with solution-processed (spray pyrolysis and sol-gel spin-coating) oxides are also provided. Chapter 3 gives an overview of the state of the art concerning p-type transparent conductive oxides and describes two of the most promising p-type oxide semiconductors designed to be applied as channel layers of TFTs: copper oxide and tin monoxide. The structural, electrical and optical properties presented by these two materials are highlighted and our results with sputtered thin films are shown. Chapter 4 is devoted to low-temperature processed dielectrics, which are also crucial materials for transparent electronic devices. Besides presenting the generic material requirements that should be met in order to integrate low-temperature dielectrics with oxide semiconductors, results for sputtered tantalum- and hafnium-based oxides are presented and discussed, mostly regarding their structural and electrical properties. Special emphasis is given to amorphous multicomponent and multilayer structures based on the combination of high-κ and high-bandgap materials. Chapter 5 is dedicated to the world of n- and p-type TFTs. After an introduction, where a short historical background and a comparison between existing TFT technologies are provided, a detailed analysis of the devices integrating the oxide semiconductors and dielectrics explored in the previous chapters is presented. Besides the effect of the different compositions and (post-)processing parameters of the active layer, other important aspects are covered, such as contact-resistance analysis, effect of passivation layer and electrical stability. Most of the chapter is related to sputtered n-type oxide TFTs based on the gallium-indium-zinc oxide system; however, sputtered p-type and solution processed n-type oxide TFTs are also presented. Chapter 6 is devoted to the new concept of paper electronics, made possible with semiconductor materials that can be processed at low temperatures and that still exhibit remarkable electrical properties, such as multi-component amorphous oxides. Special emphasis is given to paper transistors, paper memories and paper batteries, with initial results for CMOS and sensor devices provided. The book concludes with Chapter 7, dedicated to the current and upcoming applications of oxide TFTs, for which transparent circuits including both NMOS and CMOS architectures, active matrices for displays and bio-sensors are highlighted.

When we agreed to write this book, our main purpose was to share our knowledge with the transparent electronics scientific community and simultaneously to attract newcomers by introducing them to this fascinating world. Besides that, we also expect to contribute with a pedagogic tool and a key element to be consulted when material or device concepts are needed, especially for students starting their university degrees.

Acknowledgments

First of all we would like to thank the European Research Council through the Advanced Grant (given to EF) under the project "INVISIBLE" (ERC-2008-AdG 228144) Advanced Amorphous Multicomponent Oxides for Transparent Electronics, directly related to this topic, as well as the support given by the European Commission and the partners involved in the following projects: Multiflexioxides (NMP3-CT-2006-032231), the first fully running project on Transparent Electronics in Europe, and more recently ORAMA (NMP3-LA-2010-246334) and POINTS (NMP3-SL-2011-263042), which are related to multifunctional oxide-based electronic materials and printable organic-inorganic transparent semiconductor devices, respectively. Considerable gratitude is also due to the EU project (262782-2 APPLE CP-TP), dealing with printed paper products for functional labels and electronics, as well as to the project SMART-EC (258203-ICT-2009.3.3) dealing with electrochromic oxides enabling technology applications.

We would like also to thank the fruitful collaborations done directly with companies and institutes such as the SAIT-SAMSUNG project "STABOXI", related to the passivation of a-GIZO TFTs, the Electronic and Telecommunications Research Institute of South Korea (ETRI) with the project IT R&D program MKE - 2006-S079-03, "Smart window with transparent electronic devices" and to Saint Gobain Recherche France, for the project related to oxides for glazing.

This work was partially funded by the Portuguese Science Foundation (FCT-MCTES) through a multiannual contract with I3N and with projects related to oxide semiconductors such as: POCI/CTM/55945/2004 (Development of transparent and conductive oxide semiconductors p-type: from synthesis to devices); POCI/CTM/55942/2004 (Transparent thin film transistors based on zinc oxide to be used in flexible displays); PTDC/CTM/73943/2006 (Multifunctional Oxides: a novel approach for low temperature integration of oxide semiconductors as active and passive thin films in the future generation of electronic systems); PTDC/EEA-ELC/64975/2006 (High mobility transparent amorphous oxide semiconductors thin film transistors for active matrix displays); PTDC/CTM/103465/2008 (Integrated memory paper using oxide based channel thin film transistors); PTDC/EEA-ELC/099490/2008 (From electronic paper to paper electronic – Paper_@); PTDC/CTM/099124/2008 (Electrochromic thin film transistors for smart windows applications); QREN Nº 3454 (New nanoxide composites for advanced fabrication of targets for passive and active Opto/Micro/Nano-electronics applications); CMU-PT/SIA/0005/2009 (Self-organizing power management for photo-voltaic power plants) and ERA-NET/0005/2009 (Multifunctional zinc oxide-based nanostructures: from materials to a new generation of devices).

We would also like to thank the Calouste Gulbenkian Foundation for the Stimulus to Research Award 2008, "Nanotransistors of oxide semiconductors" (given to PB) and the Luso-American Development Foundation for the project in 2005 "Low-temperature sputter deposition exploration/ optimization of multi-component, amorphous and nanostructure heavy metal cation oxides for TFT and TTFT channel layer applications" and also the several scholarships given to the authors and members of the research team.

The authors also wish to express their gratitude to past deans and especially the present dean of Faculdade de Ciências e Tecnologia da Universidade Nova de Lisboa, Prof. Fernando Santana, who ensured that we enjoyed the excellent working conditions that existed at CENIMAT (FCT-UNL) and CEMOP (UNINOVA), especially the well-equipped laboratories that made this work possible.

We would like to thank Sarah Tilley, Project Editor for Chemistry at John Wiley & Sons, Ltd for her guidance and help during the editorial process.

We finish by acknowledging those who made this work possible, especially the group Microelectronic and Optoelectronic Materials of CENIMAT and CEMOP for their hard work, unlimited patience, dedication, creativity, expertise, knowledge and "emotional support" when the obstacles we faced every day were not so "transparent". Without them this work would have been impossible.

The Microelectronic and Optoelectronic Materials Team (July 2011).

1 Introduction

1.1 Oxides and Transparent Electronics: Fundamental Research or Heading Towards Commercial Products?

Transparent electronics is emerging as one of the most promising technologies for future electronic products, as distinct from the traditional silicon technology. The fact that circuits based on conventional semiconductors such as silicon and conductors such as copper can be made transparent by using different materials, the so-called transparent semiconducting and conducting oxides (TSOs and TCOs, respectively), is of great importance and allows for the definition of innovative fields of application with high added value. The viability of this technology depends to a large extent on the performance, reproducibility, reliability and cost of the transparent transistors. Transistors are the key components in most modern electronic circuits, and are commonly used to amplify or to switch electronic analog and digital signals. Besides the high-performance silicon transistors used in microprocessors or amplifiers, designated by metal-oxide-semiconductor field-effect transistors (MOSFETs) and requiring processing temperatures exceeding 1000°C, other types of transistors are available for large area electronics, where lower temperatures and costs are required. Perhaps the most relevant are the thin-film transistors (TFTs), which are intimately associated with liquid crystal displays (LCDs), where they allow one to switch each pixel of an image *on* or *off* independently.

The most immediate demonstration of transparent electronics would be the realization of a transparent display, something that has been envisaged for a long time, at least from the 1930s when H.G. Wells imagined it in his science fiction novel *The Shape of Things to Come* (Figure 1.1a, see color plate section). Nowadays, with the advent of TSOs and TCOs,

---

Transparent Oxide Electronics: From Materials to Devices, First Edition. Pedro Barquinha, Rodrigo Martins, Luis Pereira and Elvira Fortunato.

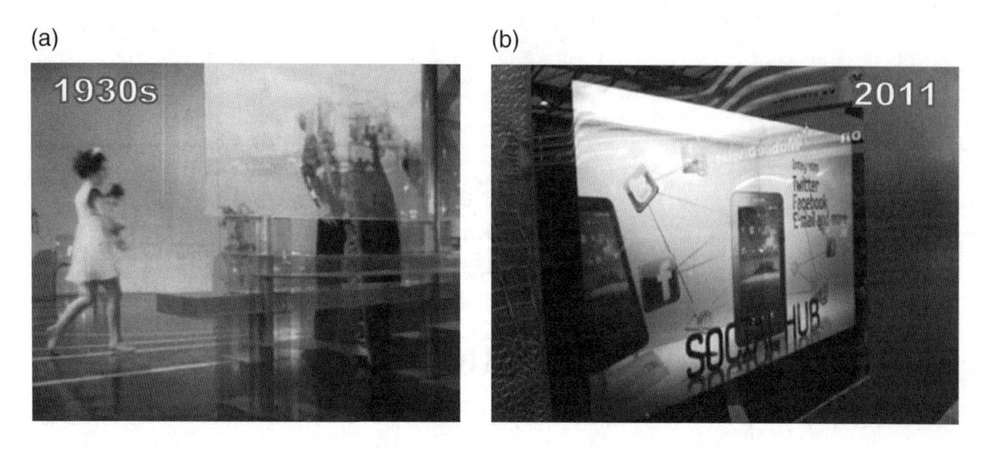

Figure 1.1 *Transparent displays: a) early vision, in H.G. Wells' 1930s novel* The Shape of Things to Come *[1]; b) Samsung's 22" transparent LCD panel now being mass-produced in 2011 [2]. Reproduced from [2] Copyright (2011) Samsung Corp.*

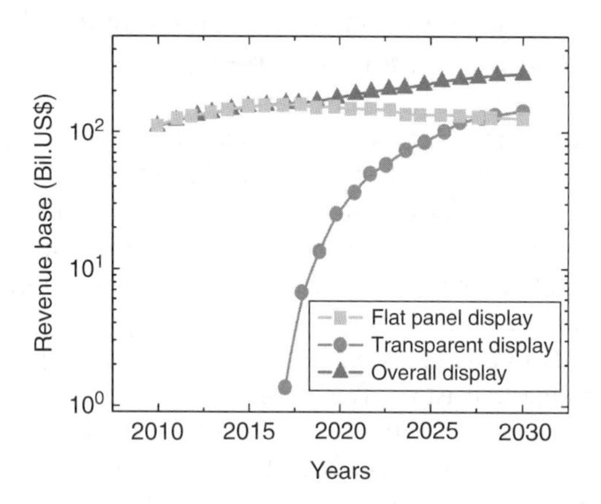

Figure 1.2 *Transparent display technology evolution and global display market. Adapted with permission from [3] Copyright (2011) DisplayBank.*

which besides transparency also allow for low temperature, low processing costs and high performance, transparent displays have become truly conceivable. In fact, even if much fundamental research will continue to be required so as to understand all the peculiarities of these materials and improve their performance and stability, the first commercial products within the transparent electronics concept have already started to be mass produced, such as the 22-inch transparent LCD panels by Samsung, in March 2011 (Figure 1.1b).

The market for transparent displays is emerging now and its future looks quite promising, as revealed by the "Transparent Display Technology and Market Forecast" report by Displaybank, which predicts a $ 87.2 biliion market by the year 2025 (Figure 1.2, see color plate section) [3].

1.2 The Need for Transparent (Semi)Conductors

Materials exhibiting both high optical transparency in the visible range of the electromagnetic spectrum and high electrical conductivity (σ) are not common when considering the categories of conventional materials, such as metals, polymers and ceramics. For instance, metals are generally characterized by having a high σ but being opaque, while ceramics are seen as electrical insulating materials which due to their typically large bandgap (E_G) can be optically transparent. However, certain ceramic materials can simultaneously fulfill the requirements of high σ and optical transparency: these are designated by transparent conducting oxides (TCOs), where typically the main free carriers are electrons (n-type materials) [4]. Physically, this can be achieved if the ceramic material has $E_G > \approx 3$ eV, a free carrier concentration (N) above $\approx 10^{19-20}$ cm^{-3} and a mobility (μ) larger than ≈ 1 cm^2 V^{-1} s^{-1}, which can be verified for metallic oxides such as ZnO, In_2O_3 and SnO_2 [5]. Due to the relatively low μ of TCOs when compared with classical semiconductors such as single crystalline silicon, which has $\mu > 400$ cm^2 V^{-1} s^{-1}, TCOs generally need to be degenerately doped if a high σ is envisaged. As with silicon, doping can be achieved by the introduction of extrinsic substitutional elements in the host crystal structure, such as elements with different valences that are introduced in the cationic sites [4, 6, 7]. Doping can also be achieved by intrinsic structural defects, such as oxygen vacancies and/or metal interstitials. This structural imperfection, or in other words the deviation from stoichiometry, which always occurs when TCOs are deposited, is the fundamental reason behind the electrical conduction of these materials: to maintain charge neutrality, in n-type (p-type) materials, the defects give rise to electrons (holes) that depending on the defects' energy levels within the E_G of the oxide can be available for the conduction process, increasing N and consequently σ [6].

The effect of oxygen defining the final properties of materials was readily observed in the early days of this research area. In fact, the first reported TCO, by Badeker in 1907, was obtained after exposing an evaporated cadmium film to an oxidizing atmosphere: the resulting material, CdO, was transparent but maintained a reasonably high σ, resembling a metal [8]. In the 1920–1930s, Cu_2O and ZnO were also investigated and researchers found experimentally that a large range of σ, exceeding six orders of magnitude, could be obtained by changing the oxygen partial pressure [9–13]. Oxygen concentration can have more implications than simply changing N and σ. As an example, in tin oxide it is reported that a large oxygen deficiency leads to the change of the tin oxidation state from +4 to +2, i.e., SnO_2 is transformed into SnO. This can totally change the electrical properties of the resulting material: for instance, as with most of the TCOs, SnO_2 is an n-type semiconductor, while SnO can present p-type behavior [4, 14].

However, even if σ can be significantly modulated by the concentration of intrinsic defects, with regard to the objective of obtaining a TCO with a high σ, extrinsic doping has to be used, with aluminum-doped zinc oxide (AZO) or tin-doped indium oxide (ITO) constituting some of the most well-known examples of these n-type materials. Even if optimal doped TCOs present σ values ($\approx 10^4$ Ω^{-1} cm^{-1} [15]) which are almost two orders of magnitude lower than those typically obtained in the cooper metal used in integrated circuits, this level of σ signifies that appreciable electrical conduction can be achieved in TCOs, allowing one to target a large range of applications, as will be shown below.

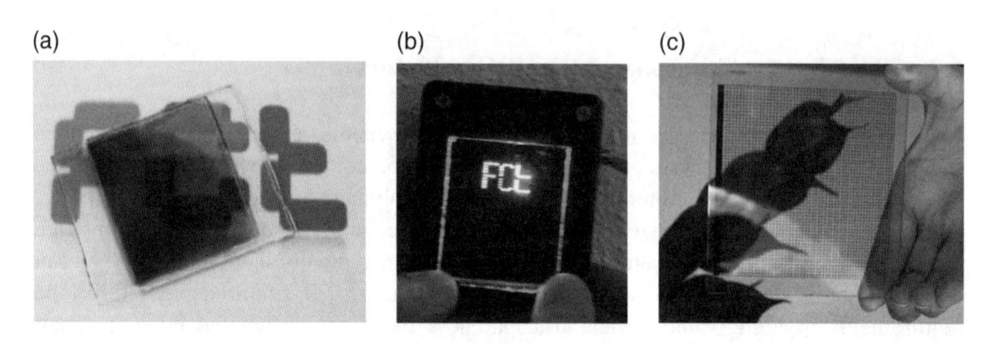

Figure 1.3 Some applications of TCOs at CENIMAT: a) electrochromic windows; b) passive matrix LED display; c) see-through solar cell.

Although work such as that of Badeker was based essentially on pure scientific interest, the continuous advances in the understanding of solid state physics and of processing and characterization tools that occurred during the first half of the 20th century allowed for substantial technological progress in TCOs research. This resulted in the improvement of the properties of materials and soon a large range of applications for them began to be envisaged. The first large-scale use of TCOs occurred during World War II, when antimony-doped tin oxide (SnO_2:Sb or ATO) was deposited by spray pyrolysis to be used as a transparent defroster for aircraft windshields [16]. During recent decades, making use of optimized TCO properties such as high σ, high transparency in the visible range, high reflectivity in the infrared, high mechanical hardness or high sensitivity to gas pressure, these materials have been extensively used as transparent electrodes in solar cells, liquid crystal displays (LCDs) and electrochromic windows, heating stages for optical microscopes, transparent heat reflectors in windows, abrasion and corrosion-resistant coatings, antistatic surface layers on temperature control coatings in orbiting satellites, gas sensors, among many other applications [4, 5]. Some examples of these applications are depicted in Figure 1.3 (see colour plate section).

In all of the electrical applications mentioned above, the TCO is an electrically passive element, i.e. it works as an electrode. Hence, with regard to electrical properties, most optimization efforts are focused on achieving the maximum possible σ, which requires a large N. However, a new class of applications requiring TCOs with considerably different electrical properties has recently emerged. In fact, the idea of producing ultra-violet (UV) detectors and diodes or even fully transparent TFTs requires the N of TCOs to be substantially decreased, in order to be able to use them as proper semiconductors, i.e. as active elements in devices [5, 17]. For instance, note that the usage of a TCO with a large N as the active layer of a TFT would result in a useless device, because the semiconductor could not be fully depleted, hence it would not be possible to switch-off the TFT. To distinguish these transparent oxides from the highly conducting TCOs, the low σ and N materials can be designated by transparent semiconducting oxides (TSOs). The properties tuning of TSOs can be made using the same principles as those discussed above for TCOs, i.e. either by intrinsic or extrinsic doping. For instance, larger oxygen concentrations during deposition should result in fewer oxygen vacancies, hence less free electrons in an n-type TSO, while extrinsic doping with elements that introduce acceptor-like levels and/or that increase E_G can also lead to similar results [6, 18].

Most of the TCOs and TSOs studied so far are n-type. However, p-type oxides are needed to extend the possibilities of transparent electronics, for instance by making possible the fabrication of complementary logic circuits. Besides the early experiments performed with poor-transparency Cu_2O in the early 1930s, the first reported p-type oxide was NiO, in 1993 [19]. Although p-type conduction was achieved, poor average visible transmittance (AVT) of 40 % and low σ ($\approx 7\,\Omega^{-1}\,cm^{-1}$) were obtained. In 1997 Kawazoe *et al.* presented a strategy for identifying oxides combining p-type conductivity with good optical transparency [20]. The authors suggested that the candidate materials should have tetrahedral coordination, with cations having a closed shell with comparable energy to those of the *2p* levels of oxygen anions, and the dimension of crosslinking of cations should be reduced. They selected $CuAlO_2$ to demonstrate the concept, and p-type conduction and reasonable transparency could in fact be achieved. The paper published by Kawazoe *et al.* had a significant impact on the research of p-type oxides, with various work being reported during the following years based on similar theoretical principles, mostly employing delafossite structure materials such as $SrCu_2O_2$ or $CuGaO_2$ [21–23]. Although the maximum σ and μ achieved with these p-type oxides are at present three to four (σ) and one to two (μ) orders of magnitude lower than the ones of optimized n-type TCOs, the achieved values begin to be suitable for their application as TSOs. Given this, different transparent optoelectronic devices employing both p- and n-type TSOs have been demonstrated, such as near-UV-emitting diodes composed of heteroepitaxially grown TSOs (p-type $SrCu_2O_2$ and n-type ZnO) [24] and UV-detectors composed of single crystalline p-type NiO and n-type ZnO [25]. However, to achieve reasonable optical and electrical properties, p-type TSOs generally require larger processing temperatures than n-type oxides, and significant research is still needed in order to surpass the temperature and performance limitations of these materials so as to fabricate transparent p-type materials compatible with the low temperature processed n-type TSOs. However, it will be shown in Chapter 5 that recent research developed at CENIMAT already allows one to obtain good performance p-type oxide TFTs with a maximum processing temperature of 200°C.

1.3 Reaching Full Transparency: Dielectrics and Substrates

To reach the target of fully transparent devices, oxides with very large electrical resistivity ($\rho > 10^{10}\ \Omega$ cm) are also required. In a transparent TFT, for instance, these oxides work as dielectric layers, insulating electrically the gate electrode from the semiconductor. The choice of the appropriate dielectric comprehends both physical requirements, such as the band offsets with the semiconductor and the level of leakage current allowable, as well as process related ones, such as compatibility with the remaining device materials in terms of deposition temperature or etching selectivity. Higher dielectric constant (κ) allows one to preserve a high capacitance with thicker dielectrics, which is especially relevant for low-temperature processed thin films, where leakage currents are normally higher. Moreover, the surface of the dielectric should be highly smooth and the material should have an amorphous structure, since high roughness and polycrystalline structure lead to increased interface defects and grain boundaries can act as paths for carrier flow, increasing leakage current and leading eventually to the dielectrics' breakdown. Generally, dielectrics with high-κ exhibit low-E_G and vice-versa [26]. Hence, to obtain a better match between the desirable structural and electrical properties, amorphous multicomponent dielectrics based on mixtures of high-κ materials, such as Ta_2O_5 or HfO_2, with high-E_G materials, such as SiO_2 or Al_2O_3, are proposed by the authors [27, 28].

Finally, substrates also have to be considered and oxides are again a solution, as glass is certainly the most versatile rigid substrate for transparent electronics, combining important properties such as low cost, smooth surface and ability for large area deposition. Moreover, if flexibility is required, polymers or even paper based on nanofibrills should be used.

References

[1] http://www.technovelgy.com.
[2] http://www.gizmag.com/samsungs-transparent-lcd-display/18283/picture/132495/.
[3] DisplayBank, "Transparent Display Technology and Market Forecast," 2011.
[4] H. Hartnagel, A. Dawar, A. Jain, and C. Jagadish (1995) *Semiconducting Transparent Thin Films*. Bristol: IOP Publishing.
[5] J.F. Wager, D.A. Keszler, and R.E. Presley (2008) *Transparent Electronics*. New York: Springer.
[6] D.P. Norton, Y.W. Heo, M.P. Ivill, K. Ip, S.J. Pearton, M.F. Chisholm, and T. Steiner (2004) ZnO: growth, doping and processing, *Materials Today* **7**, 34–40.
[7] H.Q. Chiang (2007) "Development of Oxide Semiconductors: Materials, Devices, and Integration," in *Electrical and Computer Engineering*. vol. PhD thesis Oregon: Oregon State University.
[8] K. Bädeker (1907) Über die elektrische Leitfähigkeit und die thermoelektrische Kraft einiger Schwermetallverbindungen, *Annalen der Physik* **327**, 749–66.
[9] B. Gudden (1924) Elektrizitatsleitung in kristallisierten stoffen unter ausschluss der metalle ergeb, *Exakten Naturwiss* **3**, 116–59.
[10] H. Dunwald and C. Wagner (1933) Tests on the appearances of irregularities in copper oxidule and its influence on electrical characteristics, *Z. Phys. Chem. B-Chem. Elem. Aufbau. Mater*. **22**, 212–25.
[11] H.H. von Baumbach, H. Dunwald, and C. Wagner (1933) Conduction measurements on copper oxide, *Z. Phys. Chem. B-Chem. Elem. Aufbau. Mater*. **22**, 226–30.
[12] C. Wagner (1933) Theory of ordered mixture phases. III. Appearances of irregularity in polar compounds as a basis for ion conduction and electron conduction, *Z. Phys. Chem. B-Chem. Elem. Aufbau. Mater*. **22**, 181–94.
[13] H.H. von Baumbach and C. Wagner (1933) Die elektrische leitfahigkeit von Zinkoxyd und Cadmiumoxyd, *Z. Phys. Chem. B* **22**, 199–211.
[14] Y. Ogo, H. Hiramatsu, K. Nomura, H. Yanagi, T. Kamiya, M. Hirano, and H. Hosono (2008) p-channel thin-film transistor using p-type oxide semiconductor, SnO, *Applied Physics Letters* **93**, 032113-1–032113-3.
[15] T. Minami (2005) Transparent conducting oxide semiconductors for transparent electrodes, *Semicond. Sci. Technol*. **20**, S35–S44.
[16] R.G. Gordon (2000) Criteria for choosing transparent conductors, *MRS Bull*. **25**, 52–7.
[17] H. Ohta and H. Hosono (2004) Transparent oxide optoelectronics, *Materials Today* **7**, 42–51.
[18] Y. Kwon, Y. Li, Y. W. Heo, M. Jones, P.H. Holloway, D.P. Norton, Z.V. Park, and S. Li (2004) Enhancement-mode thin-film field-effect transistor using phosphorus-doped (Zn,Mg)O channel, *Applied Physics Letters* **84**, 2685–7.
[19] H. Sato, T. Minami, S. Takata, and T. Yamada (1993) Transparent conducting p-type NiO thin films prepared by magnetron sputtering, *Thin Solid Films* **236**, 27–31.
[20] H. Kawazoe, M. Yasukawa, H. Hyodo, M. Kurita, H. Yanagi, and H. Hosono (1997) P-type electrical conduction in transparent thin films of $CuAlO_2$, *Nature* **389**, 939–42.
[21] J. Tate, M.K. Jayaraj, A.D. Draeseke, T. Ulbrich, A.W. Sleight, K.A. Vanaja, R. Nagarajan, J.F. Wager, and R.L. Hoffman (2002) p-type oxides for use in transparent diodes, *Thin Solid Films* **411**, 119–24.

[22] K. Ueda, T. Hase, H. Yanagi, H. Kawazoe, H. Hosono, H. Ohta, M. Orita, and M. Hirano (2001) Epitaxial growth of transparent p-type conducting $CuGaO_2$ thin films on sapphire (001) substrates by pulsed laser deposition, *Journal of Applied Physics* **89**, 1790–3.

[23] A. Kudo, H. Yanagi, H. Hosono, and H. Kawazoe (1998) $SrCu_2O_2$: A p-type conductive oxide with wide band gap, *Applied Physics Letters* **73**, 220–2.

[24] H. Ohta, K. Kawamura, M. Orita, M. Hirano, N. Sarukura, and H. Hosono (2000) Current injection emission from a transparent p-n junction composed of p-$SrCu_2O_2$/n-ZnO, *Applied Physics Letters* **77**, 475–7.

[25] H. Ohta, M. Hirano, K. Nakahara, H. Maruta, T. Tanabe, M. Kamiya, T. Kamiya, and H. Hosono (2003) Fabrication and photoresponse of a pn-heterojunction diode composed of transparent oxide semiconductors, p-NiO and n-ZnO, *Applied Physics Letters* **83**, 1029–31.

[26] J. Robertson (2002) Electronic structure and band offsets of high-dielectric-constant gate oxides, *MRS Bull.* **27**, 217–21.

[27] P. Barquinha, L. Pereira, G. Goncalves, R. Martins, D. Kuscer, M. Kosec, and E. Fortunato (2009) Performance and Stability of Low Temperature Transparent Thin-Film Transistors Using Amorphous Multicomponent Dielectrics, *J. Electrochem. Soc.* **156**, H824–H831.

[28] L. Pereira, P. Barquinha, G. Goncalves, A. Vila, A. Olziersky, J. Morante, E. Fortunato, and R. Martins (2009) Sputtered multicomponent amorphous dielectrics for transparent electronics, *Phys. Status Solidi A-Appl. Mat.* **206**, 2149–54.

2

N-type Transparent Semiconducting Oxides

2.1 Introduction: Binary and Multicomponent Oxides

Over recent years, intense research has been carried out on n-type transparent conducting and semiconducting oxides (TCOs and TSOs, respectively). Whether extrinsically doped or not, zinc oxide and indium oxide have been two of the most commonly used binary compounds in transparent electronics.[a] More recently, ternary and quaternary compounds such as indium-zinc oxide or gallium-indium-zinc oxide have also started to be explored, offering the opportunity to combine low processing temperatures, amorphous structures and remarkable optical and electrical performance. Far from pretending to be an exhaustive review of the literature on this topic, the following pages provide a brief glance of the generic properties of n-type TCOs and TSOs, from binary to quaternary compounds.

2.1.1 Binary Compounds: the Examples of Zinc Oxide and Indium Oxide

ZnO is perhaps the metal oxide with the largest field of application. Some examples are rubber manufacture, the concrete industry, the medical industry, cigarette filters, food additives, pigments in paints, different types of coatings, Amongst many others [1, 2]. ZnO-based varistors have also been well known for a long time [3] and ZnO has great potential to be used in other applications such as UV light emitters, spin functional devices, gas sensors, surface acoustic wave guides or as a transparent conductor and semiconductor material in the emerging field of transparent electronics [4].

[a] Tin oxide is another core binary compound in transparent electronics. Since it can also exhibit p-type conduction, its properties are reviewed briefly in Chapter 3.

Transparent Oxide Electronics: From Materials to Devices, First Edition. Pedro Barquinha, Rodrigo Martins, Luis Pereira and Elvira Fortunato.

ZnO crystallizes with a hexagonal wurtzite structure (Figure 2.1a, see color plate section), with lattice constants of a=3.24Å and c=5.19Å [5]. ZnO exhibits a direct bandgap (E_G) of 3.2–3.4eV, which can be tuned by substitutional doping on the cation site, for instance with cadmium or magnesium so as to decrease or increase E_G, respectively [4]. When degenerately doped, the bandgap of ZnO can also be broadened due to the Burstein-Moss shift, since the lowest energy states above conduction band minimum (CBM) are already occupied and absorption can only occur for higher energy states as the free carrier concentration (N) increases [5, 6]. The intrinsic defects that are mostly considered with ZnO are oxygen vacancies, interstitial zinc and interstitial hydrogen. From these, oxygen vacancies constituting defect levels lying approximately 0.01–0.05eV below CBM (within E_G) are the most relevant for n-type conduction, being that its concentration is quite similar to the N observed in single crystals [4, 7]. However, neutral oxygen vacancies can also create deep defect levels that can trap electrons and are responsible for phenomena such as persistent photoconductivity [4, 7, 8].

ZnO thin films have been produced using a large variety of techniques, such as sputtering [9–11], pulsed laser deposition (PLD) [12, 13], evaporation [14], chemical vapor deposition (CVD) [15], spray pyrolysis [16], sol-gel [17] and ink-jet [18], amongst others. Although the obtained properties are highly dependent not only on the technique but also on the processing parameters (see, e.g., [10]), thin films suitable for a large range of applications can be prepared even with room temperature processing. However, even at low processing temperatures and regardless of the deposition technique and deposition parameters used, ZnO films always tend to exhibit a polycrystalline structure, which can have a deleterious effect on the carrier transport properties and inhibit large area applications due to the lack of uniformity and reproducibility of such structures.

Although works exist regarding tentative p-type doping in ZnO, using dopants that introduce deep acceptor levels in ZnO, such as nitrogen or phosphorous, stable and reproducible properties are difficult to achieve [19]. This arises as a consequence of self-compensation mechanisms, i.e., of the redistribution of electronic state occupancy due to self-creation of an intrinsic defect that counterbalances the effect of the intentionally introduced acceptor level, in order to reduce the overall energy of the system [8, 20].

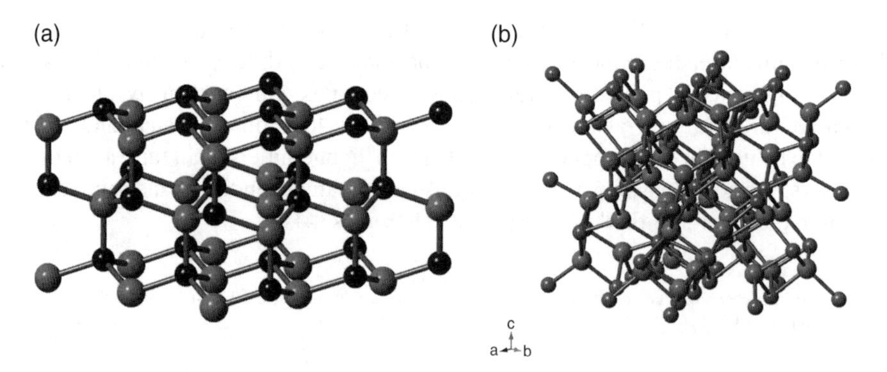

Figure 2.1 *Crystalline structure commonly adopted by a) ZnO, hexagonal (wurtzite) and b) In_2O_3, cubic (bixbyite). The small spheres represent the metallic cations, while the large spheres represent the oxygen anions.*

Regarding In_2O_3, this crystallizes according to the cubic structure of the mineral bixbyite (Figure 2.1b), with a lattice parameter of 10.12Å [5]. Although amorphous thin films can be obtained when deposited at very low temperatures (depending on the processing conditions), they readily crystallize under the cubic system mentioned above when subjected to temperatures of around 150°C [21–24]. In In_2O_3 light is absorbed by both indirect and direct interband transitions, which correspond to E_G around 2.7 and 3.5–3.7 eV, respectively [5, 9]. Bandgap widening due to Burstein-Moss shift is also extremely relevant for degenerately doped In_2O_3, with shifts larger than 0.6 eV being verified with the increase of N when the Fermi level (E_F) is above CBM [22, 25]. As for ZnO, oxygen vacancies are also assumed to be the main sources of the shallow donor levels that yield the characteristic n-type conduction to In_2O_3. These donor levels are generally very close to CBM, in the range of 0.008–0.03 eV, depending on the donor concentration, with degeneracy beginning at a donor density of 1.48×10^{18} cm^{-3} [5].

Similar techniques to those mentioned for ZnO can be used to process In_2O_3 thin films, some of them even at room temperature [22, 26–32]. Concerning applications, the most relevant is certainly the usage of tin-doped In_2O_3 (ITO) as a transparent electrode [33, 34]. This arises as a consequence of the very high electrical conductivity (σ) that is possible to achieve with sputtered ITO, in some cases above 1×10^4 Ω^{-1} cm^{-1}. However, the recent developments in aluminum- or gallium-doped zinc oxide (AZO or GZO, respectively) allow one to obtain comparable TCO performance for ZnO-based TCOs, using similar deposition techniques, even with room temperature processing [33, 35–37]. This is highly important because zinc abundance in the Earth's crust is more than two orders of magnitude larger than indium (132 and 0.1 ppm, respectively [9]), which results in a higher cost for indium-based materials.

As happens with the other widely studied n-type oxide, SnO_2, both ZnO and In_2O_3 are composed by metallic cations and oxide anions with ns^0 and $2s^22p^6$ valence electron configurations, respectively, with $n=4$ for ZnO and $n=5$ for In_2O_3 (and SnO_2). The empty metallic s-orbitals constitute the CBM, while the valence band maximum (VBM) is composed by the filled oxygen $2p$-orbitals. If these materials were stoichiometric, E_F would be in the middle of E_G, but the intrinsic and/or extrinsic defects take E_F near or within the conduction band. The nature of the CBM, derived primarily from large radii and spherical s-orbitals, allows one to have a good pathway for electron transport, since the orbitals of neighboring cations can easily overlap [38].

Table 2.1 presents a comparison between the typical electrical and optical properties found in ZnO and In_2O_3 thin films, with SnO_2 being also included for reference.

Table 2.1 *Typical optical and electrical properties found for thin films of ZnO, In_2O_3 and SnO_2, measured at 300 K. μ values in parentheses correspond to the typical μ obtained in single crystals.*

	E_G (eV)	σ (Ω cm)$^{-1}$	N (cm^{-3})	μ (cm^2 V^{-1} s^{-1})	References
ZnO	3.2–3.4 (dir)	$<10^4$	$<10^{21}$	5–50 (200–400)	[7–9, 39–41]
In_2O_3	2.7 (ind), 3.5–3.7 (ind)	$<10^4$	$<10^{21}$	10–50 (160)	[7–9, 39, 42]
SnO_2	3.6–4.3 (dir)	$<10^3$	$<10^{20}$	5–30 (240)	[8, 9, 39, 43]

For all of the materials average transmittance in the visible range (AVT) around 80–90% and E_G above 3 eV are achieved, in agreement to the requisites of a transparent material. In single crystals, mobility (μ) is of the same order of magnitude for all the oxides.[b] Even if In_2O_3 has larger *5s*-orbitals that provide better overlapping than the *4s*-orbitals of ZnO, the higher metal-atom number densities and shorter metal-metal distances of the latter tend to equilibrate the overall electrical properties achieved in both single crystalline semiconductors [7].[c] It is also observed that μ measured in single crystals is considerably higher than that obtained in deposited thin films. This would be expected, since thin films of these oxides generally exhibit a polycrystalline structure, regardless of the deposition technique and processing conditions used to fabricate them. As such, grain boundary scattering limits carrier transport considerably, reducing μ [9, 45]. In addition, note that in deposited thin films N is generally much larger than in undoped single crystals, due to the intrinsic defects created during deposition that can act as shallow donor levels. As shown by Ellmer [9], if single crystals of ZnO are intentionally doped to achieve $N>10^{20}$ cm^{-3}, μ in single crystals and polycrystalline thin films with the same N are quite similar, because at this N range μ is essentially controlled by ionized impurity scattering regardless of the material structure. Given this background, it could be plausible to assume that by tuning the deposition conditions of thin films in order to achieve a very low N, a large μ could be obtained, which would represent an ideal condition for a TSO to be employed as a channel layer in a TFT. However, this trend is not observed for polycrystalline oxide semiconductors, since for very low N the energy associated with the grain boundaries is too high, so electrons cannot surmount them, i.e., carrier transport starts to be dominated by the energy barriers at the grain boundaries, which can only be surpassed if a larger N is used [7]. Hence, even if ZnO and In_2O_3 have different structural properties, their electrical properties at the N ranges of interest for TCOs and TSOs are essentially controlled by the same mechanisms, ionized impurity scattering and grain barrier inhibited transport, respectively. This way, the typical electrical properties of both ZnO and In_2O_3 polycrystalline thin films can be considered, in general, to be quite similar.

2.1.2 Ternary and Quaternary Compounds: the Examples of Indium-Zinc Oxide and Gallium-Indium-Zinc Oxide

Even if the TCOs and TSOs mentioned above are innovative materials when compared with covalent semiconductors, allowing one to explore some unique applications, they always present a polycrystalline structure. In addition, although results in the laboratory show that it is possible to obtain good properties with low processing temperatures [34, 36, 37], most of the commercial applications of these materials rely on high (post-)deposition temperatures in order to achieve optimal performance. This is particularly relevant for ITO, which is currently one of the most widely used TCOs, for which temperatures above 200–300°C are typically used for commercial applications. Higher temperatures directly affect the cost and time required to process the materials, besides limiting the type of

---

[b] The highest deviation occurs for single crystalline ZnO, where a large $\mu \approx 400$ cm^2 V^{-1} s^{-1} was reported by Tsukazaki *et al.* [41]. To achieve this, the authors use a $Mg_xZn_{1-x}O$ buffer layer to reduce structural defects arising from lattice mismatch and chemical dissimilarity to substrate materials.

[c] Even if SnO_2 is not explored here as a n-type TSO, note that the analysis of its electrical properties is more complicated than for ZnO and In_2O_3, because the difference in the thermodynamic stability of the Sn^{2+} and Sn^{4+} is very low and so highly resistive SnO phases may be formed in this material, for instance when too many oxygen vacancies are created [5, 44].

substrates that are possible to use. Furthermore, due to the polycrystalline structure, carrier transport for oxides with low N (TSOs) is severely limited by grain boundary effects, as explained before. Besides this, polycrystalline materials are not desirable for large area applications, since it is hard to assure uniform and reproducible grain distribution and size, a problem that is well known in silicon technology [46]. Hydrogenated amorphous silicon (a-Si:H) is perhaps the best example of the successful implementation of an amorphous semiconductor in history, even if the electrical properties are considerably worse than those of polycrystalline silicon (poly-Si). Regarding oxide semiconductors, the first report on an amorphous material dates from the 1950s, when Denton *et al.* showed that glasses containing a large amount of V_2O_5 could present some electrical conductivity [47]. Several subsequent works followed the same theoretical principles, by employing different variable-valence transition metal oxides, but since carrier transport was dominated by a variable-range hopping (VRH) mechanism, the resulting μ was rather low, around 10^{-4} cm^2 V^{-1} s^{-1} [38].

As stated before, In_2O_3 thin films can present an amorphous structure, but only when produced at low temperature and under a very narrow range of processing conditions. But contrary to what happens with silicon, the electrical properties of polycrystalline and amorphous In_2O_3 films are quite similar. In fact, this was observed by Bellingham *et al.* in 1990 for films with $N>10^{20}$ cm^{-3} [22]. For this N range, the authors found that carrier transport was essentially dominated by ionized impurity scattering both for polycrystalline and amorphous films and the structural disorder of the latter had a negligible effect on the electrical properties.

The material design concept introduced in 1995–96 by Hosono and co-workers revolutionized the field of amorphous oxide semiconductors: the authors proposed that multicomponent oxides composed of post-transition cations with a $(n-1)d^{10}s^0$ electronic configuration are amorphous and present similar μ to the polycrystalline materials in the degenerated state, with values around 10 cm^2 V^{-1} s^{-1}. This was observed experimentally with various materials such as Cd_2GeO_4 implanted with H^+ or Li^+ ions, $AgSbO_3$ and Cd_2PbO_4 [48–51]. Another striking feature of these materials is that they remain amorphous even when annealed at temperatures of 500°C. Moreover, it was shown that free carriers could be generated either by ion implantation or by oxygen desorption after annealing (always preserving the amorphous structure), transforming highly resistive films with activation energies above 1 eV into highly conducting films with negligible activation energy, meaning that E_F could be taken from a deep bandgap to above CBM.

The primary reason for the excellent properties of these amorphous multicomponent oxide semiconductors can be understood by analyzing the differences in the composition of the CBM between covalent (silicon) and ionic (oxide) semiconductors (Figure 2.2, see color plate section). In crystalline silicon, CBM is composed primarily of strongly directive and anisotropic sp^3 orbitals (Figure 2.2a), hence, when moving to an amorphous silicon structure there are significant changes in the bond angles (Figure 2.2b), creating a very large concentration of localized states with energy levels inside the bandgap. This results in severely degraded carrier transport in the amorphous state, which starts to be controlled essentially by hopping between localized tail-states, with band conduction never being achieved. A totally different situation is verified for oxide semiconductors: in this case, CBM is composed by the large spherical isotropic *ns* orbitals of the metallic cations (Figure 2.2c). If the radii of these orbitals is made larger than the inter-cation distance,

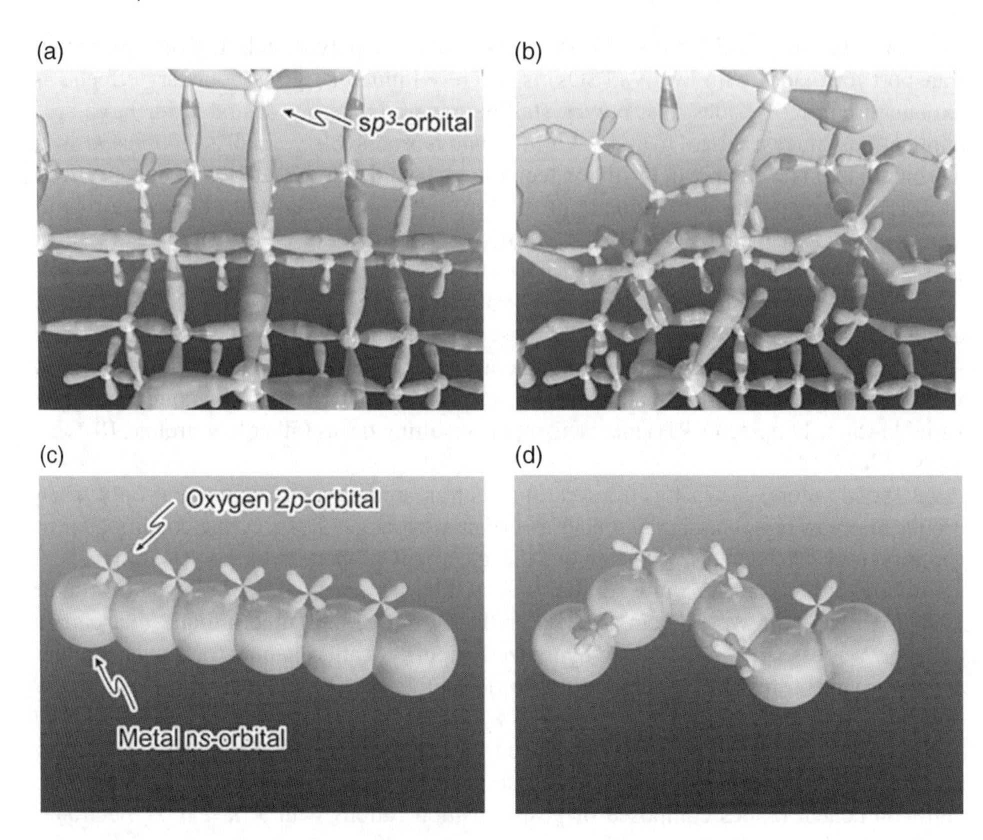

Figure 2.2 Schematics proposed by Nomura et al. of the orbitals composing the CBM on covalent semiconductors with sp^3 orbitals and ionic semiconductors with ns orbitals ($n \geq 4$): a) covalent crystalline; b) covalent amorphous; c) ionic crystalline; d) ionic amorphous. Reproduced with permission from [52] Copyright (2004) Macmillan Publishing Ltd.

which can be achieved for $n > 4$, the neighboring orbitals always overlap, despite the degree of disorder of the material (Figure 2.2d). This means that even in the amorphous state, oxide semiconductors always have a well defined carrier path in the CBM and large μ can be achieved [38, 52].

Despite the novel and exciting properties exhibited by the initial multicomponent oxide semiconductors, they had somewhat limited capabilities because in some cases good σ could only be achieved after ion implantation, while in others the proposed materials were composed by multivalent ions that during the change of their oxidation state (for instance, from Pb^{4+} to Pb^{2+}) consumed a large fraction of the electrons generated via the formation of oxygen vacancies [50]. Hopefully, a large range of elements in the periodic table exhibit the $(n-1)d^{10}s^0$ electronic configuration required to obtain an amorphous semiconductor according to this model, including zinc, indium and gallium. These constitute the most widely explored cations for amorphous multicomponent oxide semiconductor fabrication, in the form of indium-zinc oxide (IZO) and gallium-indium-zinc oxide (GIZO, Figure 2.3, see color plate section). In IZO and GIZO, In^{3+} cations are the main elements of the CBM, like in In_2O_3, but the incorporation of zinc (and gallium) in significant concentrations

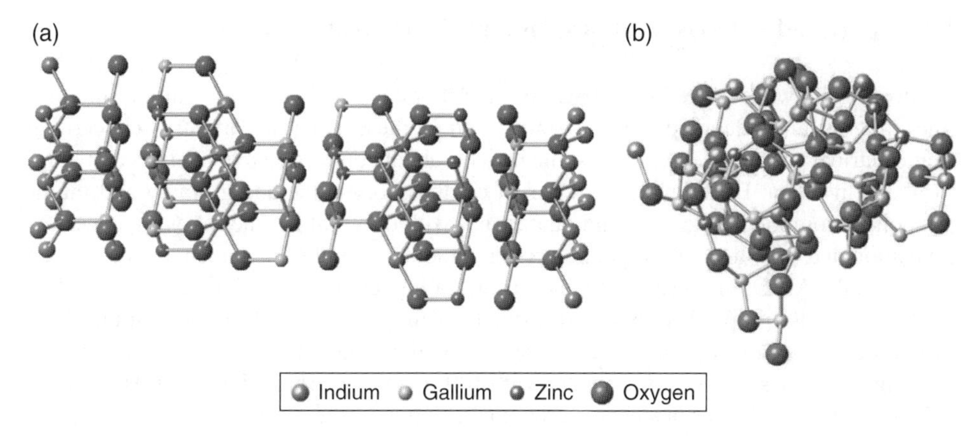

Figure 2.3 *Schematic structures of a) crystalline and b) amorphous GIZO.*

prevents the crystallization that easily occurs for In_2O_3. For room temperature deposited IZO, it is reported that the films are amorphous for a broad range of deposition conditions, at least in the range of 60/40 to 84/16 In/Zn cation % (atomic) [53]. For IZO films with 50/50 In/Zn cation % the processing conditions start to be important in order to define the structure of the thin films and polycrystalline or amorphous structures are observed by different authors [53–55]. Depending on the composition and annealing atmosphere, IZO films are reported to be amorphous up to 600°C [55]. Even if deposited at room temperature, sputtered IZO thin films already present electrical and optical properties quite similar to ITO films produced at higher temperatures [56]. However, the application of IZO as a TSO seems to be limited, because N cannot be easily decreased below 10^{17} cm^{-3} [38]. This can be solved by adding gallium to IZO, since as Ga^{3+} has a high ionic potential (+3 valence and smaller ionic radius than In^{3+} and Zn^{2+}), this element can establish strong bonds with oxygen, preventing excessive free carrier generation due to oxygen vacancies [38]. Furthermore, given the higher structural disorder achieved with the addition of an extra cation, in GIZO a broader range of amorphous compositions can be explored. In fact, as demonstrated by Orita *et al.*, zinc can even be made the predominant cation in GIZO without losing the amorphous structure [57]. Even if in this case the CBM is mainly derived from zinc *4s*-orbitals rather than the larger *5s*-orbitals when In^{3+} is the predominant cation, good electrical properties can still be achieved.

Besides indium-containing multicomponent oxide semiconductors, indium-free possibilities are also being studied, such as zinc-tin oxide (ZTO) [58–60] and gallium-zinc-tin oxide (GZTO) [44, 61]. In these materials, where either Zn^{2+} or Sn^{4+} are the predominant metal cations, TSO properties close to those achieved with GIZO can be achieved, but this generally requires processing temperatures that are considerably larger, typically above 300°C. However, the exploration of this route is highly important due to the higher cost of indium relative to other post-transition metals. The results obtained at CENIMAT on sputtered GZTO films and their application as active layers in TFTs will be shown in sections 2.3 and 5.2.2.

2.2 Sputtered n-TSOs: Gallium-Indium-Zinc Oxide System

As most of the work related to n-TSOs in our laboratory deals with the gallium-indium-zinc oxide system, a detailed analysis is presented for these materials, primarily concerning their electrical properties, including some insights about conduction mechanisms and long-term stability data. Different ceramic target compositions – including binary and multi-component compounds (ternary and quaternary, with different atomic ratios) – as well as deposition and post-deposition parameters are studied, namely the percentage of oxygen content in the Ar+O_2 mixture ($\%O_2$) and the annealing temperature (T_A), being their effect on the materials' properties discussed throughout this section.[d] All the depositions were carried out in a rf magnetron sputtering system without intentional substrate heating, on Corning 1737 glass and Si wafers coated with 100 nm thick thermal SiO_2. In order to assure reliable results for the various characterization techniques, the films were produced with a thickness ≈200–250 nm, although thinner films (≈40 nm) were also deposited for selected samples to study the effect of thickness on the electrical properties.

2.2.1 Dependence of the Growth Rate on Oxygen Content in the Ar+O_2 Mixture and Target Composition

The growth rate of sputtered films is highly dependent on the composition of the material and on the deposition parameters used (and naturally, on the sputtering system itself). Before initiating a detailed discussion about the properties of the oxide semiconductor materials, it is important to see how their growth rates are affected by the deposition parameters and composition. This is depicted in Figure 2.4 (see color plate section), regarding $\%O_2$ and target composition.

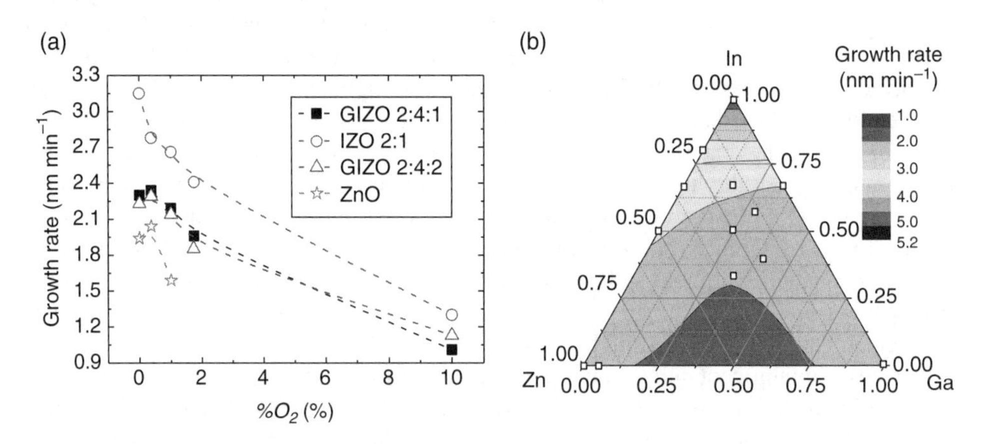

Figure 2.4 Dependence of the growth rate on a) $\%O_2$, and b) target composition.

[d] All the annealing treatments described here were performed in a tubular furnace, in air atmosphere, during 1 h. Effects of other deposition parameters, such as deposition pressure (p_{dep}) and rf power density (P_{rf}), are also briefly discussed in this section. Unless otherwise stated, films were produced with p_{dep}=0.7 Pa and P_{rf}=1.1 W cm^{-2}.

The deposition of oxides by sputtering in a pure argon atmosphere, even if starting from ceramic targets, generally results in films with high oxygen deficiency. This is essentially related to the threshold energy required to sputter a particle, which is higher for a metal-oxide than for a metal [62]. Additionally, molecular species sputtered from a target heading towards the substrate, such as a ZnO molecule, can be dissociated inside the plasma, increasing the chances of non-stoichiometric film formation [63]. Hence, a small percentage of oxygen, typically less than 10%, is usually mixed with the argon atmosphere in order to have a better control of the stoichiometry of the growing film [5, 64]. Figure 2.4a shows the effect of increasing $\%O_2$ on the growth rate, for films produced from targets with different compositions. Regardless of the composition, growth rate decreases as $\%O_2$ increases, which can be attributed to multiple factors:

- Resputtering (i.e., sputter-etch) of the growing film due to bombardment of highly energetic oxygen ions [5, 65]. This is especially relevant for higher $\%O_2$ and as will be seen later, this phenomenon can significantly affect other film properties, including its composition.
- Change of the surface conditions of the target [5, 63]. For instance, with a ZnO target, if a pure argon atmosphere is used, the target surface will be less oxidized, favoring the sputtering of zinc atoms rather than ZnO aggregates. This results in higher growth rates, since the binding energy of zinc is lower than the one of ZnO, i.e., sputtering a ZnO molecule from the target requires more energy [5]. Additionally, the energy transfer from an Ar^+ incident ion to the ejected target material is larger when the mass of the ejected particle is closer to that of the ion, resulting in a greater sputtering rate in the absence of oxygen [64]. Different surface conditions of the target can be readily seen by visual inspection of its surface after being involved in sputtering processes with different $\%O_2$.
- For low $\%O_2$ the oxidation reaction takes place essentially on the substrate, which is reflected in a small or even negligible decrease on the growth rate when compared with sputtering in a pure argon atmosphere. However, for higher $\%O_2$ target oxidation starts to play an important role, decreasing the growth rate due to the arguments given in the previous point [5, 66].

Figure 2.4b shows a ternary diagram with the growth rates of all the oxide semiconductor compositions studied in the gallium-indium-zinc oxide system, deposited using the same deposition conditions: $\%O_2=0.4\%$, $p_{dep}=0.7$ Pa and $P_{rf}=1.1$ W cm^{-2}. Comparing the binary compounds, In_2O_3 exhibits the highest growth rate, more than twice of ZnO, while Ga_2O_3 is slightly higher than ZnO. The differences should be related with the different mass (atomic number) and binding energy of the elements composing the ceramic targets as well as with the crystal structure of their surfaces, which are known to be the target parameters mostly affecting the sputtering yield [67, 68]. For instance, it was shown by Stuart and Wehner that sputter yields vary periodically with the element's atomic number, with the yield increasing consistently as the electronic *d* shells of the materials are filled [67]. For the multicomponent oxides, it is found that growth rate decreases when compared with In_2O_3[e] and gets lower for higher concentrations of zinc and gallium, i.e., it gets closer to the growth rates of isolated ZnO and Ga_2O_3, respectively.

---

[e] Note that for all the multicomponent compositions studied herein In/(In+Zn) and In/(In+Ga) atomic ratios are always ≥0.50, i.e., indium has always the same or larger atomic concentration as zinc and gallium.

2.2.2 Structural and Morphological Properties

Oxide semiconductors present a multitude of different structures and morphologies which directly affect other properties and hence dictate to a large extent the possible applications envisaged for a given material. Composition and T_A dependence were studied for all the produced oxide semiconductors. For selected GIZO compositions a more complete study was performed, comprising the effect of the various deposition parameters ($\%O_2$, p_{dep} and P_{rf}) on the structural and morphological properties but no significant or meaningful trends were found for those parameters, at least within the range of deposition conditions studied herein. On the contrary, although not explored here, it is reported that the structure and morphology of ZnO are severely affected by the deposition conditions mentioned above, especially regarding grain size, internal stress and surface roughness [37, 63, 69–71].

Figure 2.5 shows the X-ray diffraction (XRD) results obtained for a) In_2O_3, b) ZnO and c) IZO and GIZO thin films annealed at different temperatures but deposited under the same deposition conditions ($\%O_2$=0.4%, p_{dep}=0.7 Pa and P_{rf}=1.1 W cm^{-2}), on Si/SiO_2

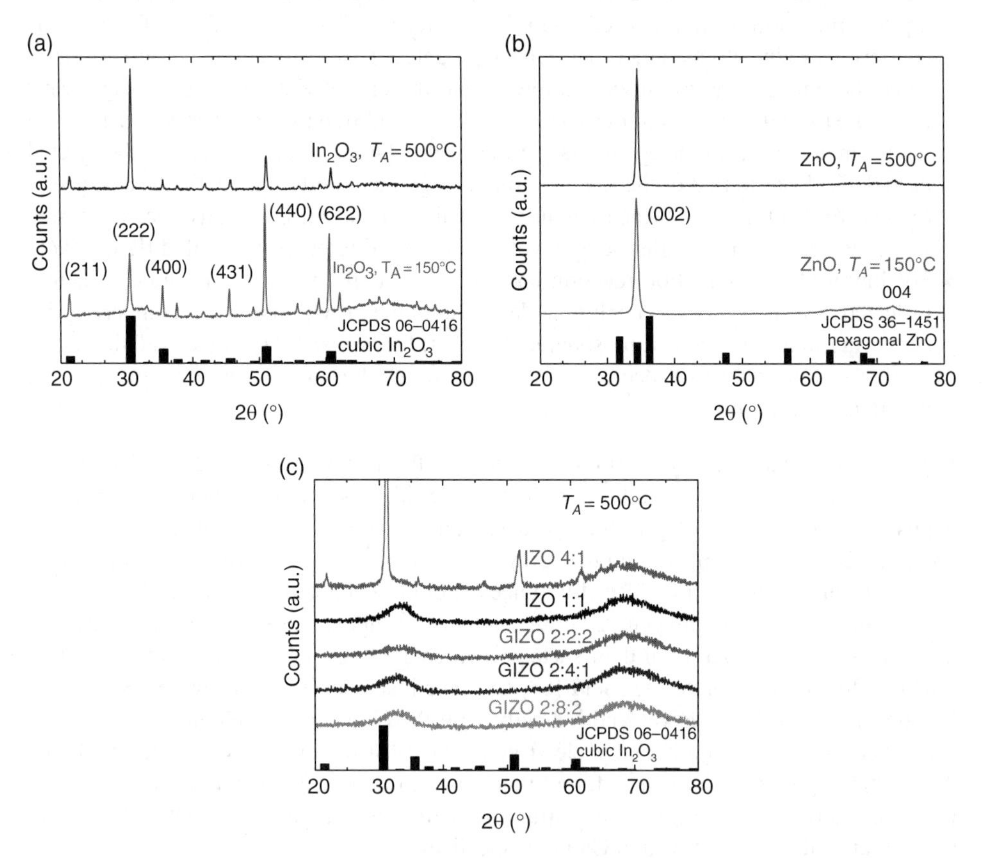

Figure 2.5 XRD diffractograms of a) In_2O_3, b) ZnO and c) IZO/GIZO thin films annealed up to 500°C. Films deposited on Si/SiO_2 substrates under the same conditions, with thickness around 200 nm. The bars show the peak positions and relative intensities for cubic In_2O_3 (JCPDS 06-0416) and hexagonal ZnO (JCPDS 36–1451) in a), c) and b), respectively.

substrates. For all the XRD analyses presented here, some decoupling for the silicon substrate was applied in order to reduce its large peak close to 70°, resulting in a broad and low intensity peak rather than an intense and sharp peak that would overshadow films' features. The data is normalized to the maximum peak intensity.

Typically, low-temperature deposited In_2O_3 films are amorphous but have a very low crystallization temperature, around 150–170°C [23, 24, 72]. Figure 2.5a shows that the In_2O_3 films annealed at 150°C are already polycrystalline, but still present some amorphous contribution, as evident from the halo peak around 32° that exists on the background of the diffractogram. Structures comprising amorphous and polycrystalline components are common for In_2O_3 films annealed (or processed) below 200°C [72, 73]. Enhanced crystallization and a predominance of (222) reflection occurs for higher T_A, and at 500°C the films exhibit sharp crystalline peaks that clearly correspond to the cubic bixbyite In_2O_3 structure. For In_2O_3 films (or ITO, which crystallize in the In_2O_3 system) deposited at low temperatures, generally the predominant planes are (222) and (400) [5], with (222) being the preferred orientation for films deposited using an oxygen containing atmosphere, which is the case here [72, 74].

By the Scherrer formula [75, 76], a crystallite size of 27 nm for the (222) orientation of In_2O_3 film annealed at 500°C is obtained, which is close to the values reported in the literature extracted by this methodology for In_2O_3 and ITO films [31, 66].

ZnO films (Figure 2.5b) are polycrystalline even before any annealing treatment. The films crystallize with a hexagonal wurtzite structure, with a preference for the (002) plane of this structure, i.e., the crystals grow along the c-axis, perpendicularly to the substrate, forming highly textured films. This preferential orientation, observed in most of the low-temperature deposited ZnO films reported in the literature [70, 77–79], is attributed to the fact that in the hexagonal structure the c-plane of the ZnO crystallites corresponds to the densest packed plane with the minimal surface energy [67, 69]. Films annealed at higher temperatures do not reveal any change on this preferred orientation, but the decreased width of the (002) peak suggests an improvement of the crystallinity. In fact, the crystallite size increases from 14 to 20 nm, as T_A increases from 150 to 500°C. Similar grain sizes were obtained for sputtered ZnO films by other authors using the Scherrer formula [64, 75, 80].

For the other binary compound used here, Ga_2O_3, amorphous films are always obtained, even if annealed at 500°C (not shown). In the literature, Ga_2O_3 sputtered films were consistently found to crystallize only when annealed at higher temperatures, typically above 800°C [81, 82].

Concerning the multicomponent oxides, as-deposited IGO, IZO and GIZO films are found to be amorphous regardless of their composition, exhibiting only a broad peak centered at $2\theta \approx 32$–34°, typical of amorphous films [23, 83–85]. By Scherrer analysis on this broad peak, possible crystallite sizes have only around 2 nm, which is consistent with the values typically reported in the literature for multicomponent oxides, validating their designation as amorphous materials [54, 59, 86–88]. Even after $T_A = 500$°C (Figure 2.5c), most of the films remain amorphous, with the only exception being the IZO 4:1 composition, which undergoes crystallization between 400 and 500°C. This film crystallizes under the same cubic bixbyite structure as the isolated In_2O_3 films, with the predominant planes and the relative intensities of the peaks being essentially the same in both cases (compare with Figure 2.5a), although the IZO films present smaller crystallite sizes, around 18 nm. The peaks are slightly shifted to higher values of 2θ for the IZO 4:1 composition due to the distortions in the In_2O_3 crystal lattice caused by zinc. Based on these results, three general rules can be deduced concerning the multicomponent oxides structure:

- Mixing different oxides, such as In_2O_3 and ZnO, inhibits or at least increases the crystallization temperature, due to the disorder introduced in the structure (compare Figure 2.5a and b with c).
- The crystallization in multicomponent oxides is favored for compositions with higher concentration of one of the composing elements, being the obtained crystal lattice based on the oxide of the predominant element[f] (compare IZO 4:1 with IZO 1:1 in Figure 2.5c).
- Multicomponent oxides with a higher number of elements (in this case, GIZO) are harder to crystallize, due to increased disorder in the structure.

The fact that these multicomponent oxides present amorphous structures, which are stable in a broad range of compositions and temperatures, is of major importance for the application of these films on devices. In fact, an amorphous structure is advantageous concerning not only process related issues, like uniformity, reproducibility and low-temperature deposition, but also regarding device performance, since amorphous films are generally much smoother than polycrystalline ones, improving interface properties. In fact, by analysis of atomic force microscopy (AFM) data, root-mean-square (RMS) roughness around 3–4 nm is obtained for ZnO films, while values below 2 nm are achieved for IZO/GIZO.

The differences in smoothness and morphology between ZnO and IZO/GIZO films are clearly visible in the scanning electron microscopy (SEM) images presented in Figure 2.6.

ZnO presents a typical textured surface (Figure 2.6a), with nanocrystals in the range of 70 nm, measured using simple imaging techniques. Note that when compared with the crystallite size determined by the Scherrer formula, these values are around three times higher, revealing that the grain sizes are underestimated by using the Scherrer formula. This effect was already verified for sputtered ZnO films [75, 89] and also for other oxides such as SnO_2 [90] and is generally attributed to the different grain size estimation methodology inherent to the techniques. While in SEM the grain size is measured by the distances between the visible grain boundaries, in XRD it is determined by the extent of the crystalline regions that diffract the X-rays coherently, which is a more stringent criterion [89]. Other reasons pointed out in the literature to justify these differences are instrumental broadening of full width at half maximum (FWHM), stress-induced and mosaic structure-induced broadening and neglect of strain [76]. The inset in Figure 2.6a shows the columnar growth of the ZnO film, being visible that the first deposited layers are composed of grains with smaller size and during the growth of subsequent layers larger crystallites start to coalesce perpendicularly to the substrate. Under certain deposition conditions, especially when using amorphous substrates, this initial layer may even form a very thin amorphous film, since new coming particles have, in general, higher tendency to relax in the same phase of nearby particles [91]. The GIZO surface reveals a totally different morphology, with very smooth and small feature sizes, less than 10 nm, measured using the same imaging methodology as for ZnO. The fact that these small features are present in these multicomponent oxides makes some authors to designate them by amorphous/nanocrystalline materials [7, 59], although no distinct peaks are visible in the XRD data.

---

[f]Although not studied here, IZO films where zinc is the predominant element are reported to crystallize with the wurtzite ZnO tructure [84, 86].

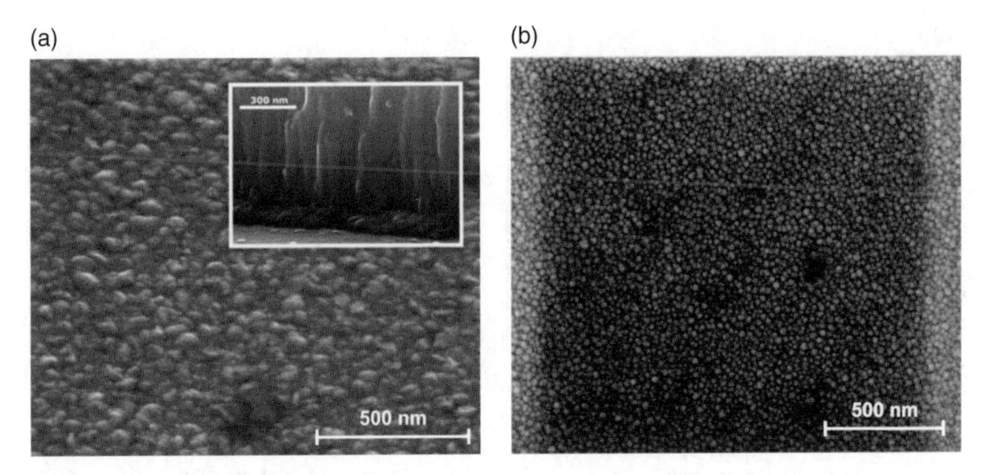

Figure 2.6 SEM images of a) ZnO and b) GIZO 2:2:1 surfaces, for thin films annealed at 500°C.

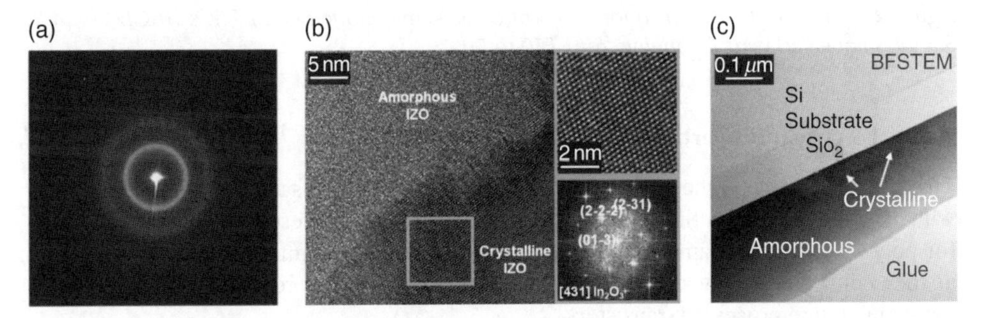

Figure 2.7 TEM analyses of oxide semiconductor thin films annealed at 500°C: a) electron diffraction pattern of a GIZO 2:2:1 film; b) electron micrograph of an IZO 4:1 cross section, revealing the presence of amorphous and crystalline domains. The images on the right side present a magnification of the crystalline domain (top) and the electron diffraction pattern of the same area (bottom); c) cross-section BFSTEM image of the IZO 4:1 film.[g]

Transmission electron microscopy (TEM) analyses were also performed on 500°C annealed films to confirm the structural results presented above. For GIZO films, the electron diffraction patterns reveal diffuse and broad electron diffraction rings, typical of amorphous films that exhibit some short-range order (Figure 2.7a) [92, 93]. However, for IZO 4:1 films annealed at 500°C, the electron diffraction pattern clearly shows evidence of a crystalline structure, which corresponds to the cubic In_2O_3 phase (Figure 2.7b). It is also interesting to note that the 250 nm thick IZO 4:1 film is not fully crystallized after 1 hour at 500°C and that the crystallization process is initiated on the IZO/SiO_2 interface, as revealed by the Bright-Field Scanning TEM (BFSTEM) image (Figure 2.7c).

[g]TEM analysis presented throughout this chapter was performed at University of Barcelona under the framework of Multiflexioxides European project.

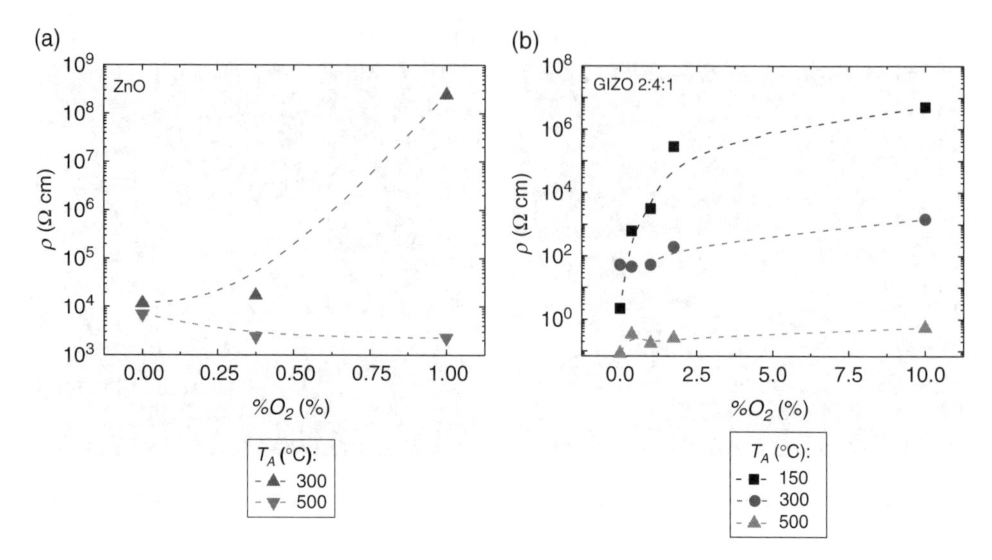

Figure 2.8 *Effect of $\%O_2$ on ρ for different oxide semiconductors: a) ZnO; b) GIZO 2:4:1. Samples annealed with T_A ranging from 150 to 500°C.*

2.2.3 Electrical Properties

This is naturally one of the most important topics, as these TSOs are studied in CENIMAT with the main intent of acting as active layers in TFTs. Data is presented and discussed both for polycrystalline and amorphous oxides, with the latter assuming particular relevance, as they constitute the most innovative and attractive materials regarding applications on low-temperature processed transistors.

2.2.3.1 Effect of oxygen content in the $Ar+O_2$ mixture

Besides doping, oxygen vacancies derived from stoichiometry deviations on the deposited films are the main contributors for electrical conduction in oxides. These oxygen vacancies form shallow electron donor levels close to the conduction band and are readily ionized near room temperature, with each doubly charged oxygen vacancy contributing two free electrons, preserving charge neutrality [22, 24, 94, 95]. Hence, the variation of $\%O_2$ is one of the most effective ways to control the electrical properties of oxide semiconductors. Figure 2.8 shows electrical resistivity (ρ) dependence on $\%O_2$ for a polycrystalline and an amorphous material, ZnO (Figure 2.8a) and GIZO 2:4:1 (Figure 2.8b), respectively.

For both cases, the main trend is for ρ to increase with $\%O_2$, which is consistent with the introductory idea given above, since the additional oxygen atoms supplied using a higher $\%O_2$ can be incorporated in the film, filling the high number of oxygen vacancies that are created on films sputtered in a pure argon atmosphere. This is clearly visible for films annealed at lower temperatures, such as 150°C[h] (Figure 2.8b), for which ρ increases more

[h] Note that for ZnO these values are not presented, as it was not possible to measure them using the four point probe setup, due to their high ρ. Based on the specifications of the semiconductor parameter analyzer and on the sample geometry, a ρ value higher than 10^8 Ω cm is expected for these highly-resistive samples.

than six orders of magnitude by changing $\% O_2$ from 0 to 10.0%. The highest ρ variation occurs for $\% O_2$ between 0 and 1.8%, and although some oxygen vacancies may still be compensated above this $\% O_2$ value, the increase on ρ between 1.8 and 10.0 $\% O_2$ can also be attributed to a different mechanism: for high $\% O_2$ the intense substrate bombardment by highly energetic oxygen ions can cause considerable resputtering of the deposited film, inducing structural defects on its growing surface, raising the film's ρ. The bombardment effect was found to be of extreme relevance on sputtered films by different authors [31, 62, 96–98]. Additionally, since the growth rate decreases so drastically for $\% O_2$=10.0%, regardless of the target/film composition (Figure 2.4a), it is also proposed that for this value of $\% O_2$ the films' compactness is affected, which may also contribute to increase ρ. Experimental evidence of this will be shown in section 2.2.4.2, using spectroscopic ellipsometry data. Another factor that can contribute to the increased ρ with $\% O_2$ in GIZO films is the reduction of the indium content of the films when high $\% O_2$ is used, as verified by X-ray fluorescence (XRF) and energy-dispersive X-ray spectroscopy (EDS).[i] As will be shown in this chapter, (G)IZO films with lower indium content have higher ρ.

Another important aspect shown in Figure 2.8 is that the increase of T_A tends to decrease the films' ρ and also the influence that $\% O_2$ has on it, although the main ρ-$\% O_2$ trend described above is still verified (a dedicated section about the effect of T_A can be found in section 2.2.3.3). For instance, with GIZO 2:4:1 annealed at 300°C, ρ increases less than two orders of magnitude within the studied $\% O_2$ range, against more than six orders at 150°C, and at 500°C the ρ variation is even smaller. However, for ZnO films annealed at 300°C the increase of ρ with $\% O_2$ is still quite significant, even if in this case $\% O_2$ is only increased up to 1.0 %. This should primarily be related with the polycrystalline structure of the material and the consequent existence of grain boundaries, which act like preferential paths for absorption/desorption of elements to/from the film, such as oxygen atoms. Another important factor that could justify the high ρ of ZnO films with $\% O_2$=1.0 %, even when annealed at 300°C, is the higher sensitivity of this oxide to the bombardment by highly energetic particles, such as oxygen ions. This effect can lead to the creation of defects on the films, like crystallographic damage and internal stress, raising ρ [62, 63, 97]. Nevertheless, after annealing at 500°C, grain size gets larger and the additional energy supplied by the increased T_A should allow to attenuate the effect of the oxygen bombardment damage, which is reflected in a considerably lower influence of $\% O_2$ on ρ.

It is also important to analyze the effect of $\% O_2$ on other electrical parameters directly related with ρ, namely carrier concentration (N) and mobility (μ). Although the films whose results are discussed in Figure 2.8 were not measurable with the hall effect system for all the range of $\% O_2$, due to their high ρ, low N and μ (at least the ones annealed at 150°C, where the $\% O_2$ effect dominates the films' properties), by using films produced under similar processing conditions but with a different target composition, such as IZO 2:1, consistent hall effect measurements were obtained. The results achieved for this multicomponent amorphous oxide are presented in Figure 2.9. First, as already seen in Figure 2.8, ρ increases with $\% O_2$, although here both the absolute ρ values and their modulation with $\% O_2$ are considerably smaller than for GIZO 2:4:1. This is expectable

[i]The decrease of indium content with $\%O_2$ also depends on the GIZO composition, but values around 5 % are obtained for GIZO 2:4:1 composition when $\%O_2$ is increased from 0.4 to 10.0 %.

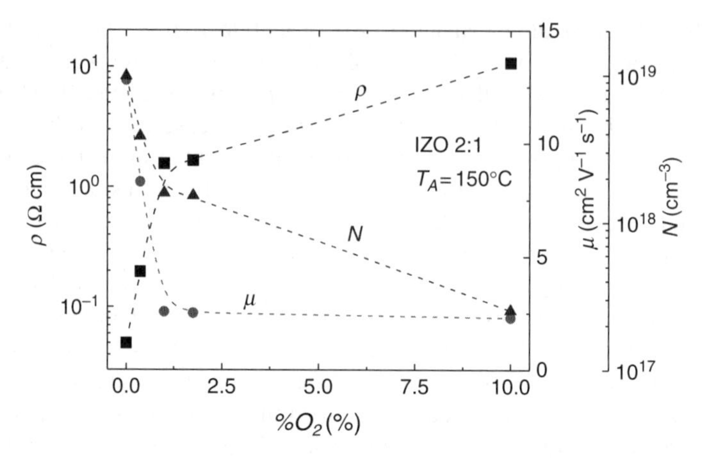

Figure 2.9 *Effect of $\%O_2$ on ρ, N and μ for IZO 2:1 thin films annealed at 150°C.*

since GIZO has one more cation than IZO and so, assuming similar conduction mechanisms for both materials, GIZO is able to incorporate more oxygen in its structure. Furthermore, gallium creates stronger bonds with oxygen than indium or zinc due to its high ionic potential (+3 valence and small ionic radius), which greatly contributes to a higher ρ of GIZO [38]. The increase of ρ with $\% O_2$ is accompanied by a decrease of N, reinforcing the idea that free electrons originate essentially from oxygen vacancies. Other important aspect of Figure 2.9 is that μ decreases as N decreases ($\% O_2$ increases), at least for $N > 10^{18}$ cm^{-3}. This trend, which is opposite to that verified for conventional crystalline semiconductors where carrier scattering with dopant ions controls μ, is attributed to the different conduction mechanisms of oxides. On these oxides, there is a random distribution of metallic cations around the conduction band edge, so charge transport gets more efficient (i.e., μ increases) for higher N, because the extra carriers allow E_F to surpass the potential barriers derived from the structural randomness. This μ-N tendency and its magnitude are naturally dependent on the considered N range, and consequently on the processing parameters and target composition, but a clear inversion is only observed for very high N, above 10^{20} cm^{-3}, where ionized impurity scattering becomes relevant (conduction mechanisms and their dependence on N are discussed in section 2.2.3.4).

2.2.3.2 *Effect of composition (binary, ternary and quaternary compounds)*

Figure 2.10a (see color plate section) shows the dependence of ρ on composition for oxide films with different In/(In+Zn) and In/(In+Ga) atomic ratios, annealed at 300 and 500°C and deposited under the same conditions ($\%O_2$=0.4%, p_{dep}=0.7 Pa and P_{rf}=1.1 W cm^{-2}). For the films annealed at 500°C, the trends of μ and N on the gallium-indium-zinc oxide system can be seen in Figure 2.10b and c.[j]

---

[j] Although it was already shown that larger dependence of the electrical properties on the processing parameters is verified for films annealed at lower T_A, 300 and 500°C films are presented on these figures as they allow to have a larger number of measurable films, hence a clearer relation between composition and electrical properties. Despite this, note that for the films measurable with lower T_A, the trends described for 300 and 500°C are also verified, but with even larger magnitudes.

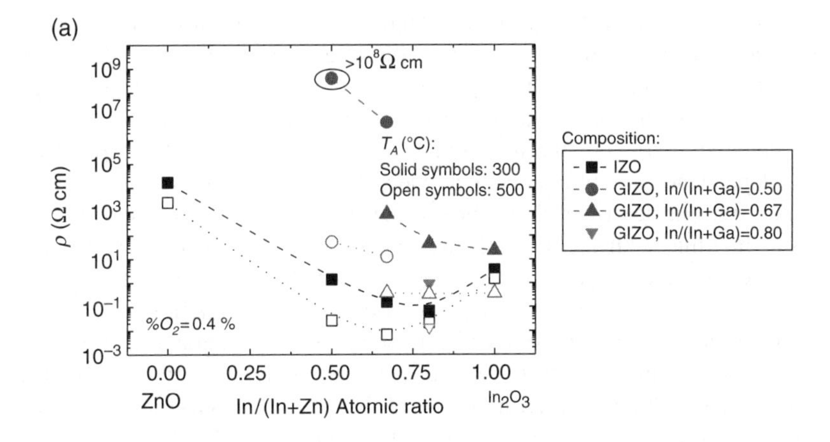

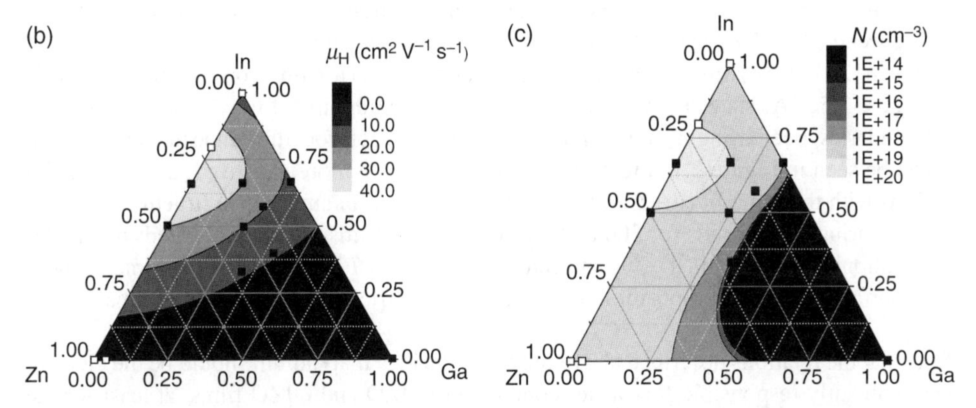

Figure 2.10 Effect of target composition on the electrical properties of oxide semiconductor thin films produced under the same deposition conditions: a) Effect of In/(In+Zn) and In/(In+Ga) atomic ratios on ρ, for TA=300 and 500°C; b) and c) are ternary diagrams for gallium-indium-zinc oxide system, for films annealed at 500°C, showing μ and N, respectively. The filled and empty squares denote amorphous and polycrystalline films, respectively.

Several conclusions can be draw from these figures:

- Comparing the binary compounds, ZnO and In_2O_3, the latter presents considerably lower ρ, which should be ascribed with the larger orbitals of indium (*5s*) than zinc (*4s*) and also with the larger grain sizes of In_2O_3. However, these advantages of In_2O_3 should be attenuated by the higher metal number density and shorter metal-metal distances of ZnO [7]. For Ga_2O_3 (not shown) produced under these deposition conditions highly resistive films are always obtained (>10^8 Ω cm), even after a 500°C annealing. This should be related to the above mentioned higher ionic potential of gallium when compared with indium or zinc, which is responsible for strong bonds between gallium and oxygen.
- For the multicomponent oxides, the general trend is for ρ to decrease as the In/(In+Zn) atomic ratio is increased. This is in agreement with the fact that the In_2O_3 films present a considerably lower ρ (higher N and μ) than ZnO films, with the multicomponent oxides

reflecting this to an extent dependent on the relative concentrations of indium and zinc. Additionally, when comparing films without gallium with In/(In+Zn)=0.80 and 1.00 (i.e., IZO and pure In_2O_3 films), ρ is higher for the latter, which is attributed to the polycrystalline structure of In_2O_3, where grain boundaries trap the free carriers, increasing the potential barriers between the grains and degrading carrier transport [63].

- The dependence of multicomponent oxide films on the gallium content follows a similar trend to the one described above for zinc, i.e., higher gallium contents (lower In/(In+Ga)) result in films with higher ρ. However, as Ga_2O_3 has an even higher ρ than ZnO, the effect of adding gallium to multicomponent films based on In_2O_3 is even higher than the effect of adding zinc, i.e., ρ is more affected by In/(In+Ga) than by In/(In+Zn) atomic ratio. This is also clearly visible on the ternary diagrams of Figure 2.10b and c, where for amorphous (G)IZO films μ and N are more affected (decreased) by gallium than by zinc addition. Furthermore, contrarily to what happens for films without gallium, when In/(In+Ga)=0.67 ρ decreases for all the range of In/(In+Zn), since the films always remain amorphous, even when no zinc exists in the structure (i.e., In/(In+Zn)=1.00).
- The trends described above for multicomponent oxides only fail for the 500°C IZO 4:1 (In/(In+Zn)=0.80) film, because crystallization occurs for this composition at this T_A (Figure 2.5c). As stated above, crystallization leads to trapping of free carriers at the grain boundaries, decreasing N and μ and increasing ρ. In fact, for films annealed at 500°C, it can be seen in Figure 2.10a that IZO 4:1 films are even more resistive than some GIZO films with the same In/(In+Zn) atomic ratio but having an amorphous structure.
- Comparing the ρ trends for 300 and 500°C annealed films, the latter T_A leads to smaller variations of ρ with In/(In+Zn) and In/(In+Ga), i.e., as T_A gets higher it assumes a more important role than composition in defining the electrical properties of the films.

Note that the relations described above reinforce the idea that indium should be the element predominantly responsible for carrier conduction in IZO and GIZO films, at least for the multicomponent compositions studied here, where indium exists in equal or superior atomic concentration as zinc or gallium. In fact, for the amorphous materials, μ always increases for higher N, disregarding whether the higher N values are achieved by the decrease of gallium or zinc concentrations. This is related to the above mentioned potential barriers associated with the random distribution of zinc and gallium close to the bottom of the conduction band, composed primarily of indium atoms. As N increases, potential barriers derived from structural randomness also decrease, since the higher N is associated with lower zinc and/or gallium content. Additionally, with high N, E_F moves above these potential barriers, enhancing μ. Therefore, as a general idea extracted from this analysis, it can be said that for the compositional range studied here higher indium concentrations enhance the carrier transport in multicomponent oxides, as long as enough concentrations of gallium and/or zinc exist in the films to inhibit their crystallization. Nevertheless, this structural change is only significant for films annealed at temperatures of 500°C.

To end this section, a final remark about the μ-N relation should be made: note that the direct relation found here for the multicomponent amorphous oxides happens for a large range of N values, at least from 10^{16} to more than 10^{19} cm^{-3}, based on the data presented in Figure 2.10b and c. However, this holds true since a high T_A (500°C) was used, which makes the electrical properties less dependent on the deposition conditions but clearly shows that composition has a major role in dictating the end properties of oxide semiconductor thin films. Although the same μ-N trend is valid for IZO and GIZO films annealed at lower T_A, it only occurs clearly for $N>10^{18}$ cm^{-3}, as seen earlier during the analysis of the $\%O_2$ effect. This point will be discussed in more detail in section 2.2.3.4.

2.2.3.3 Effect of annealing temperature

What was verified in the previous sections of this chapter regarding the T_A effect on the electrical properties was essentially that, depending on the deposition conditions and target compositions, increasing T_A increases or decreases ρ with different magnitudes and attenuates the differences between the electrical properties of films produced under different conditions and/or compositions. Since all the annealing treatments were done in air, increased adsorption or desorption of oxygen is expected to happen as T_A gets higher. Whether incorporation or removal of oxygen on the films' structure happens, mostly depends on the relation between the oxygen concentration in the annealing atmosphere and in the films: for instance, for polycrystalline materials, it is reported that the concentration of oxygen states at the grain boundaries (the preferential paths for impurities diffusion) relatively to the ambient oxygen pressure over the film determines whether adsorption or desorption takes place to reach an equilibrium state [5]. Although no grain boundaries exist in amorphous multicomponent oxides, their surfaces also strongly interact with the surrounding atmosphere even at room temperature, either by physisorption, chemisorption or desorption processes [99, 100]. Given that the properties of oxide semiconductors strongly depend on their oxygen deficiency/excess, it is not strange that they are significantly affected by these annealing treatments. Besides the oxygen concentration changes, annealing can also promote structural rearrangement and reduce the internal stress effects of the films [44, 101, 102]. Furthermore, as presented above, annealing can also lead to the crystallization of as-deposited amorphous structures or to the improvement of the crystalline quality of as-deposited polycrystalline materials, which naturally also affects the electrical properties.

Figure 2.11 illustrates the different effects that T_A between 150 and 500°C has on IZO 2:1 films produced with different $\%O_2$. N variation is chosen for this plot, as it should be the electrical parameter more directly related with the oxygen content of the films, since oxygen vacancies are one of the main sources of free carriers in oxides. Note that these films remain amorphous for all the range of analyzed temperatures.

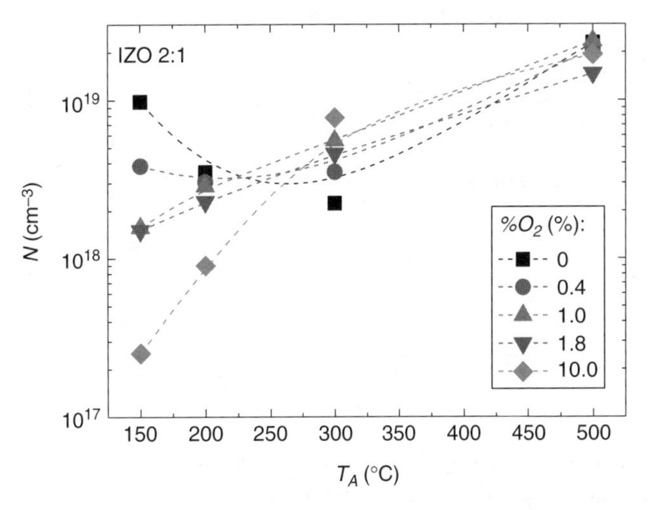

Figure 2.11 *Effect of T_A on N for IZO 2:1 thin films produced with different $\%O_2$.*

Three different behaviors can be clearly seen as T_A increases from 150 to 300°C:

- N decreases for films deposited in pure argon atmosphere ($\%O_2=0\%$), since oxygen adsorption should take place to compensate the high deficiency of this element on the films.
- N increases for films deposited with $\%O_2 \geq 1.0\%$, which could be attributed not only to oxygen desorption due to the high oxygen concentration in the film relatively to the annealing atmosphere, but also to structural relaxation, as films deposited with higher $\%O_2$ should be more prone to bombardment effects, specially when produced with $\%O_2=10.0\%$.
- N doesn't present a significant variation for $\%O=0.4\%$, meaning that a better equilibrium between oxygen concentration in the films and in the atmosphere should exist right from 150°C and also that the films should be structurally more stable.

If T_A is increased to 500°C, N increases for all the films. For this temperature, the oxygen content of the films should be in equilibrium with the annealing atmosphere and a better structural arrangement between indium, zinc and oxygen atoms should be obtained for all the range of $\%O_2$, but still preserving for all the cases an amorphous structure. Hence, all the films present very similar electrical properties at 500°C.

It is also interesting to note that the deposition conditions generally providing the best compromise in terms of electrical performance ($\%O_2=0.4\%$) are also the ones leading to smaller T_A effects. However, for the production of oxide semiconductors intended to be used as highly conducting electrodes (TCOs), which are also vital components of a transparent TFT, different processing conditions need to be used, to assure the lowest possible ρ without severely compromising the transparency. To achieve these sorts of properties, films are usually grown under a ballistic transport plasma regime, i.e., with higher P_{rf} and lower p_{dep} than the films presented so far [7]. Figure 2.12 shows the effect of T_A on such films, produced from an IZO 5:1 target.

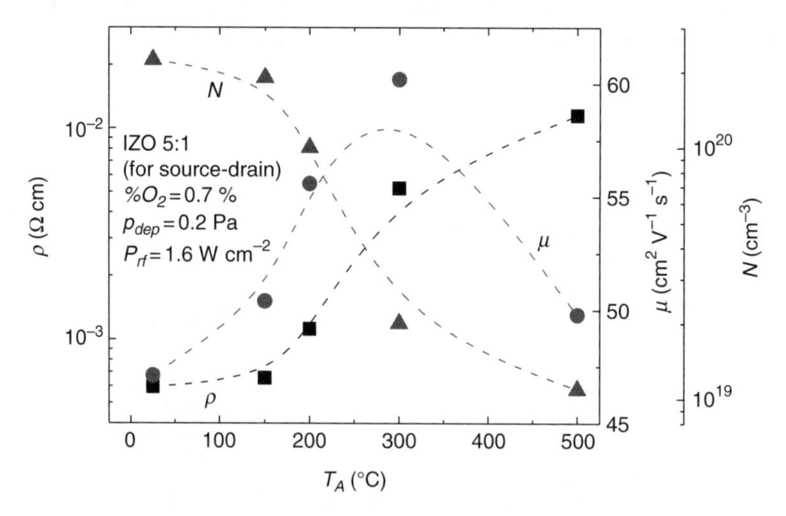

Figure 2.12 *Effect of T_A on ρ, μ and N for highly conducting IZO 5:1 thin films (for electrode application), produced using the same deposition conditions.*

As-deposited films exhibit the lowest ρ and the highest N and μ presented so far in this chapter, clearly showing that these deposition conditions and target composition are optimized to obtain films to be used as electrodes. Given the high oxygen deficiency of the films, as T_A increases oxidation takes place, which increases ρ and decreases N. Concerning μ, it assumes a different relation with N than the one previously verified for the multicomponent oxides, i.e. μ increases as N decreases. This is attributed to the different N range of these films: for IZO, it is reported that ionized impurity scattering starts to dominate carrier transport when N is above 10^{20} cm^{-3} [53]. Hence, for the N range of the IZO films presented in Figure 2.12, E_F is well above the CBM and enhanced carrier transport (higher μ) starts to be achieved for lower N, since less scattering centers derived from the ionization of native point defects affect the movement of carriers [36, 78]. Additionally, μ can also be increased due to the better structural rearrangement of the amorphous structure as T_A increases. This is even more plausible considering the low p_{dep} and high P_{rf} used here, which can lead to more significant film bombardment effects. For the highest T_A analyzed here (500°C), all the electrical properties are deteriorated (μ and N decrease, ρ increases) regarding the application of this material as a transparent electrode, similar to what was reported by other authors for IZO 5:1 films [23, 102, 103]. Several phenomena can justify this result:

- The crystallization of the film. When this happens, potential barriers associated with charge trapping and scattering effects at the grain boundaries start to be relevant, resulting in lower μ [5]. This is even more critical for materials with low crystallite size, which is the case of IZO annealed at this temperature, since the created depletion regions can easily overlap in adjacent grains, degrading charge transport. Nevertheless, as the material has still a relatively high N, above 10^{19} cm^{-3}, the grain boundary regions should only be partially depleted, so their negative effect on charge transport may be less severe [104].
- The grain boundaries constitute preferential paths for oxygen to penetrate in the film, increasing the oxygen concentration in the film, which is reflected in higher ρ and lower N. Furthermore, this oxygen diffusion process modulates the inter-grain boundary barrier height, affecting μ [105].
- As N is now smaller than for the films annealed at lower temperatures, the degenerated ionized impurity scattering mechanism, dominant for higher N, may now be replaced by a mechanism where the electronic transport is affected by the random distribution of zinc cations around the more ordered indium cations that primarily compose the CBM (note that IZO crystallizes according to the In_2O_3 system), which together with the grain boundaries create potential barriers around the conduction band edge [106]. This leads to a lower μ than for films annealed at 300°C, where higher N and amorphous structures are obtained.

Regarding the binary compounds, In_2O_3, ZnO and Ga_2O_3, the first crystallizes at very low temperatures (at 150°C it is already polycrystalline, see Figure 2.5a) and charge transport starts to be limited by the existence of grain boundaries. Hence, ρ increases significantly (from 10^{-2} to $\approx 1\,\Omega$ cm) when moving from non-annealed films to films annealed at temperatures of 150°C and above. Given the existence of grain boundaries, the films' oxidation is facilitated, thus N (and μ) decreases with T_A, from $\approx 10^{19}$ cm^{-3} (non-annealed) to $\approx 10^{18}$ cm^{-3} ($T_A \geq 150$°C). Although hall-effect measurements were not possible on ZnO films, ρ variation with T_A was opposite to the verified for In_2O_3 and of much larger

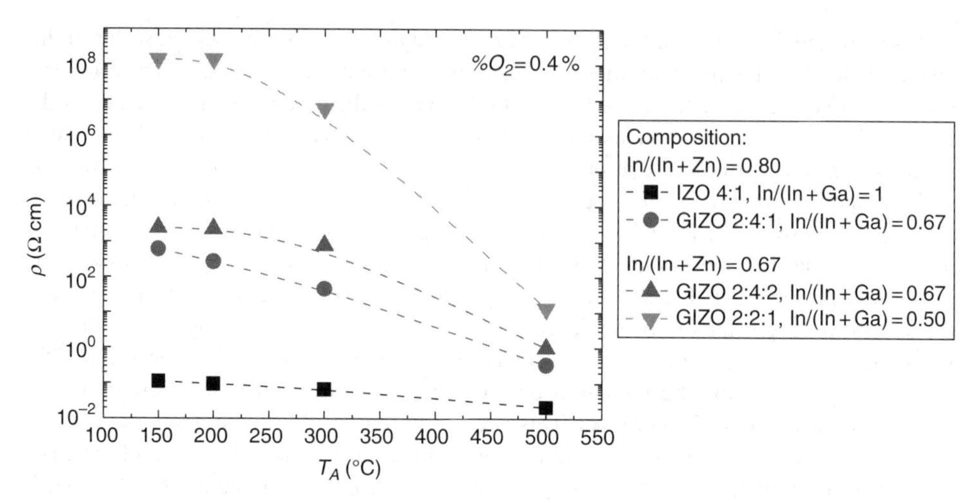

Figure 2.13 *Effect of T_A on ρ for In_2O_3, IZO 4:1 and GIZO 2:4:1 thin films produced using the same deposition conditions.*

magnitude, even if the as-deposited films are already polycrystalline: for ZnO, ρ decreases from $>10^8$ Ω cm (as-deposited) down to 10^3 Ω cm (500°C), for films produced under the same deposition conditions as In_2O_3 ones. For Ga_2O_3 films, even after being annealed at 500°C, ρ was still non-measurable. This result, together with the compositional dependence on the electrical properties of the oxide semiconductors presented in section 2.2.3.2, makes it understandable that the multicomponent oxides based on In_2O_3 present different variations with T_A depending on their composition, with films with larger zinc and mainly gallium contents being more sensitive to T_A (Figure 2.13). In addition, even if the (G)IZO films are amorphous at $T_A = 500$°C, it should be plausible to assume that some In_2O_3-rich agglomerates start to be formed at this T_A (given that In_2O_3 is always the first phase to crystallize for the films analyzed here), contributing to the large reduction of ρ at $T_A = 500$°C verified not only in Figure 2.13 but also in previously presented data, such as in Figures 2.8b and 2.10a.

2.2.3.4 *Additional considerations about the conduction mechanisms in oxide semiconductors*

The composition and the conditions under which the oxide semiconductor films are prepared have a significant effect on their electrical properties, as shown above. In addition, some brief discussion was provided regarding the mechanisms possibly affecting charge transport in oxide semiconductors, both in amorphous and polycrystalline films. In this section a larger quantity of experimental data is compiled in order to discuss the relevant conduction mechanisms in more detail.

- *Dependence of ρ and μ on N*

As E_F is shifted inside the bandgap (or even above the conduction band edge) depending on the value of *N*, it is not strange that other electrical parameters such as ρ or μ show a marked dependence on *N*. Figure 2.14 shows these relations, for binary and multicomponent oxide films deposited with different conditions and target compositions, with $T_A = 150$ and 500°C.

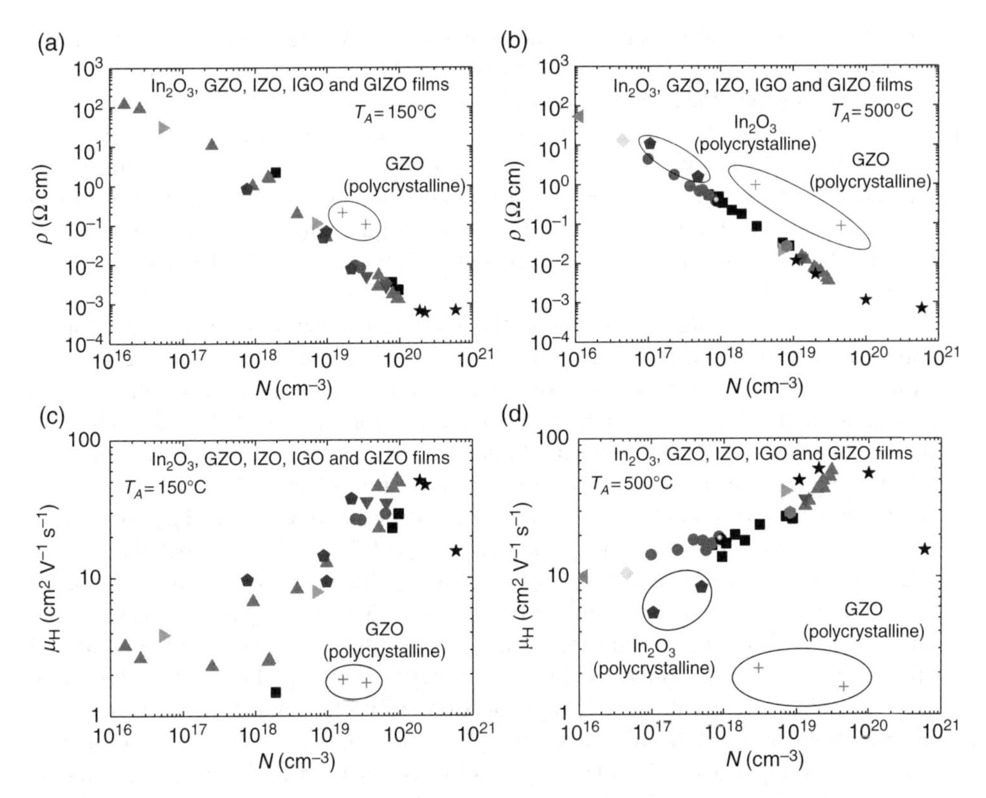

Figure 2.14 *Dependence of ρ and μ on N for oxide semiconductor thin films produced under different deposition conditions, using different target compositions and annealed at 150 and 500°C. Each symbol represents a specific target composition. a) $\rho(N)$, T_A = 150°C; b) $\rho(N)$, T_A = 500°C; c) $\mu(N)$, T_A = 150°C; d) $\mu(N)$, T_A = 500°C.*

As seen in Figure 2.14a and b, ρ sharply decreases with N, following a linear relationship in a double logarithmic plot. This trend is not surprising, since ρ in oxide semiconductors directly decreases with the concentration of intrinsic donors originated from lattice defects – such as oxygen vacancies or metal atoms on interstitial lattice sites – or with the introduction of extrinsic dopants with different oxidation numbers than the atoms of the host lattice [9]. Hence, the ρ-N relation found for the analyzed oxide semiconductors reinforces the idea that these materials behave as impurity semiconductors, where the intrinsic or extrinsic impurities create donor or acceptor levels that control the obtainable electrical properties. Similar trends can be found in the literature for intentionally doped and undoped oxides, based on ZnO and In_2O_3, for different ranges of N [9, 22, 74]. Although this is the main result given by these plots, it is important to consider some considerations regarding the actual data deviations from the ideal ρ-N linear relation:

- For very high N, close to 10^{21} cm^{-3}, ρ doesn't decrease anymore with N, since a degenerated ionized impurity scattering mechanism starts to dominate the transport properties. This is made even clearer by inspecting Figure 2.14c and d, where it can be seen that μ is significantly decreased for films with such a high N. As pointed out earlier, ionized impurity scattering should be dominant for this N range, since the large number of scattering centers generate an electrostatic field that remains effective for large distances, deflecting the charge carriers [5].

- The as-deposited polycrystalline material data presented on these plots, relative to gallium-doped ZnO (5 wt% of Ga_2O_3, denoted GZO) films, deviates from the linear ρ-N relation followed by the amorphous multicomponent oxides, following instead a parallel shifted ρ-N line, both for 150 and 500°C annealed films. Similar splits are found in the literature for single-crystalline and polycrystalline ZnO films, for $N \approx 10^{18}$ cm^{-3}, with the latter exhibiting larger ρ due to the effect of grain boundaries [9]. However, when $N > 10^{20}$ cm^{-3}, which is the N range where most of the oxide semiconductors are optimized for the application as transparent electrodes, even the data of polycrystalline and single-crystalline structures merge into one single line, since the depletion regions associated with the grain boundaries at this N range are so narrow that the electrons are able to tunnel through the barriers, so charge transport is mostly dominated by scattering at ionized impurities [9, 37]. On the other hand, when comparing ρ values for polycrystalline and amorphous In_2O_3 films, both with $N > 10^{20}$ cm^{-3}, Nakazawa *et al.* obtained two parallel ρ-N lines, with amorphous films exhibiting higher ρ for a given N [74]. The authors attributed this to the different states of structural defects of both types of films. Given this background, the higher ρ values obtained here for polycrystalline ZGO films when compared with amorphous (G)IZO ones having the same N are justified essentially by the combined effect of grain boundary (small) and ionized impurity scattering (high) effects typical of polycrystalline oxides with this N range [5]. Contrarily, electron movement is relatively insensitive to structural disorder in an amorphous structure primarily composed of large indium cations, typical of (G)IZO films. In addition, even when comparing polycrystalline oxide semiconductors based on In_2O_3 and ZnO structures, the latter are generally slightly more resistive for a given N, which agrees with the experimental data obtained here [63]. Nevertheless, note that when produced under optimal conditions, GZO films can present very low ρ, around 3×10^{-4} Ω cm, which is achieved with $N \approx 10^{21}$ cm^{-3}, when the films are deposited under a ballistic transport plasma regime [37].
- Even considering only the amorphous multicomponent oxides, there is considerably more dispersion on the ρ-N data for 150°C than for 500°C annealed films, especially when $N < 10^{18}$ cm^{-3}. This result suggests a model where carriers are relatively insensitive to structural disorder above this N but are more prone to be affected by higher densities of defects for lower N, when E_F starts to move out of the conduction band, being that these defects are dependent on the processing conditions and composition. When T_A is increased, a large part of these defect levels are annihilated, allowing for a more linear ρ-N relation. This is in agreement with the results presented in previous sections, where it was seen that the electrical properties start to depend less on the deposition conditions as T_A increases.

Confirmation and more insight about the dispersion phenomena mentioned above are obtained by analyzing the μ-N plots (Figure 2.14c and d). For the 150°C annealed films, the two different regimes occurring below and above $N = 10^{18}$ cm^{-3} are very clear, with the films below this threshold N value presenting reduced μ, which remains almost unchanged down to $N = 10^{16}$ cm^{-3}, consistent with the higher defect level density below the conduction band. As will be seen below, GIZO films, even if annealed at 500°C, have thermally activated (non-degenerated) conduction for $N = 10^{16}$ cm^{-3}, meaning that E_F is below the mobility edge for this N range [83]. Above $N = 10^{18}$–10^{19} cm^{-3}, μ sharply increases with N, which is consistent with the existence of potential barriers associated with the random distribution of gallium and zinc atoms. These barriers should have a relatively small height, as the conduction is essentially dominated by the large indium *5s* spherical orbitals, which easily overlap even in a disordered structure, while the zinc and gallium cations assure that this structure does not crystallize and prevent excessive free carrier generation [38]. As N increases above 10^{20} cm^{-3}, ionized impurity scattering dominates charge transport, which results in lower μ, as discussed above. For (G)IZO films annealed at 500°C, a continuous increase of μ with N is verified right from $N = 10^{16}$ cm^{-3}, meaning that most of the defects affecting the films with low N are extinguished after the annealing treatment and E_F can reach the conduction band edge for considerably lower N than films annealed at lower temperatures. Another interesting feature of Figure 2.14c and d is that μ is increased to a much lower extent by the annealing treatment for the N region comprised around 10^{19}–10^{20} cm^{-3}. Assuming that on this range of N the most important scattering effects on these amorphous oxides are related with the structural randomness, it is expectable that the annealing treatment leads to some structural relaxation, improving μ. However, as this improvement on μ is relatively small (around 10–20 %) when compared to the μ increase occurring for lower N, it can be concluded that structural disorder does not affect significantly the electrical properties of (G)IZO films.

Data is also depicted for In_2O_3 films which are already polycrystalline at 150°C. Given the N values of these low-temperature annealed In_2O_3 films (between 10^{18} and 10^{19} cm^{-3}), carrier transport is not significantly affected either by grain boundaries or by degenerated ionized impurities, hence similar μ-N trends are verified for In_2O_3 and (G)IZO films, reinforcing the idea that indium is the main element contributing to electronic transport. However, as T_A is increased to 500°C N decreases below $N = 10^{18}$ cm^{-3}, leading to grain boundary inhibited transport. Concerning the GZO films, as they have higher N than crystalline In_2O_3 films, carriers are more prone to be scattered at the higher number of ionized impurities, such as oxygen vacancies and gallium ions, decreasing their μ as N increases [9, 107], with this trend being valid both for low and high T_A. Nevertheless, note that a combined effect of ionized impurity scattering and grain boundary scattering is also plausible for the lower N films, with $N \approx 10^{19}$ cm^{-3} or below [5].

To finish the analysis of Figure 2.14, it is also interesting to note that after annealing at higher temperature, when the properties are less affected by the deposition conditions, the data in Figure 2.14b and d can be divided into well defined groups, with each one of these groups corresponding to a specific target composition, a range of μ and N. This constitutes valuable information about the flexibility of the (G)IZO system, being useful to select the necessary target composition in order to achieve the required electrical properties for each application.

- *Temperature dependence for polycrystalline binary compounds*

Important details about the conduction mechanisms of semiconductors can be extracted by measuring their electrical properties for different temperatures [108]. As an example for binary compounds, measurements for polycrystalline In_2O_3 films annealed at 500°C are presented in Figure 2.15. σ shows very small thermal activation ($E_a \approx 0.014$ eV) and N is not significantly changed with temperature, which is consistent with a close to degenerate

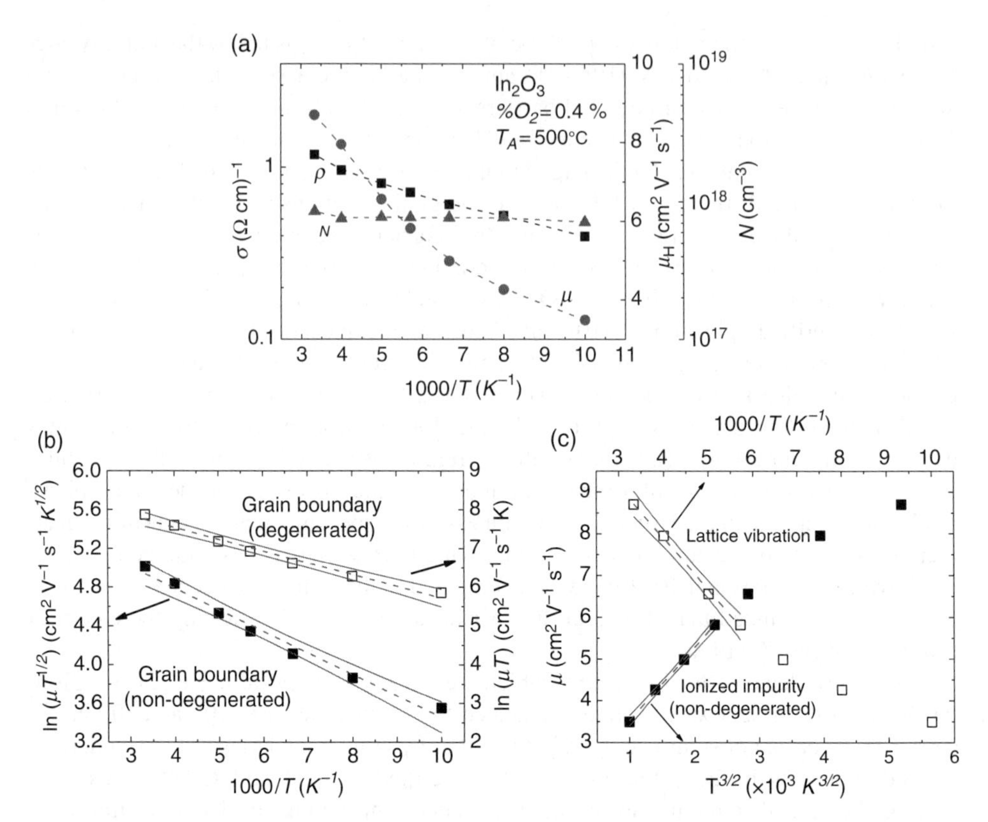

Figure 2.15 *Temperature dependence of the electrical properties for In_2O_3 thin films annealed at 500°C: a) Arrhenius plot for σ, μ and N; b) Relation between μ and T using the grain boundary scattering model for degenerated and non-degenerated cases; c) Relation between μ and T using the ionized impurity and lattice vibration scattering models.*

Table 2.2 *Different scattering centers and their μ-T relation for polycrystalline oxide semiconductors [107, 109].*

Scattering center	μ-T relation
Grain boundary (degenerated)	$\ln(\mu T)\tilde{}-T^{-1}$
Grain boundary (non-degenerated)	$\ln(\mu T^{1/2})\tilde{}-T^{-1}$
Lattice vibration	$\mu\tilde{}T^{-1}$
Ionized impurity (degenerated)	μ independent of T
Ionized impurity (non-degenerated)	$\mu\tilde{}T^{3/2}$

state. On the contrary, μ shows a considerable dependence on T, meaning that complete degeneracy is not obtained. To help in analyzing the μ-T trend and identifying the relevant conduction mechanisms, Table 2.2 presents some of the most important scattering mechanisms for polycrystalline oxide semiconductors.

The μ data was fitted for all the possible mechanisms listed on this table (Figure 2.15b and c). Starting with the grain boundaries effect, reasonably linear fits ($R^2 \approx 0.975$) are achieved for both degenerate and non-degenerate relations. The fact that both models yield the same fitting results may be explained by the N value of the film, $\approx 8\times10^{17}$ cm^{-3} at room temperature, which is close to the threshold for degeneracy on this material (around 1.48×10^{18} cm^{-3}). Based on this, it can be predicted that the analyzed film has a donor level slightly below the CBM, but without overlapping it [5, 42]. However, some deviations from an ideal fit are obtained in both cases, which could be indicative that grain boundaries are not the main scattering centers involved, or that another scattering mechanism plays simultaneously an important role. In fact, in polycrystalline oxide semiconductors with this range of N, μ is typically governed by the contribution of different mechanisms, namely bulk properties and grain boundary effects [5]. In addition, as pointed out above, grain boundaries are expected to make some contribution to the conduction mechanism for low N (comparatively to the N values typically obtained in TCOs, above 10^{20} cm^{-3}), specially in materials with small grain size, due to their large depletion regions that directly affect the grain boundaries' potential barriers height [104, 109]. To test the possibility that electronic conduction has simultaneous contributions, μ data is fitted for the ionized impurity and lattice vibration models in Figure 2.15c, which were seen by other authors to affect non-degenerated single crystalline In_2O_3 [42]. Clearly, linear relations are not obtained for all the temperature ranges, but each scattering mechanism seems to be dominant for different temperatures. Based on this, it is proposed that ionized impurities are the main scattering centers for low temperatures, while lattice vibrations dominate for higher temperatures. Similar mechanisms are found in the literature for GZO films [107]. Hence, it seems that at least three different scattering centers affect charge transport for the analyzed In_2O_3 films, namely those associated with ionized impurities (low T), lattice vibrations (high T) and grain boundaries. Nevertheless, for a valid and accurate attribution of these different mechanisms to the different temperature ranges, a more detailed analysis is required, including more data points and the analysis of films with different N.

- *Temperature dependence for amorphous ternary/quaternary compounds*

Figure 2.16 shows the temperature dependence of the electrical properties of IZO 5:1 thin films annealed at 150°C, which are intended to be used as the electrodes of the TFTs. Given the temperature independency of the large N, above 10^{20} cm^{-3}, E_F should already be above E_C, hence the conduction is essentially degenerated, as happens typically in TCOs. However, μ decreases for higher temperatures, which can be due to a lattice scattering mechanism, perhaps phonon or alloy related, as proposed by Leenheer *et al.* for IZO films with similar N [53]. For IZO films with higher N, $\approx 6\times10^{20}$ cm^{-3}, which can be obtained by using $\%O_2=0\%$, this μ-T dependency is not verified but μ is considerably decreased (see Figure 2.14c) when compared with the one of the films presented in Figure 2.16.[k] Thus, complete degeneracy associated with an ionized impurity scattering mechanism is only achieved for very high N values. These results are consistent with those obtained by Leenheer *et al.*, where complete μ-T independency only occurred for $N\approx 5\times0^{20}$ cm^{-3} [53].

Concerning the GIZO analysis, films produced in pure argon atmosphere or with very small $\%O_2$ ($\leq 0.4\%$) were prepared using different target compositions, namely GIZO

[k] IZO films processed under these conditions ($\%O_2=0\%$) are not suitable to be used as transparent electrodes, given that their AVT is severely degraded ($\approx 50\%$).

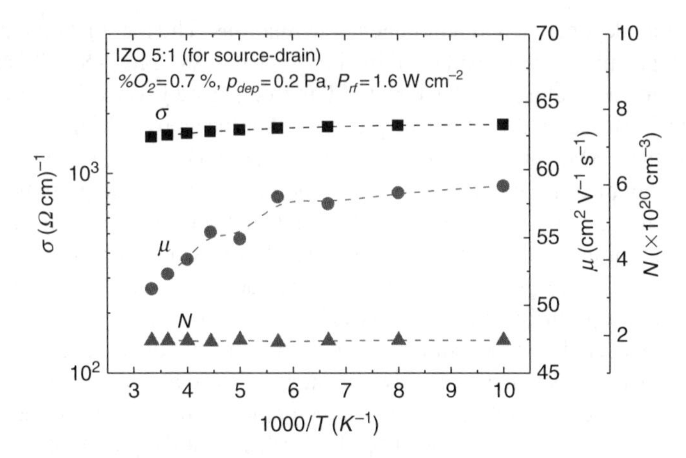

Figure 2.16 *Temperature dependence of the electrical properties (Arrhenius plot for σ, μ and N) for IZO 5:1 thin films annealed at 150°C, intended to be used as electrodes in TFTs.*

2:4:1, 2:2:1 and 2:2:2, and different T_A, 300 and 500°C. This set of films provided reasonably low ρ and high μ to be measurable by Hall-effect and a large range of N that allows one to understand the effects of E_F shifting. Figure 2.17 shows the obtained results for these GIZO films (In_2O_3 results are also plotted for comparison). It can be seen that fully thermal activated behavior (i.e., ρ, μ and N changing significantly with T, having $E_a \approx 0.05$ eV) is only obvious for $N<10^{17}$ cm^{-3}, suggesting that above this N value E_F exceeds the mobility edge. However, μ is thermally activated even for $N=10^{19}$ cm^{-3}, which suggests that some potential barriers exist above this mobility edge, in the vicinity of the conduction band. These potential barriers are associated with the structural randomness, mainly due to the random distribution of Ga^{3+} and Zn^{2+} ions, which form non-localized tail states around the conduction band edge. Thus, when E_F is located at energies corresponding to these potential barriers, carrier transport (i.e., μ) is enhanced when the temperature is increased.

In order to provide a clear explanation of the proposed model, different regimes can be considered:

- For $N \approx 10^{20}$ cm^{-3}, $E_F > E_C$ and μ shows little dependence on temperature. This suggests that E_F is above a certain threshold energy (E_{th}) where carrier transport is not affected by the potential barriers associated with structural randomness. Thus, conduction is essentially degenerated, with the small μ-T dependence for higher T being possibly ascribed to lattice scattering effects, as mentioned above for highly conducting IZO films.
- As N is decreased, first to 10^{19} cm^{-3} and then to 10^{18} cm^{-3}, μ changes with T and decreases when compared with the values obtained for $N=10^{20}$ cm^{-3}. Hence, even if $E_F > E_C$ can still be verified, carrier transport is limited because $E_F < E_{th}$, i.e., the electrons are affected by the potential barriers. This type of conduction resembles a percolation mechanism, where the carriers move through a distribution of potential barriers, being in effect more significant as E_F moves to lower energies, towards the conduction band bottom.
- For the GIZO film with $N=10^{17}$ cm^{-3} an interesting effect is seen which, at first sight, could contradict the model shown above, since μ for this film is higher than for the films

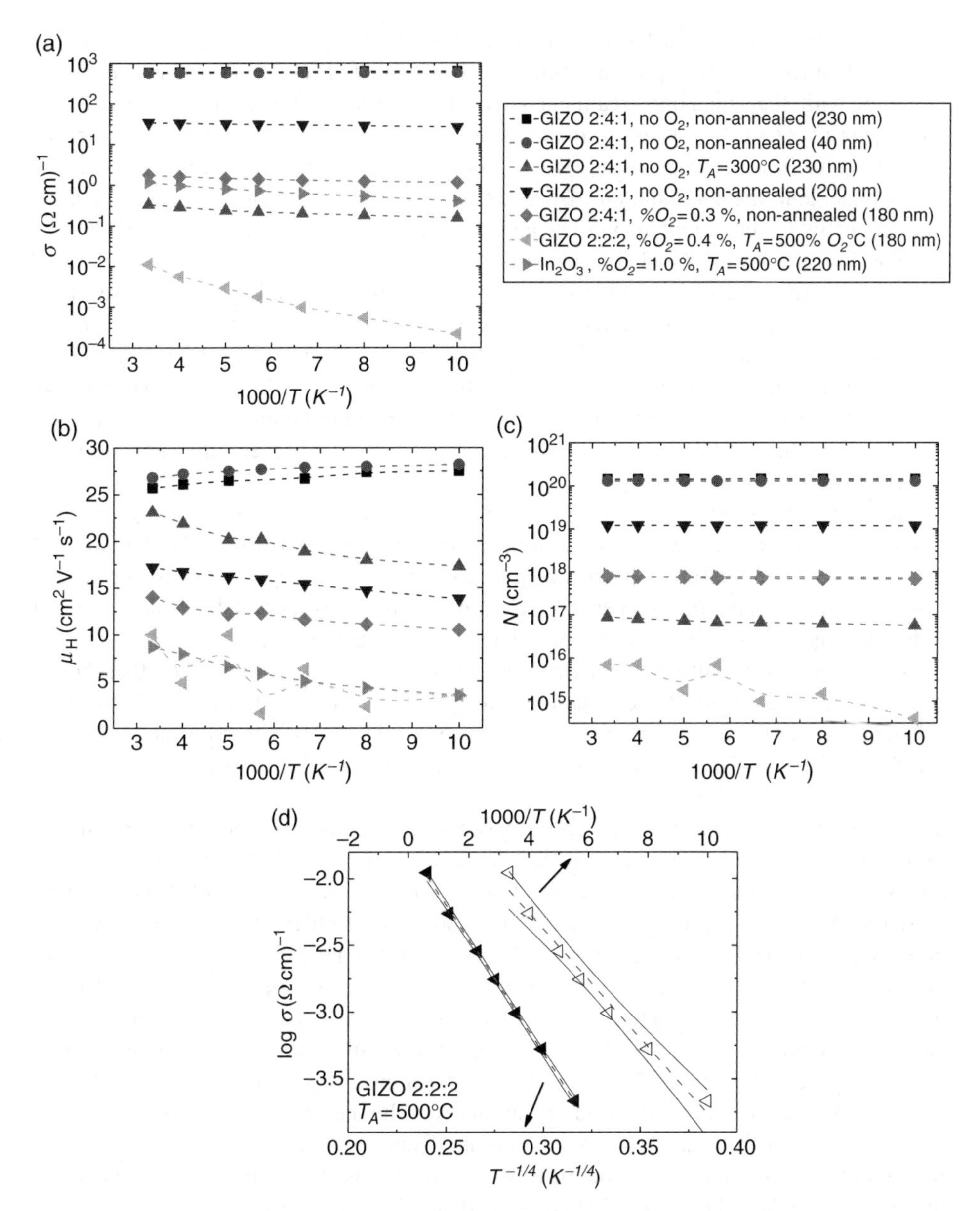

Figure 2.17 *Temperature dependence of the electrical properties for GIZO thin films produced with different compositions, processing conditions and annealed at different temperatures, in order to assure different N values. Arrhenius plots for a) σ, b) μ and c) N. In d) the σ data for the GIZO 2:2:2 film is plotted against 1000/T and $T^{-1/4}$, to show the possibility of a percolation conduction mechanism.*

with N=10^{18} and 10^{19} cm^{-3}, even if E_F is now located very close to E_C or even slightly below it, since N starts to present some dependence on T. The reason for this behavior is related with the annealing treatment: since the as-grown film is highly deficient in oxygen (a $\%O_2$=0 % was used), the annealing treatment promotes the film's oxidation, resulting in lower N. But as suggested above, annealing can also result in structural rearrangement and annihilation of defects arising from the growth process, which is reflected in lower potential barriers and decreased trap states density [110]. Hence, despite having lower N and consequently being more prone to the effects of structural disorder and/or defects (E_F located at lower energies), a considerable decrease in the density of those scattering centers occurs with annealing, improving the quality of the film and the transport properties. The fact that μ presents some more thermal activation than for the higher N films is consistent with this, since E_F is located at lower energy levels. This result is in agreement with the data presented in Figure 2.14c and d, where it can be seen that annealed films with $N \approx 10^{17}$ cm^{-3} can present higher μ than non-annealed films with $N \approx 10^{19}$ cm^{-3}.

- Finally, for films with $N \approx 10^{16}$ cm^{-3}, even if annealed at 500°C, both N and μ are thermally activated and μ is the lowest among all the analyzed GIZO films. For this N range, E_F should be located below E_C, on energies corresponding to high trap densities, related with structural randomness, heavy distortion of indium-oxygen-metal bonding angles and defects which are not totally annihilated by the annealing treatment [44, 111, 112]. However, definite Hall voltage signals are obtained, which is an indication that the tail states created just below the conduction band are not localized and carrier mean free path is much larger than the chemical bond distances, which is opposite to what happens on a-Si:H, where the Hall sign double anomaly is verified [83, 106, 113].

Since percolation between the potential barriers formed by the random distribution of cations seems to be the dominant conduction mechanism for GIZO, the σ data was also represented as a function of $T^{1/4}$, which describes the percolation model if the potential barrier height has Gaussian-type distributions [106]. Linear fits for σ-$T^{1/4}$ and σ-$1000/T$ are presented in Figure 2.17d for the films with the lowest σ, where thermal activation is more evident. The results show that considerably improved fits are obtained for σ-$T^{1/4}$, reinforcing that a percolation conduction model is valid for GIZO. Note that a σ-$T^{-1/4}$ relation may also be in accordance with a VRH mechanism, typical of disordered or highly doped semiconductors [106]. In the VRH mechanism the transport is limited by localized states and the charge carriers hop from a neutral atom to another neutral atom situated at the same energy level, even if located at very interatomic distances away [5]. VRH shouldn't be a relevant mechanism here, since the measured N gives consistent Hall voltage signals, i.e., the carriers shouldn't be localized [114] as predicted by this model.

Similar results were reported for amorphous, polycrystalline and single-crystalline GIZO films, although with some differences on the threshold N where the transition between the different regimes takes place, which can be attributed to the different growing methods and resulting quality of the films [83, 106]. Also, even if not defining whether hopping or percolation is the dominant conduction mechanism, Leenheer *et al.* obtained analogous μ-T relations in the IZO system [53].

Additionally, note that even if the IZO and GIZO films with high N presented in Figures 2.16 and 2.17 have similar N values, μ is considerably higher for IZO. This clearly reinforces one of the main ideas exposed throughout this chapter, i.e., that (post-)deposition conditions (for instance, the low p_{dep} used for IZO and the fact that GIZO films are not annealed) and composition play a very important role defining the electrical properties of oxide semiconductors when high T_A is not used. Nevertheless, even if exhibiting lower μ than IZO films, the GIZO films of Figure 2.17 still present a large μ, considerably higher (≈60 %) than polycrystalline In_2O_3 films (also represented in Figure 2.17) having similar N. This is even more striking considering that the GIZO films were not annealed and the In_2O_3 films had an annealing treatment at 500°C that enhanced their crystallinity. This result clearly shows the advantage in terms of electrical performance of using multicomponent amorphous materials, at least for this range of N, where In_2O_3 transport is limited by grain boundaries, lattice scattering and ionized impurities, as discussed above.

2.2.3.5 *Effect of thickness (d_s)*

Figure 2.17 also shows another interesting feature of amorphous oxide semiconductors that has not yet been discussed: the electrical properties are essentially the same for GIZO 2:4:1 films with different thicknesses, around 230 and 40 nm. This situation is contrary to what happens in most polycrystalline materials, where ρ variation with thickness only saturates for thicknesses above 200–300 nm [5]. Below these values, ρ typically shows a large dependence on thickness: for instance, in GZO films ρ can be increased more than one order of magnitude when thickness is decreased from 200 to 100 nm [5]. This ρ improvement for higher thickness in polycrystalline materials is mostly attributed to improved crystallinity, lower number of defects and less surface scattering effects for thicker films [5]. Even when considering amorphous covalent materials, such as a-Si:H, where the grain boundary problems are absent, the electronic quality of the films is reported to improve (in terms of density-of-states) as the thickness increases [115], which is related with the fact that dangling bonds created during the growth process are mostly distributed around 100 nm from the top-surface [116]. The insensitivity of GIZO electrical properties to thickness should result from its amorphous structure composed of large overlapping cationic orbitals and strong ionic bonds, as opposed to the amorphous structure with large density of dangling bonds of covalent amorphous semiconductors. Hence, despite the thickness of GIZO, an overall structure with low density of defects is always created, much lower than the one typically found on a-Si:H, for which the tail states density is reported to be two to three orders higher than for GIZO [111]. Nevertheless, when GIZO thickness is decreased to a few nm surface effects such as the interaction with atmospheric oxygen should naturally start to become relevant and can dominate the overall properties of the thin films.

2.2.3.6 *Electrical stability measurements*

Selected films (≈200 nm thick) deposited with different $\%O_2$, target compositions and T_A were electrically characterized over 18 months. During this period, the films were kept under normal room conditions, i.e., exposed to room temperature and atmospheric pressure, but stored in the dark to avoid possible light induced effects. Figure 2.18 shows the most relevant results obtained.

The results are correlated with what was observed in the previous sections of this chapter:

- Concerning the GIZO films produced with different $\%O_2$ and annealed at a low T_A, 150°C (Figure 2.18a), it is verified that the most stable films are those produced using low $\%O_2$ (0.4 %). For films produced without oxygen ρ increases with time, while for

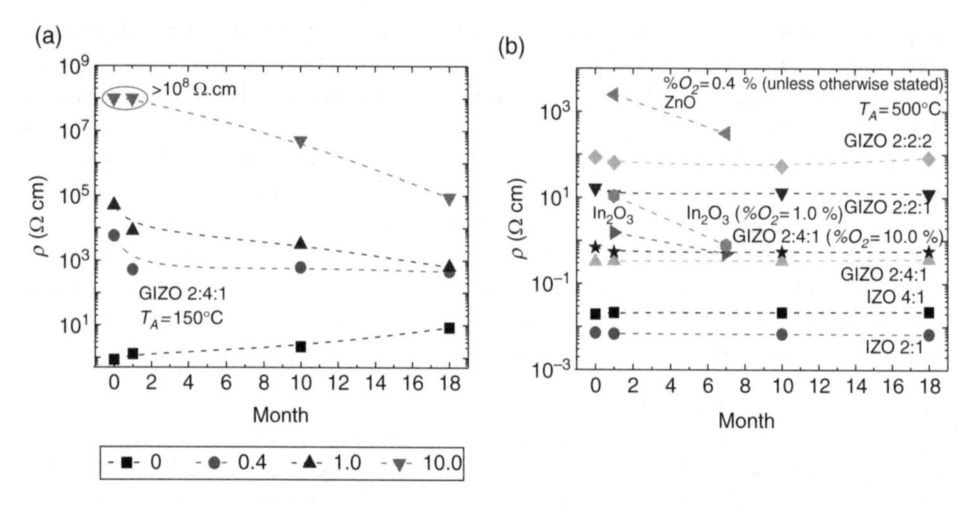

Figure 2.18 *ρ measurements in an 18 month time-scale after the fabrication of the oxide semiconductor thin films: a) Effect of %O_2 for GIZO 2:4:1 thin films annealed at 150°C; b) Effect of composition for thin films annealed at 500°C.*

films produced with higher %O_2 the opposite happens. Furthermore, the most significant variation occurs for the films deposited with very high %O_2, for which ρ decreases more than three orders of magnitude during the 18 months test. These results resemble those presented for instance in Figure 2.11, where the variation of N as a function of T_A was analyzed for IZO films processed with different %O_2. The effect here should be essentially the same, i.e., the materials always tend to an equilibrium point, where the discrepancies between their electrical properties, arising from the different processing conditions, are reduced. Here, this evolution occurs in a much larger time scale, due to the totally different amounts of energy supplied in each case: while here the films were annealed only at 150°C and kept at room temperature for 18 months, which is consistent with low energy exchange processes, in the case of Figure 2.11 the properties of the films were changed almost instantaneously for higher T_A, because the involved energy, supplied thermally to the materials, was much higher.

- The previous point is in agreement with the data depicted in Figure 2.18b. In this case, it can be seen that films produced with different target compositions and annealed at 500°C do not show significant variations on ρ during the 18 months test. This is true for the multicomponent oxides, which are amorphous or at most crystallize close to this temperature, with very small grain sizes (IZO 4:1). For the binary compounds In_2O_3 and ZnO, which are both polycrystalline even at 150–170°C, notorious variations are still observed after annealing at 500°C, which should mostly be related with the presence of grain boundaries where several oxygen adsorption/desorption processes occur. The instability of these binary compounds is reinforced by the different variations observed for In_2O_3 films processed with %O_2=0.4 and 1.0 %, which do not happen for the multicomponent amorphous oxides when annealed at high temperature (as an example, compare in Figure 2.18b the properties of GIZO 2:4:1 films produced with %O_2=0.4 and 10.0 %, where even for the latter good stability is obtained).

However, note that considering an even larger time scale for low temperature annealed (or non-annealed) films does not necessarily lead to exactly the same properties as those instantaneously obtained after a high T_A treatment. Processes involving severe structural changes, such as significant atomic rearrangement and amorphous-crystalline transition, or desorption of strongly bonded oxygen atoms, generally require that certain threshold energy values are surpassed, and these cannot be supplied by simply leaving the film at room temperature for a long period of time.

Itagaki *et al.* also analyzed the shelf life stability on non-annealed IZO films with different compositions over three months [117]. The results presented show that during this period ρ can decrease by around four orders of magnitude for large zinc contents (In/(In+Zn) atomic ratio ≈0.20), but only around one order of magnitude for large indium contents (In/(In+Zn) atomic ratio ≈0.80). Most of these changes occur during the first month, which is in agreement with the data presented above, at least for the less resistive samples.

Based on the data collected for IZO and GIZO films during the 18 months study and on Itagaki's results, it is proposed that for low temperature annealed (or non-annealed) multicomponent amorphous oxide films stability is enhanced when deposited under ideal conditions, mostly regarding $\%O_2$ that should be low (≈0.4%), and for indium-richer compositions, but increased T_A brings improved stability even for deposition conditions/compositions that deviate from these. However, note that even when fulfilling the requirements for good stability listed above, it is expected that for very thin films, adsorption/desorption processes at the films' surface can play a very significant role and considerable changes in the electrical properties can happen through time.

Naturally, these are only preliminary conclusions about the stability of these materials. The described trends must be confirmed and the mechanisms responsible for them must be clearly understood before any definitive conclusions are made.

2.2.4 Optical Properties

Given that optical effects deal with transitions between bands and/or energy levels inside the bandgap, optical measurements are of great interest when it comes to analyzing semiconductors [118]. In addition, high optical transmittance is generally desirable for oxide semiconductors, as one of the main applications of these materials is in the emerging area of transparent electronics. This section presents a brief analysis of the optical properties achieved for the produced oxide semiconductors, obtained by transmittance measurements and spectroscopic ellipsometry. The dependence of these properties on $\%O_2$, composition and T_A is discussed.

2.2.4.1 General considerations about the optical measurements

For most of the amorphous semiconductors, non-direct optical transitions are allowed and the exponent x in Tauc's relation $\alpha^x \sim (E-E_G)$ assumes the value 0.5 [119]. However, for In_2O_3 it was reported that electron momentum is conserved and the shapes of the absorption edges are similar when considering amorphous and crystalline structures [22]. Therefore, the direct bandgap model with $x=2$, which is typically applied for polycrystalline In_2O_3 films, also provides a good description of the optical absorption band edge for amorphous In_2O_3 films. Given that IZO, IGO and GIZO are amorphous materials primarily composed

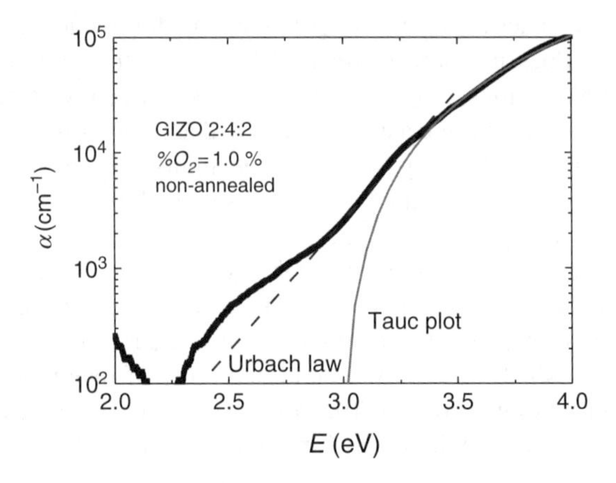

Figure 2.19 *α-E plot for a non-annealed GIZO thin film, showing Tauc and Urbach models fittings to the experimental data.*

of In_2O_3, this model is also extended to their analysis. This methodology was followed by different authors to extract E_G on multicomponent oxides based on In_2O_3 [23, 53, 85, 120], although others adopt the typical $\alpha^{1/2}$-E plot of covalent amorphous materials for the determination of E_G [57, 121, 122]. Since the E_G determination by α^2-E yielded results closer to those obtained by spectroscopic ellipsometry, this was the methodology followed here, but both $x=0.5$ and $x=2$ yielded reasonably good linear fits to the experimental data at the onset of absorption rise. Differences in E_G as large as 0.5–0.6 eV were found when using $x=0.5$ or $x=2$ (higher E_G for $x=2$), although the trends described below are essentially the same for both methodologies.

Optical analysis can also provide valuable information about the defect states of a given material. In particular, oxide semiconductors have important subgap absorptions related with deep and shallow levels that have a considerable effect on electrical properties, and hence it is important that they be analyzed [110, 121]. However, to fit different models to the experimental data obtained by transmittance and reflectance measurements, one has to be aware of the limitations of the spectrometer and of the sample itself. Figure 2.19 shows an example of an α-E plot for a GIZO film, where Tauc and Urbach models are fitted to the experimental data [118, 121].

Although the Tauc plot provides a good fit to the experimental results above the bandgap (in the strong absorption region), α data fitted with the Urbach law is close to the reliability limit of the experimental setup. It is clear that some subgap absorption is present, but detailed comparisons between different films or the correctness of fitting the Urbach law to the tail-like absorption reveal a large degree of inaccuracy. Hence, to correctly evaluate the subgap optical properties some other techniques should be used, such as constant photocurrent method (CPM) or modeling by spectroscopic ellipsometry. Although this last technique was also used here, it served essentially to confirm results and a detailed modulation of subgap states was for now out of the scope of our work. Results on GIZO subgap defect states studied by optical analyzes and first-principle calculations can be found in [123].

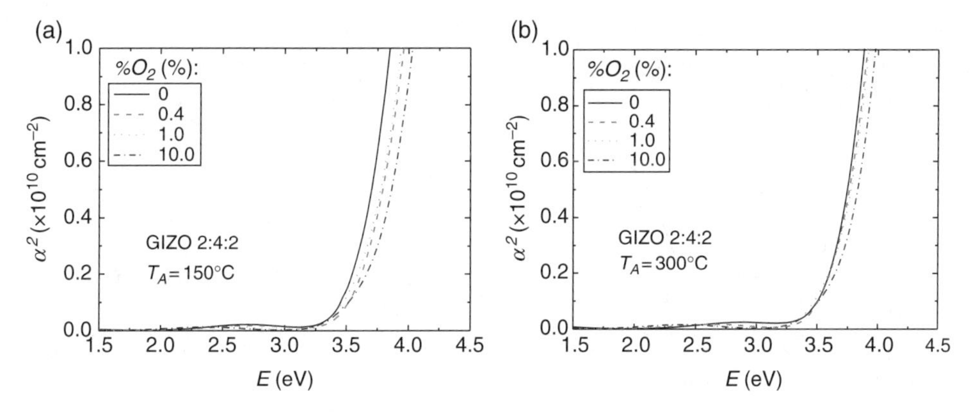

Figure 2.20 *α^2-E plots as a function of $\%O_2$ for GIZO 2:4:2 thin films annealed at a) 150 and b) 300°C.*

2.2.4.2 *Effect of oxygen content in the Ar+O_2 mixture*

Figure 2.20 shows the effect of $\%O_2$ on the α^2-E plots for films produced using a GIZO 2:4:2 target, annealed at 150 (Figure 2.20a) and 300°C (Figure 2.20b).

Two effects are immediately observable: E_G increases with $\%O_2$ (see also Table 2.3) and although the same trend is verified for both T_A, at 300°C the effect is less pronounced, which agrees with the results regarding the electrical properties, where fewer discrepancies are also found for higher T_A. The increase of E_G with $\%O_2$ is tentatively attributed to different factors:

- Higher $\%O_2$ leads to an increase of gallium at the cost of indium concentration. Given that Ga_2O_3 has a considerably higher E_G than In_2O_3 or ZnO, this can be reflected in a higher E_G on the final GIZO structure (see also the discussion on composition below).
- Higher $\%O_2$ allows one to fill a higher number of absorptive defect states, associated with oxygen deficiency. This way, the levels available for optical absorption are located at higher energies as $\%O_2$ increases, raising E_G. A similar mechanism is proposed by Suresh *et al.* for GIZO films [88].

Based on the results presented in the previous section, films produced with higher $\%O_2$ are expected to have lower N. For TCO films, this typically leads to lower E_G, in agreement to the well-known Burstein-Moss shift [6]. However, an opposite trend is verified for the results discussed here, because the GIZO 2:4:2 films presented in Figure 2.20 are not degenerate. Hence, E_G variation should be governed by mechanisms such as the ones proposed above. Similar E_G-$\%O_2$ trends are verified for other compositions and/or p_{dep} and P_{rf} conditions that yield films with low N, usable as active layers on TFTs. IZO films with higher N, used for electrodes, will be analyzed in section 2.2.4.4. In Figure 2.20 it is also noticeable that for all the films α^2 does not increase abruptly, which is consistent with subgap absorption due to the band-tails, typical on amorphous materials [118]. The trends observed with $\%O_2$ on GIZO 2:4:2 were confirmed by spectroscopic ellipsometry analysis, as presented in Figure 2.21 for non-annealed films. The extinction coefficient (k) does not have an abrupt increase with E, confirming the amorphous structure of the material (Figure 2.21a). Also, the shift of the absorption edge with $\%O_2$ agrees with the results

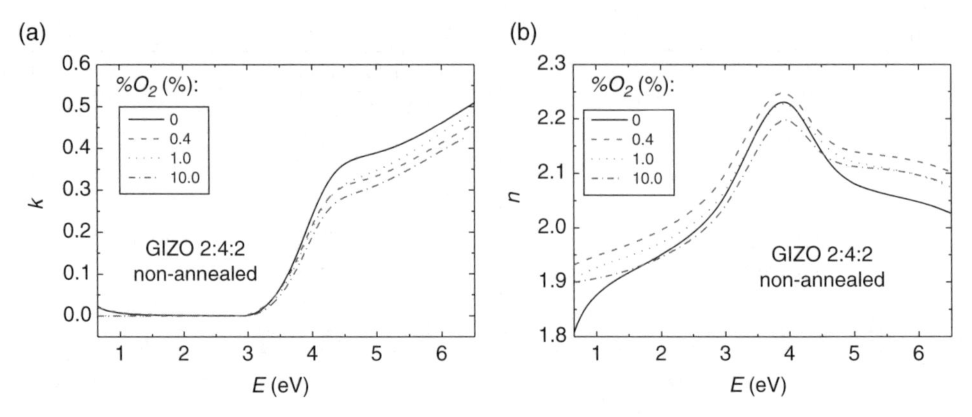

Figure 2.21 *Spectroscopic ellipsometry data, showing a) k-E and b) n-E plots as a function of* $\%O_2$ *for non-annealed GIZO 2:4:2 thin films.*

Table 2.3 E_G *values extracted from* α^2*-E plots and from spectroscopic ellipsometry data (identified by SE), for GIZO 2:4:2 thin films produced with different* $\%O_2$*, annealed at 150 and 300°C.*

$\%O_2$ (%)	E_G (eV)	
	150°C	300°C
0	3.47 (SE), 3.54	3.61
1.0	3.51 (SE), 3.62	3.64
10.0	3.60 (SE), 3.70	3.68

presented in Figure 2.20. Concerning Figure 2.21b, the contribution of the Drude model is visible in the low energy part of the *n-E* spectrum: for the film produced without oxygen a large *N* is obtained, affecting the relative permittivity at this energy range and consequently the value of *n* [5]. For higher $\%O_2$ the contribution of the Drude model is negligible, due to a considerably lower *N* of the films [5]. As suggested by the previous electrical characterization results, the degree of compactness also seems to be affected by $\%O_2$, which is verified by the peak value of *n*, that reaches a maximum for $\%O_2=0.4\,\%$. The minimum *n* value is obtained for $\%O_2=10.0\ \%$, suggesting lower compactness for films produced under these conditions, presumably due to the deleterious substrate bombardment effects. The E_G calculated from the data in Figures 2.20 and 2.21 are presented in Table 2.3, showing that spectroscopic ellipsometry yields lower values. This is related to the fact that the simulation model used here takes into account the absorption due to tail-states in order to calculate E_G.

2.2.4.3 Effect of composition (binary and multicomponent oxides)

In materials composed by two or more compounds, E_G is expected to assume a value intermediate to the ones of the composing compounds, varying in proportion to their concentration on the final material. However, the variation is rarely linear, mainly because

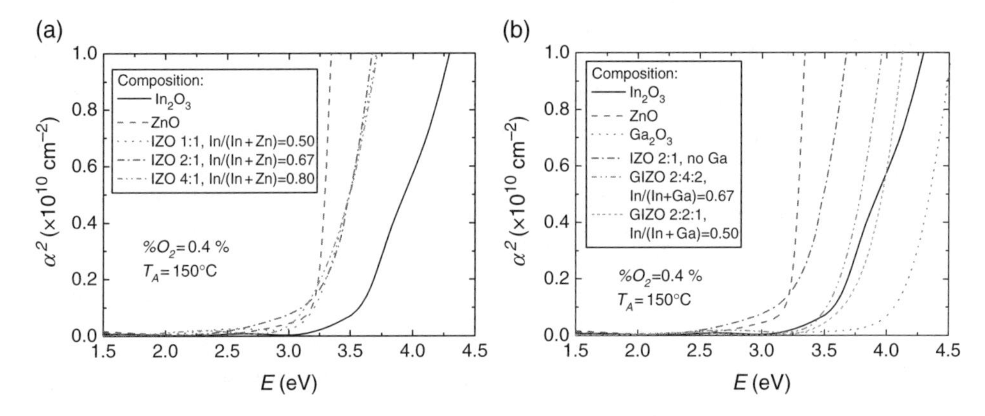

Figure 2.22 *α^2-E as a function of a) In/(In+Zn) and b) In/(In+Ga) atomic ratios for oxide semiconductor thin films annealed at 150°C. Data for the binary compounds (In_2O_3, ZnO and Ga_2O_3) is also included for reference.*

Table 2.4 *E_G values extracted from α^2-E plots for different binary and multicomponent compositions, for thin films annealed at 150°C.*

Composition	In_2O_3	ZnO	Ga_2O_3	IZO 2:1, (In/(In+Ga))=1	GIZO 2:4:2 (In/(In+Ga))=0.67	GIZO 2:2:1 (In/(In+Ga))=0.50
E_G (eV)	3.52	3.24	4.16	3.33	3.64	3.81

of the formation of band tails due to the random perturbation introduced in the host lattice by the minority atoms. This is the typical case of germanium-silicon alloys [118]. To see if the same applies for multicomponent oxides, E_G evolution is analyzed for different In/(In+Zn) (Figure 2.22a) and In/(In+Ga) (Figure 2.22b) atomic ratios. The extracted E_G values are presented in Table 2.4.

It can be seen that within the range of In/(In+Zn) ratios analyzed here E_G does not present significant changes, being the values located between 3.29 and 3.33 eV. These values are similar to the ones reported in literature for IZO films with moderate-to-high ρ. For films with $N > 10^{20}$ cm^{-3} E_G can increase above 3.5 eV due to Burstein-Moss shift [23, 53]. The fact that a systematic trend for E_G is not verified with the In/(In+Zn) atomic ratio should be ascribed to the insensitivity of the electronic structure to the metal composition, as proposed by Leenheer *et al.* [53]. Additionally, a very large E_G variation would not be expected, given the proximity of In_2O_3 and ZnO bandgaps [124]. As expected, even if gallium is incorporated in the structure (i.e., for GIZO), E_G is still relatively unaffected by the In/(In+Zn) atomic ratio. However, In/(In+Ga) atomic ratio has an important effect on E_G, with a shift of ≈0.5 eV being verified when In/(In+Ga) is decreased from 1 to 0.50. Since Ga_2O_3 has a considerably larger E_G than In_2O_3 and ZnO, the increase in Ga_2O_3 content widens the bandgap in the amorphous GIZO films [57]. Similar dependences for GIZO films were found by Kang *et al.* [124].

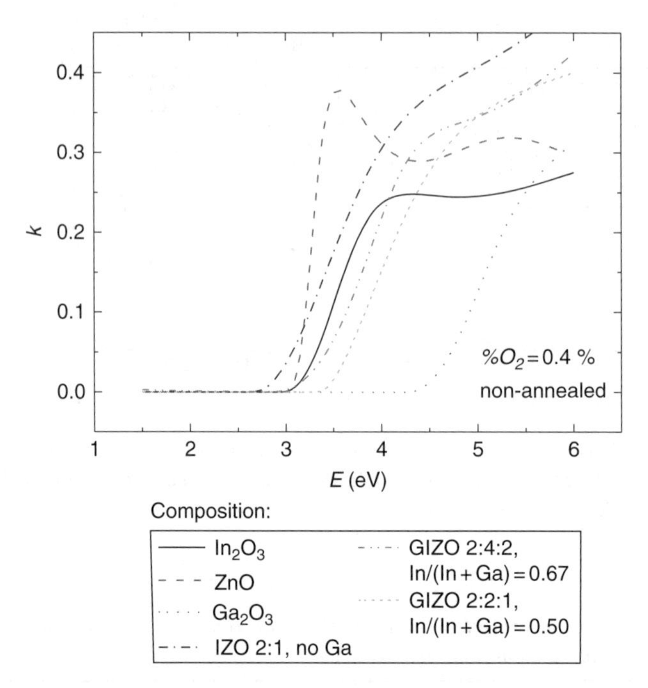

Figure 2.23 *Spectroscopic ellipsometry data, showing k-E plots for oxide semiconductor films with different compositions. Non-annealed films.*

Spectroscopic ellipsometry analysis reinforces the trends discussed above. In fact, the k-E plots represented in Figure 2.23 for non-annealed films with different compositions are in agreement with Figure 2.22b, being verified that the addition of gallium to the IZO structure significantly shifts the absorption edge toward higher E. Furthermore, given that all the films analyzed in Figure 2.23 are amorphous except ZnO (note that these are non-annealed films), it is noticeable a sharper increase of k with E for ZnO films, consistent with the fact that in polycrystalline materials band transitions occur for a well defined E, while in amorphous materials they occur for a broader E range.

2.2.4.4 *Effect of annealing temperature*

Besides reducing the discrepancy between the obtainable properties using different deposition conditions, such as $\%O_2$ (Figure 2.20), increased T_A may also promote important structural and chemical modifications that are reflected in the optical properties. Figure 2.24 shows the effect of T_A on In_2O_3 and ZnO thin films.

For In_2O_3 (Figure 2.24a) a considerably sharper α^2 rise can be seen when moving from 150 to 200°C, which is coincident with the crystallographic changes occurring on the material: if at 150°C the structure is essentially polycrystalline but still has a significant amorphous contribution, the amorphous phase tends to disappear for higher T_A, being that this evolution also accompanied by changes on the preferential orientations of the crystals, as seen in the XRD analysis (Figure 2.5). However, E_G is only changed from 3.52 to 3.63 eV, reinforcing the idea that the properties of ionic semiconductors based on large cations are

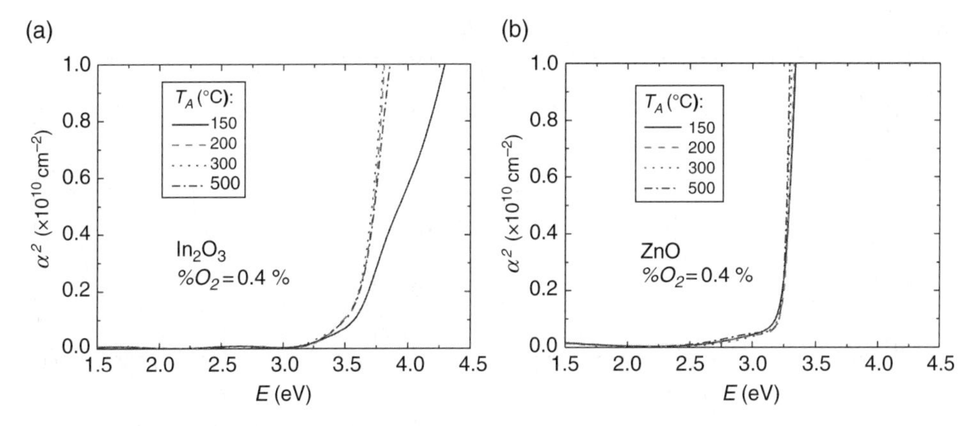

Figure 2.24 α^2-E plots as a function of T_A for a) In_2O_3 and b) ZnO thin films, produced under the same deposition conditions.

considerably less sensitive to structural disorder than covalent semiconductors. Concerning ZnO (Figure 2.24b), which always presents a polycrystalline structure with the same preferential orientation, a slightly sharper rise is verified for higher T_A, being consistent with the improvement of the crystalline quality (larger grain size). Both materials present average transmittance in the visible range (AVT) of 80–85 % and the obtained E_G are in agreement to what is typically found in the literature.

For the multicomponent oxides, different behaviors are observed as T_A increases, depending on the N range. Two illustrative cases are shown in Figure 2.25:

- For materials with low N, such as GIZO 2:4:2, transmittance spectra and consequently α^2-E plots and E_G are not significantly changed with T_A (Figure 2.25a). The only visible change occurs for films annealed at 500°C, for which N increases from $<10^{14}$ (estimated from ρ measurements) to 10^{17} cm^{-3} (measured). These results are consistent with the trend presented for $\%O_2$, since these materials are not degenerated (although at 500°C they should be close to the onset of degeneracy), making the Burstein-Moss rule non-applicable. For the 500°C annealed films, structural relaxation and annihilation of trap states close to the conduction band edge, associated with random ionic distribution and heavy distortion of indium-oxygen-metal bonds, can also contribute to the slight decrease of E_G. For this high T_A, even if In_2O_3 phases are not yet visible in XRD data, some In_2O_3–rich agglomerates may start to form,[1] contributing to a decrease of E_G ($E_{G\ In2O3} < E_{G\ GIZO}$). AVT is maintained around 84 % for all the analyzed T_A.
- On the other hand, significant changes are verified for materials with high N, such as the IZO films produced for electrode application (Figure 2.25b). As seen before (Figure 2.12), N decreases from $\approx 2\times10^{20}$ to $\approx 1\times10^{19}$ cm^{-3} with T_A, but even the lowest N value assures degeneracy. Given the high N involved, a considerable decrease is verified on the transmittance in the near infrared region, i.e., the infrared absorption edge is shifted towards the visible region as N increases. This effect is well predicted by Drude's theory

[1]Note that for the IZO and GIZO compositions analyzed in this work In_2O_3 is always the first phase to crystallize.

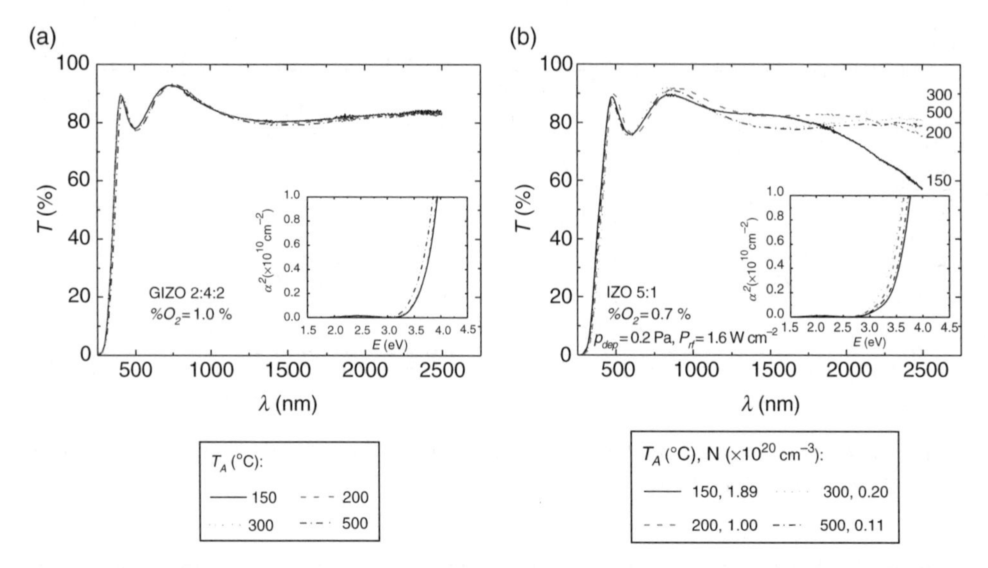

Figure 2.25 *Effect of T_A on the optical transmittance in the UV-Visible-NIR regions for multicomponent oxide semiconductor thin films produced under different processing conditions and compositions, in order to achieve different N ranges: a) GIZO 2:4:2 thin films; b) IZO 5:1 thin films. The insets show the α^2-E plots obtained from the transmittance data.*

for free electrons in metals and is associated with the plasma oscillation of the free carriers that screen the incident electromagnetic wave via transitions within the conduction band [5, 36]. AVT values are also affected by annealing (and consequently by N), being consistently changed between 76 and 79% as N is decreased from 1.9×10^{20} (150°C) to 2.0×10^{19} (300°C). Another different effect from the previously analyzed oxides is that the absorption edge is blue shifted for higher N, in accordance with the Burstein-Moss effect, since in degenerated semiconductors the lowest states in the conduction band are blocked as N increases, turning optical absorption only possible for higher energies [83, 120].

Typical Burstein-Moss shifts require that E_G varies linearly with $N^{2/3}$, when considering a three dimensional parabolic band material [53]. Figure 2.26 shows that the IZO films from Figure 2.25b obey this relation until a T_A of 300°C, with the same being verified for GIZO 2:8:2 films, which always have $N>10^{18}$ cm^{-3}. The deviations from linearity can be due to the error in fitting the α^2-E plot, electron–ion and electron–electron scattering effects which can narrow E_G even when N increases in a degenerate semiconductor [5, 22, 125]. Also, structural changes arising due to the increase of T_A may also affect the optical properties. In fact, when these structural changes are extreme, leading to crystallization (which happens for the IZO films), E_G can deviate significantly from the linear E_G-$N^{2/3}$ relation, as shown by the isolated points in the plot below. A similar effect on E_G was also found by Gonçalves *et al.* in IZO films annealed at 500°C [85]. For GIZO 2:8:2 films, even if crystallization does not occur for T_A=500°C, the formation of In_2O_3-rich agglomerates contributes to decrease E_G, as suggested above.

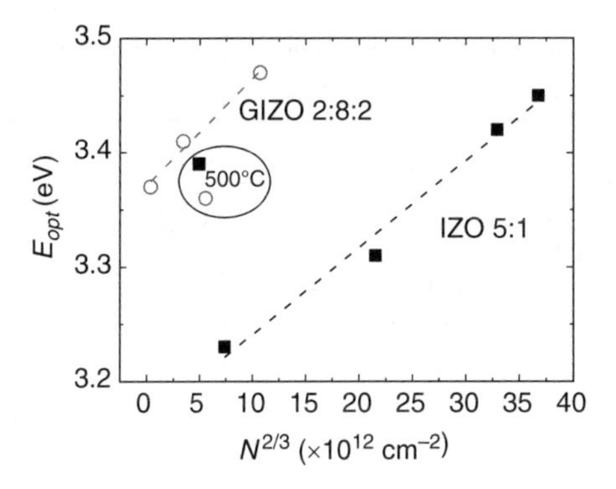

Figure 2.26 E_G-$N^{2/3}$ *plots for IZO 5:1 and GIZO 2:8:2 thin films, with different N values obtained by varying* T_A. *The linear fit shows good agreement of the data with the Burstein-Moss rule for degenerate semiconductors.*

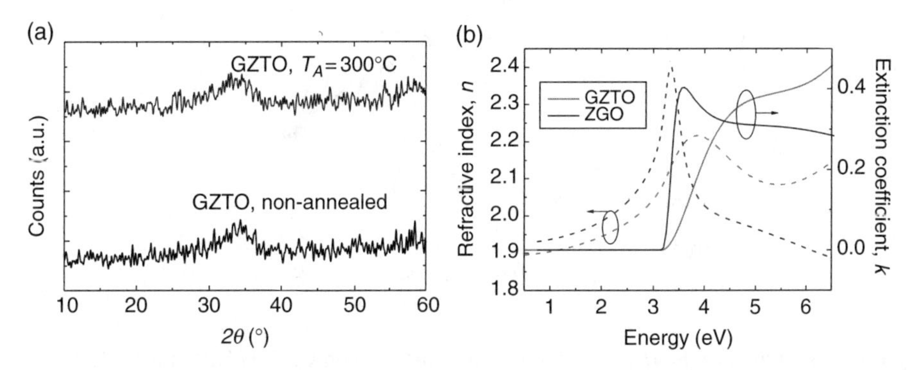

Figure 2.27 Structural (a, XRD) and optical (b, spectroscopic ellipsometry) analysis of co-sputtered GZTO films. Adapted with permission from [61] Copyright (2008) American Institute of Physics.

2.3 Sputtered n-TSOs: Gallium-Zinc-Tin Oxide System

Given the high cost of indium, the search for multicomponent oxides without this element is also a research topic of great interest. As mentioned in section 2.1.2, ZTO and GZTO are seen as the most promising choices for indium-free amorphous oxide semiconductors. At CENIMAT, initial results with GZTO were obtained by co-sputtering from Ga-doped ZnO and metallic tin targets [61]. XRD data obtained from films deposited on glass substrates shows that even after annealing at 300°C the films are amorphous (Figure 2.27a). AFM analysis reveals a very smooth surface, with a RMS roughness of 0.6 nm. Optical characterization by spectroscopic ellipsometry also reinforces the disordered structure of the GZTO films, given the considerably less abrupt increase of k when compared with polycrystalline GZO films (Figure 2.27b).

Elemental analysis using X-ray photoelectron spectroscopy (XPS) reveals a surface enrichment in gallium and tin as T_A increases (Table 2.5). Based on the binding energies

Table 2.5 *Atomic percentages for Zn, Ga, Sn and O obtained from XPS data, for co-sputtered GZTO films.*

T_A (°C)	Ga	Zn	Sn	O (oxide)	O (hydroxide)
RT	1.6	24.6	6.6	35.5	31.7
200	1.6	24.6	7.2	40.4	26.2
300	2.5	21.2	11.8	45.3	19.2

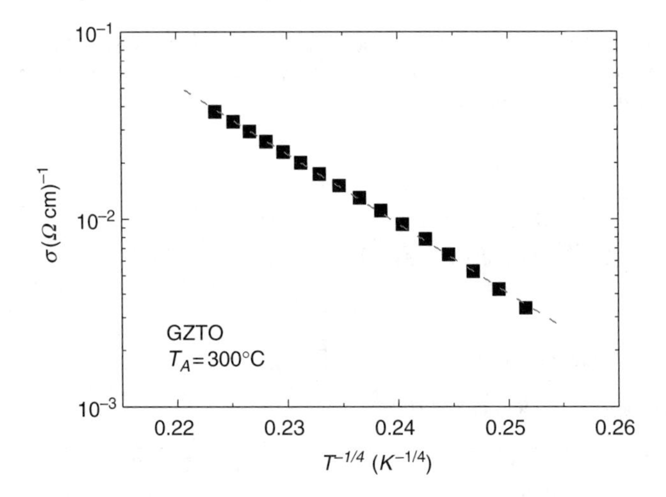

Figure 2.28 *Temperature dependence of σ for co-sputtered GZTO films annealed at 300°C.*

(for Ga and Zn) and modified Auger parameter (for Sn) of the different elements identified in Table 2.5, they correspond to the oxidized forms Ga_2O_3, ZnO and SnO_2. Another result visible from XPS data is the increase of the ratio O(oxide)/O(hydroxide) as T_A increases. Besides that, the spectra charge shifts decrease as T_A increases (9.5, 8.5, and 4 eV for RT, 200 and 300°C, respectively), suggesting that the number of hole traps in the samples decrease with T_A.

Regarding electrical properties, as-deposited GZTO films are highly resistive ($\rho \approx 10^{10}$ Ω cm) and do not present a clear semiconductor behavior in σ-T measurements. After annealing, values in the range of 10^2 Ω cm are obtained and the good linear fits obtained in the σ-$T^{-1/4}$ plots (Figure 2.28) suggest a percolation conduction model, similarly to what was verified for GIZO films (see section 2.2.3.4).

Even if much more work needs to be carried out to optimize GZTO films regarding their processing conditions and composition and despite the higher required T_A when compared with GIZO films, this initial study has already allowed one to fabricate good performing transistors, as will be shown in section 5.2.2.

Figure 2.29 a) SEM, b) AFM and c) XRD data for ZTO thin films deposited by spray-pyrolysis at 400°C.

2.4 Solution-Processed n-TSOs

Even if nowadays most of the TSO thin films are processed by physical routes such as sputtering, over the last five years research on solution-processed TSOs has gained an increased relevance. Low cost, high throughput and large-area coverage are some of the main advantages envisaged for solution-processing. As with sputtered films, multicomponent materials such as ZTO or GIZO are those more intensively explored with regard to low temperature device applications, primarily for TFTs. In CENIMAT, both ZTO and GIZO are being studied, using spray-pyrolysis (ZTO) and sol-gel spin-coating (ZTO and GIZO). This section gives a brief overview regarding our initial results.

2.4.1 ZTO by Spray-pyrolysis

ZTO thin films were prepared by using equal molar ratios of tin (II) chloride ($SnCl_2$) as a source of Sn and zinc acetate dehydrate {$Zn(CH_3COO)_2.2H_2O$} as a source of Zn. These were dissolved in hydrochloridic acid (HCl) and diluted in methanol, and after that mixed and stirred to prepare the final spray solutions. Films with thickness ranging from 10 to 30 nm were sprayed on pre-heated (400°C) Corning 1737 glass and glass coated with ITO/ATO substrates. After deposition, samples were kept at the pre-heating temperature for 30 min. Further detail can be found in [126].

Films deposited on ITO/ATO (the same substrates used to fabricate TFTs) were analyzed by SEM and AFM (Figures 2.29a and b, see color plate section), being observed as a

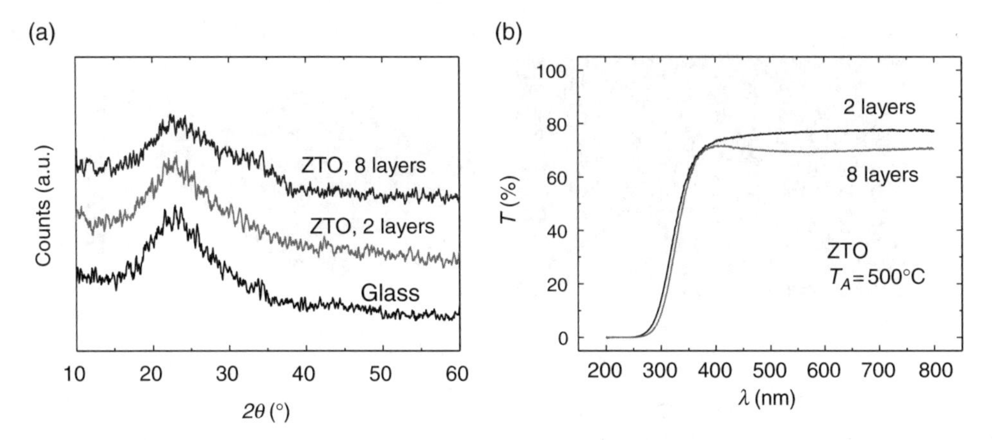

Figure 2.30 *a) XRD and b) optical transmittance data for ZTO thin films deposited by spin-coating, with T_A=500°C.*

surface microstructure consisting of granular shaped clusters distributed homogeneously. On these glass/ITO/ATO/ZTO structures, RMS roughness was found to be 8.6 nm. Figure 2.29c shows the XRD patterns obtained for ZTO films on glass. Small diffraction peaks are obtained, corresponding to tin oxide [126]. More recently, amorphous ZTO films were obtained by decreasing the processing temperature to 200°C.

Further characterization of ZTO thin films is currently being performed, but working TFTs were already fabricated at 200 and 400°C (see section 5.2.4.1).

2.4.2 ZTO by Sol-gel Spin-coating

The same sources of tin and zinc used in section 2.4.1 were used here ($SnCl_2$ and $Zn(CH_3COO)_2.2H_2O$, respectively). In this case, the precursor solution was prepared with 2-methoxyethanol and ethanolamine was added to the mixed solution, which was stirred at 50°C in air for 1h. After the spin-coating process (on glass substrates) the films were annealed at 500°C in air for 1h [127]. In order to obtain films with different thicknesses, multiple layers were sequentially spin-coated and dried. Figure 2.30a presents diffractograms obtained in grazing incidence geometry for ZTO films on glass with two and eight layers (thickness ≈20 and 80 nm), showing that the ZTO films are amorphous. Depending on the composition and processing conditions of solutions and thin films, reports of amorphous solution-processed ZTO films after T_A=600°C can be found in the literature [128]. Optically, ZTO films with two and eight layers exhibit AVT ≈ 76 and 70 %, respectively.

2.4.3 GIZO Sol-gel by Spin-coating

Given the good results achieved with sputtered GIZO films, most of our work with solution-processed TSOs has also been devoted to this material. The effects of composition (by changing the zinc concentration in the precursor solution, obtaining molar ratios of Ga:In:Zn of 1:3:1, 1:3:2 and 1:3:3) and T_A were studied, in order to obtain TSOs that would allow for high performance TFTs.

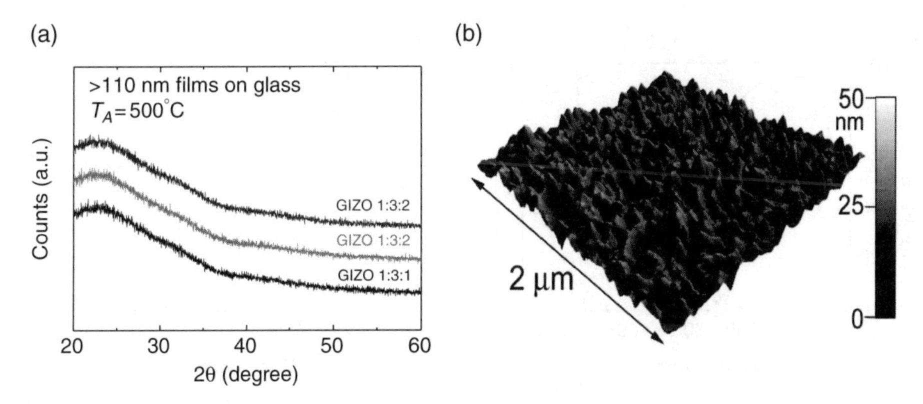

Figure 2.31 *a) XRD and b) AFM data for GIZO thin films deposited by spin-coating, with T_A=500°C. AFM data refers to GIZO 1:3:1 films. Adapted with permission from ref [129] Copyright (2010) American Institute of Physics.*

Zinc acetate dihydrate, indium acetate, gallium nitrate hydrate, and ethanolamine in anhydrous 2-methoxyethanol were used to prepare the precursor solutions. The solutions were stirred at 50°C in air for 1 h and aged for 24 h. After the spin-coating process the films were annealed at temperatures ranging from 300 to 500°C in air for 1h. Further experimental details can be found in [129]. Thick films (>110 nm) deposited on glass were analyzed by XRD, revealing that regardless of their composition all the films are amorphous, even after T_A=500°C (Figure 2.31a, see color plate section). Note that as it happens with sputtered GIZO films, different crystallization temperatures can be found in the literature for solution-processed GIZO films (ranging from ≈95 to >600°C), depending on their composition and preparation methods of solutions and thin films [130–132].

To simulate the conditions to be used on TFTs, similar but thinner films (≈20–30 nm thick) were also deposited on glass coated with ITO/ATO. AFM analysis reveals that the films have a RMS roughness of ≈2–3 nm (Figure 2.31b). These results are also consistent with the SEM analysis, where a smooth surface was observed.

The need to reduce processing costs and expand the range of applications of TSOs demands lower processing temperatures. However, this can be a challenging task with solution-processed films, as generally high temperatures are required to thermally decompose the metal oxide precursors and by-products in the thin films. Thermogravimetry (TG), differential scanning calorimetry (DSC) and Fourier transform infrared (FTIR) spectroscopy were performed to infer the critical temperatures for the formation of GIZO films. Initial weight losses verified in the DSC analysis are primarily attributed to solvent evaporation, being the large exothermic peak at 235°C associated with initial alloying of metal hydroxides (Figure 2.32a) [133]. However, the formation of GIZO films is only completed above 300°C. In fact, FTIR analysis shows a broad and weak IR peak at ≈3380 cm^{-1} for films with T_A=300°C (Figure 2.32b), which is associated to the O-H stretching vibration, presumably indicating the presence of metal hydroxides that at this temperature could not be converted in oxides [134].

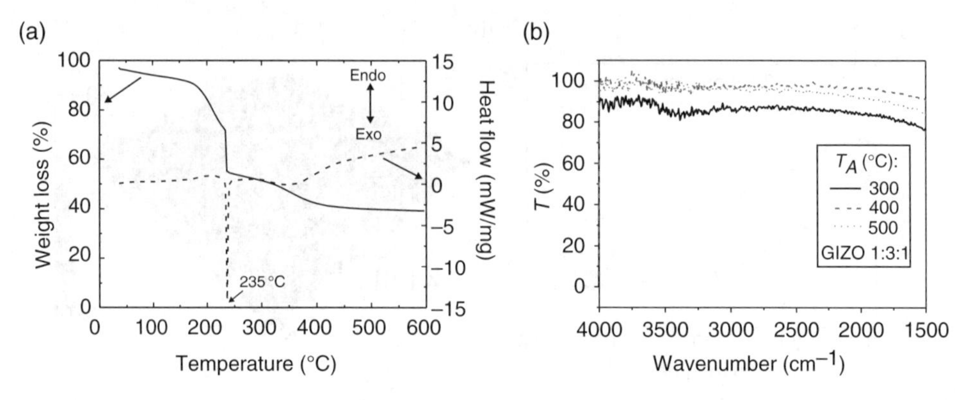

Figure 2.32 *a) TG-DSC and b) FTIR data for GIZO 1:3:1 thin films deposited by spin-coating.*

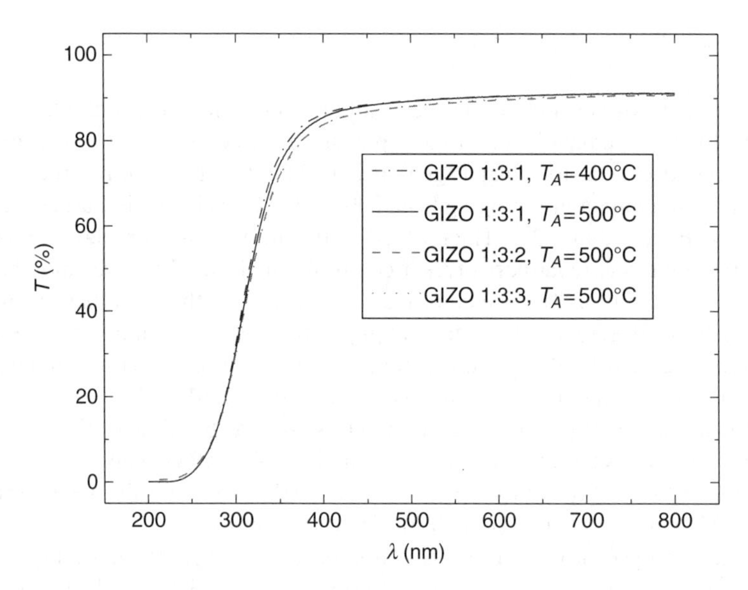

Figure 2.33 *Effect of composition and T_A on the optical transmittance of spin-coated GIZO films.*

Considering the experimental conditions studied herein, the effect of composition and T_A on the optical properties of GIZO films is negligible (Figure 2.33). The films are highly transparent, exhibiting an AVT ≈ 90 % and E_G≈3.4 eV.

The GIZO films reported here were used to fabricate TFTs on ITO/ATO coated glass substrates, as will be shown in Chapter 5, section 5.2.4.3.

References

[1] G H. Kim, B.D. Ahn, H.S. Shin, W. H. Jeong, H.J. Kim, and H. J. Kim (2009) Effect of indium composition ratio on solution-processed nanocrystalline InGaZnO thin film transistors, *Applied Physics Letters* **94**, 233501–1–233501-3.

[2] J.H. Shin, Y.J. Cho, and D.K. Choi (2009) Effect of RF Power on the Structural, Optical and Electrical Properties of Amorphous InGaZnO Thin Films Prepared by RF Magnetron Sputtering, *J. Korean Inst. Met. Mater.* **47**, 38–43.

[3] D.R. Clarke (199) Varistor ceramics, *J. Am. Ceram. Soc.* **82**, 485–502.

[4] D.P. Norton, Y.W. Heo, M.P. Ivill, K. Ip, S.J. Pearton, M.F. Chisholm, and T. Steiner (2004) ZnO: growth, doping & processing, *Materials Today* **7**, 34–40.

[5] H. Hartnagel, A. Dawar, A. Jain, and C. Jagadish (1995) *Semiconducting Transparent Thin Films*. Bristol: IOP Publishing.

[6] E. Burstein (1954) Anomalous optical absorption limit in InSb, *Physical Review* **93**, 632–3.

[7] J.F. Wager, D.A. Keszler, and R.E. Presley (2008) *Transparent Electronics*. New York: Springer.

[8] R.L. Hoffman (2002) "Development, Fabrication, and Characterization of Transparent Electronic Devices," in *Electrical and Computer Engineering*. Master's thesis Oregon: Oregon State University.

[9] K. Ellmer (2001) Resistivity of polycrystalline zinc oxide films: current status and physical limit, *J. Phys. D-Appl. Phys.* **34**, 3097–3108.

[10] P.F. Carcia, R.S. McLean, M.H. Reilly, and G. Nunes (2003) Transparent ZnO thin-film transistor fabricated by rf magnetron sputtering, *Applied Physics Letters* **82**, 1117–19.

[11] E.M.C. Fortunato, P.M.C. Barquinha, A. Pimentel, A.M.F. Goncalves, A.J.S. Marques, R.F.P. Martins, and L.M.N. Pereira (2004) Wide-bandgap high-mobility ZnO thin-film transistors produced at room temperature, *Applied Physics Letters* **85**, 2541–3.

[12] B.J. Jin, S. Im, and S.Y. Lee (2000) Violet and UV luminescence emitted from ZnO thin films grown on sapphire by pulsed laser deposition, *Thin Solid Films* **366**, 107–10.

[13] B.J. Jin, S.H. Bae, S.Y. Lee, and S. Im (2000) Effects of native defects on optical and electrical properties of ZnO prepared by pulsed laser deposition, *Mater. Sci. Eng. B-Solid State Mater. Adv. Technol.* **71**, 301–5.

[14] Y. Nakanishi, A. Miyake, H. Kominami, T. Aoki, Y. Hatanaka, and G. Shimaoka (1999) Preparation of ZnO thin films for high-resolution field emission display by electron beam evaporation, *Appl. Surf. Sci.* **142**, 233–6.

[15] H. Sato, T. Minami, T. Miyata, S. Takata, and M. Ishii (1994) Transparent conducting ZnO thin-films prepared on low-temperature substrates by chemical-vapor-deposition using $Zn(C_5H_7O_2)_2$, *Thin Solid Films* **246**, 65–70.

[16] S.A. Studenikin, N. Golego, and M. Cocivera (1998) Fabrication of green and orange photoluminescent, undoped ZnO films using spray pyrolysis, *Journal of Applied Physics* **84**, 2287–94.

[17] M. Ohyama, H. Kozuka, and T. Yoko (1997) Sol-gel preparation of ZnO films with extremely preferred orientation along (002) plane from zinc acetate solution, *Thin Solid Films* **306**, 78–85.

[18] S.T. Meyers, J.T. Anderson, C.M. Hung, J. Thompson, J.F. Wager, and D.A. (2008) Keszler, Aqueous Inorganic Inks for Low-Temperature Fabrication of ZnO TFTs, *J. Am. Chem. Soc.* **130**, 17603–9.

[19] J.Z. Wang, E. Elamurugu, V. Sallet, F. Jomard, A. Lusson, A.M.B. do Rego, P. Barquinha, G. Goncalves, R. Martins, and E. Fortunato (2008) Effect of annealing on the properties of N-doped ZnO films deposited by RF magnetron sputtering, *Appl. Surf. Sci.* **254**, 7178–82.

[20] A. Zunger (2003) Practical doping principles, *Applied Physics Letters* **83**, 57–9.

[21] J.R. Bellingham, W.A. Phillips, and C.J. Adkins (1991) Amorphous indium oxide, *Thin Solid Films* **195**, 23–32.

[22] J.R. Bellingham, W.A. Phillips, and C.J. Adkins (1990) Electrical and optical properties of amorphous indium oxide, *J. Phys.-Condes. Matter* **2**, 6207–21.

[23] N. Ito, Y. Sato, P.K. Song, A. Kaijio, K. Inoue, and Y. Shigesato (2006) Electrical and optical properties of amorphous indium zinc oxide films, *Thin Solid Films* **496**, 99–103.

[24] B.G. Lewis and D.C. Paine (2000) Applications and processing of transparent conducting oxides, *MRS Bull.* **25**, 22–7.
[25] I. Hamberg, C.G. Granqvist, K.F. Berggren, B.E. Sernelius, and L. Engström (1984) Band-gap widening in heavily Sn-doped In_2O_3, *Phys. Rev. B* **30**, 3240.
[26] H. Kim, C.M. Gilmore, A. Pique, J. S. Horwitz, H. Mattoussi, H. Murata, Z.H. Kafafi, and D.B. Chrisey (1999) Electrical, optical, and structural properties of indium-tin-oxide thin films for organic light-emitting devices, *Journal of Applied Physics* **86**, 6451–61.
[27] A. Gurlo, M. Ivanovskaya, N. Barsan, M. Schweizer-Berberich, U. Weimar, W. Gopel, and A. Dieguez (1997) Grain size control in nanocrystalline In_2O_3 semiconductor gas sensors, *Sens. Actuator B-Chem.* **44**, 327–33.
[28] V. Vasu and A. Subrahmanyam (1990) Reaction-kinetics of the formation of indium tin oxide films grown by spray pyrolysis, *Thin Solid Films* **193**, 696–703.
[29] Y. Sawada, C. Kobayashi, S. Seki, and H. Funakubo (2002) Highly-conducting indium-tin-oxide transparent films fabricated by spray CVD using ethanol solution of indium (III) chloride and tin (II) chloride, *Thin Solid Films* **409**, 46–50.
[30] G. Buhler, D. Tholmann, and C. Feldmann (2007) One-pot synthesis of highly conductive indium tin oxide nanocrystals, *Adv. Mater.* **19**, 2224–7.
[31] Y. Shigesato, S. Takaki, and T. Haranoh (1992) Electrical and structural properties of low resistivity tin-doped indium oxide films, *Journal of Applied Physics* **71**, 3356–64.
[32] Y. Shigesato, D.C. Paine, and T.E. Haynes (1993) Study of the effect of ion-implantation on the electrical and microstructural properties of tin-doped indium oxide thin films, *Journal of Applied Physics* **73**, 3805–11.
[33] T. Minami (2008) Present status of transparent conducting oxide thin-film development for Indium-Tin-Oxide (ITO) substitutes, *Thin Solid Films* **516**, 5822–8.
[34] P. Canhola, N. Martins, L. Raniero, S. Pereira, E. Fortunato, I. Ferreira, and R. Martins (2005) Role of annealing environment on the performances of large area ITO films produced by rf magnetron sputtering, *Thin Solid Films* **487**, 271–6.
[35] T. Minami (2005) Transparent conducting oxide semiconductors for transparent electrodes, *Semicond. Sci. Technol.* **20**, S35–S44.
[36] E. Fortunato, D. Ginley, H. Hosono, and D.C. Paine (2007) Transparent conducting oxides for photovoltaics, *MRS Bull.* **32**, 242–7.
[37] V. Assuncao, E. Fortunato, A. Marques, H. Aguas, F.I.M.E.V. Costa, and R. Martins (2003) Influence of the deposition pressure on the properties of transparent and conductive ZnO : Ga thin-film produced by r.f. sputtering at room temperature, *Thin Solid Films* **427**, 401–5.
[38] H. Hosono (2006) Ionic amorphous oxide semiconductors: Material design, carrier transport, and device application, *J. Non-Cryst. Solids* **352**, 851–8.
[39] H.Q. Chiang (2007) "Development of Oxide Semiconductors: Materials, Devices, and Integration," in *Electrical and Computer Engineering*. PhD thesis Oregon: Oregon State University.
[40] D.C. Look, D.C. Reynolds, J.R. Sizelove, R.L. Jones, C.W. Litton, G. Cantwell, and W.C. Harsch (1998) Electrical properties of bulk ZnO, *Solid State Communications* **105**, 399–401.
[41] A. Tsukazaki, A. Ohtomo, and M. Kawasaki (2006) High-mobility electronic transport in ZnO thin films, *Applied Physics Letters* **88**, 3.
[42] R.L. Weiher (1962) Electrical Properties of Single Crystals of Indium Oxide, *Journal of Applied Physics* **33**, 2834–9.
[43] C.G. Fonstad and R.H. Rediker (1971) Electrical Properties of High-Quality Stannic Oxide Crystals, *Journal of Applied Physics* **42**, 2911–18.
[44] H. Hosono, K. Nomura, Y. Ogo, T. Uruga, and T. Kamiya (2008) Factors controlling electron transport properties in transparent amorphous oxide semiconductors, *J. Non-Cryst. Solids* **354**, 2796–2800.
[45] M. Okude, K. Ueno, S. Itoh, M. Kikuchi, A. Ohtomo, and M. Kawasaki (2008) Effect of in situ annealed SnO_2 buffer layer on structural and electrical properties of (001) SnO_2/TiO_2 heterostructures, *J. Phys. D-Appl. Phys.* **41**, 4.
[46] C.R. Kagan and P. Andry (2003) *Thin-film transistors*. New York: Marcel Dekker, Inc.
[47] E.P. Denton, H. Rawson, and J.E. Stanworth (1954) Vanadate Glasses, *Nature* **173**, 1030–2.

[48] H. Hosono, N. Kikuchi, N. Ueda, H. Kawazoe, and K.-i. Shimidzu (1995) Amorphous transparent electroconductor $2CdO.GeO_2$: Conversion of amorphous insulating cadmium germanate by ion implantation, *Applied Physics Letters* **67**, 2663–5.
[49] H. Hosono, Y. Yamashita, N. Ueda, H. Kawazoe, and K.-i. Shimidzu (1996) New amorphous semiconductor: $2CdO.PbO_x$, *Applied Physics Letters* **68**, 661–3.
[50] H. Hosono, M. Yasukawa, and H. Kawazoe (1996) Novel oxide amorphous semiconductors: Transparent conducting amorphous oxides, *J. Non-Cryst. Solids* **203**, 334–44.
[51] H. Hosono, N. Kikuchi, N. Ueda, and H. Kawazoe (1996) Working hypothesis to explore novel wide band gap electrically conducting amorphous oxides and examples, *J. Non-Cryst. Solids* **200**, 165–9.
[52] K. Nomura, H. Ohta, A. Takagi, T. Kamiya, M. Hirano, and H. Hosono (2004) Room-temperature fabrication of transparent flexible thin-film transistors using amorphous oxide semiconductors, *Nature* **432**, 488–92.
[53] A.J. Leenheer, J.D. Perkins, M. van Hest, J.J. Berry, R. P. O'Hayre, and D.S. Ginley (2008) General mobility and carrier concentration relationship in transparent amorphous indium zinc oxide films, *Phys. Rev. B* **77**, 5.
[54] N.L. Dehuff, E.S. Kettenring, D. Hong, H. Q. Chiang, J.F. Wager, R.L. Hoffman, C.H. Park, and D.A. Keszler (2005) Transparent thin-film transistors with zinc indium oxide channel layer, *Journal of Applied Physics* **97**, 064505-1–064505-5.
[55] M.G. McDowell and I.G. Hill (2009) Influence of Channel Stoichiometry on Zinc Indium Oxide Thin-Film Transistor Performance, *IEEE Trans. Electron Devices* **56**, 343–7.
[56] G. Bernardo, G. Goncalves, P. Barquinha, Q. Ferreira, G. Brotas, L. Pereira, A. Charas, J. Morgado, R. Martins, and E. Fortunato (2009) Polymer light-emitting diodes with amorphous indium-zinc oxide anodes deposited at room temperature, *Synth. Met.* **159**, 1112–15.
[57] M. Orita, H. Ohta, M. Hirano, S. Narushima, and H. Hosono (2001) Amorphous transparent conductive oxide $InGaO_3(ZnO)_m$ (m<=4): a Zn 4s conductor, *Philosophical Magazine Part B* **81**, 501–15.
[58] R.L. Hoffman (2006) Effects of channel stoichiometry and processing temperature on the electrical characteristics of zinc tin oxide thin-film transistors, *Solid-State Electron.* **50**, 784–7.
[59] H.Q. Chiang, J.F. Wager, R.L. Hoffman, J. Jeong, and D.A. Keszler (2005) High mobility transparent thin-film transistors with amorphous zinc tin oxide channel layer, *Applied Physics Letters* **86**, 013503-1–013503-3.
[60] P. Gorrn, M. Sander, J. Meyer, M. Kroger, E. Becker, H.H. Johannes, W. Kowalsky, and T. Riedl (2006) Towards see-through displays: Fully transparent thin-film transistors driving transparent organic light-emitting diodes, *Advanced Materials* **18**, 738–41.
[61] E.M.C. Fortunato, L.M.N. Pereira, P.M.C. Barquinha, A.M.B. do Rego, G. Goncalves, A. Vila, J.R. Morante, and R.F.P. Martins (2008) High mobility indium free amorphous oxide thin film transistors, *Applied Physics Letters* **92**, 222103-1–222103-3.
[62] Q.M. Song, B.J. Wu, B. Xie, F. Huang, M. Li, H. Q. Wang, Y.S. Jiang, and Y.Z. Song (2009) Resputtering of zinc oxide films prepared by radical assisted sputtering, *Journal of Applied Physics* **105**, 044509-1–044509-9.
[63] K. Ellmer (2000) Magnetron sputtering of transparent conductive zinc oxide: relation between the sputtering parameters and the electronic properties, *J. Phys. D-Appl. Phys.* **33**, R17–R32.
[64] J.O. Barnes, D.J. Leary, and A.G. Jordan (1980) Relationship between deposition conditions and physical-properties of sputtered ZnO, *J. Electrochem. Soc.* **127**, 1636–40.
[65] J.A. Thornton and V.L. Hedgcoth (1976) Transparent conductive Sn-doped indium oxide coatings deposited by reactive sputtering with a post cathode, *Journal of Vacuum Science & Technology* **13**, 117–21.
[66] A.S. Ryzhikov, R.B. Vasiliev, M.N. Rumyantseva, L.I. Ryabova, G.A. Dosovitsky, A.M. Gilmutdinov, V.F. Kozlovsky, and A.M. Gaskov (2002) Microstructure and electrophysical properties of SnO_2, ZnO and In_2O_3 nanocrystalline films prepared by reactive magnetron sputtering, *Mater. Sci. Eng. B-Solid State Mater. Adv. Technol.* **96**, 268–74.
[67] K. Wasa, M. Kitabakate, and H. Adachi (2004) *Thin Film Materials Technology: Sputtering of Compound Materials*. New York: William Andrew, Inc. and Springer-Verlag GmbH & Co. KG.

[68] H. Andersen, H. Bay, R. Behrisch, M. Robinson, H. Roosendaal, and P. Sigmund (1981) *Sputtering by Particle Bombardment I: Physical Sputtering of Single-Element Solids*. New York: Springer-Verlag Berlin Heidelberg.
[69] J.F. Chang, H.L. Wang, and M.H. Hon (2000) Studying of transparent conductive ZnO: Al thin films by RF reactive magnetron sputtering, *Journal of Crystal Growth* **211**, 93–7.
[70] S. Masuda, K. Kitamura, Y. Okumura, S. Miyatake, H. Tabata, and T. Kawai (2003) Transparent thin film transistors using ZnO as an active channel layer and their electrical properties, *Journal of Applied Physics* **93**, 1624–30.
[71] J.-H. Jou, M.-Y. Han, and D.-J. Cheng (1992) Substrate dependent internal stress in sputtered zinc oxide thin films, *Journal of Applied Physics* **71**, 4333–6.
[72] F.O. Adurodija, L. Semple, and R. Brüning (2005) Real-time in situ crystallization and electrical properties of pulsed laser deposited indium oxide thin films, *Thin Solid Films* **492**, 153–7.
[73] H. Morikawa and M. Fujita (2000) Crystallization and electrical property change on the annealing of amorphous indium-oxide and indium-tin-oxide thin films, *Thin Solid Films* **359**, 61–7.
[74] H. Nakazawa, Y. Ito, E. Matsumoto, K. Adachi, N. Aoki, and Y. Ochiai (2006) The electronic properties of amorphous and crystallized In_2O_3 films, *Journal of Applied Physics* **100**, 8.
[75] N.H. Kim and H.W. Kim (2004) Room temperature growth of zinc oxide films on Si substrates by the RF magnetron sputtering, *Materials Letters* **58**, 938–43.
[76] B. Cullity (1956) *Ellements of X-ray Diffraction*. USA: Addison-Wesley Publishing Company, Inc.
[77] J. Yoo, J. Lee, S. Kim, K. Yoon, I.J. Park, S.K. Dhungel, B. Karunagaran, D. Mangalaraj, and J.S. Yi (2004) "High transmittance and low resistive ZnO : Al films for thin film solar cells," in *EMRS Symposium on Thin Film Chalcogenide Photovoltaic Materials*, Strasbourg, 213–17.
[78] U. Ozgur, Y.I. Alivov, C. Liu, A. Teke, M.A. Reshchikov, S. Dogan, V. Avrutin, S.J. Cho, and H. Morkoc (2005) A comprehensive review of ZnO materials and devices, *Journal of Applied Physics* **98**, 103.
[79] Dhananjay and S.B. Krupanidhi (2007) Low threshold voltage ZnO thin film transistor with a $Zn_{0.7}Mg_{0.3}O$ gate dielectric for transparent electronics, *Journal of Applied Physics* **101**, 123717-1– 123717-6.
[80] K. Ellmer, R. Cebulla, and R. Wendt (1997) "Polycrystalline ZnO- and ZnO: Al-layers: Dependence of film stress and electrical properties on the energy input during the magnetron sputtering deposition," in *Symposium on Polycrystalline Thin Films - Structure, Texture, Properties and Applications III, at the 1997 MRS Spring Meeting*, San Francisco, Ca, 245–50.
[81] J. Kim and K. Yoon (2009) Influence of post-deposition annealing on the microstructure and properties of Ga_2O_3:Mn thin films deposited by RF planar magnetron sputtering, *Journal of Materials Science: Materials in Electronics* **20**, 879–84.
[82] P. Marie, X. Portier, and J. Cardin (2008) Growth and characterization of gallium oxide thin films by radiofrequency magnetron sputtering, *Physica Status Solidi (a)* **205**, 1943–6.
[83] A. Takagi, K. Nomura, H. Ohta, H. Yanagi, T. Kamiya, M. Hirano, and H. Hosono (2005) Carrier transport and electronic structure in amorphous oxide semiconductor, a-$InGaZnO_4$, *Thin Solid Films* **486**, 38–41.
[84] K. Tominaga, T. Takao, A. Fukushima, T. Moriga, and I. Nakabayashi (2002) Amorphous ZnO-In_2O_3 transparent conductive films by simultaneous sputtering method of ZnO and In_2O_3 targets, *Vacuum* **66**, 505–9.
[85] G. Goncalves, P. Barquinha, L. Raniero, R. Martins, and E. Fortunato (2008) Crystallization of amorphous indium zinc oxide thin films produced by radio-frequency magnetron sputtering, *Thin Solid Films* **516**, 1374–6.
[86] M.P. Taylor, D.W. Readey, C.W. Teplin, M. van Hest, J. L. Alleman, M.S. Dabney, L.M. Gedvilas, B.M. Keyes, B. To, J.D. Perkins, and D.S. Ginley (2005) The electrical, optical and structural properties of $In_xZn_{1-x}O_y$ ($0 <= x <= 1$) thin films by combinatorial techniques, *Meas. Sci. Technol.* **16**, 90–4.
[87] P.T. Erslev, H.Q. Chiang, D. Hong, J.F. Wager, and J.D. Cohen (2008) Electronic properties of amorphous zinc tin oxide films by junction capacitance methods, *J. Non-Cryst. Solids* **354**, 2801–4.
[88] A. Suresh, P. Gollakota, P. Wellenius, A. Dhawan, and J.F. Muth (2008) Transparent, high mobility of InGaZnO thin films deposited by PLD, *Thin Solid Films* **516**, 1326–9.

[89] R. Cebulla, R. Wendt, and K. Ellmer (1998) Al-doped zinc oxide films deposited by simultaneous rf and dc excitation of a magnetron plasma: Relationships between plasma parameters and structural and electrical film properties, *Journal of Applied Physics* **83**, 1087–95.

[90] R.E. Presley, C.L. Munsee, C.H. Park, D. Hong, J.F. Wager, and D.A. Keszler (2004) Tin oxide transparent thin-film transistors, *J. Phys. D-Appl. Phys.* **37**, 2810–13.

[91] H.H. Hsieh and C.C. Wu (2007) Amorphous ZnO transparent thin-film transistors fabricated by fully lithographic and etching processes, *Applied Physics Letters* **91**, 013502-1–013502-3.

[92] R. Martins, P. Barquinha, A. Pimentel, L. Pereira, and E. Fortunato (2005) Transport in high mobility amorphous wide band gap indium zinc oxide films, *Phys. Status Solidi A-Appl. Mat.* **202**, R95–R97.

[93] C.G. Choi, S.J. Seo, and B.S. Bae (2008) Solution-processed indium-zinc oxide transparent thin-film transistors, *Electrochem. Solid State Lett.* **11**, H7–H9.

[94] P.F. Carcia, R.S. McLean, and M.H. Reilly (2005) Oxide engineering of ZnO thin-film transistors for flexible electronics, *J. Soc. Inf. Disp.* **13**, 547–54.

[95] I. Hamberg and C.G. Granqvist (1986) Evaporated Sn-doped In_2O_3 films - basic optical-properties and applications to energy-efficient windows, *Journal of Applied Physics* **60**, R123–R159.

[96] O. Kappertz, R. Drese, J.M. Ngaruiya, and M. Wuttig (2005) Reactive sputter deposition of zinc oxide: Employing resputtering effects to tailor film properties, *Thin Solid Films* **484**, 64–7.

[97] S. Zafar, C.S. Ferekides, and D.L. Morel (1995) Characterization and analysis of ZnO:Al deposited by reactive magnetron sputtering, *Journal of Vacuum Science & Technology A: Vacuum, Surfaces, and Films* **13**, 2177–82.

[98] K. Tominaga, K. Kuroda, and O. Tada (1988) Radiation effect due to energetic oxygen-atoms on conductive Al-doped ZnO films, *Jpn. Journal of Applied Physics Part 1 – Regul. Pap. Short Notes Rev. Pap.* **27**, 1176–80.

[99] D. Hong and J.F. Wager (2005) Passivation of zinc-tin-oxide thin-film transistors, *J. Vac. Sci. Technol. B* **23**, L25–L27.

[100] D. Kang, H. Lim, C. Kim, I. Song, J. Park, Y. Park, and J. Chung (2007) Amorphous gallium indium zinc oxide thin film transistors: Sensitive to oxygen molecules, *Applied Physics Letters* **90**, 192101-1–192101-3.

[101] T. Sasabayashi, N. Ito, E. Nishimura, M. Kon, P.K. Song, K. Utsumi, A. Kaijo, and Y. Shigesato (2003) "Comparative study on structure and internal stress in tin-doped indium oxide and indium-zinc oxide films deposited by r.f. magnetron sputtering," in *3rd International Symposium on Transparent Oxide Thin Films for Electronics and Optics (TOEO-3)*, Tokyo, 219–23.

[102] B. Yaglioglu, Y.J. Huang, H.Y. Yeom, and D.C. Paine (2006) A study of amorphous and crystalline phases in In_2O_3-10wt.% ZnO thin films deposited by DC magnetron sputtering, *Thin Solid Films* **496**, 89–94.

[103] G. Goncalves, E. Elangovan, P. Barquinha, L. Pereira, R. Martins, and E. Fortunato (2007) Influence of post-annealing temperature on the properties exhibited by ITO, IZO and GZO thin films, *Thin Solid Films* **515**, 8562–6.

[104] F.M. Hossain, J. Nishii, S. Takagi, A. Ohtomo, T. Fukumura, H. Fujioka, H. Ohno, H. Koinuma, and M. Kawasaki (2003) Modeling and simulation of polycrystalline ZnO thin-film transistors, *Journal of Applied Physics* **94**, 7768–77.

[105] J.W. Orton, B.J. Goldsmith, M.J. Powell, and J.A. Chapman (1980) Temperature-dependence of intergrain barriers in polycrystalline semiconductor-films, *Applied Physics Letters* **37**, 557–9.

[106] K. Nomura, T. Kamiya, H. Ohta, K. Ueda, M. Hirano, and H. Hosono (2004) Carrier transport in transparent oxide semiconductor with intrinsic structural randomness probed using single-crystalline $InGaO_3(ZnO)_5$ films, *Applied Physics Letters* **85**, 1993–5.

[107] B. Du Ahn, S.H. Oh, H.J. Kim, M.H. Jung, and Y.G. Ko (2007) Low temperature conduction and scattering behavior of Ga-doped ZnO, *Applied Physics Letters* **91**, 252109-1–252109-3.

[108] D.K. Schroder (2006) *Semiconductor Material and Device Characterization*, 3rd ed. New Jersey: John Wiley & Sons, Inc.

[109] D.H. Zhang and H.L. Ma (1996) Scattering mechanisms of charge carriers in transparent conducting oxide films, *Applied Physics A: Materials Science & Processing* **62**, 487–92.

[110] K. Nomura, T. Kamiya, H. Yanagi, E. Ikenaga, K. Yang, K. Kobayashi, M. Hirano, and H. Hosono (2008) Subgap states in transparent amorphous oxide semiconductor, In-Ga-Zn-O, observed by bulk sensitive x-ray photoelectron spectroscopy, *Applied Physics Letters* **92**, 202117-1–202117-3.
[111] H.H. Hsieh, T. Kamiya, K. Nomura, H. Hosono, and C.C. Wu (2008) Modeling of amorphous $InGaZnO_4$ thin film transistors and their subgap density of states, *Applied Physics Letters* **92**, 133503-1–133503-3.
[112] M. Kimura, T. Nakanishi, K. Nomura, T. Kamiya, and H. Hosono (2008) Trap densities in amorphous-$InGaZnO_4$ thin-film transistors, *Applied Physics Letters* **92**, 133512-1–133512-3.
[113] J.M. Holender and G.J. Morgan (1992) The double-sign anomaly of the Hall coefficient in amorphous silicon: Verification by computer simulations, *Philosophical Magazine Letters* **65**, 225–31.
[114] J. Robertson (2008) Disorder, band offsets and dopability of transparent conducting oxides, *Thin Solid Films* **516**, 1419–25.
[115] S. Martin, C.S. Chiang, J.Y. Nahm, T. Li, J. Kanicki, and Y. Ugai (2001) Influence of the amorphous silicon thickness on top gate thin-film transistor electrical performances, *Jpn. Journal of Applied Physics Part 1 - Regul. Pap. Short Notes Rev. Pap.* **40**, 530–7.
[116] S. Yamasaki, U.K. Das, T. Umeda, J. Isoya, and K. Tanaka (2000) Creation and annihilation mechanism of dangling bonds within the a-Si:H growth surface studied by in situ ESR technique, *Journal of Non-Crystalline Solids* **266–269**, 529–33.
[117] N. Itagaki, T. Iwasaki, H. Kumomi, T. Den, K. Nomura, T. Kamiya, and H. Hosono (2008) Zn-In-O based thin-film transistors: Compositional dependence, *Phys. Status Solidi A-Appl. Mat.* **205**, 1915–19.
[118] J.I. Pankove (1971) *Optical Processes in Semiconductors*. New York: Dover Publications, Inc.
[119] N.F. Mott and E.A. Davis (1979) *Electronic Processes in Non-Crystalline Materials*. Oxford: Clarendon.
[120] Y.G. Cao, L. Miao, S. Tanemura, M. Tanemura, Y. Kuno, Y. Hayashi, and Y. Mori (2006) Optical properties of indium-doped ZnO films, *Jpn. Journal of Applied Physics Part 1 – Regul. Pap. Brief Commun. Rev. Pap.* **45**, 1623–8.
[121] T. Kamiya, K. Nomura, and H. Hosono (2009) Electronic structure of the amorphous oxide semiconductor a-$InGaZnO_{4-x}$: Tauc-Lorentz optical model and origins of subgap states, *Phys. Status Solidi A-Appl. Mat.* **206**, 860–7.
[122] B. Kumar, H. Gong, and R. Akkipeddi (2005) High mobility undoped amorphous indium zinc oxide transparent thin films, *Journal of Applied Physics* **98**, 5.
[123] T. Kamiya, K. Nomura, M. Hirano, and H. Hosono (2008) Electronic structure of oxygen deficient amorphous oxide semiconductor a-$InGaZnO_4$: Optical analyses and first-principle calculations, *physica status solidi (c)* **5**, 3098–3100.
[124] D. Kang, I. Song, C. Kim, Y. Park, T.D. Kang, H.S. Lee, J. W. Park, S.H. Baek, S.H. Choi, and H. Lee (2007) Effect of Ga/In ratio on the optical and electrical properties of GaInZnO thin films grown on SiO_2/Si substrates, *Applied Physics Letters* **91**, 091910-1–091910-3.
[125] C.G. Huang, M.L. Wang, Q. L. Liu, Y.G. Cao, Z.H. Deng, Z. Huang, Y. Liu, Q. F. Huang, and W. Guo (2009) Physical properties and growth kinetics of co-sputtered indium-zinc oxide films, *Semicond. Sci. Technol.* **24**, 095019-1–095019-6.
[126] S. Parthiban, E. Elangovan, K.J. Saji, P.K. Nayak, T. Busani, E. Fortunato, and R. Martins (2011) *Internal Report*.
[127] P.K. Nayak, J.V. Pinto, G. Gonçalves, R. Martins, and E. Fortunato (2011) *IEEE Journal of Display Technology*, in press.
[128] Y. J. Chang, D. H. Lee, G. S. Herman, and C. H. Chang, High-Performance, Spin-Coated Zinc Tin Oxide Thin-Film Transistors, *Electrochemical and Solid-State Letters* **10**, H135–H138.
[129] P.K. Nayak, T. Busani, E. Elamurugu, P. Barquinha, R. Martins, Y. Hong, and E. Fortunato (2010) Zinc concentration dependence study of solution processed amorphous indium gallium zinc oxide thin film transistors using high-k dielectric, *Applied Physics Letters* **97**, 183504-1–183504-3.
[130] Y. Ya-Hui, S.S. Yang, K. Chen-Yen, and C. Kan-San (2010) Chemical and Electrical Properties of Low-Temperature Solution-Processed In-Ga-Zn-O Thin-Film Transistors, *Electron Device Letters, IEEE* **31** 329–31.

[131] K. Yong-Hoon, H. Min-Koo, H. Jeong-In, and P. Sung Kyu (2010) Effect of Metallic Composition on Electrical Properties of Solution-Processed Indium-Gallium-Zinc-Oxide Thin-Film Transistors, *Electron Devices, IEEE Transactions on* **57**, 1009–14.

[132] Y. Wang, S. Liu, X. Sun, J. Zhao, G. Goh, Q. Vu, and H. Yu (2010) Highly transparent solution processed In-Ga-Zn oxide thin films and thin film transistors, *Journal of Sol-Gel Science and Technology* **55**, 322–327.

[133] G.H. Kim, B.D. Ahn, H.S. Shin, W.H. Jeong, H.J. Kim, and H.J. Kim (2009) Effect of indium composition ratio on solution-processed nanocrystalline InGaZnO thin film transistors, *Applied Physics Letters* **94**, 233501.

[134] G.H. Kim, H.S. Shin, B.D. Ahn, K.H. Kim, W.J. Park, and H.J. Kim (2009) Formation Mechanism of Solution-Processed Nanocrystalline InGaZnO Thin Film as Active Channel Layer in Thin-Film Transistor, *J. Electrochem. Soc.* **156**, H7–H9.

3

P-type Transparent Conductors and Semiconductors

3.1 Introduction

The enormous success of n-type oxides and their application to transparent conductive oxides (TCOs) as "invisible electrodes" and as semiconductors (TSOs) in active channel layers used in the production of thin film transistors (TFTs) has raised interest in p-type oxides for both types of applications. This is either as a contact (for instance, in organic based devices whose semiconductors with the best performance are p-type or as a contact for the p-layer in a large variety of p-type semiconductors used in optoelectronics, such as solar cells) or as semiconductors to be applied to produce p-type TFTs. The main goal is to have ambipolar devices similar to Complementary Metal Oxide Semiconductors (CMOS), the device that is both the fastest and has the lowest energy consumption, as required when complex and integrated electronic blocks are to be fabricated.

However, until now there have been few reports on p-type TCOs, most of them within the last 10 years, while there has been no report on p-type oxide TFTs with performance similar to that of n-type oxide TFTs. As far as TCOs are concerned, many questions regarding the reproducibility of highly conductive and transparent thin films, as required for optoelectronics, remain unanswered. Moreover, as far as TSOs are concerned, we are facing exactly the same problems presented by organic TFTs but now in an opposite way, since most organic TFTs with high electrical performances described in the literature are p-type [1–4], while the oxide are n-type. Organic materials have been studied for more than 50 years and the significant improvements in semiconducting properties associated with the discovery of electroluminescence in organic diode structures make these materials

Transparent Oxide Electronics: From Materials to Devices, First Edition. Pedro Barquinha, Rodrigo Martins, Luis Pereira and Elvira Fortunato.

excellent candidates for low-cost electronic and optoelectronic integrated systems [5, 6]. One of the main advantages of organic materials is the low temperature fabrication which allows for lightweight flexible displays, while the main disadvantage remains the lifetime of the devices, mostly related to ageing effects and the stability of the devices for long periods of working. However, organic TFTs have been the topic of intense research over recent decades. In spite of that, the overall performance parameters of these devices are still poor compared to oxide materials, in terms of the low mobilities ($<2\,cm^2\,V^{-1}\,s^{-1}$) [7], poor stability and device-to-device variability as well as degraded performance, processability, and environmental stability with regard to to oxygen and environmental moisture, where the humidity is a key problem parameter as far as the stability of the device is concerned. Moreover, the n-type organic semiconductors have much lower mobilities (10^{-2} to $10^{-1}\,cm^2\,V^{-1}\,s^{-1}$), [8–11] limiting their field of application, such as in CMOS, which leads one to look at alternative solutions.

At present, almost all of the reported oxide TCOs and TSOs for TFTs are based on n-type oxides [12–17]. For p-type oxides, the carrier conduction path (valence band) is mainly formed from the oxygen *p* asymmetric orbitals, which severely limits the carrier mobility. Thus, p-type oxides have very low carrier mobility compared to their n-type counterparts, which is the main obstacle to obtaining high performance TCOs and p-channel oxide TFTs. Recently much attention has been given to Cu based semiconductors, of which the delafossite family $CuMO_2$ (M=Al, Ga, In, Y, Sc, La, etc.) is the most important. The simple binary oxides based on ZnO and NiO have been also studied as promising p-type TCOs and TSOs, but without achieving relevant results.

In the following sections the discussion will be focused on one overview of the state of the art concerning p-type TCOs and a deep analysis of TSOs centred mainly on two emergent and promising p-type materials based on copper oxide and tin monoxide, since up to now the only working oxide based p-type TFTs with reasonable electrical performances have been based on these two materials.

3.2 P-type Transparent Conductive Oxides

The use of highly conductive and highly transparent thin films in the visible range of the spectrum is of great relevance for a broad range of optoelectronic device applications, ranging from displays to solar cells. Up to now, most of the known TCOs have been n-type [15, 16, 18], which raises problems related to band offset when these TCOs are used for contacting p-layers, either organic or inorganic (non-full ohmic contact is achieved). Moreover, carriers' recombination is an issue since we have holes to be collected via an electrode where the major free carriers are electrons. One way to improve the efficiency of such optoelectronic devices, especially for solar cells based on thin film technology, is by using p-type TCOs. Also, for applications such as organic light emitting diodes (OLED) the use of p-type TCOs will lead to tremendous gains in their emitting performance. The same happens for the novel generation of transparent p-n oxide [19] based devices or functional sensors and optoelectronic systems working in the UV wavelength range. However, it has been very difficult to produce p-type TCOs. This difficulty has been related to the electronic structure of oxides: the strong localization of the upper edge of the valence band to oxide ions [19]. Therefore, any selection of a p-type conducting oxide must include modification of the energy band structure so as to reduce localization behavior, which in turn requires new insight into the relationship between electronic structure and the properties of oxide materials. The second condition to be considered in the selection of the candidate oxide is the type of structure selected and what it is the binding energy between the cation and oxide ion, which will condition the type of free carriers interaction (direct or indirect band gap transitions). To overcome the inherent difficulties in producing p-type TCO, Kawazoe and his co-workers [20] proposed the chemical design of the p-type conducting and transparent oxides, where the main candidates were Cu-delafossites, such as $CuGaO_2$ and $CuAlO_2$ oxide systems, in their polycrystalline form. Much effort has also been made to prepare p-type layers based on ZnO doped with group V elements (N, P, As, Sb) [20], but such compounds are not stable due to inherent defects such as oxygen vacancies, which act as a hole killer [21]. Despite this limitation, many studies report the fabrication of p–n junctions based on TCO layers, such as: $CuAlO_2$ (p-type)/ZnO (n-type) [22], $CuCrO_2$:Mg (p-type)/ZnO (n-type) [23], Cu_2O (p-type)/ZnO (n-type) [24] and ZnO:Sb (p-type)/ZnO (n-type) [25].

In Table 3.1 we show the evolution of the room temperature resistivity of p-type TCOs thin films since 1995, exhibiting optical transmittances that are always above 80%. The table reveals that only in the last two to three years has a strong jump in the resistivity been achieved, mainly due to the systems Ti-O-Pd [26]; Cu-Nb-O_x [27] and TiO_x [28], where the lowest room temperature resistivity recorded was for the $TiO_{1.5}$ system, which was of 3×10^{-4} Ω cm that corresponds to a hole concentration and Hall mobility of 5×10^{20} cm^{-3} and 30 cm^2 V^{-1} s^{-1}, respectively. These values surpass even those achieved in the n-type system constituted by the $TiO_{1.5}$:Nb system where an electron concentration, a Hall mobility and a room temperature resistivity of 2.2×10^{20} cm^{-3}, 3.0 cm^2 V^{-1} s^{-1} and 1.3×10^{-3} Ω cm, respectively was recorded [28].

Table 3.1 *Electro-optical characteristics of some p-type TCOs with high optical transmittance (>80 %) in the visible range.*

Material	T (%)	ρ (Ω cm)	n (cm^{-3})	μ (cm^2 V^{-1} s^{-1})	E_g (eV)	Year
NiO:Li [30]	80	1.43	–	–	3.6 – 4.0	1997
$SrCu_2O_2$: K [31]	~80	4.8×10^{-2}	6.1×10^{17}	0.46	3.30	1998
Cu-Ga-O_2 [32]	80	20				2000
$CuGaO_2$ [33]	80	6.3×10^{-2}	1.7×10^{18}	0.23	3.60	2001
$CuAlO_2$ [34]	~80	0.22	4.4×10^{17}	–	3.75	2003
SnO_2:Al [35]	80	2.78×10^{1}	6.7×10^{18}	25.90	4.10	2004
ZnO:N [36]	80	0.85	6.6×10^{17}	7.9	–	2005
ZnO:N [36]	80	1.35×10^{-3}	7.6×10^{13}	111	–	2005
ZnO-Cr:N [36]	80	1.82	4.8×10^{17}	23.60	–	2005
$CuAlO_2$ [37]	–	0.39	1.2×10^{18}	–		2005
$CuCrO_2$:Mg [38]	80	1.00	2.0×10^{19}	0.20	3.11	2008
$CuCrO_2$ [39]	>80	1.00	1.2×10^{19}	0.25	2.58 – 2.79	2008
ZnMgO [40]	–	0.22	9.6×10^{17}	1.40	–	2009
TiO_2:Pd [26]	–	1×10^{-3}	–	–	~1.25	2011
Cu-Nb-O [27]	80	6.2×10^{-4}	2.1×10^{21}	0.36	2.6	2011
$TiO_{1.5}$ [28]	80	3.4×10^{-4}	5×10^{20}	30		2011

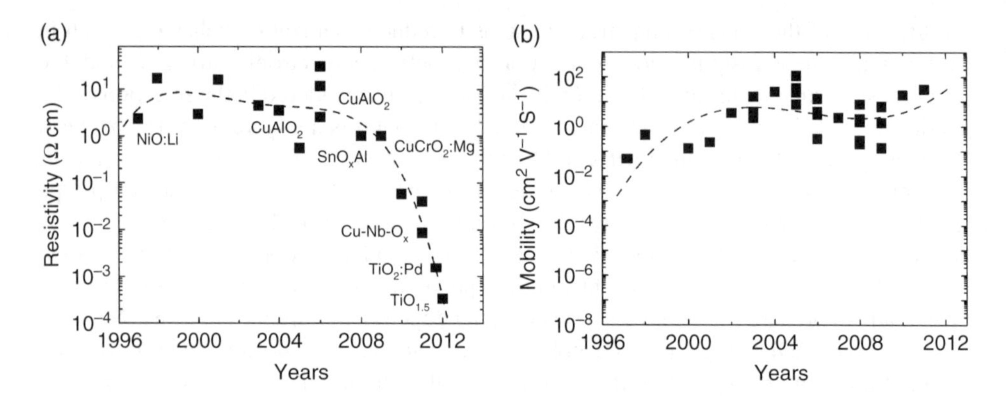

Figure 3.1 *a) Sketch of the room temperature resistivity of the p-type TCO thin films with transmittance above 80 % evolution since 1996; b) Evolution of the mobility since 1996 of the mobility of p-type TCO.*

The explanation for these high values is attributed to the high level of defects, which could induce localized states formed close to the valence band, similar to what happens in the Zn-O system [29].

The resistivity now recorded for the $TiO_{1.5}$ system is slightly higher than the value measured for anatase Nb doped titanium oxide films ($Ti_{1-x}Nb_xO_2$) by Furubayashi *et al.* ($2\text{–}3\times10^{-4}$ Ω cm) [41], where high hole carrier mobilities were recorded. Nevertheless this value is not shown in the diagram of the evolution of the p-type TCO resistivity depicted above since a cross checking measurement in order to determine the type of carrier such as the Seebeck effect is not mentioned by the authors. In Figures 3.1a and b we also highlight some of the most highly transparent p-type TCO systems studied over the years, showing the corresponding evolution of the resistivity and mobility recorded at room temperature. There, we can see the impressive improvement achieved concerning the decrease of the resistvity of the films produced, preserving a high hole carrier mobility, as required for TCOs applications.

Overall we can say that the data depicted in Figure 3.1 show that high performance p-type TCO has been achieved recently thanks to a better understanding of the oxide's electronic structure, which has enabled one to design different oxide structures, by exploiting the role of cation species in determining the proper percolation path of hole carriers, similar to that which has been done with n-type oxides [42]. By doing so, we can "design" the structure and oxide composition by aiming to target high thin film performances, independent of its type of structure order (amorphous or polycrystalline) as is done with n-type oxides [43].

3.3 Thin Film Copper Oxide Semiconductors

The first evidence of semiconductor properties in a metal oxide was found in copper oxide, more precisely Cu_2O, cuprous oxide, in 1917 [44] by Kennard *et al.* (Figure 3.2a). Solid-state devices based on Cu_2O semiconductors are known for more than 90 years even before the era of germanium and silicon devices. Rectifier diodes based on this semiconductor were used industrially as early as 1926 [45] (Figure 3.2b) and most of the theory of semiconductors was developed using the data on Cu_2O devices [46–48].

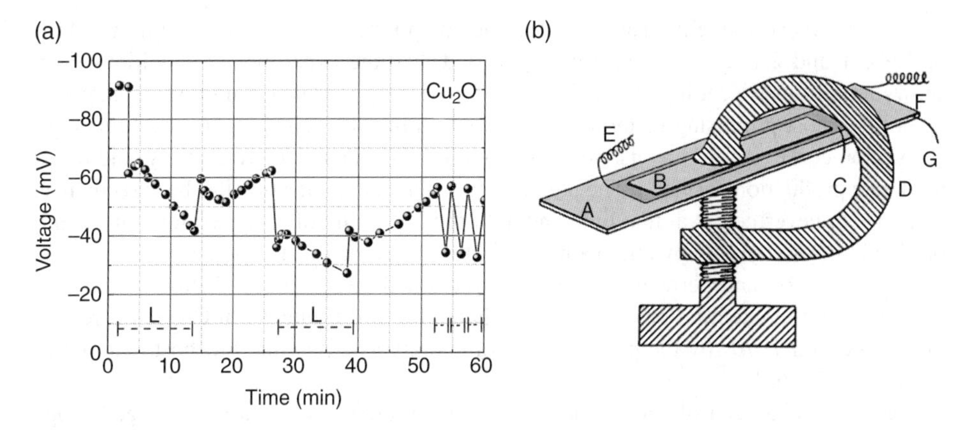

Figure 3.2 a) Evidence of the photo electromotive effect of Cu_2O. A series of readings of the potential at intervals of a minute, the dotted line marked L showing the periods of illumination. Apparently illumination produces, besides the sudden effect, a slow drift of potential followed by recovery in the dark but lagging by two to three minutes. Adapted with permission from ref [44] Copyright (1917) American Physical Society. b) Example of the first rectifier assembly. A is oxidized surface of copper strip; C is the piece of lead used for contact; B is mica insulation; and F is terminal connected to mother copper. Adapted with permission from ref [45] Copyright (1926) AAAS.

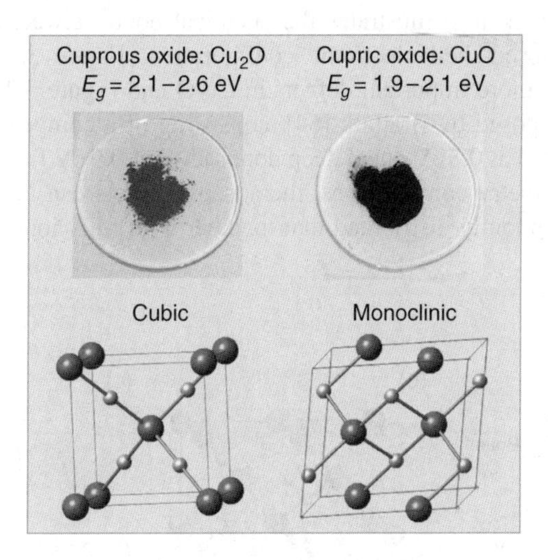

Figure 3.3 Copper (I) oxide or cuprous oxide is one of the principal oxides of copper. The compound can appear either yellow or red, depending on the size of the particles. Cu_2O crystallizes in a cubic structure. Copper (II) oxide or cupric oxide belongs to the monoclinic crystal system. It is a black solid with an ionic structure.

Oxides of copper are known to show p-type conductivity and are attracting renewed interest as promising semiconductor materials for a wide range of optoelectronic devices. There are two common forms of copper oxide: cuprous oxide or cuprite (Cu_2O) and cupric oxide or tenorite (CuO). Figure 3.3 (see color plate section) shows a comparison between these two compounds in terms of optical properties and crystal structure.

Both the CuO (monoclinic) and Cu_2O (cubic) are p-type semiconductors with a band gap of 1.9–2.1 and 2.1–2.6 eV respectively [49] and in some experimental conditions Cu_2O shows mobilities exceeding 100 cm^2 V^{-1} s^{-1} [50–53]. Besides that, Cu_2O was regarded as one of the most promising materials for application in solar cells [54–56] due to its high-absorption coefficient in the visible region, non-toxicity, abundant availability, and low to moderate production cost [57]. The potential for solar cell application has been recognized since 1920, nevertheless at that time and until the beginning of space exploration, energy production from the Sun by photovoltaic effect was just a curiosity. In the following years doping of silicon and germanium was discovered and these semiconductors became the standard and the attention paid to Cu_2O declined. However, interest in Cu_2O was revived during the mid-1970s by the photovoltaic community as a possible low cost material for solar cells [58–62].

The p-type character of Cu_2O is attributed to the presence of negatively charged copper vacancies (V_{Cu}), which introduce an acceptor level at about 0.3 eV above the valence band (V_B) [63]. Some authors also proposed the co-existence of both intrinsic, acceptor and donor levels with a ratio slightly larger than 1 and less than 10 [58, 64]. The nature of the donor levels is not completely clear and is even controversial, where the simplest candidates are oxygen vacancies [65]. In contrast to the majority of metal oxides, in which the top of V_B is mainly formed from localized and anisotropic O 2p orbitals, which leads to a low hole mobility due to hopping conduction, the top of the V_B is composed of fully occupied hybridized orbitals (Cu 3d and O 2p) with Cu d sates dominating the top of the V_B. Figure 3.4a (see color plate section) illustrates the chemical bond between an oxide ion and a cation that has a closed-shell electronic configuration; Figure 3.4b shows a pictorial representation of the more important defects in Cu_2O and Figure 3.4c depicts the simple electronic model proposed by Brattain [64], consisting of a compensated semiconductor with one acceptor level at 0.3 eV and a deep donor level at 0.9 eV from V_B.

For quasi-stoichiometry compositions, the acceptor levels can be related to point and interstitial defects randomly distributed that together with fully ionized copper vacancies

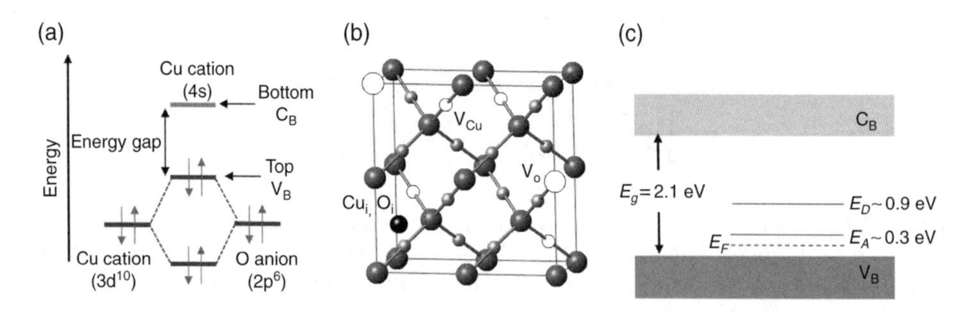

Figure 3.4 *a) Chemical bond between an oxide ion and a cation that has a closed-shell electronic configuration; b) representation of the more important defects in Cu_2O; c) a simple electronic model proposed by Brattain, with a compensated semiconductor with one acceptor level at 0.3 eV and a deep donor level at 0.9 eV from V_B. Adapted with permission from ref [87] Copyright (2010) American Institute of Physics.*

(V_{Cux}^{y}) and oxygen charged or neutral interstitials (O_i^y) produce band deformation close to the valence band, promoting the formation of negatively charged acceptor-like states (see Figure 3.4b). For instance, V_{Cux} associated with the removal of a Cu^{2+}, together with two electrons from the copper cationic sub-lattice, promotes the formation of holes (*h*, empty states) in the valence band, according to the generic quasi-chemical reactions [66]:

$$V_{Cu}^{*} = V_{Cu}^{-} + h^{+} \tag{3.1a}$$

or

$$V_{Cu}^{*} = V_{Cu}^{2-} + 2h^{+}, \tag{3.1b}$$

whose oxidation reaction due to the oxygen in the form of gas is given by [66]:

$$O_2^g = 2O_0^x + V_{Cu}^{-} + V_{Cu}^{2-} + 3h^{+}, \tag{3.2}$$

where the concentration of copper and oxygen vacancies should satisfy the relations [66]:

$$\left[\bar{V}_{Cu}\right] = \left[V_{Cu}^{-}\right] + \left[V_{Cu}^{2-}\right] \tag{3.3a}$$

and

$$\bar{V}_O = \left[V_O^x\right] + \left[V_O^{+}\right] + \left[V_O^{2+}\right], \tag{3.3b}$$

As there also exists a donor like level associated mainly with oxygen vacancies (V_O^{2-}) and the films' composition is out of stoichiometry, in general terms the quasi-chemical reaction can be translated by [66]:

$$(CuO)_{1-C} + (Cu_2O)_C \leftrightarrow Cu^{2+} + Cu^{+} + yV_{cu}^{2-} + zV_{Cu}^{-} + wO_{iCu}^{2-} + aO_O^{2-} + bV_O^{2+} + \frac{x}{2}O_2, \tag{3.4}$$

where O_{iCu}^{2-} corresponds to charged anti-structure defects created by the incorporation of *O* atoms into some Cu interstices or sub-lattice sites (they act as acceptor like states); O_O^{2-} corresponds to charged defects created in the *O* sub-lattice; which leads to the formation of $y+z+w$ holes and $a+b$ electrons, where $y+z+w >> a+b$, function of [66]:

$$\Delta x = \frac{\left[\bar{O}_i\right] + \left[\bar{O}_{iCu}\right] + \left[\bar{V}_{Cu}\right] - \left[\bar{C}u_i\right] - \left[\bar{C}u_O\right] - \left[\bar{V}_O\right]}{N_0}, \tag{3.5}$$

where $\bar{O}_i$, $\bar{O}_{iCu}$ and $\bar{C}u_i$, $\bar{C}u_O$ are respectively the average oxygen and copper interstices or substitutionals in both sub-lattices; N_0 is the number of lattice sites in each sub-lattice per cubic centimeter.

The conduction process in these structures is believed to occur by hopping controlled by the ions existing in different valance states, whose electronic conductivity is controlled by the fraction of reduced transition metal ions [$C=(Cu^{2+}+Cu^{+})/Cu_{total}$], leading to p-type conductivity (σ) with hole mobility (μ_H) able to exceed 100 cm^2 V^{-1} s^{-1} [52]. The p-type character of Cu_2O is attributed to the presence of negatively charged copper vacancies (V_{Cu}), which introduce an acceptor level about 0.3 eV above V_B [63].

High quality Cu_2O thin films have been grown by several methods, such as sputtering [67–70], pulsed laser deposition [71–73], molecular beam epitaxy [74], electrochemical deposition [75–78], sol-gel [79], plasma evaporation [80], chemical vapour deposition [81] and thermal oxidation [57, 82, 83]. Of all these methods sputtering is a relatively cost effective process that can be used for large area deposition and thermal oxidation is the simplest, inexpensive and most conventional method for obtaining copper oxide. It is well known that Cu oxidizes easily at low temperatures. This characteristic has impeded the application of Cu in integrated circuits. However, the high oxidation rate of Cu and high reduction rate of its oxides at low temperature can be exploited for some potential applications, such as thin film transistors, as we will show in Chapter 5.

Despite the high quality of CuO thin films with mobilities achieving values of 256 cm^2 V^{-1} s^{-1} [50] it has been difficult to make TFTs with high channel mobilities. TFTs prepared from Cu_2O have shown very poor performance (field-effect mobilities and on/off current ratio were below 1 cm^2 V^{-1} s^{-1} and 10^2 respectively), mainly because of the difficulty in controlling the hole density in the channel layer [51]. While conventional passive applications require a hole density to be as high as possible for optimal conductivity, a TFT requires only moderate hole density in the semiconductor channel layer in order to allow for conductance modulation by an applied electric field: the definition of *field effect* [84].

3.3.1 Role of Oxygen in the Structure, Electrical and Optical Performance

3.3.1.1 Structure evaluation

The x-ray diffraction (XRD) patterns obtained from the as-deposited Cu_xO films are shown in Figure 3.5 as a function of oxygen partial pressure (O_{PP}[a]). It is perceptible from the figure that the crystalline phase is strongly dependent on the variation in O_{PP}. The films deposited with 0 % O_{PP} are confirmed to be metallic cubic Cu phase with a strongest orientation along (111). When the O_{PP} is increased to 9 %, the crystallinity is changed to cubic Cu_2O phase [85] with a strongest orientation along (111). When the O_{PP} is increased further to $\geq$ 25 % O_{PP}, the films start crystallizing with monoclinic CuO phase [85] with a strongest orientation along (111). It is noteworthy that the strongest intensity (111) and other diffraction peaks are decreasing with an increasing O_{PP}. This situation suggests an amorphization process, which is further substantiated by the result of almost amorphous nature in the case of the films deposited with 100 % O_{PP}.

It was found that in the case of Cu_xO the annealing is not effective in inducing any change in the copper oxide phases. However, the increasing intensity and the strongest orientation along (111) as the annealing time increases, confirmes the improvement in crystallinity of the as-deposited films. Typical XRD patterns obtained from films annealed

[a] O_{pp} is defined as the ratio between the oxygen pressure present in the chamber, function of the gas flow allowed to the chamber and the total pressure present in the chamber (argon partial pressure plus the oxygen partial pressure), without throttling the output exhaust gas flow of the chamber.

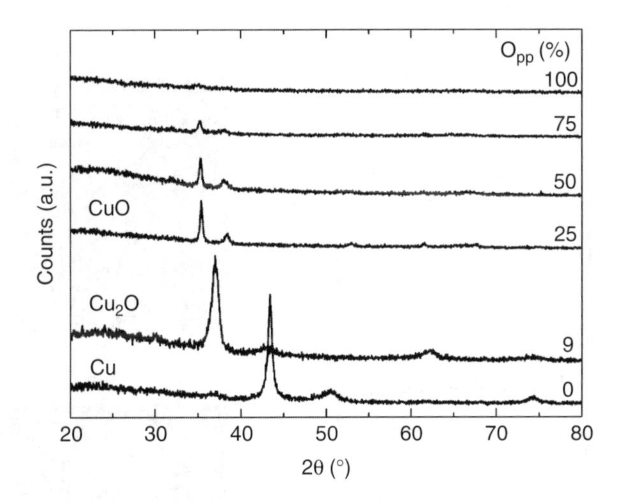

Figure 3.5 *X-ray diffraction patterns for Cu_xO films produced with an O_{PP} between 0 and 100%, as-deposited, with a thickness of 200 nm deposited on a glass substrate.*

for 10 hours are shown comparatively in Figure 3.6 (see color plate section) with one of as-deposited films. It is found that the annealing is not effective in inducing any change in the Cu_2O phase. The grain size (d_g) inferred from the Scherrer formula [86] for as-deposited films is 8.30 nm, increasing with the annealing time to a maximum of 15.72 nm for films annealed for 10 hours. To substantiate the increase in d_g the films were analyzed by AFM (Atomic Force Microsocopy) and the obtained microstructures are shown as an inset in Figure 3.6. The surface roughness increased from 4.6 to 10.4 nm corroborating the d_g obtained from XRD data.

3.3.1.2 Electrical properties evaluation

Normally, in order to fabricate high quality Cu_2O semiconductor thin films to be used as an active layer in devices, it is necessary to improve the film quality and consequently the electrical properties of the material. Generally, in order to achieve high quality Cu_2O thin films as far as the electrical properties are concerned, the depositions involve high temperatures, which is always higher than 500°C. Recently it was shown by Li *et al.* [50] that is possible to growth high quality Cu_2O thin films by reactive magnetron sputtering using a Cu target at temperatures of the order of 200°C, followed by a thermal treatment of 600°C. Li and co-workers found that with the introduction of a buffer layer, the crystallinity increases associated with an increase on the grain size and under oxygen optimum growing conditions they achieved mobilities of the order of 256 $cm^2\ V^{-1}\ s^{-1}$ with a hole concentration of $10^{14}\ cm^{-3}$ as shown in Figure 3.7 [50].

In order to reduce the deposition and/or the annealing steps Fortunato *et al.* [87] have recently produced Cu_2O by reactive rf magnetron sputtering at room temperature (RT), followed by annealing temperatures of only 200°C.

Figure 3.8 represents the resistivity of Cu_xO as a function of the O_{PP}. The Cu_2O film (9% O_{PP}) exhibits the higher bulk resistivity (ρ) of $2.7\times10^3\ \Omega$ cm, which has been gradually

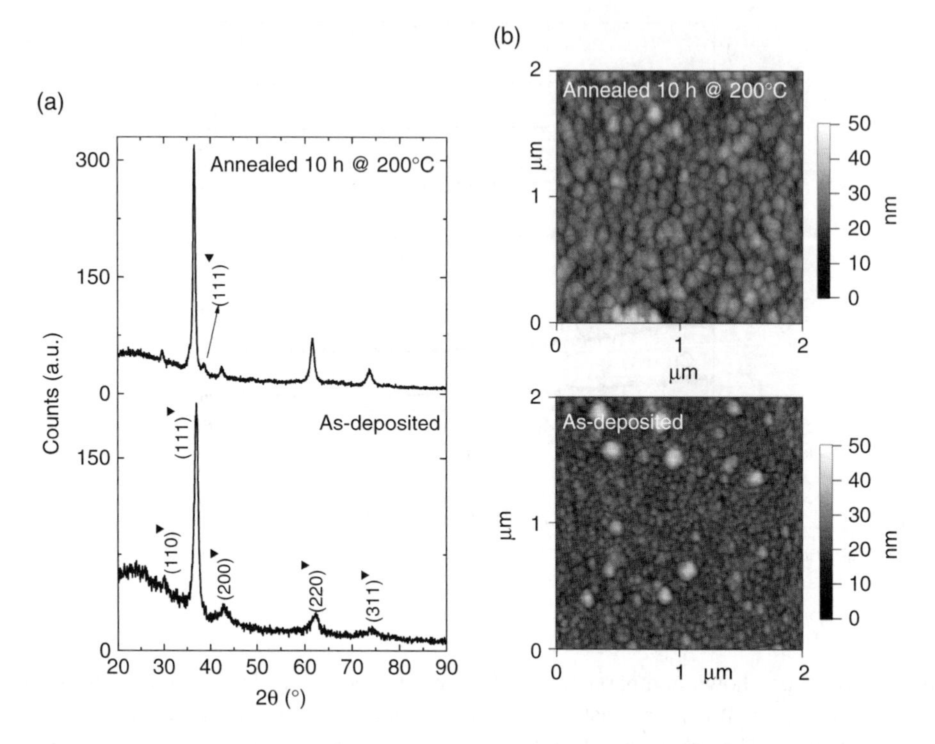

Figure 3.6 *a) X-ray pattern of the Cu_2O film annealed at 200°C in air for 10 hours in comparison with the as-deposited film. (Symbol representations: ▼-Cu_2O phase; ▶-CuO phase); b) AFM images of the corresponding films. Adapted with permission from ref [87] Copyright (2010) American Institute of Physics.*

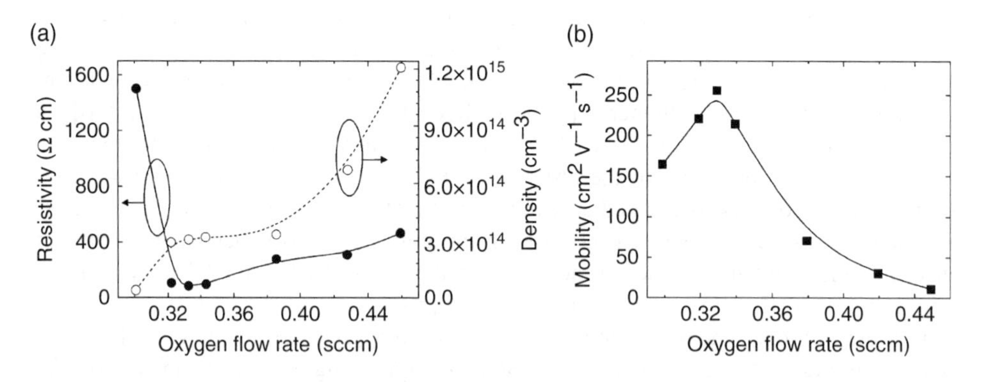

Figure 3.7 *a) Resistivity and hole density and b) mobility of Cu_2O thin films with LTB-Cu_2O as a function of oxygen flow rates. Adapted with permission from [50] Copyright (2009) Elsevier Ltd.*

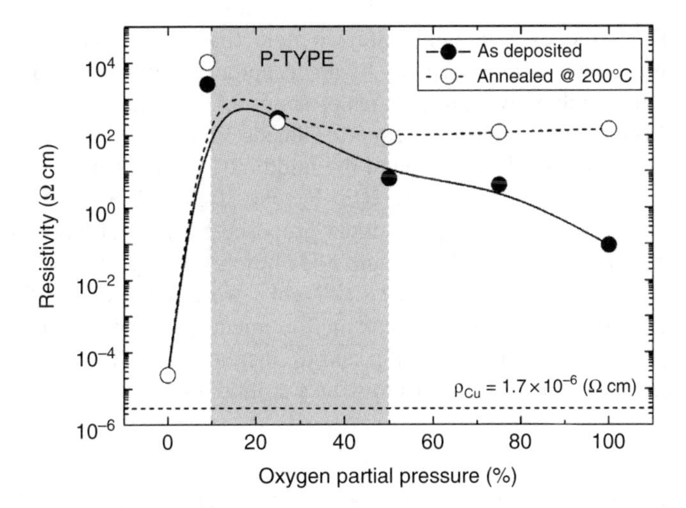

Figure 3.8 *Dependence of ρ measured by four-point probe on O_{PP}, for as-deposited and annealed Cu_xO films. It is also indicated the ρ of metallic Cu.*

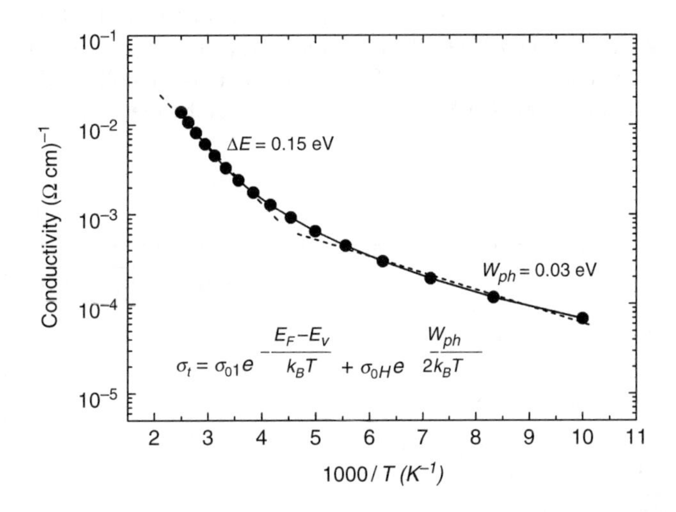

Figure 3.9 *Temperature dependence of dark conductivity for a Cu_2O thin film with 270 nm. The equation represents the conduction mechanisms for this type of semiconductors. For higher temperatures the conductivity is controlled by an acceptor center located 0.3 eV above the top of V_B (V_{Cu}) while for low temperatures hopping controls the conductivity. Adapted with permission from ref [87] Copyright (2010) American Institute of Physics.*

decreased with the increase in O_{PP}, in the case of CuO films. However, the ρ of 2.93×10^2 Ω cm obtained for 25% O_{PP} is increased to 9.09×10^{-2} Ω cm for the increase in O_{PP} to 100%.

The dependence of the conductivity (σ) on the inverse of the absolute temperature (T^{-1}) is shown in Figure 3.9 for the films produced with 9% O_{PP}. These films exhibit the Cu_2O phase and present a typical semiconductor behavior (negative slope of the plot σ versus T^{-1}). From the experimental data we note that $\sigma(T)$ data are fitted by two conduction paths controlled respectively by an acceptor energy level E_A and by phonons of energy W_{ph}, to which activation energies of about 0.15 eV and 0.03 eV are associated respectively. Taking into account the equation inside Figure 3.9 [88], where the Fermi energy level (E_F) is assumed to be around the middle of energy gap between E_A and the top energy of V_B, we estimate $E_A \approx 2\Delta E \approx 2(E_F - E_V) \approx 0.30$ eV and $W_{ph} \approx 0.06$ eV, which is in agreement with the conduction path model proposed by other authors [64, 89, 90]. A p-type resistivity (ρ) of $2.7 \times 10^3\ \Omega$ cm, and a Hall mobility of $\mu_H \cong 0.65\ cm^2\ V^{-1}\ s^{-1}$ and a hole carrier concentration (P) of $3.7 \times 10^{15}\ cm^{-3}$ were obtained at RT. The p-type conduction was also confirmed by Seebeck measurements performed in the same sample. After annealing at 200° C for 10 hours, ρ and μ_H increase to $1 \times 10^4\ \Omega$ cm and $18.5\ cm^2\ V^{-1}\ s^{-1}$, respectively, while P decreases to values around $3 \times 10^{13}\ cm^{-3}$.

3.3.1.3 Optical properties evaluation

The optical band gap (E_{op}) for directly allowed transitions of the Cu_2O thin films was calculated using Tauc's law: $\alpha^2 = (h\nu - E_{op})$, where α represents the absorption coefficient, h the Planck's constant and ν the photon frequency. The obtained value was 2.39 eV, which is slightly higher than the conventional value of 2.1 eV for Cu_2O [91]. Figure 3.10 shows the dependence of E_{op} as a function of the oxygen partial pressure for Cu_xO. For low O_{PP} E_{op} was 2.39 eV while for higher O_{PP} the E_{op} stabilizes at an average value of 2.05 eV. These values are in agreement with the fact that for low O_{PP} we have monolithic Cu_2O phase while for higher O_{PP} we have cubic CuO phase, as was observed by the XRD results [57].

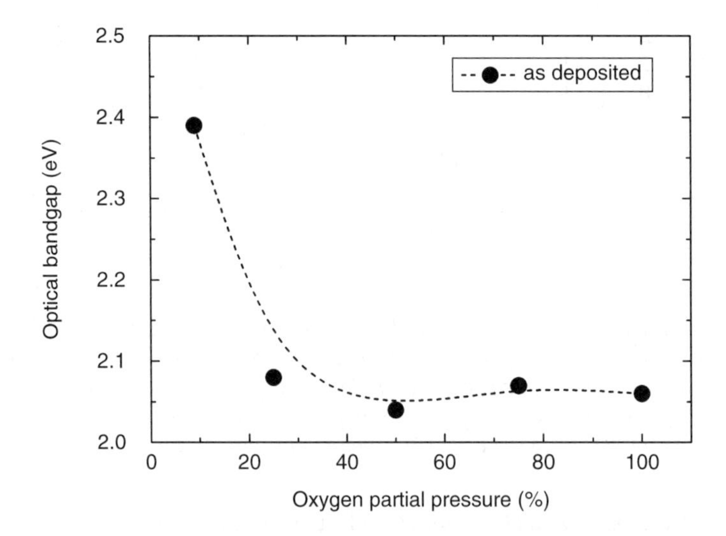

Figure 3.10 *Dependence of the optical band gap as a function o the oxygen partial pressure for CuxO thin films as-deposited. After annealing in air, no significant changes were observed.*

In summary, we can say that the structure and physical properties exhibited by CuO_x ($1 \leq x \leq 2$) thin films deposited on glass by reactive RF magnetron sputtering are highly dependent on the oxygen partial pressure percentage (O_{PP}) used during the deposition process. The metallic copper phase obtained from 0% O_{PP} has been crystallized with Cu_2O phase for the increase in O_{PP} to 9%, which was then changed to CuO phase ($O_{PP} \leq 50\%$). On the other hand, the crystallite size calculated from XRD data varied between 8.3 to 23.4 nm, respectively. The lattice parameter (a) is estimated to be ~3.60 Å for single Cu phase and ~4.20 Å for the Cu_2O phase. The films deposited with O_{PP} of 9%, 35% and 50% have p-type conductivity with ρ around 2.70×10^3, 2.93×10^2 and 6.39 Ω cm, respectively. The ρ and μ_H are decreased with the increasing O_{PP}, related to phase change from metallic Cu to Cu_2O and CuO, and also influenced by change in crystallinity. The films produced with highest O_{PP} showed the lowest resistivity, with ρ around 4.21 and 9.09×10^{-2} Ω cm, respectively. The annealing increases the film crystallinity and the roughness of the films produced with 9% O_{PP} (Cu_2O). The μ_H also increases from 0.65 to 18.5 $cm^2\ V^{-1}\ s^{-1}$, by enhancing the annealing temperature.

3.4 Thin Film Tin Oxide Semiconductors

Tin oxides are also known as wide band gap oxide semiconductors that can be used in several optoelectronics applications. Tin oxide is present into two well-known forms: tin monoxide SnO and tin dioxide SnO_2. SnO_2 (and impurity doped SnO_2) is a typical functional material with multiple applications, including transparent conducting oxides [92, 93], low emission windows coatings [94] and solid state gas sensing material [95]. In recent decades, due to its technological applications SnO have been used in a variety of applications such as: anode materials for lithium rechargeable batteries [96, 97], coatings [98], catalysts for several acids [99] and precursor for the production of SnO_2 [100]. Recently, SnO has received particular attention because of the difficulty in obtaining stable and high quality p-type semiconductors based on ZnO [101], NiO [102, 103] and Cu_2O [104]. The p-type conductivity observed in SnO results from the formation of a substantial amount of native acceptors and two types are considered for native-acceptor-like defect [85]: Sn vacancies and O interstitials.

The unit cells of SnO_2 and SnO are shown in Figure 3.11 (see color plate section). SnO_2 has a tetragonal structure and the unit cell contains two tin and four oxygen atoms. Each tin atom is at the centre of six oxygen atoms placed approximately at the corners of a regular octahedron, while each oxygen atom is surrounded by three tin atoms at the corners of an equilateral triangle. On the other hand, SnO has a specific electronic structure associated with the presence of divalent tin, Sn(II), in a layered crystal structure with a Sn-O-Sn sequence and a van der Waals gap between Sn layers ~2.52 Å. Oxygen atoms are tetrahedrally bonded to Sn ones. The Sn atoms are situated at the top of regular square-based pyramids that are based on oxygen atoms with Sn-O distances equal to 2.224 Å [105].

SnO_2 is an intrinsic n-type semiconductor and the electrical conduction results from the existence of defects, which may act as donors or acceptors. These defects are generally due to oxygen vacancies or interstitial tin atoms and are responsible for making electrons available at the conduction band.

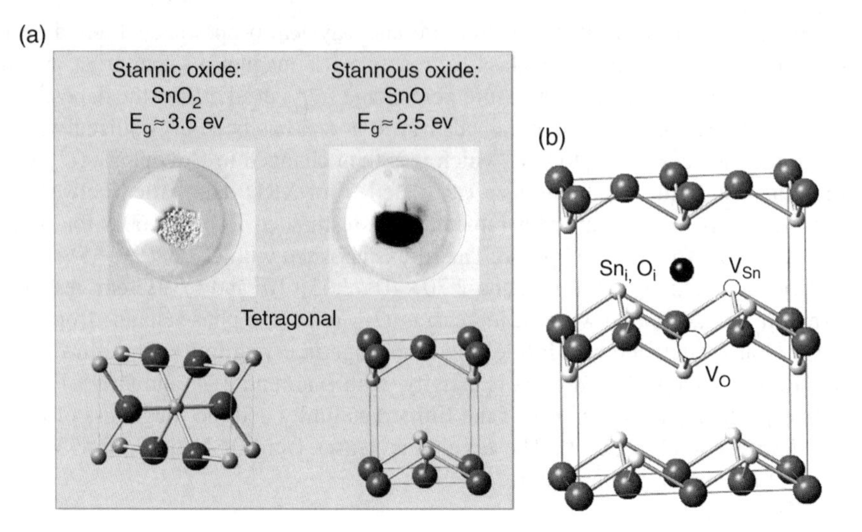

Figure 3.11 a) Unit cell of the crystal structure of SnO_2 and SnO; b) Sketch of the unit cell of tin oxide showing the presence of interstitial atoms and vacants of oxygen and tin in the structure.

In spite of the recent developments concerning p-type oxide semiconductors aimed at fabricating TFTs, the results achieved so far refer to devices typically processed at temperatures above 575°C and are limited by the low hole (h) mobilities [51, 106]. The valence band maxima (VBM) of oxide semiconductors are mainly formed from localized and anisotropic O 2p orbitals, which lead to a low h mobility due to a percolation/hopping conduction [107]. However, if we consider the case of stoichiometric or non-stoichiometric SnO_α, this can be changed. In this case the higher-energy region of VBM contains Sn 5s, Sn 5p and O 2p components nearly equally, but very near the upper part of VBM, the contributions of Sn 5s and O 2p are predominant, and so, they define the way how free holes are generated. This can lead to an overall enhancement in h mobility. On the other hand, in the conduction band, the O 2p component is relatively small and the states near the bottom of CBM are mainly formed by Sn 5p. This is advantageous in reducing the localization of the valence band edge [32].

Consideration of the hole transport paths leads us to expect that an oxide in which VBM is made of spatially spread s orbitals would have a large hole mobility and would be a better p-type oxide semiconductor. The origin of p-type conductivity of SnO is mainly attributed to Sn vacancies and O interstitials [85] (see Figure 3.11b). If there is excess oxygen in the film, some cations will be transformed into Sn^{3+} in order to maintain electrical neutrality. This process can be considered as Sn^{2+} capturing a hole and forming weak bonded holes. These holes are located inside the bandgap near the top of the valence band and serve as acceptor states in the energy band structure.

Under this condition we expect that the hole transport requires structures mainly dominated by SnO_x with some embedded Sn cations in which VBM is made of pseudo-closed ns^2 orbitals to form hybridized orbitals [108], as is indicated in Figure 3.12. The origin of the p-type conductivity of SnO_x is mainly attributed to Sn vacancies and O interstitials [85], which when fully ionized (respectively V_{Sn}^{2-} and O_i^{2-}) produces band deformation close to VBM and so, to the formation of acceptor like states, negative charged, located very close to valence band tails (see also Figure 3.12). That is, the formation of an energy band level localized close to the top of VBM which, for temperatures above the absolute zero, are partly filled by electrons coming from the valence band, according to one of the quasi-chemical stoichiometric reactions [109]:

$$SnO \leftrightarrow Sn^{2+} + 2V_{Sn}^{-} + \frac{1}{2}O_2(1h/\mathrm{V_{Sn}}) \tag{3.6a}$$

$$SnO \leftrightarrow Sn^{2+} + V_{Sn}^{2-} + \frac{1}{2}O_2\ (2h/\mathrm{V_{Sn}}); \tag{3.6b}$$

$$\begin{gathered} SnO \leftrightarrow Sn^{2+} + V_{Sn}^{-} + O_{Sn}^{-} + \frac{1}{2}O_2 + V_O^{+} \\ (\mathrm{2h/(V_{Sn} + anti\ structured\ charged\ O\ defect,\ O_{Sn})}) \end{gathered} \tag{3.6c}$$

$$SnO \leftrightarrow Sn^{2+} + V_{Sn}^{-} + O_{Sn}^{-}(2\,h/\mathrm{V_{Sn}} + \mathrm{O_{Sn}}) \tag{3.6d}$$

In general, for non-stoichiometric films where donor-like centres such as oxygen vacancies are also present, the following reaction is expected [66]:

$$SnO_x \leftrightarrow Sn^{3+} + Sn^{2+} + yV_{Sn}^{2-} + zV_{Sn}^{-} + wO_{Sn}^{2-} + aO_O^{2-} + bV_O^{2+} + \frac{x}{2}O_2 \tag{3.7}$$

which leads to the formation of $y+z+w$ holes and $a+b$ electrons, where $y+z+w >> a+b$.

In general terms, the deviation of stoichiometry, when all possible non-associated native atomic defects are present, can be described by the formula [66]:

$$\Delta x = \frac{[\bar{O}_i]+[\bar{O}_{Sn}]+[\bar{V}_{Sn}]-[\bar{Sn}_i]-[\bar{Sn}_O]-[\bar{V}_O]}{N_0} \tag{3.8}$$

where

$$\bar{V}_{Sn} = [V_{Sn}^{x}]+[V_{Sn}^{-}]+[V_{Sn}^{2-}]+[V_{Sn}^{3-}] \tag{3.9a}$$

and

$$\bar{V}_O = [V_O^{x}]+[V_O^{+}]+[V_O^{2+}] \tag{3.9b}$$

are the average total concentration of vacancies in the tin and oxygen sub-lattices, respectively; N_0 is the number of lattice sites in each sub-lattice per cubic centimeter, which will determine

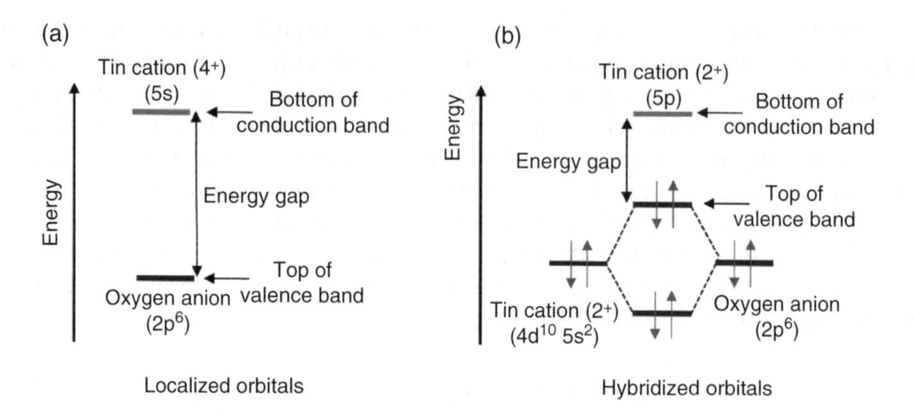

Figure 3.12 *Comparison between the band structure of a) SnO_2 and b) SnO. In the case of SnO the top of the valence band consist of hybridized orbital of O 2p and Sn 5s. Adapted with permission from ref [110] Copyright (2010) American Institute of Physics.*

the sub-band gap localization. Here we have to take into account that cation vacancies form more easily than O interstitials in most metal oxides with densely packed structures, where the open spaces surrounded by Sn^{2+} may facilitate the formation of O interstitials.

This localized band formation sets the conduction mechanism, as suggested by the SnO layered crystal structure [106]. Thus, we expect that the p-type conductivity in SnO_x is mainly originated by Sn vacancies, as demonstrated by Togo *et al.* [85] (see Figure 3.11b). Thus, TFTs based on p-type SnO_x are expected to fulfill these requirements due to the particular nature of band structure. Contributions from Sn 5s states to VBM could offer appreciable hole mobility in this material, without using high processes temperature. The comparison between the band structure of SnO_2 and SnO is presented in Figure 3.12. In the case of SnO the top of the valence band consist of hybridized orbital of O 2p and Sn 5s.

From what has been said, it is relevant to know the morphology, structure and composition of the films processed, which are highly influenced by the process conditions used, as it is in the case of films processed by rf magnetron sputtering where the oxygen partial pressure, the residence time of species within the deposition chamber, the distance between target and substrate and the power density used are extremely determinant of the films' properties, complemented by the annealing environment and temperatures used and the annealing times that the samples sustain. In the following pages we present the morphology, structure, composition, electrical and optical data of tin oxide films processed at different oxygen partial pressures, aiming to observe the overall properties of the tin oxides processed by sputtering technique.

3.4.1 Structure, Composition and Morphology of Tin Oxide Films

The SnO_x films evaluated here were produced by rf magnetron sputtering using two systems with different geometries with which are associated residence times of species in the ratio of 2:1. Moreover, all of the films were produced at room temperature (RT) using a metallic target and O_{pp} between 0 and 40 %, followed by annealing treatment.

The p-type conduction behavior was only observed in a reduced process O_{pp} window, and differently in both systems analyzed. In the system presenting the longest residence time, the p-type process window is achieved using O_{pp} between around 5.0–12 %, while in the system presenting half of this residence time, the window is almost reduced by 50 % (O_{pp} between around 2.5 % and 4.5 %). That is, a shift towards low oxygen contained in the chamber during the deposition process is achieved in the second system that also has the highest exhaust pumping speed and so requires a higher throttling than the first.

For the system with the longest residence times, the films deposited with O_{pp} below 5 % (or < 2.5 %, for the second system) were metallic with poor adhesion to the substrate. Those deposited above 12 % (or > 4.5 %, for the second system) were highly resistive, showing n-type conduction after annealing above 200°C. This is well understood based on tin oxide's two well known forms: SnO (p-type conduction) and SnO_2 (n-type conduction). Since a metallic tin target was used for sputtering, very low O_{pp} was not sufficient to fully oxidize the film. It is believed that O_{pp} in the range of 5–12 % (or 2.5 %–4.5 %) are suitable to form the SnO phase and for $O_{pp} \geq 12$ % (4.5 %) favors SnO_2 formation. Moreover, in order to expand the process window as much as possible, it is also desirable to use systems with long residence times. Otherwise, high controlled process systems are required.

3.4.1.1 *Structure evaluation*

Figure 3.13 shows the X-ray diffractograms of the as-deposited (amorphous) and annealed SnO_x (polycrystalline) films deposited in the system with the longest species residence time for an O_{pp} between 7 and 21 %. After annealing in atmosphere air at 200°C crystallization occurs and a mixture of both tetragonal β-Sn and α-SnO phases are found for O_{pp} =7.1 and 11.3 %, which is caused by the incomplete Sn oxidation and/or metal segregation due to oxygen disproportionation. It was reported that disproportionate distribution of internal oxygen at intermediate temperatures leads to metallic Sn segregation during the heating and annealing processes, which explains the appearance of the metallic phase [111]. It is also well known that tin oxide exists in two main forms: stannous SnO and stannic SnO_2. SnO generally crystallizes in a tetragonal structure, but depending on the conditions of preparation it may also crystallize in an orthorhombic phase [112]. On the other hand, stannic tin oxide (SnO_2) crystallizes in a rutile structure usually for higher temperatures. In this study the SnO_2 phase does not appear, mainly due to the deposition conditions used, which are more favorable for the SnO phase.

Figure 3.14 a shows the XRD pattern of 120 nm thick films processed in the second system (exhibiting the lowest species residence time) using Cu K_α line. It shows that at RT films have large portion of metallic tin with very small SnO phase. After annealing up to 200°C for 30 minutes in air, part of metallic tin seem to be oxidized to SnO phase. The XRD patterns obtained (in the similar conditions) from SnO powder and metallic Sn are also shown in Figure 3.14a.

In order to evaluate the structure, composition and the ratio of SnO and SnO_2 phases present in the p-type SnO_x films processed, ^{119}Sn Mössbauer spectroscopic studies were

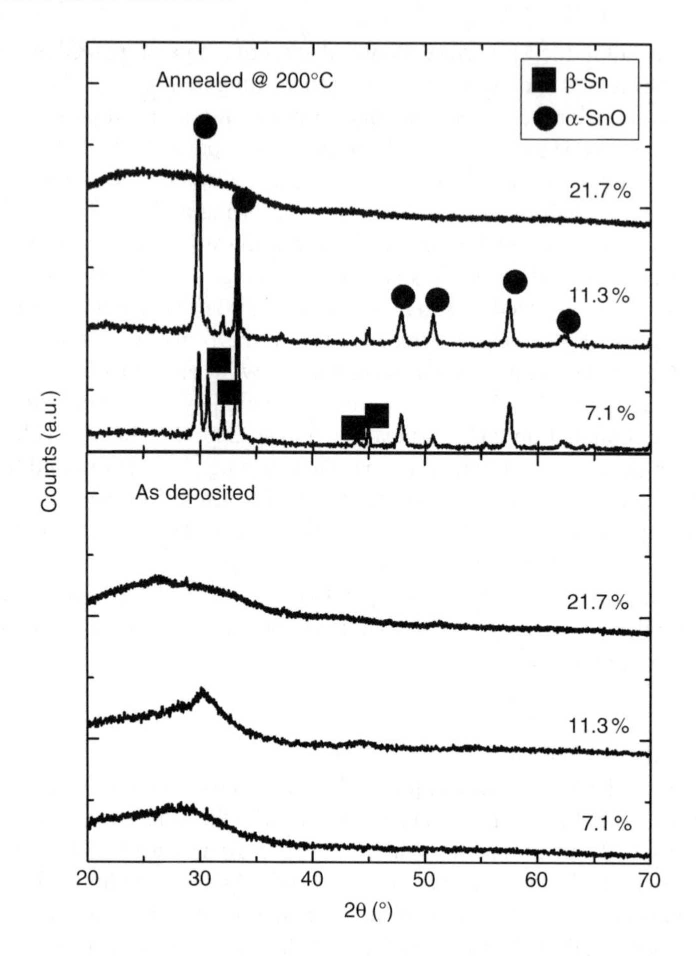

Figure 3.13 *XRD patterns for a* SnO_x *film used at the channel layer, as-deposited and annealed at 200°C in air atmosphere, with a thickness of 200 nm deposited on a glass substrate. Adapted with permission from ref [110] Copyright (2010) American Institute of Physics.*

performed, whose results are shown in Figure 3.14b. Transmission Mössbauer spectra of bulk metallic Sn, SnO and SnO_2 samples (Figure 3.14b, Table 3.2) were obtained in order to compare the corresponding hyperfine parameters with those of the phases detected in the films by conversion-electron ^{119}Sn Mössbauer spectroscopy (CEMS). The films were analyzed by CEMS at room temperature placing them inside a proportional backscatter detector RIKON-5 (Wissel) in flowing 5% CH_4-95% He gas mixture. Both CEMS and transmission Mössbauer spectra were collected using a conventional constant acceleration spectrometer and a 5 mCi $Ca^{119m}SnO_3$ source. The velocity scale was calibrated using a ^{57}Co (Rh) source and an α-Fe foil. The Sn isomer shifts (IS) are given relative to $BaSnO_3$ reference material at 295 K and obtained by adding 0.031 mm s^{-1} to the IS relative to the source. The spectra were fitted to Lorentzian lines using a non-linear least-squares method.

The Mössbauer analysis performed in these samples agrees with the XRD data, showing that in the as-deposited films most of the Sn is present as amorphous SnO and metallic β-Sn. Minor amounts of SnO_2, only detected by CEMS, are also present. The spectra of the bulk samples are identical to those reported in the literature for β-Sn, α-SnO and SnO_2, respectively [113]. The α-SnO spectrum shows a small contamination of SnO_2 at lower Doppler velocities, which is a common contamination of α-SnO when exposed to air and is easily detectable by Mössbauer spectroscopy due to the higher recoilless fraction of Sn^{4+} in SnO_2 as compared to Sn^{2+} in α-SnO [114]. The typical asymmetry of the α-SnO doublet, the high velocity peak having a higher relative area than the low-velocity peak, is also observed [115, 116].

The IS and IS quadrupole splitting (QS) of this SnO_2 deduced from the spectra are slightly different from those of bulk SnO_2 probably due to the poor crystallinity of this

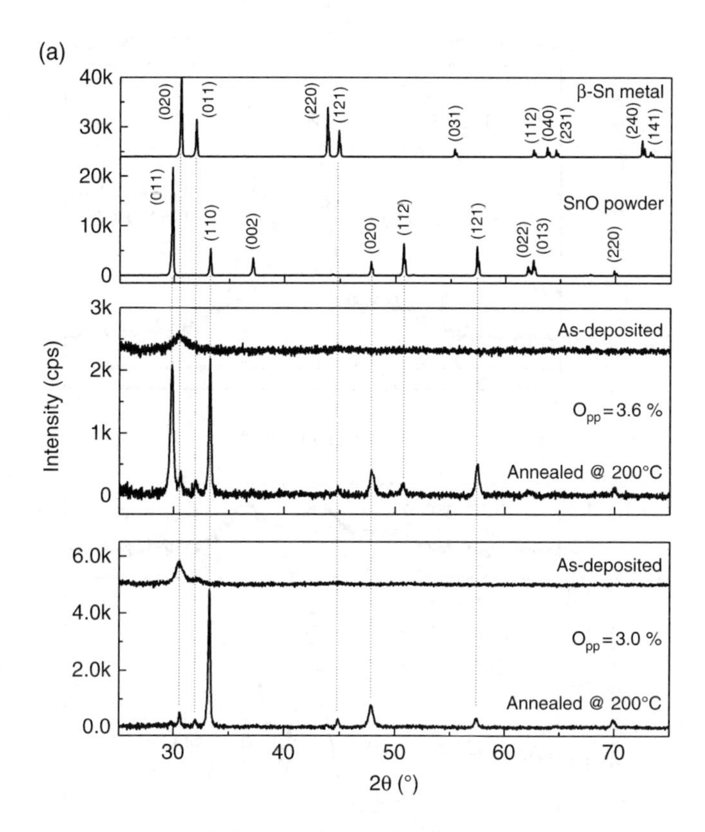

Figure 3.14 *a) XRD patterns of deposited SnO_x thin films. b) Transmission ^{119}Sn Mössbauer spectra of bulk samples and CEMS spectra of as-deposited and annealed SnO_x films deposited at 3.0% and 3.6% O_{pp}. The Mossbauer spectroscopy was done at ITN by Prof. J.C. Waerenborgh.*

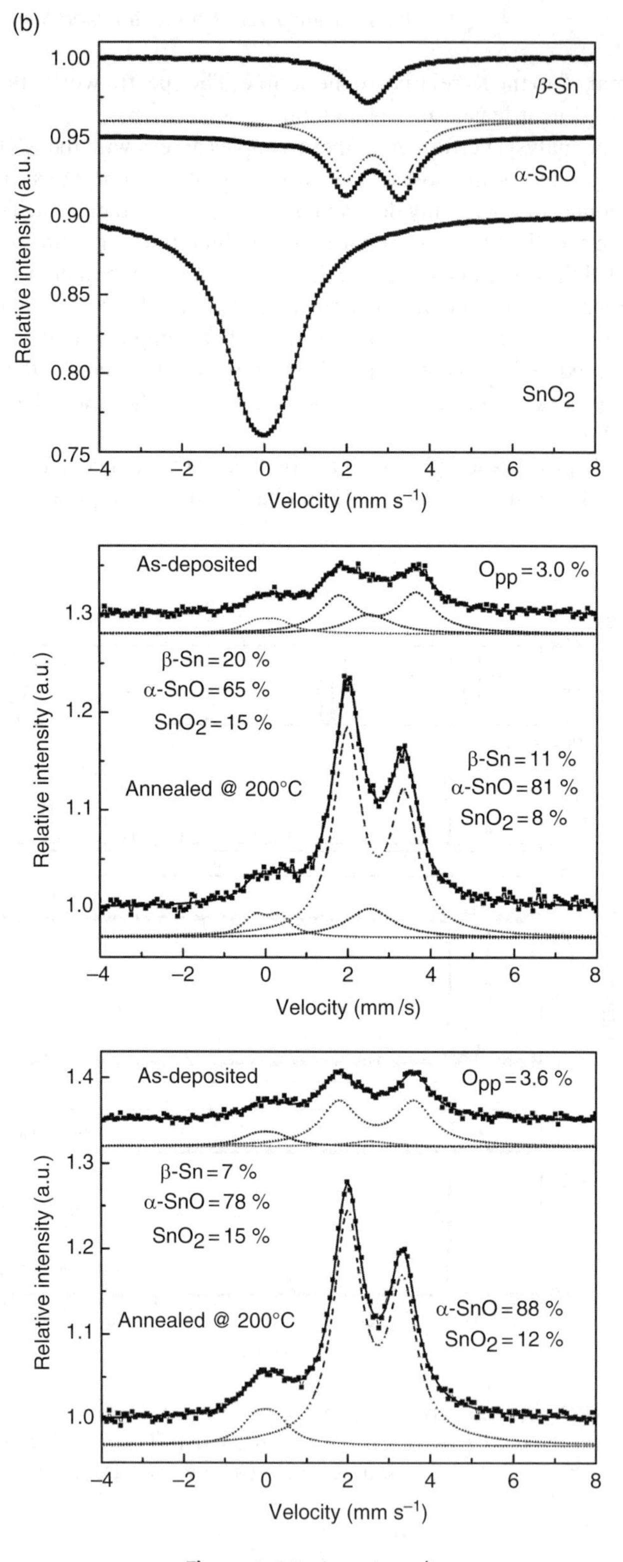

Figure 3.14 *(continued)*

oxide in the film. Even though the recoilless factors of β-Sn, SnO and SnO_2 in films are not known precisely, these factors are not expected to be very different for the same species in the different samples. So the fraction of Sn atoms in each phase should follow the same variation, with annealing or with O_{pp} during deposition, as the relative areas of the corresponding contributions to the spectra. Not surprisingly, the fraction of Sn present as β-Sn is lower in the film deposited at higher O_{pp} (3.6%), than in those deposited at 3.0%. Annealing up to 200°C not only crystallizes amorphous SnO to form α-SnO, but also oxidizes β-Sn to α-SnO. Thus, it suggests that the oxide thin film semiconductor is mainly composed of SnO phase with a smaller contribution from metallic tin and hence we can represent the channel material composition as SnO_x with $1<x<2$.

The CEMS spectra of the samples were fitted by three contributions with typical IS of Sn^{4+}, Sn^{2+} and metallic Sn. For all spectra the estimated IS of the absorption peak due to metallic Sn (Table 3.2) is very similar to that of β-Sn, confirming the presence of this phase in the films, as corroborated by XRD. The observed QS and the line widths of Sn^{2+} in the as-deposited samples are higher than the corresponding parameters for bulk α-SnO. After annealing the decrease of IS and QS leads to values, which within experimental error, are close to those of crystalline α-SnO. This suggests that the Sn^{2+} oxide present in the as-deposited film is amorphous and after annealing up to 200°C crystalline α-SnO is formed. This is in agreement with the XRD data, which shows this phase in the annealed samples while in the as-deposited films only β-Sn is detected. The estimated QS and line widths in the amorphous SnO are higher than in crystalline α-SnO denoting a more asymmetric environment and a higher diversity of near neighbor configurations of Sn^{2+} in the amorphous lattice. Furthermore, no asymmetry is observed for the Sn^{2+} oxide doublet

Table 3.2 *Estimated parameters from the transmission ^{119}Sn Mössbauer spectra of bulk samples and CEMS spectra of films taken at 295 K.*

Sample	IS [mm s^{-1}]	QS [mm s^{-1}]	Γ [mm s^{-1}]	Sn phase	I [%]
SnO_2 bulk	0.01	0.56	1.34	–	100
α-SnO bulk	2.67–	1.34	0.98	α-SnO	95
	0.03	0.58	0.77	SnO_2	5
β-Sn metal	2.56	–	1.03	–	100
Film	2.76	1.87	1.04	SnO	65
O_{pp}=3.0 %,	0.11	0.52	0.84	SnO_2	15
RT	2.56	–	1.3	β-Sn	20
Film	2.7	1.38	0.79	SnO	81
O_{pp}=3.0 %,	0.08	0.59	0.7	SnO_2	8
200°C, 30 min	2.56	–	1.35	β-Sn	11
Film	2.73	1.82	1.02	SnO	78
O_{pp}=3.6 %,	0.11	0.53	0.81	SnO_2	15
RT	2.56	11	1.3	β-Sn	7
Film	2.7	1.34	0.82	SnO	88
O_{pp}=3.6 %,	0.06	0.48	0.85	SnO_2	12
200°C, 30 min					

IS (mm/s) isomer shift relative to metallic $BaSnO_3$ at 295 K; QS (mm s^{-1}) quadrupole splitting; Γ (mm s^{-1}) line-width; I relative area. Estimated errors ≤0.02 mm s^{-1} for IS, QS, Γ and <2% for I.

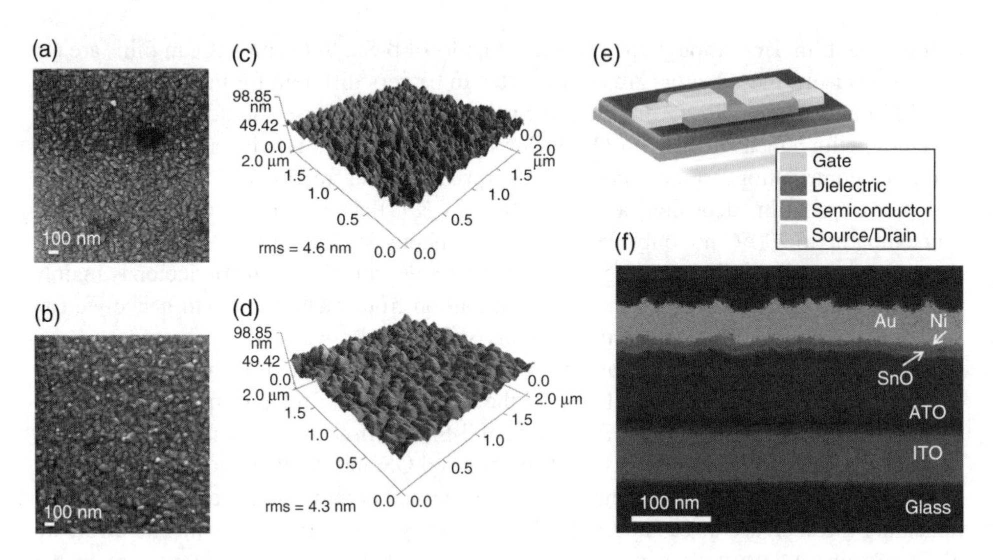

Figure 3.15 *SEM images of post-annealed tinoxide used as channel layers prepared at O_{pp} of a) 3.0 % and b) 3.6 %, for TFT applications. Corresponding AFM images of the TFT channel layers are shown in c) and d). e) Schematic illustration of SnO_x TFT structure employed for the present study. f) Cross-sectional SEM image of the SnO_x channel layer.*

in the as-deposited sample, as expected for an amorphous phase. As crystalline α-SnO is formed, asymmetry of the corresponding doublet becomes evident.

3.4.1.2 Morphology evaluation

In order to understand the surface morphology of the tin oxides used to produce the channel layer in p-type TFT, we performed scanning electron microscopy (SEM) and AFM of the oxide semiconductors referred to in Figure 3.14 and Table 3.2, after annealing. The surface microstructures obtained from the films deposited at 3.0 % and 3.6 % O_{pp} are shown in Figure 3.15 (see color plate section). Even though there is no difference in the shape of the grains as evidenced by both techniques, the size of the grains obtained from the films deposited at O_{pp}=3.0 % is slightly lower than those deposited at O_{pp}=3.6 %. The root mean square roughness (4.6 nm) of the films deposited at O_{pp}=3.0 % is slightly higher than that deposited at 3.6 % (4.3 nm). Cross-sectional SEM image obtained from the TFT prepared at 3.0 % O_{pp} (Figure 3.15f) shows perfect coverage of SnO_x channel layer over the insulator surface with highly compact, uniform and homogeneous thin film without visible defects.

3.4.2 Electrical and Optical Properties of Tin Oxide Films

3.4.2.1 Electrical properties evaluation

It is essential to study the electrical properties of the tin oxide thin films processed using the two systems with different residence times (2:1), in order to understand the charge carrier transport properties and the type of carriers involved (electrons and holes). The

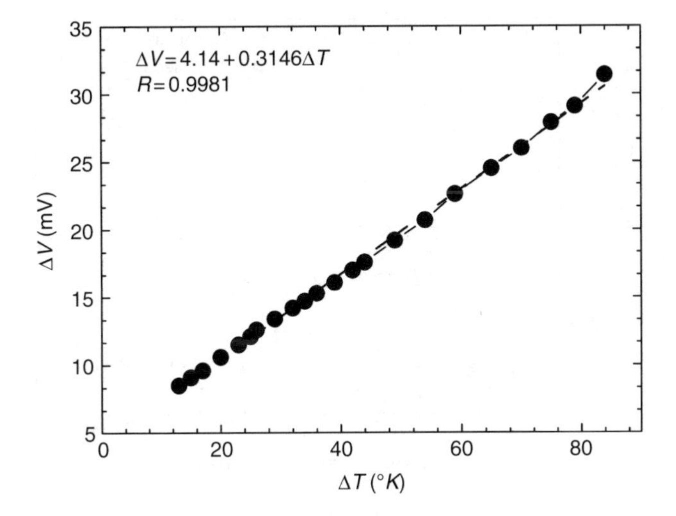

Figure 3.16 *Thermoelectric data for the SnO_x samples.*

Hall Effect and Seebeck measurements are the most commonly used techniques to identify carrier mobility and the charge carrier type. In contrast to n-type oxides, Hall measurement does not give reliable results for p-type oxides, primarily because of the very small drift mobility of holes. Moreover, when both electrons and holes are present in a semiconductor sample, both charge carriers experience a Lorentz force in the same direction. So, both electrons and holes will pile up at the same side of the sample and consequently the measured Hall voltage depends on the relative mobilities and concentrations of holes and electrons. Hall mobility for an ambipolar semiconductor is thus given by:

$$\mu_{Hall} = -\frac{n\mu_n^2 - p\mu_p^2}{n\mu_n + p\mu_p} \tag{3.10}$$

where n, p, μ_n and μ_p represent electron density, hole density, electron drift mobility and hole drift mobility respectively. This suggests a considerable reduction in Hall mobility in an ambipolar semiconductor compared to the actual drift mobilities of the charge carriers.

On the other hand, the Seebeck effect is the conversion of temperature differences directly into electricity, by establishing between the cold and hot side of the film evaluated a voltage gradient. The magnitude of the voltage gradient created by this effect is on the order of several microvolts per Kelvin difference and the signal give us the main type of free carriers involved in the charge transport mechanism of the film under analysis.

As far as thermoelectric power measuremnts are concerned, both set of films under analysis where evaluated. The data obtained present a high Seebeck coefficients, consistent with p-type semiconductor behavior. Figure 3.16 shows the plot of the temperature difference (ΔT) against the measured thermovoltage (ΔV) of a SnO_x thin film produced in

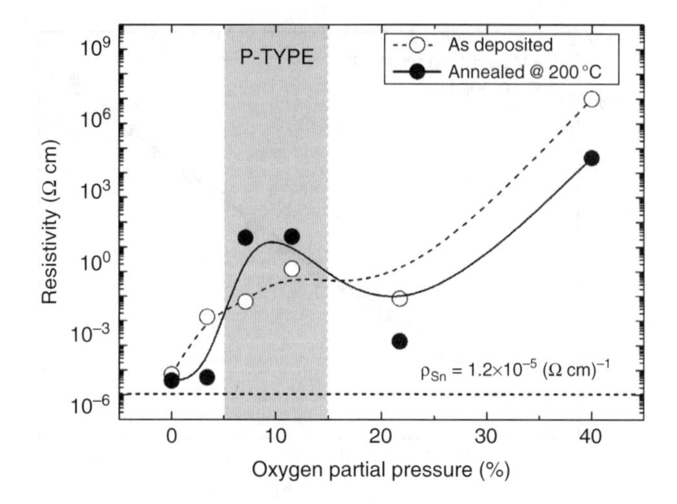

Figure 3.17 *Dependence of ρ measured by four-point probe on O_{pp}, for as-deposited and annealed SnO_x films. The dashed line represents the ρ of as deposited SnO_x, Sn metallic rich.*

the system with the highest residence time, where the slope of the linear fitting is the Seebeck coefficient (S), according to the relation:

$$\Delta V = S\Delta T \tag{3.11}$$

which exhibits a $S \approx 314.6\,\mu V°C^{-1}$.

Figure 3.17 shows the dependence of ρ as a function of O_{pp} for as-deposited tin oxide films using the system with the largest species' residence time. As the O_{pp} increases ρ also increases and two different electronic conduction behaviors are observed after annealing, depending on the O_{pp} range. In fact, although all the as-deposited SnO_x films exhibit n-type conductivity (measured by Hall effect), after annealing only the films produced with an O_{pp} between 5 and 12% present p-type characteristics. In this O_{pp} range ρ increases after annealing, while for films processed outside this O_{pp} region ρ decreases. These different behaviors are reinforced by X-ray photoelectron spectroscopy (XPS) measurements (not presented in this chapter), where an increase on the O to Sn ratio between 40% and 60% (depending on the processing conditions) is observed for the p-type samples after annealing. By Hall Effect measurements it is also verified that on the p-type films the carrier concentration can be adjusted between $\approx 10^{16}$ and 10^{18} cm^{-3}, corresponding to a maximum μ_H of 4.8 cm^2 V^{-1} s^{-1}.

The p-type characteristics are also confirmed by thermoelectric measurements as depicted in Figure 3.16, indicating that the majority carriers in SnO_x are holes. This is caused by the fact that for the O_{pp} range mentioned above the p-type conductivity is controlled by O atoms in interstitials or vacancies in the Sn sublattice acting as acceptors, while for higher O_{pp} the n-type behavior is due to the existence of Sn atoms in interstitials and O vacancies [117–120]. For lower O_{pp} (< 5%) a metallic behavior is observed, with an average ρ of 4.4×10^{-5} Ω cm, typical of metallic Sn (1.2×10^{-5} Ω cm). In fact, X-ray analysis

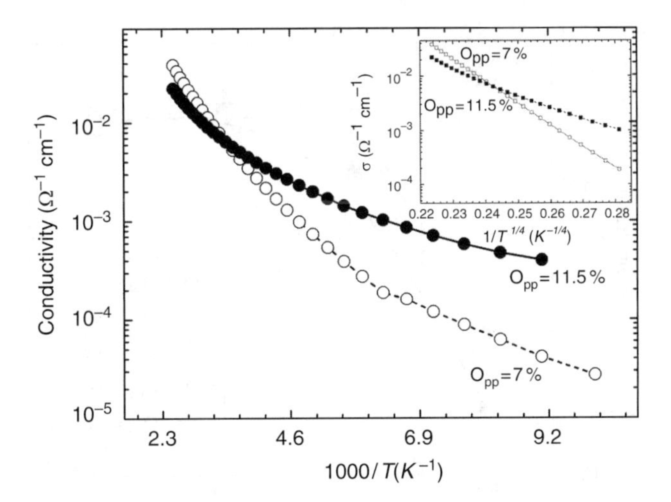

Figure 3.18 *Dependence of σ on T^{-1} for samples fabricated at $O_{pp} \cong 7\,\%$ and $O_{pp} \cong 11.5\,\%$. The inset shows the same plot for the same samples for σ versus $T^{-1/4}$. Adapted with permission from ref [110] Copyright (2010) American Institute of Physics.*

confirms that for $O_{pp} < 5\,\%$ β-Sn is the predominant phase. As was observed by other authors [106] p-type SnO_x is only obtained at a very narrow region of growth conditions, where the species residence time is a key parameter, since SnO_2 is formed for higher O_{pp}, whereas metallic Sn precipitates at lower O_{pp}. Since the growth of SnO requires a reducing atmosphere and the vapor pressure of SnO and Sn is high, the deposition at room temperature can be a promising advantage, since inhibits the re-evaporation of deposited thin films.

Figure 3.18 shows the Arrhenius plots for samples processed using the system with the highest species' residence time, with $O_{pp} \cong 7\,\%$ and $O_{pp} \cong 11.5\,\%$, which do not follow a simple thermally activated transport mechanism. The data could be fitted by a model with two conduction paths: one at high temperatures ($T \geq 200\,K$), controlled by a broad acceptor band located at energy E_a above the VBM and energy phonon assisted ($E_{\omega ph}$) and another conduction path at low temperatures, where the transport is fully governed by phonons assisting the conduction mechanism [121]. This is translated by the relation:

$$\sigma \cong \sigma_0 e^{\left(-\frac{E_a + E_{\omega ph}}{k_B T}\right)} + \sigma_{0H} e^{\left(-\frac{E_{\omega ph}}{k_B T}\right)} \tag{3.12}$$

where σ_0 and σ_{oH} are the conductivity pre-exponential factors of the two conduction pathways referred to above and k_B is the Boltzmann constant. Under this condition we could estimate $E_a \cong 0.10\text{-}0.90\,eV$ while $E_{\omega ph} \cong 0.05\text{-}0.03\,eV$, values that agree well with the expected values from the model proposed by Mott [121] when there exists a localized disorder. However, better linear fits are obtained in the log σ versus $T^{-1/4}$ plot in the whole temperature range evaluated in this work (inset to Figure 3.18), not related to variable-range hopping but rather to percolation conduction. In this case, the random

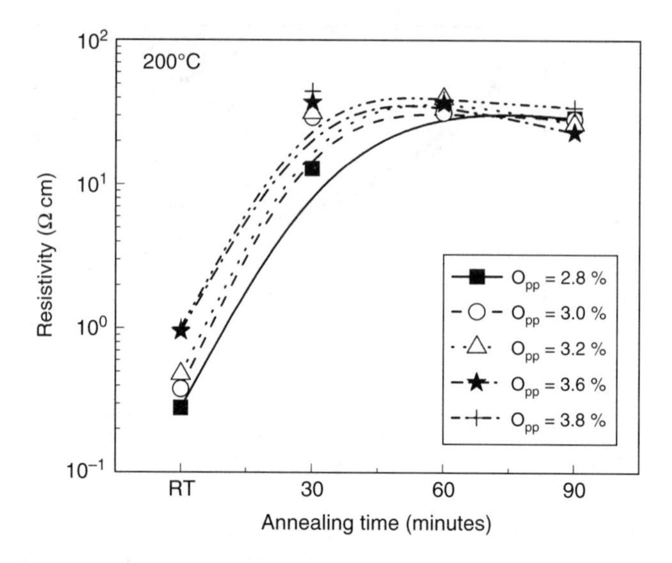

Figure 3.19 *Variation in the resistivity of SnO_x films as a function of annealing time.*

distribution of Sn^{2+} ions modulates the electronic structure around the valence band edge and may form statistical potential barrier distribution with 0.10 eV high and a few tens of milli-electron volts width, agreeing with the values predicted by Kamiya *et al.* [107].

Concerning the electrical analysis of films processed at RT using the system with the lowest species' residence time, the Hall Effect coefficient shows fluctuations in the sign and magnitude. Typical carrier mobilities were of the order of 10^{-1} cm^2 V^{-1} s^{-1}. Because of the ambipolar nature of the RT deposited films, these results suggest that Hall measurement of SnO_x films highly underestimates the actual carrier mobility. After annealing up to 200°C for 30 minutes, films prepared with O_{pp} of 2.5–4.5 % have shown positive Hall coefficient consistently. There was no change in Hall coefficient sign even though the magnitudes were showing small fluctuations. This suggests a considerably large density of holes compared to the density of electrons. Figures 3.19 and 3.20 show the resistivity, mobility and carrier concentration variation of as-prepared and annealed SnO_x films.

As observed in the Mossbauer spectra of RT deposited films (see Figure 3.14b), these films consist of metallic tin, SnO and SnO_2 phases. In this condition, both electrons and holes can contribute to the charge transport in the system. At 200°C annealing in air, oxygen atoms are incorporated in the films, which oxidizes with metallic tin leading to SnO phase formation with reduction in electron density and an increase in hole density. This may be the reason for initial increase in the resistivity of the films. In this case, the material has large density of holes compared to the density of electrons which results in positive Hall voltage with typical Hall mobility of 2 cm^2 V^{-1} s^{-1}.

Annealing at much higher temperatures causes again a decreasing tendency in material resistivity (not shown here). This time, it is because of the phase transformation of the material from SnO (p-type) to SnO_2 (n-type). Actual carrier mobility must be much higher than the measured Hall mobility value.

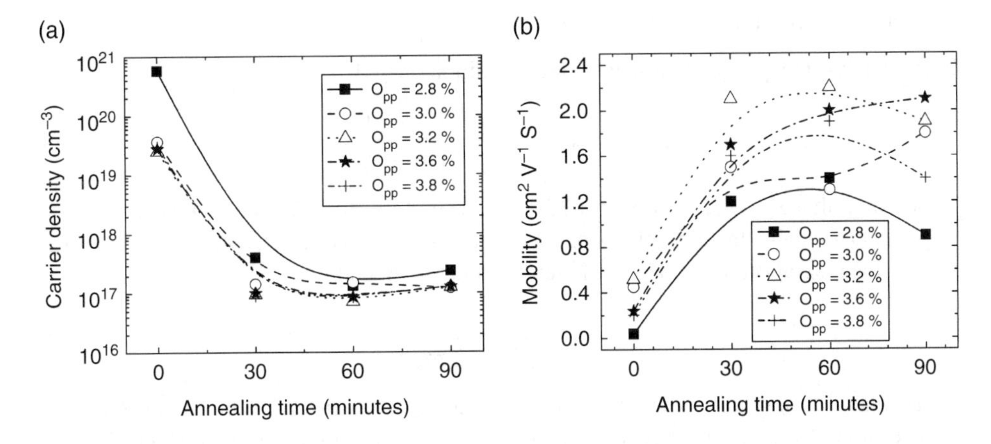

Figure 3.20 *a) Variation in the carrier density and b) mobility of SnO_x films as a function of annealing time.*

3.4.2.2 Capacitance measurements

The capacitance as a function of the applied voltage analysis allow us to gain information about the semiconductor relaxation time and the role of bulk and interface defects in the electrical performances of the oxide semiconductors used as an active channel layer in devices such as TFTs.

CV measurements have been done in TFT (see Chapter 5) using the oxide process window reported in Figure 3.14, by short-circuiting the drain and source terminals of devices processed using the system exhibiting the lowest residence time. By doing so, we enhance the overall capacitance measured allowing for a better discrimination in the data achieved. Under these conditions, we assume an electrical model consisting of a contact resistance R_C (equal for drain and source regions, since we have a symmetric device) in series with the combination of two *RC* resonators containing the channel resistance that varies dynamically, from a value close to R_I (that represents the charge modulated close to the dielectric/semiconductor interface) and to R_B (that represents the bulk channels resistivity not charge modulated): $R_I \leq R_{Dyn} \leq R_B$. Moreover, the capacitance between the drain/source terminal and the gate ($C_{osg} = C_{odg}$) is also dependent on the behavior of interface defects and how deep we can extend the channel modulation along overall channel thickness. $C_{osg} = C_{odg}$ corresponds to the series association between a capacitance dependent on the interface extension between the dielectric and the channel semiconductor and so, dependent also on the defects at the dielectric channel interface, C_{BS} given by:

$$C_{BS} = \frac{\rho(V)}{dV/dx} = \varepsilon\varepsilon_0 \frac{\rho(V)}{Q} \tag{3.13}$$

and C_B is the geometric capacitance, mainly determined by the extension of the dielectric layer through which carriers are induced in the channel, generating the source and drain currents (I_{St} and I_{Dt}, respectively).

Under these conditions we expect that:

1) As the frequency tends to a steady state condition, $\omega \rightarrow 0$:

$$C_{osg} = C_{odg} = C(o) = \frac{R_I C_{BS} + R_B^2 C_B}{(R_I + R_B)^2} \tag{3.14}$$

provided that $R_I >> R_B$, we obtain $C(0) \approx C_{BS}$. Therefore, for semiconductors that are not perfect (basically, the amorphous ones) it is necessary to use a frequency modulation less than the relaxation frequency of the semiconductor, given by

$$f_r = \frac{1}{2\pi R_B C_B} = \frac{1}{2\pi \rho_B \varepsilon_S \varepsilon_0} \tag{3.15}$$

In general terms, this means that the capacitance varies inversely in proportion to the extension of the depletion region (where charges are accumulated in the channel: $C(0) \approx \varepsilon\varepsilon_0 / W$ and W is the extension of such depleted region. In very thin channel layers may be limited by the channel thickness used).

Indeed the working frequency should be even less (according to the circuit), not exceeding:

$$f_l = \frac{1}{2\pi R_B (C_{BS} + C_B)} \tag{3.16}$$

Thus, we expect that frequency modulation to be associated with the shift of interface defects and their extension along the interface, with more and more states becoming visible as the frequency decreases. The only limit here is the ratio between these screen lengths and the real thickness of the semiconductor that may condition the charge penetration depth. That is, the values of the capacitance can be channel thickness conditioned for devices whose thickness is in the range of few nanometers.

2) $\omega \rightarrow \infty$:

$$C(\infty) \cong \frac{C_{BS} C_B}{C_{BS} + C_B} \tag{3.17}$$

This leads to the geometrical capacitance, given by: $C_G \approx \varepsilon\varepsilon_0/d$, where d is the thickness of the dielectric used.

3) In the transition region we may say that the slop is in proportion to the square root of trapped charges N_S.

Figure 3.21 shows the normalized CV plots for different frequencies. There is also plotted the transfer characteristics of the p-type TFT under analysis (the vertical scale on the right-hand side of the figure). Besides the hysteresis that we associate with the interface defects, the CV plots show that the overall maximum capacitance increases as the frequency decreases, as expected. Moreover, at a frequency of about 100 Hz the capacitance tends to show a flat behavior and the hysteresis voltage shift is enhanced, similar to that of the transfer IV TFT characteristics (see Chapter 5). We associate this behavior with interface localized states that only respond at frequencies below f_l as expected. That is, as the frequency decreases, more and more states will contribute once the penetration depth of the charges into the channel is also enhanced.

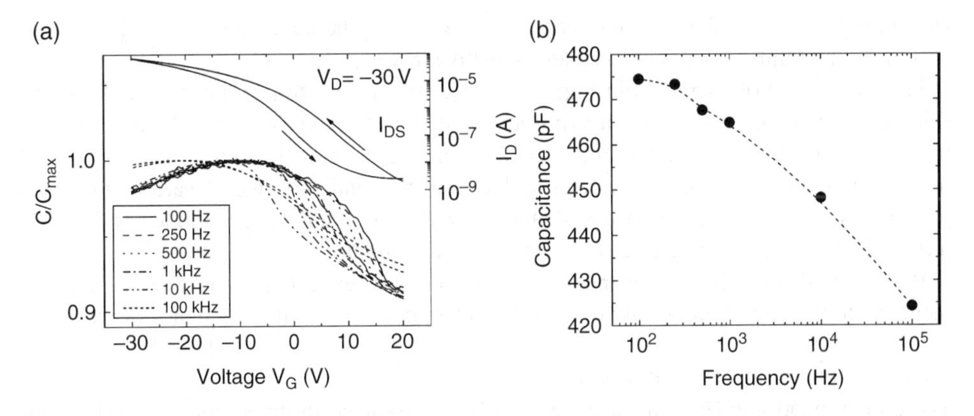

Figure 3.21 *a) Normalized CV plot for several frequencies (below), showing also the IV transfer characteristic of the TFT (above); b) The Maximum capacitance (C_{max}) achieved as a function of the frequency used.*

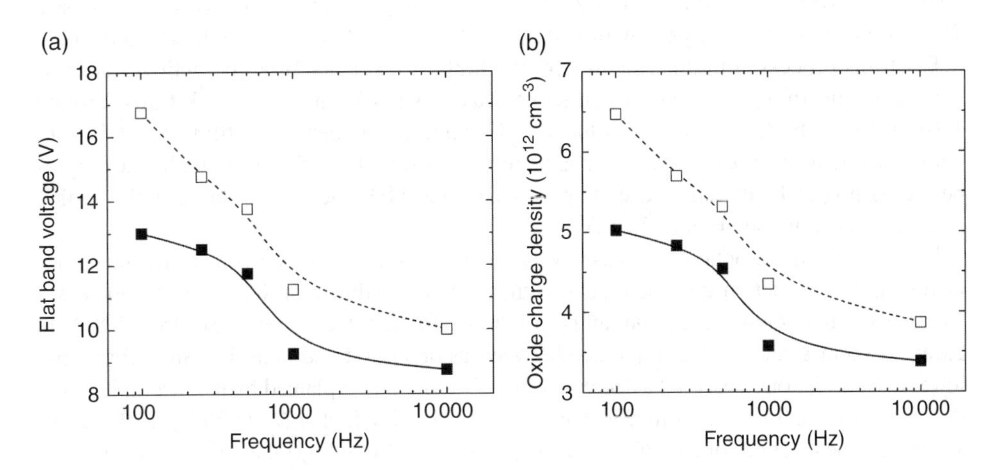

Figure 3.22 *a) Flat band Voltage dependence on the frequency; b) Oxide charge density as a function of the frequency.*

Apart from that, we note that the minimum capacitance is not fully flat. We associate this behavior with the small thickness of the semiconductor (≤12 nm) which limits the maximum dimension of the depletion region we can achieve. As shown in Figure 3.21, to "close" the channel we need to deplete the semiconductor layer by applying a positive voltage. Here we must remember that we have a series capacitance ($C_{BS}//C_B$) whose final values are dominated by the smallest single value. However, as the depletion layer increases the C_{BS} decreases. As the possible maximum width of depletion region is 12 nm (the semiconductor thickness), this means that the minimum capacitance of the system won't be much lower than the total capacitance.

Another factor that can limit the capacitance variation is the interface states. This may create a parallel capacitance (C_{ss}) associated with the depletion region capacitance (C_{BS}).

Depending on the number of interface states, the C_{ss} may be large and if larger than C_{BS} it will make the total capacitance even less sensitive to variation on C_{BS}.

From the CV plots depicted in Figure 3.21 we may also estimate the flat band voltage shift and the oxide charge density (proportional to defect density) as a function of the frequencies used, as depicted in Figure 3.22. There we indicate the two values determined for each frequency due to the hysteresis, defining the window range of variation of such parameters. For this calculation we assumed $N_a = 10^{17}$ cm^{-3} and a negligible work function difference between the gate electrode and the semiconductor.

Both plots show a similar dependence on frequency, as expected, leading to an enhancement of the flat band voltage and of the oxide charge density as the frequency decreases.

3.4.2.3 Optical properties evaluation

The optical band gap (E_{op}) of the SnO_x thin films produced in the systems with different residence times was calculated using the Tauc law: $\alpha^x = (h\nu - E_{op})$, where α represents the absorption coefficient, h the Planck's constant, ν the photon frequency and x a constant depending on the type of optical transition in the material. The E_{op} value is then obtained by linearly extrapolating the plot of $(\alpha)^x$ vs $h\nu$ and finding the intersection with the abscissa. The optical data were analyzed with $x = ½$ (direct transition) and $x = 2$ (indirect transition).

For films processed in the system with the highest residence time the optical transmittance for thin films varies with the thickness used, significantly. For thick films (around 120 nm thick) the transmittance in the visible range for as deposited films at RT is very poor, improving to about 50% after annealing. On the other, for films with thicknesses below 10 nm (as the thickness used to produce p-type TFT) the transmittance in the visible range is very high (see Figure 3.23a).

For these films the optical gap was inferred by considering that the main carrier transitions involving the conduction and valence bands are governed by direct transitions. As O_{pp} increases we observe an increase on E_{op} until an O_{pp} of 20%, after that the E_{op} decreases from 2.8 eV to 2.6 eV without a remarkable difference between as-deposited and annealed SnO_x films. For lower O_{pp} ($< 5\%$) one observed a decrease of the E_{op} which is explained by the incomplete oxidation of metallic Sn, as confirmed by the X-ray analysis. For higher O_{pp} ($> 20\%$) E_{op} decreases suggesting the presence of an indirect bandgap as was confirmed by other authors [98, 106]. The films processed using the deposition system with the lowest species' residence time exhibit optical properties quite close to those depicted in Figure 3.23 for the process window where the tin oxide films are clear p-type.

The optical transmittances of RT deposited films 120 nm thick (more than one order of magnitude higher than the thickness as active channel layer) were very poor in the visible region, mainly because of the presence of large concentration of metallic tin in the films (Figure 3.24a). Annealing up to 200°C for 30 minutes increased considerably the optical transmittance in the visible region. However, the maximum transmission was still around 50%. SnO has been reported as an indirect bandgap material with a direct bandgap value of around 2.5 eV and indirect band gap value around 1 eV. Whereas SnO_2 is a direct bandgap material with bandgap value of around 3.6 eV.

Estimated direct bandgap values are in the range of 2.6–2.75 eV for annealed films (Figures 3.24b and c). However, the presence of a narrow gap indirect band results in considerable optical absorption below these estimated values. The indirect E_g values estimated from a plot of $(\alpha h\nu)^{0.5}$ vs $h\nu$ results in 1.6–2.2 eV which is around 1–0.55 eV below the corresponding direct bandgap values (Figure 3.24d). These values are close to the reported

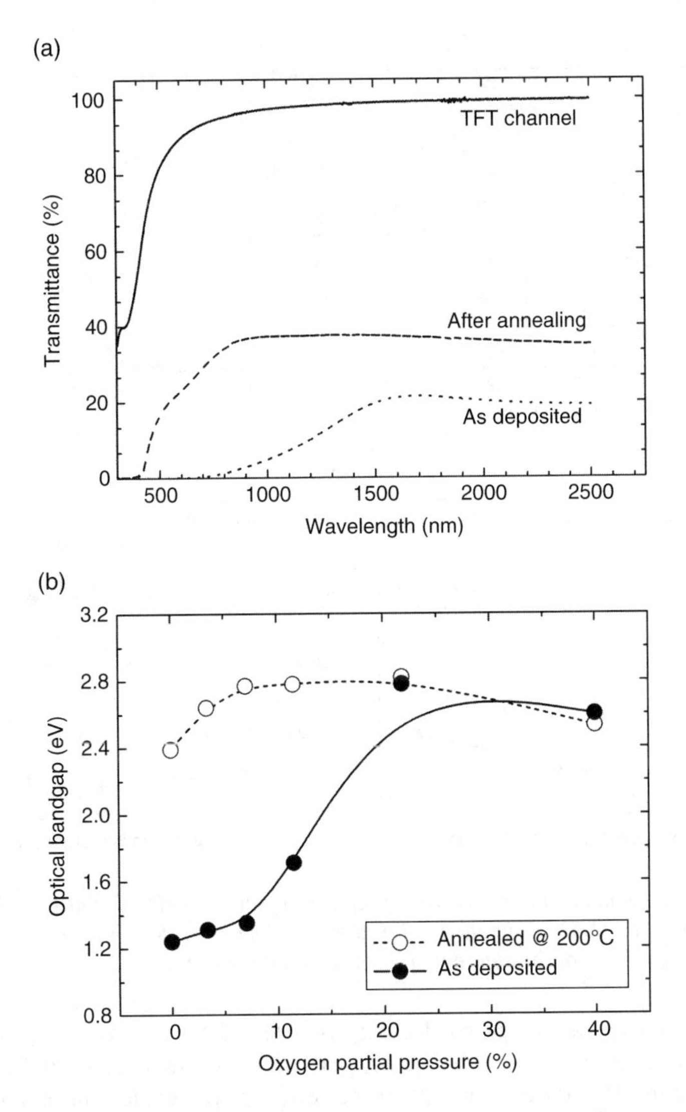

Figure 3.23 *a) Optical transmittance of SnOx films with thicknesses of 120 nm (dashed and dotted lines) and with a thickness ≤10 nm, as used as channel layer in p-type TFT; b) Dependence of the optical band gap as a function o the oxygen partial pressure for* SnO_x *thin films as deposited and after annealing at 200°C.*

bandgap values for SnO films. Small variations in these values may be understood on the basis of fractional variations of various phases (metallic tin, SnO and SnO_2) in the films.

In summary we have demonstrated that it is possible to produce transparent oxide semiconductors based on SnO_x by reactive magnetron sputtering without intentional substrate heating, that exhibit p-type conductivity after a low temperature annealing at 200°C. The SnO_x heat treated films are polycrystalline presenting a mixture of both tetragonal β-Sn and α-SnO phases. The SnO phase was identified and quantified by two independent techniques, XRD and Mossbauer spectroscopy, corroborating the p-type oxide semiconductor behavior

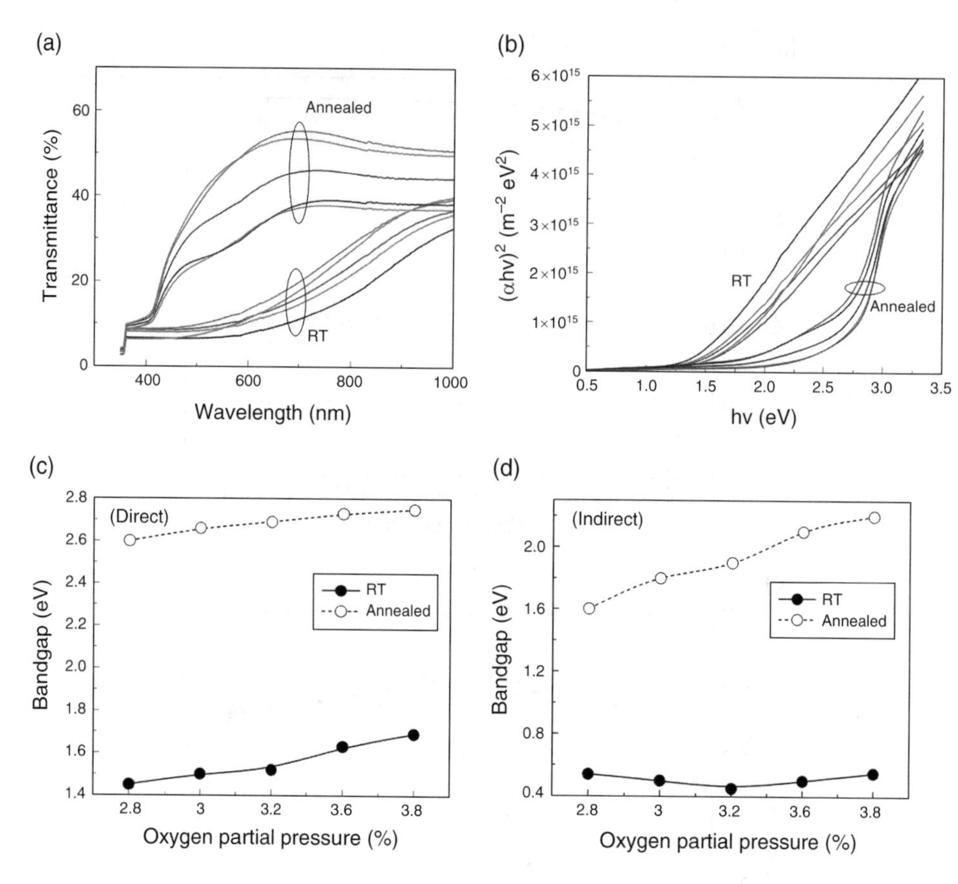

Figure 3.24 *a) Optical transmission of as prepared and annealed SnO_x films; b) Plot of $(\alpha h\nu)^2$ vs $h\nu$ for direct bandgap estimation; c) Estimated direct bandgap values; and d) indirect bandgap values. In a) and b) different lines refer to different O_{pp}.*

obtained. Moreover, the p-type conductivity is obtained for a narrow O_{pp} range, highly dependent on the species' residence time within the process system used. For the two process systems used the O_{pp} process window for achieving p-type oxide semiconductors varies between 5 to 12% and 2.5% to 4.4% respectively for systems where the ratio between species' residence time was 2:1. The hole carrier concentration (P) estimated was in the range of $\approx 10^{16}$–10^{18} cm^{-3}, ρ between 10^1 and 10^2 Ω cm and a μ_H reaching a maximum of 4.8 cm^2 V^{-1} s^{-1}. Concerning the optical properties, the p-type films present $E_{op} \approx 2.8$ eV and for 30 nm thick films the optical average transmittance is around 85%, between 400 and 2000 nm, decreasing to about 50% when the film thickness is enlarged by a factor of 4.

References

[1] C.D. Dimitrakopoulos and P.R.L. Malenfant (2002) Organic thin film transistors for large area electronics, *Advanced Materials* **14**, 99–117.

[2] A. Facchetti, M.H. Yoon, and T.J. Marks (2005) Gate dielectrics for organic field-effect transistors: New opportunities for organic electronics, *Advanced Materials* **17**, 1705–25.

[3] H. Sirringhaus (2005) Device physics of Solution-processed organic field-effect transistors, *Advanced Materials* **17**, 2411–25.
[4] H. Klauk (2010) Organic thin-film transistors, *Chemical Society Reviews* **39**, 2643–66.
[5] J.M. Shaw and P.F. Seidler (2001) Organic electronics: Introduction, *Ibm Journal of Research and Development* **45**, 3–9.
[6] S.R. Forrest and M.E. Thompson (2007) Introduction: Organic electronics and optoelectronics, *Chemical Reviews* **107**, 923–5.
[7] X.W. Zhan, A. Facchetti, S. Barlow, T.J. Marks, M.A. Ratner, M.R. Wasielewski, and S.R. Marder (2011) Rylene and Related Diimides for Organic Electronics, *Advanced Materials* **23**, 268–84.
[8] H. Yan, Z.H. Chen, Y. Zheng, C. Newman, J.R. Quinn, F. Dotz, M. Kastler, and A. Facchetti (2009) A high-mobility electron-transporting polymer for printed transistors, *Nature* **457**, 679–86.
[9] R. Schmidt, J.H. Oh, Y.S. Sun, M. Deppisch, A.M. Krause, K. Radacki, H. Braunschweig, M. Konemann, P. Erk, Z.A. Bao, and F. Wurthner (2009) High-Performance Air-Stable n-Channel Organic Thin Film Transistors Based on Halogenated Perylene Bisimide Semiconductors, J. *Am. Chem. Soc.* **131**, 6215–28.
[10] H. Usta, A. Facchetti, and T.J. Marks (2008) Air-stable, solution-processable n-channel and ambipolar semiconductors for thin-film transistors based on the indenofluorenebis(dicyanovinylene) core, J. *Am. Chem. Soc.* **130**, 8580–81.
[11] S. Handa, E. Miyazaki, K. Takimiya, and Y. Kunugi (2007) Solution-processible n-channel organic field-effect transistors based on dicyanomethylene-substituted terthienoquinoid derivative, J. *Am. Chem. Soc.* **129**, 11684–5.
[12] R. Martins, P. Barquinha, A. Pimentel, L. Pereira, and E. Fortunato (2005) Transport in high mobility amorphous wide band gap indium zinc oxide films, *Physica Status Solidi a-Applications and Materials Science* **202**, R95–R97.
[13] P. Canhola, N. Martins, L. Raniero, S. Pereira, E. Fortunato, I. Ferreira, and R. Martins (2005) Role of annealing environment on the performances of large area ITO films produced by rf magnetron sputtering, *Thin Solid Films* **487**, 271–6.
[14] J.Y. Kwon, D.J. Lee, and K.B. Kim (2011) Review Paper: Transparent Amorphous Oxide Semiconductor Thin Film Transistor, *Electronic Materials Letters* **7**, 1–11.
[15] G. Goncalves, V. Grasso, P. Barquinha, L. Pereira, E. Elamurugu, M. Brignone, R. Martins, V. Lambertini, and E. Fortunato (2011) Role of Room Temperature Sputtered High Conductive and High Transparent Indium Zinc Oxide Film Contacts on the Performance of Orange, Green, and Blue Organic Light Emitting Diodes, *Plasma Processes and Polymers* **8**, 340–5.
[16] E. Fortunato, L. Raniero, L. Silva, A. Goncalves, A. Pimentel, P. Barquinha, H. Aguas, L. Pereira, G. Goncalves, I. Ferreira, E. Elangovan, and R. Martins (2008) Highly stable transparent and conducting gallium-doped zinc oxide thin films for photovoltaic applications, *Solar Energy Materials and Solar Cells* **92**, 1605–10.
[17] T. Kamiya, K. Nomura, and H. Hosono (2010) Present status of amorphous In-Ga-Zn-O thin-film transistors, *Science and Technology of Advanced Materials* **11**, 044305-1–044305-23
[18] S. Parthiban, E. Elangovan, K. Ramamurthi, R. Martins, and E. Fortunato (2009) High near-infrared transparency and carrier mobility of Mo doped In(2)O(3) thin films for optoelectronics applications, *Journal of Applied Physics* **106**, 063716-1–063716-7.
[19] H. Kawazoe, M. Yasukawa, H. Hyodo, M. Kurita, H. Yanagi, and H. Hosono (1997) P-type electrical conduction in transparent thin films of $CuAlO_2$, *Nature* **389**, 939–42.
[20] K.U.H. Yanagi, S. Ibuki, T. Hase, H. Hosono, and H. Kawazoe (2000) Transparent p- and n-type conductive oxides with delafossite structure, *Materials Research Society Symposium Proceedings* **623**, 235–43.
[21] S.B. Zhang, S.H. Wei, and A. Zunger (2001) Intrinsic n-type versus p-type doping asymmetry and the defect physics of ZnO, *Phys. Rev. B.* **63**, 075205-1–075205-7.
[22] K. Tonooka, H. Bando, and Y. Aiura (2003) Photovoltaic effect observed in transparent p-n heterojunctions based on oxide semiconductors, *Thin Solid Films* **445**, 327–31.
[23] K. Tonooka and N. Kikuchi (2006) Preparation of transparent $CuCrO_2$: Mg/ZnO p-n junctions by pulsed laser deposition, *Thin Solid Films* **515**, 2415–18.

[24] M. Izaki, T. Shinagawa, K. T. Mizuno, Y. Ida, M. Inaba, and A. Tasaka (2007) Electrochemically constructed p-Cu_2O/n-ZnO heterojunction diode for photovoltaic device, *Journal of Physics D-Applied Physics* **40**, 3326–9.

[25] J.Z. Zhao, H.W. Liang, J.C. Sun, J.M. Bian, Q.J. Feng, L.Z. Hu, H.Q. Zhang, X. P. Liang, Y.M. Luo, and G. T. Du (2008) Electroluminescence from n-ZnO/p-ZnO : Sb homojunction light emitting diode on sapphire substrate with metal-organic precursors doped p-type ZnO layer grown by MOCVD technology, *Journal of Physics D-Applied Physics* **41**, 195110-1–195110-4.

[26] K. Sieradzka, D. Kaczmarek, J. Domaradzki, E. Prociow, M. Mazur, and B. Gornicka (2011) Optical and electrical properties of nanocrystalline TiO_2:Pd semiconducting oxides, *Central European Journal of Physics* **9**, 313–18.

[27] S. Yamazoe, S. Yanagimoto, and T. Wada (2011) Wide band gap and p-type conductive Cu-Nb-O films, *Physica Status Solidi-Rapid Research Letters* **5**, 153–5.

[28] E. Le Boulbar, E. Millon, J. Mathias, C. Boulmer-Leborgne, M. Nistor, F. Gherendi, N. Sbai, and J.B. Quoirin (2011) Pure and Nb-doped TiO(1.5) films grown by pulsed-laser deposition for transparent p-n homojunctions, *Applied Surface Science* **257**, 5380–3.

[29] M. Nistor, F. Gherendi, N. B. Mandache, C. Hebert, J. Perriere, and W. Seiler (2009) Metal-semiconductor transition in epitaxial ZnO thin films, *Journal of Applied Physics* **106**, 103710-1–103710-7.

[30] P. Puspharajah, S. Radhakrishna, and A. K. Arof (1997) Transparent conducting lithium-doped nickel oxide thin films by spray pyrolysis technique, *Journal of Materials Science* **32**, 3001–6.

[31] A. Kudo, H. Yanagi, H. Hosono, and H. Kawazoe (1998) $SrCu_2O_2$: A p-type conductive oxide with wide band gap, *Applied Physics Letters* **73**, 220–2.

[32] H. Yanagi, H. Kawazoe, A. Kudo, M. Yasukawa, and H. Hosono (2000) Chemical design and thin film preparation of p-type conductive transparent oxides, *Journal of Electroceramics* **4**, 407–14.

[33] K. Ueda, T. Hase, H. Yanagi, H. Kawazoe, H. Hosono, H. Ohta, M. Orita, and M. Hirano (2001) Epitaxial growth of transparent p-type conducting $CuGaO_2$ thin films on sapphire (001) substrates by pulsed laser deposition, *Journal of Applied Physics* **89**, 1790–3.

[34] A.N. Banerjee, R. Maity, and K.K. Chattopadhyay (2003) Preparation of p-type transparent conducting $CuAlO_2$ thin films by reactive DC sputtering, *Materials Letters* **58**, 10–13.

[35] M.M. Bagheri-Mohagheghi and M. Shokooh-Saremi (2004) The influence of Al doping on the electrical, optical and structural properties of SnO_2 transparent conducting films deposited by the spray pyrolysis technique, *Journal of Physics D-Applied Physics* **37**, 1248–53.

[36] E. Kaminska, A. Piotrowska, J. Kossut, A. Barcz, R. Butkute, W. Dobrowolski, E. Dynowska, R. Jakiela, E. Przezdziecka, R. Lukasiewicz, M. Aleszkiewicz, P. Wojnar, and E. Kowalczyk (2005) Transparent p-type ZnO films obtained by oxidation of sputter-deposited $Zn3N_2$, *Solid State Communications* **135**, 11–15.

[37] A.N. Banerjee, C.K. Ghosh, and K.K. Chattopadhyay (2005) Effect of excess oxygen on the electrical properties of transparent p-type conducting $CuAlO_{2+}$x thin films, *Solar Energy Materials and Solar Cells* **89**, 75–83.

[38] S.H. Lim, S. Desu, and A.C. Rastogi (2008) Chemical spray pyrolysis deposition and characterization of p-type CuCr1-xMgxO_2 transparent oxide semiconductor thin films, *Journal of Physics and Chemistry of Solids* **69**, 2047–56.

[39] A.C. Rastogi, S.H. Lim, and S.B. Desu (2008) Structure and optoelectronic properties of spray deposited Mg doped p-$CuCrO_2$ semiconductor oxide thin films, *Journal of Applied Physics* **104**, 023712-1–023712-11.

[40] L. Gong, Z.Z. Ye, J.G. Lu, L.P. Zhu, J.Y. Huang, and B.H. Zhao (2009) Formation of p-type ZnMgO thin films by In-N codoping method, *Applied Surface Science* **256**, 627–30.

[41] Y. Furubayashi, T. Hitosugi, Y. Yamamoto, K. Inaba, G. Kinoda, Y. Hirose, T. Shimada, and T. Hasegawa (2005) A transparent metal: Nb-doped anatase TiO^2, *Applied Physics Letters* **86**, 252101-1–252101-3.

[42] B. Szyszka, P. Loebmann, A. Georg, C. May, and C. Elsaesser (2010) Development of new transparent conductors and device applications utilizing a multidisciplinary approach, *Thin Solid Films* **518**, 3109–14.

[43] R. Martins, P. Barquinha, I. Ferreira, L. Pereira, G. Goncalves, and E. Fortunato (2007) Role of order and disorder on the electronic performances of oxide semiconductor thin film transistors, *Journal of Applied Physics* **101**, 044505-1–044505-7.
[44] E.H. Kennard and E.O. Dieterich (1917) An Effect of Light upon the Contact Potential of Selenium and Cuprous Oxide, *Physical Review* **9**, 58.
[45] L.O. Grondahl (1926) Theories of a new solid junction rectifier, *Science* **64**, 306–8.
[46] E. Duhme and W. Schottky (1930) Blocked and photo effects on the border of copper oxide against brushed metal layers, *Naturwissenschaften* **18**, 735–6.
[47] W. Schottky and F. Waibel (1933) The electron conduction of copper oxydule, *Physikalische Zeitschrift* **34**, 858–64.
[48] W. Schottky and F. Waibel (1935) The electron inroduction of copper oxyduls, *Physikalische Zeitschrift* **36**, 912–14.
[49] M.A. Rafea and N. Roushdy (2009) Determination of the optical band gap for amorphous and nanocrystalline copper oxide thin films prepared by SILAR technique, *Journal of Physics D-Applied Physics* **42**, 015413-1–015413-6.
[50] B.S. Li, K. Akimoto, and A. Shen (2009) Growth of Cu_2O thin films with high hole mobility by introducing a low-temperature buffer layer, J. *Cryst. Growth* **311**, 1102–5.
[51] K. Matsuzaki, K. Nomura, H. Yanagi, T. Kamiya, M. Hirano, and H. Hosono (2008) Epitaxial growth of high mobility Cu_2O thin films and application to p-channel thin film transistor, *Applied Physics Letters* **93**, 202107-1–202107-3.
[52] B. Li (2009) Growth of Cu_2O thin films with high hole mobility by introducing a low-temperature buffer layer, J. *Cryst. Growth* **311**, 1102–05.
[53] E. Fortin and F. L. Weichman (1966) Hall effect and electrical conductivity of CU_2O monocrystals, *Canadian Journal of Physics* **44**, 1551–61.
[54] L.C. Olsen, R.C. Bohara, and M.W. Urie (1979) Explanation for low-efficiency Cu_2O Schottky-barrier solar-cells, *Applied Physics Letters* **34**, 47–9.
[55] C. Wadia, A. P. Alivisatos, and D.M. Kammen (2009) Materials Availability Expands the Opportunity for Large-Scale Photovoltaics Deployment, *Environmental Science & Technology* **43**, 2072–7.
[56] R.J. Iwanowski and D. Trivich (1985) Cu/Cu_2O Schottky-barrier solar-cells prepared by multistep irradiation of a Cu_2O substrate by H^+ ions, *Solar Cells* **13**, 253–64.
[57] V. Figueiredo, E. Elangovan, G. Goncalves, P. Barquinha, L. Pereira, N. Franco, E. Alves, R. Martins, and E. Fortunato (2008) Effect of post-annealing on the properties of copper oxide thin films obtained from the oxidation of evaporated metallic copper, *Applied Surface Science* **254**, 3949–54.
[58] G.P. Pollack and D. Trivich (1975) Photoelectric properties of cuprous-oxide, *Journal of Applied Physics* **46**, 163–72.
[59] L. Papadimitriou, N. A. Economou, and D. Trivich (1981) Heterojunction solar-cells on cuprous-oxide, *Solar Cells* **3**, 73–80.
[60] B.P. Rai (1988) CU_2O solar-cells – a review, *Solar Cells* **25**, 265–72.
[61] L.C. Olsen, F.W. Addis, and W. Miller (1982) Experimantal and theoretical studies of CU_2O solar-cells, *Solar Cells* **7**, 247–79.
[62] A.O. Musa, T. Akomolafe, and M.J. Carter (1998) Production of cuprous oxide, a solar cell material, by thermal oxidation and a study of its physical and electrical properties, *Solar Energy Materials and Solar Cells* **51**, 305–16.
[63] H. Raebiger, S. Lany, and A. Zunger (2007) Origins of the p-type nature and cation deficiency in Cu_2O and related materials, *Phys. Rev. B* **76**, 045209-1–045209-5.
[64] W.H. Brattain (1951) The copper oxide rectifier, *Reviews of Modern Physics* **23**, 203–12.
[65] O. Porat and I. Riess (1995) Defect chemistry of Cu_2O at elevated temperatures. 2. Electrical-conductivity, thermoelectric-power and charged point-defects, *Solid State Ion.* **81**, 29–41.
[66] Z. Jarzebski (1973) *Oxide Semiconductors*, English edn, vol. 4. Oxfors: Pergamon Press.
[67] Y.S. Lee, M.T. Winkler, S.C. Siah, R. Brandt, and T. Buonassisi (2011) Hall mobility of cuprous oxide thin films deposited by reactive direct-current magnetron sputtering, *Applied Physics Letters* **98**, 192115-1–192115-3.

[68] T.H. Darma, A.A. Ogwu, and F. Placido (2011) Effects of sputtering pressure on properties of copper oxide thin films prepared by rf magnetron sputtering, *Materials Technology* **26**, 28–31.
[69] H.C. Lu, C.L. Chu, C.Y. Lai, and Y.H. Wang (2009) Property variations of direct-current reactive magnetron sputtered copper oxide thin films deposited at different oxygen partial pressures, *Thin Solid Films* **517**, 4408–12.
[70] J.Y. Park, T.H. Kwon, S.W. Koh, and Y.C. Kang (2011) Annealing Temperature Dependence on the Physicochemical Properties of Copper Oxide Thin Films, *Bulletin of the Korean Chemical Society* **32**, 1331–5.
[71] G. Yang, A.P. Chen, M. Fu, H. Long, and P. X. Lu (2011) Excimer laser deposited CuO and Cu_2O films with third-order optical nonlinearities by femtosecond z-scan measurement, *Applied Physics a-Materials Science & Processing* **104**, 171–5.
[72] A.P. Chen, H. Long, X. C. Li, Y.H. Li, G. Yang, and P.X. Lu (2009) Controlled growth and characteristics of single-phase Cu_2O and CuO films by pulsed laser deposition, *Vacuum* **83**, 927–30.
[73] W. Seiler, E. Millon, J. Perriere, R. Benzerga, and C. Boulmer-Leborgne (2009) Epitaxial growth of copper oxide films by reactive cross-beam pulsed-laser deposition, J. *Cryst. Growth* **311**, 3352–8.
[74] D.S. Darvish and H.A. Atwater (2011) Epitaxial growth of Cu_2O and ZnO/Cu_2O thin films on MgO by plasma-assisted molecular beam epitaxy, J. *Cryst. Growth* **319**, 39–43.
[75] M. Ichimura and Y. Song (2011) Band Alignment at the Cu_2O/ZnO Heterojunction, *Japanese Journal of Applied Physics* **50**, 051002.
[76] R.M. Liang, Y.M. Chang, P.W. Wu, and P. Lin (2010) Effect of annealing on the electrodeposited Cu2O films for photoelectrochemical hydrogen generation, *Thin Solid Films* **518**, 7191–5.
[77] G.C. Liu, L.D. Wang, and D.F. Xue (2010) Synthesis of Cu_2O crystals by galvanic deposition technique, *Materials Letters* **64**, 2475–8.
[78] K.P. Musselman, A. Wisnet, D.C. Iza, H.C. Hesse, C. Scheu, J.L. MacManus-Driscoll, and L. Schmidt-Mende (2010) Strong Efficiency Improvements in Ultra-low-Cost Inorganic Nanowire Solar Cells, *Advanced Materials* **22**, E254–E258.
[79] A.Y. Oral, E. Mensur, M.H. Aslan, and E.B. Basaran (2004) The preparation of copper(II) oxide thin films and the study of their microstructures and optical properties, *Materials Chemistry and Physics* **83**, 140–4.
[80] K. Santra, C.K. Sarkar, M.K. Mukherjee, and B. Ghosh (1992) Copper-oxide thin-films grown by plasma evaporation method, *Thin Solid Films* **213**, 226–9.
[81] S.Y. Lee, S.H. Choi, and C.O. Park (2000) Oxidation, grain growth and reflow characteristics of copper thin films prepared by chemical vapor deposition, *Thin Solid Films* **359**, 261–7.
[82] V. Figueiredo, E. Elangovan, G. Goncalves, N. Franco, E. Alves, S.H.K. Park, R. Martins, and E. Fortunato (2009) Electrical, structural and optical characterization of copper oxide thin films as a function of post annealing temperature, *Physica Status Solidi a-Applications and Materials Science* **206**, 2143–8.
[83] J. Liu and D. F. Xue (2008) Thermal oxidation strategy towards porous metal oxide hollow architectures, *Advanced Materials* **20**, 2622–7.
[84] A.C. Tickle (1969) *Thin-Film Transistors – A New Approach to Microelectronics*. New York: John Wiley & Sons, Inc.
[85] A. Togo, F. Oba, I. Tanaka, and K. Tatsumi (2006) First-principles calculations of native defects in tin monoxide, *Phys. Rev. B* **74**, 195128.
[86] P. Scherrer (1918) Göttinger Nachrichten Gesell, **2,** 98.
[87] E. Fortunato, V. Figueiredo, P. Barquinha, E. Elamurugu, R. Barros, G. Goncalves, S.H.K. Park, C.S. Hwang, and R. Martins (2010) Thin-film transistors based on p-type Cu_2O thin films produced at room temperature, *Applied Physics Letters* **96**, 192102-1–192102-3.
[88] N.F. Mott (1968) Metal-insulator transition, *Reviews of Modern Physics* **40**, 677–83.
[89] A. Mittiga, F. Biccari, and C. Malerba (2009) Intrinsic defects and metastability effects in Cu(2) O, *Thin Solid Films* **517**, 2469–72.
[90] M. Tapiero, Zielinge J.P., and C. Noguet (1972) Electrical conductivity and thermal activation-energies in CU_2O single-crystals, *Physica Status Solidi a-Applied Research* **12**, 517–20.

[91] Y. Nakano, S. Saeki, and T. Morikawa (2009) Optical bandgap widening of p-type Cu(2)O films by nitrogen doping, *Applied Physics Letters* **94**, 022111-1–022111-3.
[92] A.L.D.H.L. Hartnagel, A.K. Jain, C. Jagadish (1995) *Semconducting Transparent Thin Films*, Institute of Physics Publishing.
[93] H.H.D. Ginley, D.C. Paine (2010) *Handbook of Transparent Conductors*, Springer.
[94] G.Y. Zhao, X. Zhi, Y. Ren, and T. Zhu (2009) "Preparation of Sb:F:SnO_2 Films and Their Application in low-E Glass," in *Eco-Materials Processing and Design X.* vol. 620–622, H. Kim, J. F. Yang, T. Sekino, and S. W. Lee, eds, 5-8.
[95] K.I.A.J. Watson (1994) *Stannic Oxide Gas Sensor: principles and applications*, CRC Press.
[96] H. Li, X. J. Huang, and L.Q. Chen (1998) Direct imaging of the passivating film and microstructure of nanometer-scale SnO anodes in lithium rechargeable batteries, *Electrochem. Solid State Lett.* **1**, 241–3.
[97] D. Aurbach, A. Nimberger, B. Markovsky, E. Levi, E. Sominski, and A. Gedanken (2002) Nanoparticles of SnO produced by sonochemistry as anode materials for rechargeable lithium batteries, *Chem. Mat.* **14**, 4155–63.
[98] R.Y. Korotkov, R. Gupta, P. Ricou, R. Smith, and G. Silverman (2008) Atmospheric plasma discharge chemical vapor deposition of SnOx thin films using various tin precursors, *Thin Solid Films* **516**, 4720–7.
[99] W.S. Baker, J.J. Pietron, M.E. Teliska, P.J. Bouwman, D.E. Ramaker, and K.E. Swider-Lyons (2006) Enhanced oxygen reduction activity in acid by tin-oxide supported Au nanoparticle catalysts, *Journal of the Electrochemical Society* **153**, A1702–A1707.
[100] P. Ifeacho, T. Huelser, H. Wiggers, C. Schulz, and P. Roth (2007) Synthesis of SnO2-x nanoparticles tuned between 0 <=×<= 1 in a premixed low pressure H-2/O-2/Ar flame, *Proceedings of the Combustion Institute* **31**, 1805–12.
[101] J.Z. Wang, R. Martins, N.P. Barradas, E. Alves, T. Monteiro, M. Peres, E. Elamurugu, and E. Fortunato (2009) Intrinsic p Type ZnO Films Deposited by rf Magnetron Sputtering, *Journal of Nanoscience and Nanotechnology* **9**, 813–16.
[102] H. Shimotani, H. Suzuki, K. Ueno, M. Kawasaki, and Y. Iwasa (2008) p-type field-effect transistor of NiO with electric double-layer gating, *Applied Physics Letters* **92**, 242107-1–242107-3.
[103] S. Takami, R. Hayakawa, Y. Wakayama, and T. Chikyow (2010) Continuous hydrothermal synthesis of nickel oxide nanoplates and their use as nanoinks for p-type channel material in a bottom-gate field-effect transistor, *Nanotechnology* **21**, 134009-1–134009-4.
[104] L.Y. Liang, Z.M. Liu, H.T. Cao, and X.Q. Pan (2010) Microstructural, Optical, and Electrical Properties of SnO Thin Films Prepared on Quartz via a Two-Step Method, *Acs Applied Materials & Interfaces* **2**, 1060–5.
[105] I. Lefebvre, M. A. Szymanski, J. Olivier-Fourcade, and J.C. Jumas (1998) Electronic structure of tin monochalcogenides from SnO to SnTe, *Phys. Rev. B* **58**, 1896–1906.
[106] Y. Ogo, H. Hiramatsu, K. Nomura, H. Yanagi, T. Kamiya, M. Hirano, and H. Hosono (2008) p-channel thin-film transistor using p-type oxide semiconductor, SnO, *Applied Physics Letters* **93**, 032113-1–032113-3.
[107] T. Kamiya and H. Hosono (20100 Material characteristics and applications of transparent amorphous oxide semiconductors, *NPG Asia Mater* **2**, 15–22.
[108] Y. Ogo, H. Hiramatsu, K. Nomura, H. Yanagi, T. Kamiya, M. Kimura, M. Hirano, and H. Hosono (2009) Tin monoxide as an s-orbital-based p-type oxide semiconductor: Electronic structures and TFT application, *Physica Status Solidi a-Applications and Materials Science* **206**, 2187–91.
[109] Z.M. Jarzebski (1973) *Oxide Semiconductors*. Oxford: Pergamon Press.
[110] E. Fortunato, R. Barros, P. Barquinha, V. Figueiredo, S.H.K. Park, C.S. Hwang, R. Martins (2010) Transparent p-type SnOx thin film transistors produced by reactive rf magnetron sputtering followed by low temperature annealing, *Applied Physics Letters*, **97**, 052105-1–052105-3.
[111] J. Geurts, S. Rau, W. Richter, and F. J. Schmitte (1984) SnO films and their oxidation to SnO_2 – Raman-scattering, IR reflectivity and X-ray-diffraction studies, *Thin Solid Films* **121**, 217–25.
[112] W.K. Choi, H. Sung, K.H. Kim, J.S. Cho, S.C. Choi, H.J. Jung, S.K. Koh, C M. Lee, and K. Jeong (1997) Oxidation process from SnO to SnO_2, *Journal of Materials Science Letters* **16**, 1551–4.

[113] G. Shenoy and F. Wagner (1978) *Mössbauer Isomer Shifts*. Amsterdam: North Holland Publ. Co..
[114] D.E. Conte, A. Aboulaich, F. Robert, J. Olivier-Fourcade, J.C. Jumas, C. Jordy, and P. Willmann (2010) $Sn_{(x)}[BPO_{(4)}]_{(1-x)}$ composites as negative electrodes for lithium ion cells: Comparison with amorphous $SnB_{(0.6)}P_{(0.4)}O_{(2.9)}$ and effect of composition, *Journal of Solid State Chemistry* **183** 65–75.
[115] R.H. Herber (1983) Mossbauer lattice temperature of tetragonal (P4 NMM) SnO, *Phys. Rev. B* **27**, 4013–17.
[116] M.S. Moreno and R.C. Mercader (1994) Mossbauer study of SnO lattice-dynamics, *Phys. Rev. B* **50**, 9875–81.
[117] E. Leja, J. Korecki, K. Krop, and K. Toll (1979) Phase-composition of Snox thin-films obtained by reactive sputtering, *Thin Solid Films* **59**, 147–55.
[118] E. Leja, T. Pisarkiewicz, and A. Kolodziej (1980) Electrical-properties of nonstoichIometric tin oxide-films obtained by the DC reactive sputtering method, *Thin Solid Films* **67**, 45–8.
[119] T.M. Uen, K.F. Huang, M.S. Chen, and Y.S. Gou (1988) Preparation and characterization of some tin oxide-films, *Thin Solid Films* **158**, 69–80.
[120] K.F. Huang, T.M. Uen, Y.S. Gou, C.R. Huang, and H.C. Yang (1987) Temperature-Dependence of transport-properties of evaporated indium tin oxide-films, *Thin Solid Films* **148**, 7–15.
[121] N.F. Mott and E.A. Davis (1979) *Electronic Processes in Non-crystalline Materials*, 2nd edn, Oxford: Clarendon.

4

Gate Dielectrics in Oxide Electronics

4.1 Introduction

During this last decade, oxide semiconductors have been extensively studied with regard to their application as channel material in thin-film transistors (TFTs) because of their superior electrical properties as compared with amorphous silicon. The possibility of low temperature processing combined with a wide band gap made these devices suitable for transparent and flexible electronics. Materials such as zinc oxide (ZnO), indium-zinc oxide (IZO), zinc-tin oxide (ZTO) and gallium-indium-zinc-oxide (GIZO) and have been the subject of intensive study over this last decade [1–8].

Huge developments were achieved in oxide semiconductors but the dielectric has been somehow forgotten or, at least, been treated as secondary, despite being as important as the channel layer in a TFT as its properties determine the charge accumulation at the dielectric/semiconductor interface. However, the existence of defects at this interface, as well as within the dielectric itself determines to a large extent the device's static and dynamic performance and reliability, in parameters such as field effect mobility, leakage current, and stability under electrical stress.

When thinking about a well established technology such as silicon based TFTs, the dielectrics normally used have been silicon nitride (Si_3N_4), in amorphous silicon [9, 10], and silicon oxide (SiO_2), in polycrystalline silicon devices [11, 12]. The technological maturity of the deposition technique used (essentially PECVD) and the good insulating and interface properties supported this option in silicon thin film TFTs. The processing temperature is normally between 200–300°C and did not represent a problem since the semiconductor is also processed within this range of temperatures (or even higher, in the case of polycrystalline

Transparent Oxide Electronics: From Materials to Devices, First Edition. Pedro Barquinha, Rodrigo Martins, Luis Pereira and Elvira Fortunato.

silicon). However, migration to low cost substrates such as the polymeric ones raised some difficulties since it forced a decrease in deposition temperature for the dielectric layers that may compromise their insulting properties and reliability [13–15].

Silicon oxide has also been considered as the preferred dielectric in oxide TFTs. Due to its low dielectric constant (κ) the films used for high performance TFTs must be the thinnest possible, requiring highly compact and defect free SiO_2 films. Most of the studies done so far on oxide TFTs have used thermal SiO_2 in doped silicon wafers (acting as a gate electrode) that ensures good insulating characteristics and can withstand an electric field above 10 MV cm^{-1} without reaching breakdown.

Another material that has been widely used as a gate dielectric in oxide TFTs is ATO, which consists of alternate layers of aluminum oxide (Al_2O_3) and titanium oxide (TiO_2) deposited by Atomic Layer Deposition (ALD) on an ITO layer that acts as a gate electrode. This way it is possible to have transparent devices using a dielectric that has good insulating characteristics and a high breakdown electric field. Besides, the higher dielectric constant of the Al_2O_3 and TiO_2 compared with SiO_2 results in a better gate capacitance that allows for reduced operation voltage.

However, these are just clever approaches to overcome perhaps the most challenging difficulty in producing a low temperature TFT, which is related to the gate dielectric performance as none of these techniques are suitable for low temperature processing. Thermal silicon oxide is obtained above 1000°C in silicon substrates while good ALD films normally require deposition temperatures higher than 300°C. Even if recent developments decreased the ALD deposition temperature, the low deposition rate and difficulties in up-scaling for large areas are still obstacles to industrial application.

As far as the state of the art is concerned, the ideal solution has not yet been found, and contrary to the semiconductor layer that is more or less restricted to a small group of oxides, hard work still lies ahead of us if we are to find the most suitable dielectric layer on the route to low temperature transparent and flexible electronics.

4.2 High-κ Dielectrics: Why Not?

Since the first microprocessor, the Intel 4004, was introduced in 1971, electronic circuits have been constantly integrating a high density of increasingly smaller transistors in order to achieve high performance, better functionality and low costs. This evolution was predicted in 1965 by Gordon Moore, one of the co-founders of Intel, who stated that the number of transistors in a chip would double about every two years. To have an idea of the numbers involved in this evolution, note that the Intel 4004 had a clock speed of 108 kHz, 2300 transistors and a manufacturing technology of 10 μm, while the Intel Penryn from 2007 exhibits a clock speed above 3 GHz and about 820 million transistors fabricated with 45 nm technology. Although the semiconductor industry currently faces technological problems with regard to further scale down, such as the limitations of lithographic tools and severe short-channel and quantum effects when the channel lengths of the transistors are reduced to the dimensions of a few silicon atoms, perhaps the most important issue that has arisen in recent times is related to the dielectric layer. In single crystalline silicon technology, thermally grown SiO_2 is used as the dielectric and the successful Si/SiO_2 combination is probably what contributes mostly to the remarkable properties exhibited by metal-oxide-semiconductor field-effect transistors (MOSFETs). Si/SiO_2 represents an almost perfect interface, because thermal SiO_2 is not actually deposited but rather grown by the reaction of oxygen with silicon. Besides, it has excellent insulating properties with a gap around 9 eV and a high breakdown electric field, typically above 10 MVcm^{-1} [16].

Today's demand for miniaturization requires that in the new generation of transistors with channel dimensions below 45 nm the thickness of the SiO_2 layer must be lower than 1 nm, which represents a critical issue even for such an excellent insulator as thermal SiO_2, because gate leakage current dramatically increases due to quantum tunneling effects [17]. This creates a technological limitation that cannot be easily surpassed without looking for new device structures or dielectric materials [18]. One of the best possibilities for overcoming this limitation involves the usage of materials with a higher dielectric constant than SiO_2, the so-called high-κ dielectrics. This way, thicker insulating films can be used, while maintaining the same capacitance per unit area. These materials began to be intensively studied in the mid-1990s and are currently being used to fabricate Intel microprocessors with features below 45 nm in size.

The importance of high-κ material on TFTs is also relevant. It is intended that a good device presents a high current when in on-state and, simultaneously, low operation voltages. One way to achieve this is by increasing the gate capacitance, as good quality crystalline semiconductor thin films cannot be obtained at low temperatures. To increase gate capacitance two options are available: either we use a high-κ material or we decrease the thickness of conventional TFT gate insulators. However, thickness reduction may be a limiting option: if these oxide TFTs are thought to be deposited on low cost substrates such as polymers, all layers must processed at low temperatures (typically below 150°C) and high leakage current may arise due to poor insulating characteristics obtained under these conditions, even when dealing with "well ranked" good insulators such as SiO_2. Despite high-k dielectrics have been previously thought to replace SiO_2 in crystalline silicon technology they are also one option for oxide based TFTs, increasing the gate capacitance, while maintaining a physical thickness able to ensure low gate leakage current. Thus, this is a valid option for opening the way for the concept of transparent and flexible electronics.

4.3 Requirements

The importance of high-κ dielectrics is obvious for present and future electronic circuits. However, some drawbacks exist, such as increased parasitic capacitances [19] and low band gap, since for most metal oxide dielectrics it is inversely proportional to κ (Figure 4.1a) [20]. Moreover, despite preventing direct tunneling as thicker high-κ layers ensure the same capacitance per unit area of SiO_2, if its band gap (E_g) is too low undesirable gate leakage can still remain an issue, due to excitation of electrons or holes by Schottky emission into the dielectric conduction or valence bands, or to other defect-assisted transport mechanisms, such as Poole-Frenkel effect or hopping conduction [20, 21]. But even if the gap of the dielectric layer is considerably larger than that of the semiconductor, another requirement should be fulfilled in order to achieve good reliability: the offsets of the dielectric's valence band maximum (VBM) in a p-type transistor and conduction band minimum (CBM) in a n-type transistor should be at least 1 eV relative to those of the semiconductor (Figure 4.1b) [22].

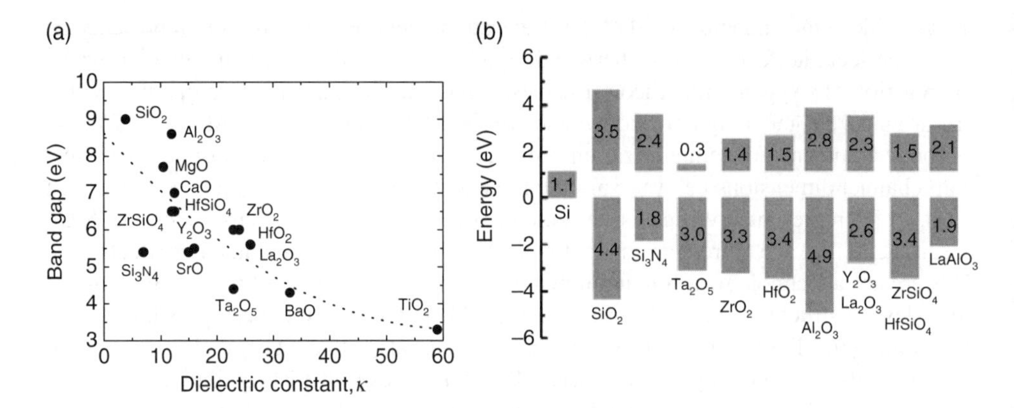

Figure 4.1 (a) Relation between the band gap and dielectric constant for some dielectrics; (b) calculated band offsets of some dielectrics on crystalline silicon. Adapted with permission from ref [22] Copyright (2002) Materials Research Society.

Another important aspect to be considered is the structure: while most of the high-κ dielectrics are polycrystalline even when processed at low temperatures, amorphous layers are preferred because grain boundaries act as preferential paths for impurity diffusion and leakage current, resulting in inferior dielectric reliability. Moreover, amorphous materials present smoother surfaces than polycrystalline ones, resulting in improved interface properties [18, 23]. The quality of the dielectric/semiconductor interface is preponderant, determining the performance of the transistors, and most high-κ materials exhibit mid-gap interface state densities one to two orders of magnitude higher than the one of thermal SiO_2 [18]. Other considerations regarding the choice of a suitable dielectric involve its thermodynamic stability with the semiconductor and the process's compatibility with current or expected tools [18].

All requirements listed above are linked to crystalline silicon technology. However, some of these requirements must be tuned for application in oxide TFTs. Concerning band gap no differences exist and a high value is desirable. This is even more important in oxides TFTs since the channel material has a E_g normally above 3 eV, raising some problems in assuring a band offset higher than 1 eV. Figure 4.2 is presents the estimation of the bands offset for some common dielectrics (high-κ included), normalized to the VBM of zinc oxide (ZnO), based on charge neutrality level (CNL) data published by Robertson *et al.* and King *et al.* for dielectrics and oxide semiconductors [24–26]. The horizontal line defines the minimum 1 eV limit for the conduction band offset (CBO) in ZnO.

It is possible to see that, theoretically, SiO_2 and Al_2O_3 are excellent dielectrics from a band offset point of view as their large gap ensures values well above 1 eV both at the conduction and valence bands. However, some high-κ materials such as hafnium oxide (HfO_2) or zirconium oxide (ZrO_2) are still a good option for oxide TFTs, with a CBO above 2 eV. To be in a more comfortable position, alloying them with materials such as SiO_2 may lead to the formation of silicates and increase slightly the band gap, achieving a more favorable band offset, as happens for $HfSiO_4$. The CBO of these dielectrics is even higher in the case of tin oxide (SnO_2) and indium oxide (In_2O_3) and so even tantalum oxide (Ta_2O_5), may be an interesting option as this value is around 1 eV.

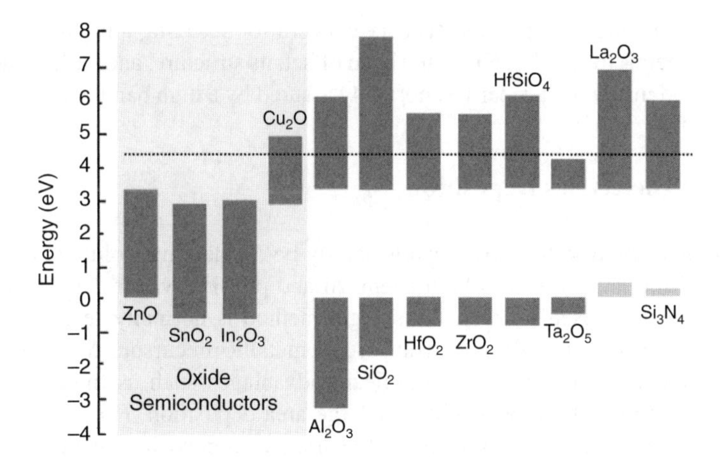

Figure 4.2 *Calculated band offset of some dielectrics. The values are normalized to the VBM of ZnO. The thinner line indicates the minimum of 1 eV for CBO.*

Table 4.1 *Summary of some relevant characteristics of some dielectrics.*

Dielectric	κ	E_g (eV)	ZnO CBO (eV)	SnO_2 CBO (eV)	In_2O_3 CBO (eV)
SiO_2	3.9	9	4.3	4.8	4.6
Al_2O_3	9	8.8	3.6	3.1	2.9
Si_3N_4	7	5.4	2.5	3.0	2.8
HfO_2	25	6	2.1	2.6	2.4
La_2O_3	30	5.7	3.4	3.9	3.7
Ta_2O_5	26	4.4	0.9	1.4	1.2
ZrO_2	24	5.8	2	2.5	2.3
$HfSiO_4$	14	6.5	2.7	3.2	3

Recently, the possibility of also having p-type TFTs based on semiconductor oxides was demonstrated [27–30]. This way, it is also important to understand which dielectric can be considered for this type of devices. Copper monoxide (Cu_2O) is one of the materials that can be used as p-type semiconductor [29, 30] and the bands offset for this material are also presented in Figure 4.2 for this material. In this case the valence band offset (VBO) should be considered and, theoretically, all the dielectrics listed are suitable for p-type oxide TFTs.

On the other hand, when seeking low temperature processing the dielectric thermal stability is not such a critical issue. Besides, the channel layer is also an oxide material and these tend not to react between them at low temperatures. However, there are reports of interdifusion of elements at temperatures as low as 300°C and this must be taken into account [31]. Concerning the structure, amorphous layers are again desirable in order to reduce leakage paths through grain boundaries and improve interface properties [16, 32]. This is because pure crystalline structures are not possible at low temperatures and because most of the semiconductor layers used tend also to be amorphous, as this class of materials has already proved to be the best candidates for oxide TFTs.

In summary, we may say that the primary concern for selecting a high-κ material (or a dielectric in general) for oxide TFTs is the band offset, its structure, and surface morphology besides the good insulating properties, normally assured by a high band gap (see Table 4.1).

4.4 High-κ Dielectrics Deposition

The deposition of high-κ dielectrics can generally be divided into solution and gas-phase processes, and the latter includes both chemical and physical vapor deposition methods. Concerning solution based process, the sol–gel method is normally selected and can be used to deposit dielectric thin films by using organometallic precursors. Materials processed by sol–gel methods generally offer significant advantages such as high purity, ease of composition control and the possibility for large area deposition. However temperatures around 400°C are needed in order to anneal the dried layer deposited on the substrate [33]. Moreover, cracks may appear during this step. In chemical vapor deposition the preference has been for atomic layer deposition (ALD – Atomic Layer Deposition) [34, 35] and chemical vapor deposition of organo-metallic precursors (MOCVD – metal organic chemical vapor deposition) [36, 37]. These techniques allow for dense thin films close to stochiometry, nice step coverage and good insulating characteristics, since growth is controlled layer by layer. Even though MOCVD has great potential the major problem is related to impurities such as carbon that require high temperature in order to be removed, above 350°C [38–40]. In ALD contamination is not an issue but the deposition rate is too low and the process temperature is still above 250°C [41, 42].

Concerning the physical techniques, sputtering, pulsed laser deposition (PLD) or thermal evaporation can be used to deposit high-κ dielectrics [43–48]. In particular sputtering is very interesting due to low contamination, easily controllable growth, the possibility of large scale deposition and low temperature processing. In crystalline silicon technology, this technique was not considered to be effective, mainly due to the high density of induced interface defects in top gate devices and the weak step cover capability. However, the possibility of fabricating TFTs at low temperatures makes them interesting for a wide range of flexible devices involving low cost substrates. Moreover, physical routes seem to be one the most promising alternatives for oxide based devices processed entirely at room temperature. Another advantage of the physical techniques is the ability to virtually deposit all kinds of materials, not depending on the existence/synthesis of liquid or gaseous precursors.

In this chapter we will focus on radio frequency magnetron sputtering (RFMS) deposition. This has been the most commonly used technique for depositing the channel layer in oxide based TFTs and therefore its usage for the dielectric layers also means that a device can be processed using the same deposition technology, which may be a desirable scenario for industrial up-scaling.

4.5 Sputtered High-κ Dielectrics in Oxide TFTs

Sputtering has been used to deposit several high-κ oxides such as HfO_2 [49–51], Ta_2O_5 [52, 53], ZrO_2 [54, 55], La_2O_3 [56], Y_2O_3 [57, 58], to mention some of those most studied. The deposition is normally done at room temperature from ceramic targets but many

Table 4.2 *Electrical characteristics of some oxide TFTs using high-κ dielectrics.*

High-κ	Semic.	Dielectric Thickness (nm)	μ ($cm^2 V^{-1} s^{-1}$)	S ($V dec^{-1}$)	V_T (V)	I_G(A)	Ref
HfO_2	GIZO	–	7.2	0.25	0.44	–	[62]
HfO_2	ZnO	250	1	–		$<10^{-9}$@4V	[63]
HfO_2	GIZO	180	8.4	0.25	–1	–	[64]
HfO_2	ZnO	200	14.7	–	2	–	[67]
Y_2O_3	GIZO	140	12	0.2	1.4	–	[59]
Ta_2O_5	GIZO	200	61.5	0.61	0.25	–	[68]
Ta_2O_5/SiO_2	GIZO	300	35	0.24			[66]
ZrO_2	GIZO	200	28	0.56	3.2	$<10^{-8}$@10V	[61]

authors report annealing treatments above 300°C in order improve the insulating characteristics and reduce the defect density of the dielectric layers [52, 55, 57]. These studies were focused on optimization of dielectric layers for crystalline silicon technology.

When considering the recent oxide semiconductor-based TFTs technology, working devices were obtained using sputtered high-κ dielectrics at low temperature such as Y_2O_3, [59, 60], ZrO_2 [61] HfO_2, [62–64] and Ta_2O_5 [65, 66]. Table 4.2 lists some of the electrical characteristics of these devices.

However, most of these high-κ dielectrics present a polycrystalline structure even when deposited at room temperature, whereas an amorphous structure is preferred so as to decrease the gate leakage and improve interface properties. Besides, as seen in section 4.3, higher-κ oxide dielectrics tend to have low E_g which can bring additional problems regarding band offsets. As a possible solution for this, a similar approach to that followed for multicomponent oxide semiconductors can be used: by mixing a high-κ/low gap oxide with a low-κ/high gap oxide, using, for instance, co-sputtering deposition, it is expected that the properties of the resulting multicomponent dielectric can be tuned by varying the relative proportions of the composing oxides. Moreover, given the induced structural disorder obtained by mixing the two oxides, an amorphous structure can be obtained in the multicomponent dielectric thin film, coupled with a moderate-to-high-κ and enlarged gap. This approach was considered in silicon technology with the main purpose of stabilizing an amorphous structure up to high temperatures [69–71]. On the following section we will show the results aimed at the optimization of these multicomponent dielectrics and envisaging oxide TFTs technology. We must emphasize that all of the films were deposited without any intentional substrate heating.

4.6 Hafnium Oxide

Hafnium oxide (HfO_2) was one of the materials that received the most attention when it was realized that SiO_2 had to be replaced by high-κ materials in advanced crystalline silicon technology. This was mainly due its dielectric constant of around 25 and the theoretical thermal stability when in contact with silicon [72]. In fact, it was later discovered that the formation of interfacial compounds (silicates or SiO_x) may occur under certain processing conditions [73–75]. Nevertheless, this material has been intensively studied since the formation of the interfacial layer, once controlled, could increase resistivity and improve the quality of the interface without significant losses in gate capacitance.

As far as E_g is concerned, despite some discrepancy in the values reported in the literature, it lies close to 6 eV [18, 76]. The band offset with silicon is greater than 1 eV, both in the conduction and valence band, which makes this material usable in both n-type and p-type transistors. This misalignment is also favorable in the case of oxide semiconductors, which makes HfO_2 a possible candidate for dielectrics also in oxide based TFTs. However, we may face a problem when using this material because HfO_2 films tend to be polycrystalline when deposited by RFMS even without any intentional substrate heating. The ease of crystallization is also common to several deposition techniques, whether physical or chemical [77–83]. This can be seen in Figure 4.3 where is shown the XRD data of HfO_2 thin films with around 200 nm sputtered at room temperature using different oxygen (O_2)/argon (Ar) flow ratios. This is considered to be the oxygen percentage (%O_2). The deposition was done from a ceramic target at pressure (P_{dep}) of 0.3 Pa and RF power density applied to the target (P_{rf}) was 7.5 W cm^{-2}. We can confirm that the films are microcrystalline where monoclinic phase predominates. This is the typical low temperature phase reported for sputtered HfO_2 and is also common to other processing techniques [77–83].

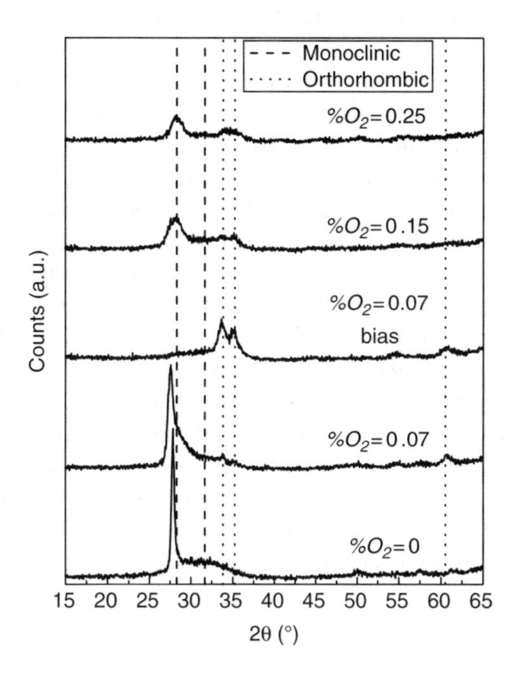

Figure 4.3 *XRD patterns of HfO_2 films deposited without substrate bias by RFMS with different %O_2. P_{dep} was 0.3 Pa and P_{rf} was 7.5 W cm^{-2}. The sample deposited with substrate bias is also presented for comparison. In this case was used a P_{bias} of 240 mW cm^{-2} applied to the substrate holder. Adapted with permission from ref [78] Copyright (2009) Wiley-VCH.*

The HfO_2 deposited by RFMS tends to be polycrystalline in a wide range of processing conditions but $\%O_2$ may change the crystalline structure [78, 84, 85]. The orthorhombic phase becomes detectable, which is meta-stable and not commonly observed in sputtered films. However, it is well known that the monoclinic to orthorhombic transformation can occur in polycrystalline bulk material under high pressure [86, 87] or after thermal annealing [88, 89]. Here, this phase transformation is attributed to the high oxygen flow that induces the substrate bombardment by negative ions reflected at the target surface that have higher energy than the sputtered species and are responsible for structural rearrangements in the deposited film [85, 90, 91]. Oxygen has a much lower atomic mass than argon. Hence, oxygen ions are much more mobile than argon and will be accelerated to higher velocities and, consequently, will also be reflected at the cathode with high velocity. The small volume and high velocity of the reflected oxygen atoms will reduce the probability of their scattering by plasma species as they cross the plasma region. As a consequence, they will deposit more energy into the substrate [85].

Estimating the grain sizes using the Scherrer analysis, we verify that it decreases as oxygen flow increases, due to the presence of both monoclinic and orthorhombic phases. Cheynet *et al.* also verified that when the orthorhombic phase is present the grain size tend to be smaller than in the monoclinic phase [92]. The small grain size may result from enhanced nucleation under ionic substrate bombardment. We can also observe that the peak centred ~28° is slightly moved to low 2θ values when adding oxygen to the gas mixture, which could indicate the presence of internal stress.

Structural modification under ionic bombardment is confirmed when using RF substrate bias where the orthorhombic phase becomes clearly predominant. This results from a more dramatic substrate bombardment since it promotes the involvement of argon ions that have high mass and are able to supply even more energy on the surface of the growing film.

By means of TEM in Figure 4.4, it may be noted that when the monoclinic phase is dominant and the grains exhibit a pyramidal shape. The upper area is polycrystalline and with near-vertical grain boundaries. As an explanation of this morphology, they seem to be formed at a nucleation site at a certain distance to the substrate and to grow up in a pyramidal shape until contact with neighbouring grains, when a columnar growth is forced. On the other hand, under substrate bias the growth of orthorhombic phase associated with columnar grains occurs. Here nucleation seems to be more intensive due to the high directional energy flow coming from ionic bombardment and the grains are forced to grow vertically and tend to be small. It is also possible to see that in both cases the HfO_2 layer near the substrate (during initial deposition stage) is amorphous. The existence of this incubation layer and grain size dependence on thickness are somehow typical of HfO_2 thin films deposition [80, 81].

Associated with a different structure, there is also a modification in surface morphology as shown in AFM microstructures (Figure 4.4). Low RMS roughness values are obtained when the orthorhombic phase dominates which is also a consequence of the different grain shape and small size.

Optical characterization may be also useful in reaching a better understanding of some of the other properties observed in HfO_2 thin films. A nondestructive optical characterization technique such as spectroscopic ellipsometry (SE) can be used to determine the band gap, detect the presence of the crystalline phase, monitor interfacial reactions or infer the presence of sub-gap states that result from defects within HfO_2 [93–95]. This can be done by a proper modelling of the experimental results, and determining parameters such as

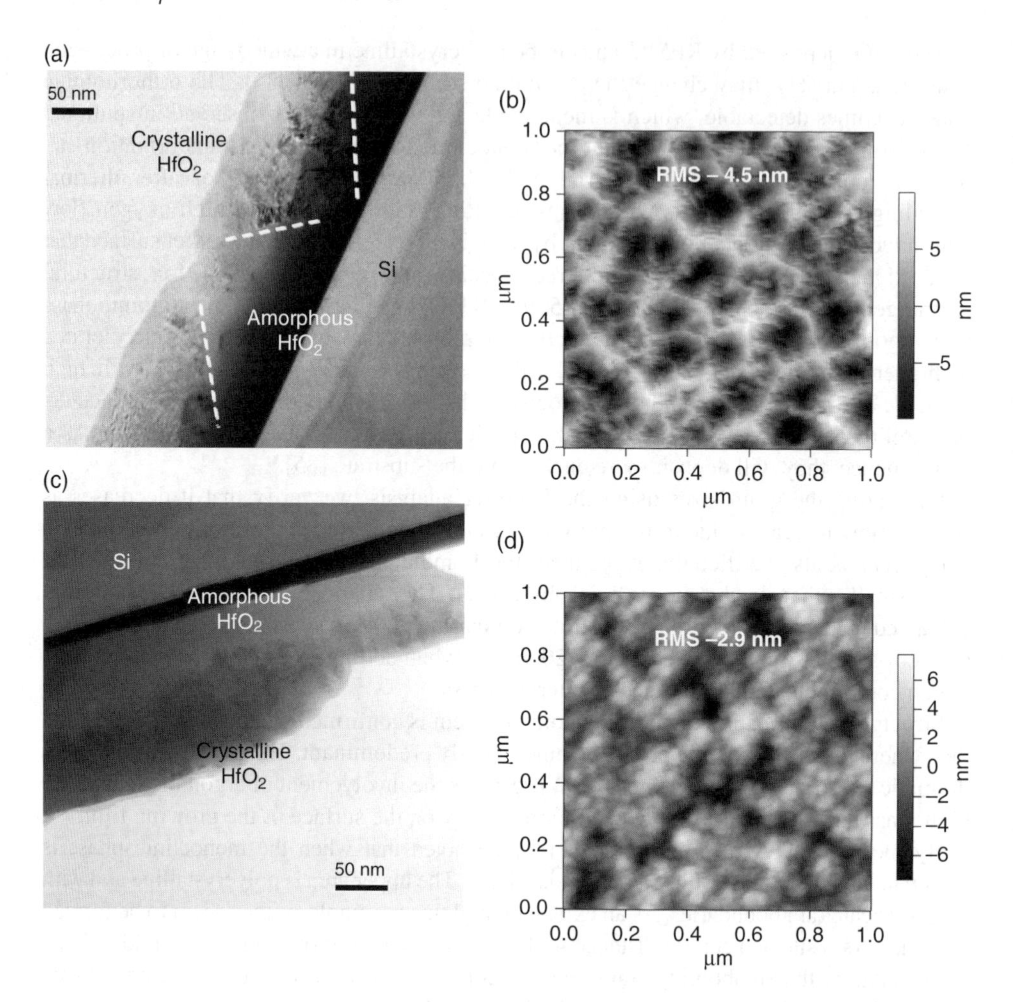

Figure 4.4 *TEM cross section and AFM surface topography ($1\times1\mu m^2$) of HfO_2 thin films deposited using a %O_2 of 0.07 at 0.3 Pa and P_{rf} of 7.5 W cm^{-2}: a) and b) Sample deposited without substrate bias; c) and d) sample deposited with P_{bias} of 240 mW cm^{-2}.[a] Adapted with permission from ref [78] Copyright (2009) Wiley-VCH.*

dielectric function, refractive index and thickness of thin films. The basis of ellipsometry is the measurement of changes in amplitude and phase between the parallel and perpendicular components of a beam of polarized light, relative to the plane of incidence, after reflection in a sample. This reflection causes a phase shift in these two components, as well as a modification of the ratio of their intensities. The experimental data are obtained by measuring the polarization after reflection, known as the Fresnel reflection coefficient, given by:

$$\rho = \frac{r_p}{r_s} = \tan\psi \cdot \exp(i\Delta) \tag{4.1}$$

[a] TEM as well as XPS analysis presented throughout this chapter was performed at Universitat de Barcelona under the framework of Multiflexioxides FP6 European project.

where r_p and r_s are the reflection coefficients for light parallel and perpendicularly polarized to the plane of incidence. The expression can be described in terms of ellipsometric angles (Δ and Ψ), being tan Ψ the ratio between the amplitude of reflection coefficients and Δ related to relative phase change.

Different dispersion formulas have been used in the modulation of dielectric materials, most of them developed for amorphous structures [93, 96]. The application to polycrystalline films, and sometimes non-stoichiometric ones, raises some difficulties. However, the dielectric function in nanocrystalline structures may, in some cases, be described by the dispersion formulas developed for amorphous materials [97, 98]. When the E_g of the film to be analyzed is within the measurement range Tauc-Lorentz dispersion has proven to be efficient in fitting the dielectric function of several materials, even if polycrystalline ones, including HfO_2 [94, 95]. It is described by [99]:

$$\varepsilon = \varepsilon_1 + \varepsilon_2$$
$$\varepsilon_2 = \begin{cases} \dfrac{1}{E}\dfrac{AE_1C(E-E_g)^2}{(E^2-E_1^2)^2+C^2E^2} \rightarrow E > E_g \\ 0 \rightarrow E < E_g \end{cases} \quad (4.2)$$

where A is the intensity of transition, E_1 the energy at maximum transition, and C the dispersion parameter, inversely proportional to the structural order. The real part of dielectric function (ε_1) is obtained by Kramers-Kronig integration of the imaginary part:

$$\varepsilon_1 = \varepsilon_\infty + \frac{2}{\pi}P\int_{Eg}^{\infty}\frac{\xi\varepsilon_2(\xi)}{\xi^2 - E^2}d\zeta \quad (4.3)$$

The refractive index and extinction coefficient can then be directly calculated from ε_1 and ε_2.

In polycrystalline materials it is common to have multiple transitions at different and well defined energies and this may be taken into account by adding more Lorentz oscillators [94, 100]. As mentioned above, it is also common in deposited dielectric layers (HfO_2 included) the existence of defects responsible for the appearance of energy states near the conduction band, which contributes to absorption below the gap energy [92, 101]. To better adapt the model to experimental results, an extra oscillator can be combined with the Tauc-Lorentz dispersion in order to account for the contribution of sub gap absorption.

The complete model (Figure 4.5) consists of four layers: the first layer is a crystalline silicon reference that simulates the substrate, then a SiO_2 reference layer is used to simulate the interfacial layer (if present), then the Tauc-Lorentz dispersion to describe the dielectric function of the bulk and finally a Bruggeman Effective Medium Approximation (BEMA), consisting in equal parts of Tauc-Lorentz dispersion and a voids reference, used to simulate the surface roughness.

After describing the fitting procedure used for HfO_2 let's now look at the refractive index (n) of sputtered films obtain by SE (Figure 4.6a). It decreases as the oxygen flow becomes high and it is possible to speculate that this is associated with the mixed phase and more disordered structure, and with small grain size. This creates more defects that result in gap

TL dispersion + voids
TL dispersion
SiO2
c-Si

Figure 4.5 Theoretical structure used for the modeling of SE results.

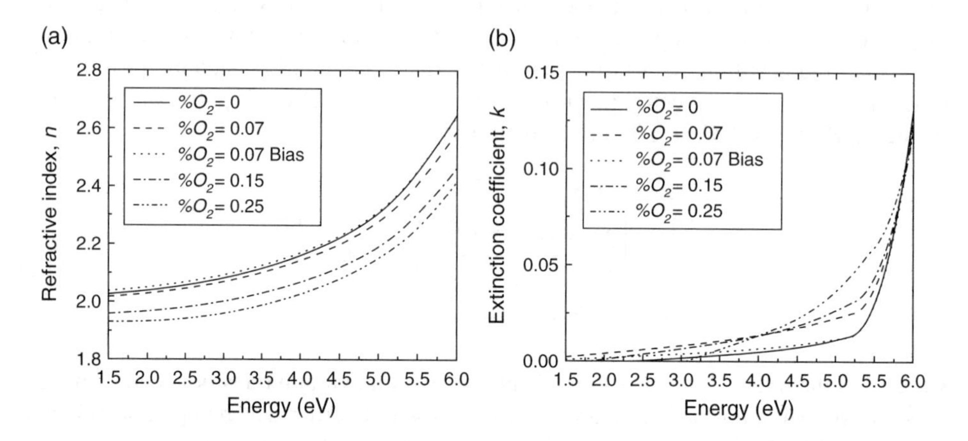

Figure 4.6 a) Refractive index and b) extinction coefficient obtained by SE for HfO_2 samples prepared with different $\%O_2$ at 0.3 Pa and P_{rf} of 7.5 W cm^{-2}. P_{bias} was 240 mW cm^{-2}.

states close to the conduction band that are responsible for the increase in the extinction coefficient (k) for energies below the absorption edge (Figure 4.6b) [101]. On the other hand, the refractive index is increased when the RF substrate bias (P_{bias}) is used. This could be due to the presence of just one phase but it is also likely to be related with oxygen and argon ionic substrate bombardment. Under argon bombardment the deposition rate (*R*) decreases by 30 % due to the re-sputtering of the substrate that eliminates weak bonds. This results in the densification of the deposited films and, consequently, in a reduction in the sub gap absorption.

The electronic structure of HfO_2 close to the band gap is rather complicated where both valence and conduction band edges are flat bands and indirect transition at 5.65 eV is followed closely in energy (5.9 eV) by a direct transition [92]. This explains the agreement in the values found in the literature for indirect and direct gap calculation methods. In these RFMS deposited films the E_g determined by the Tauc-Lorentz model lies between 5.31 and 5.34 eV. The small increase detected may be associated with small grain size as the oxygen flow increases. Cheynet *et al.* have reported an increase in E_g when the orthorhombic phase is mixed with the monoclinic, resulting in small grain size [92]. The grains are generally bigger in size and E_g is reduced when the pure monoclinic phase is present. However, Cheynet *et al.* observed a large variation in the calculated values, due to the method used to determine them. The fact that the films presented here have an amorphous phase may also affect the films' E_g determination.

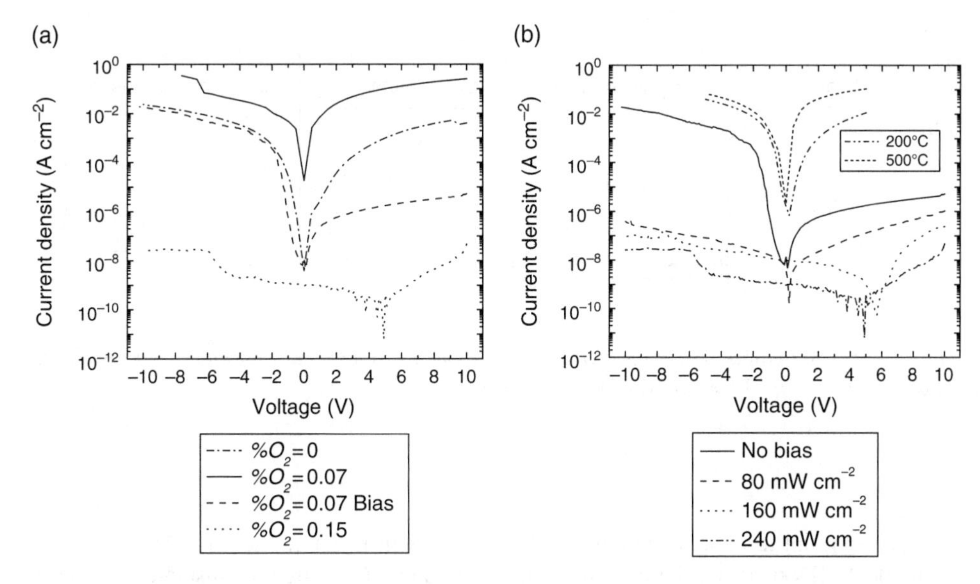

Figure 4.7 *J–V characteristics of HfO_2 samples deposited on p-type c-Si at 0.3 Pa and P_{rf} of 7.5 W cm^{-2}: a) $\%O_2$ and P_{bias} influence and b) effect of different P_{bias} and annealing temperature.*

Despite the high negative ion bombardment, the use of a minimum oxygen flow is necessary to compensate for the oxygen vacancies in the film that are inherent to RF sputtering deposition either from metallic or oxide targets. This is evidenced by the current density vs. voltage (J–V) characterization of MIS structures in crystalline silicon, using sputtered HfO_2 films (Figure 4.7). The introduction of oxygen greatly reduces the leakage current when compared with the sample deposited using just argon. However, there are no benefits in further increasing the oxygen flow above a certain valuc. As mentioned above, this is the result of high negative ion bombardment of the film and also the presence of two different phases that can create highly defective grain boundaries, due to large lattice mismatches, acting as leakage paths [16]. So, the addition of oxygen is important but needs to be adjusted to result in the optimal insulating properties.

On the other hand, when biasing the substrate (where only the orthorhombic phase exists), the leakage current is further reduced. This supports the abovementioned idea that the coexistence of two phases contributes to the increase in leakage current. However, it also confirms that the films benefit from high compactness and bond reconstruction. The difference is small but high P_{bias} leads to a slight decrease in leakage current. Another important result is related with the reduction on the gate leakage current when substrate bias is used on HfO_2 thin films, contrary to what happens when compared with thermal annealing. Previous reports have already shown that thermal annealing in RFMS as-deposited polycrystalline HfO_2 did not result in much reduction in gate leakage current density [84, 102]. This shows the effectiveness of RF bias in improving the insulating properties of films deposited at room temperature, which is extremely relevant for device production in low cost substrates.

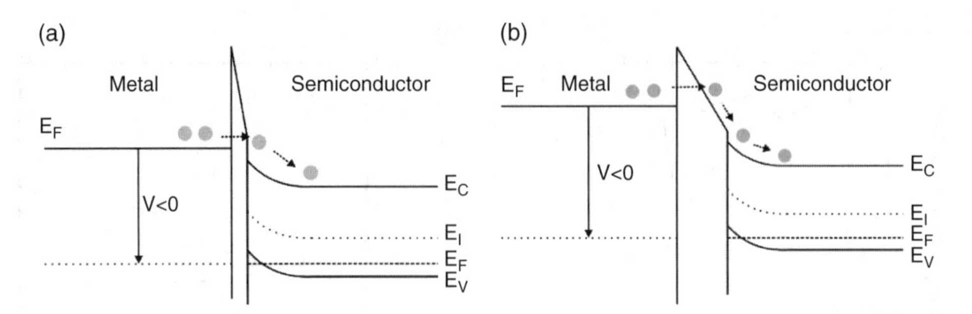

Figure 4.8 *Schematic illustration of the mechanism of electrical conduction in insulators: a) direct tunneling b) Fowler-Nordheim tunneling.*

It is important to understand how the leakage current is determined in a MIS structure. In an ideal MIS structure there is no electrical conduction through the insulator. However, ideal MIS structures do not exist and different kinds of conduction mechanisms may occur depending on the insulator's thickness and the defects within it. The direct tunneling between the metal Fermi level and the top of the semiconductor conduction band may occur in very thin insulators. For thicker layers, typically above tens of nanometers, the direct quantum tunneling current is small. However, when directly polarizing the MIS structure (negative voltage on the gate in case of p-type semiconductor), electron tunneling from the metal Fermi level to the conduction band of the insulator is possible. Once in the insulator conduction band the electrons are free to move. Accordingly, the relationship between current density (J) and electric field in the insulator (E_{el}) is described by the Fowler-Nordheim theory (see Figure 4.8):

$$J = K_1 \cdot E_{el}^2 \exp\left(-\frac{4}{3}\frac{\sqrt{2m_e}}{qh}\frac{(q\phi_B)^{3/2}}{E} \right) = K_1 \cdot E_{el}^2 \exp\left(-\frac{K_2}{E_{el}} \right) \quad (4.4)$$

where m_e is the electron effective mass in the insulator, q the electron charge and $q\phi_B$ is the barrier between the Fermi level of the metal and the top of the conduction band of the insulator. K_1 and K_2 are typical constants of the transport process. If there is a linear relationship between the ratio of the current density by the square of the electric field as a function of the inverse of electric field, this mechanism dominates the electrical conduction in a MIS structure.

The Fowler-Nordheim mechanism usually dominates the conduction in insulators at high electric fields, or when their thickness is very small, and implies that the charge carriers move freely through the insulator. This happens, for example, in the case of thermally grown silicon oxide, where the defect density is low. However, this is not the reality for most deposited dielectrics, which have a high density of structural defects, responsible for producing additional energy states close to the bands edges. These trap states dominate the current flow through a process of capture and emission of charge carriers that are injected into the insulator. This conduction mechanism is known as Poole-Frenkel,

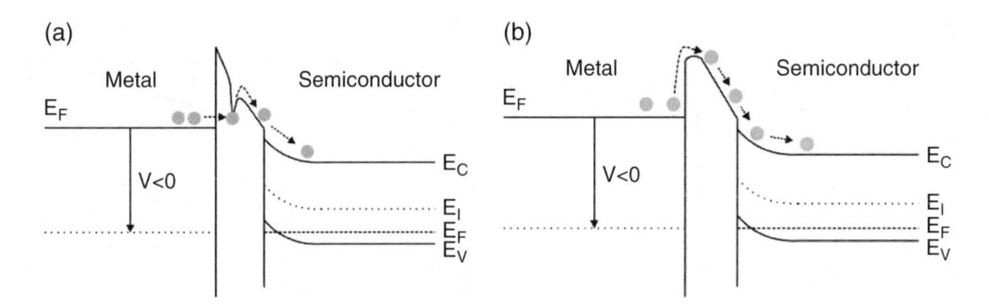

Figure 4.9 *Schematic illustration of the mechanism of electrical conduction in insulators: a) Poole-Frenkel emission and b) Schottky emission.*

where the insulator is traversed by a simple drift current density, relating to the electric field as follows:

$$J = \sigma_{PF} \cdot E_{el} \cdot \exp\left[-\frac{q}{k_B T}\left(\phi_B - \sqrt{\frac{qE_{el}}{\pi\varepsilon_0\varepsilon_{hf}}}\right)\right] = J_0 \exp\left(\frac{\beta_{PF}\sqrt{E_{el}} - q\phi_B}{zk_B T}\right) \tag{4.5}$$

where J_0 is the current density for low electric fields, k_B the Boltzmann constant, T the temperature, ε_0 the vacuum permittivity and ε_{hf} is the high-frequency dielectric constant, assumed here normally to be the square of the refractive index [103]. The compensation factor (z), which varies between 1 and 2, increasing with the density of trap states, is added to the expression in order to translate the effect of these traps. The barrier lowering coefficient or Poole-Frenkel, β_{PF} is given by:

$$\beta_{PF} = \sqrt{\frac{q^3}{\pi\varepsilon_0\varepsilon_{hf}}} \tag{4.6}$$

When there is a linear relationship between the current density/electric field ratio as a function of the square root of the electric field, this mechanism dominates the conduction through the insulator. However, it is necessary to be careful in this analysis because the Schottky-Richardson mechanism (or modified Poole-Frenkel emission), mostly controlled by the interface, has the same dependency. Although slightly different expressions, the barrier lowering coefficient (β_{SC}) is determined in the same way and, theoretically, is one half of the Poole-Frenkel coefficient (see Figure 4.9).

Given this background, the current density variation under direct polarization is shown in (Fig. 4.10a). It was not possible to determine the breakdown field for a sample deposited without bias as the leakage current density reached the meter compliance. On films produced under substrate bias this value is above 2 MV cm^{-1} and 4 MV cm^{-1} for low and high P_{bias}, respectively. Before breakdown, as the electric field increases, the current passing through the dielectric is mainly dominated by the Poole-Frenkel mechanism. Figure 4.10b) represents the ratio of current density by the electric field as a function of the square root of the latter. The experimental data show the existence of two regions in which the current density varies almost linearly with the electrical field, for samples deposited without bias. However, we consider that only the region for high electrical field is

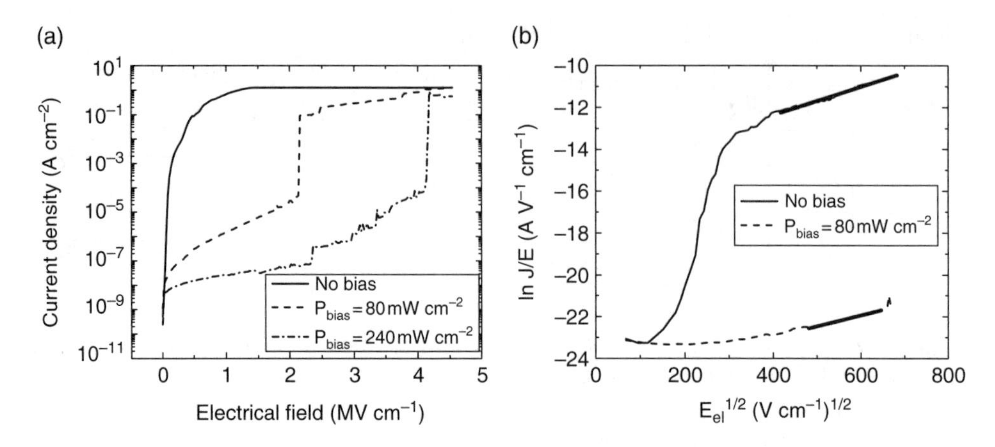

Figure 4.10 *Leakage current on HfO_2 thin films deposited under different P_{bias} at 0.3 Pa, P_{rf} of 7.5 W cm^{-2} and a %O_2 of 0.07: a) Current density function of electrical field and b) Poole-Frenkel mechanism plot and fitting on high field regime.*

Table 4.3 *Parameters obtained by adjusting the expression for the Poole-Frenkel mechanism on HfO_2 deposited under different P_{bias} at 0.3 Pa, P_{rf} of 7.5 W cm^{-2} and a %O_2 of 0.07.*

	β_{PF} (z=1) (eV cm$^{1/2}$ V$^{-1/2}$)	β_{PF} (z=2) (eV cm$^{1/2}$ V$^{-1/2}$)	β_{PF} (theoretical) (eV cm$^{1/2}$ V$^{-1/2}$)	β_{SC} (theoretical) (eV cm$^{1/2}$ V$^{-1/2}$)
No bias	1.73×10^{-4}	3.46×10^{-4}	3.79×10^{-4}	1.89×10^{-4}
8 mW cm^{-2}	1.34×10^{-4}	2.69×10^{-4}	3.78×10^{-4}	1.89×10^{-4}

attributable to the Poole-Frenkel conduction. For a sample deposited without bias a better approximation to β experimental values is achieved when the compensating factor is high, which still indicates a high density of trap states [104]. The value of ε_{hf} used in the calculation of the barrier lowering factors is estimated by the square of the refractive index (n) obtained by ellipsometry (approximately 2). For a sample deposited under substrate bias the β value is smaller than without bias (indication of trap density reduction) and closest to theoretical β_{SC}. Some authors had associated this proximity to the Schottky mechanism controlled current [105, 106]. However, Simmons *et al.* claimed this is more likely related to "anomalous" Poole-Frenkel instead as the current in relatively thick thin films is essentially bulk controlled [107] (see Table 4.3).

As we are dealing with dielectric materials it is relevant to calculate the dielectric constant for these films, using standard procedures. C–V measurements performed in the same MIS structures as those referred to above, were done at 1 MHz and lead to values that vary between 21.8 and 24.3, being the highest value obtained for the films prepared under a substrate bias, as expected for denser films.

In summary, RFMS deposited HfO_2 thin films tend to be polycrystalline where monoclinic to orthorhombic transition may occur as the oxygen flow increases. The presence of two phases has as a consequence small grain size and less dense films. Substrate bias promotes the existence of only the orthorhombic phase with small grain size but dense structure. The precise control of oxygen flow is needed in order to obtain good insulating characteristics in HfO_2 thin films. However, the data achieved show that the substrate bias results in a strong decrease in the gate leakage, proving to be a determinant deposition parameter so as to obtain electronically high grade quality gate dielectric layers deposited by RFMS. In fact, substrate bias seems to be more effective than thermal annealing in improving the films' properties in as-deposited polycrystalline films. This is an important point with regard to overall optimization of insulating layers in low cost substrates that cannot withstand high processing temperatures. Finally, we observe that the Poole-Frenkel (or trap assisted) mechanism dominates electron conduction through the insulator and the results obtained confirm the defect reduction under a substrate bias.

4.6.1 Multicomponent Co-sputtered HfO_2 Based Dielectrics

So far, we have seen that the processing conditions of HfO_2 layers must be carefully tuned in order to obtain the best insulating properties and control the film structure. However, polycrystalline films are easily obtained while amorphous structures are preferred mainly due to the absence of grain boundaries, properties' homogeneity over large areas and highly smooth surfaces. Based on this, pure HfO_2 deposited by RFMS has little interest for application in oxide TFTs, especially in bottom gate structures where high roughness may lead to bad interface properties. However, when extending the concept of multicomponent oxides from oxide semiconductors to dielectrics it is possible to get stable amorphous structures. The question is now which oxides can be mixed with HfO_2. In our approach we decided to deal with two problems at the same time. As we have seen above, band offset of HfO_2 with ZnO or SnO_2, for instance, is at the limit of 1 eV. So, the material to be chosen must allow for a slight increase in E_g and, at the same time, result in an amorphous structure, when mixed with HfO_2. Both SiO_2 and Al_2O_3 seem to be good options. First of all, when mixing HfO_2 with these oxides it is possible get stable amorphous films up to high temperatures [71, 108–110]. Second, both have high E_g, close to 9 eV, and are able to form a solid solution or even compounds that result in a gap enlargement.

When HfO_2 is co-sputtered with SiO_2 (HSiO) the films become amorphous as evidenced by the XRD patterns shown in (Figure 4.11a) meaning that crystallization of HfO_2 films during deposition is blocked. The introduction of SiO_2 decreases the crystallization enthalpy per mol, making the crystallization harder to achieve [110]. This results in a RMS surface roughness lower than that obtained for the polycrystalline films. The XRD patterns suggest that the resulting structure of HfO_2+Al_2O_3 (HAlO) co-sputtered films is also amorphous. In this case, the RMS surface roughness is even lower than for HSiO (Figures 4.11b and c). This means that the effect of the oxygen ion bombardment on the morphology of co-sputtered films is different in SiO_2 and Al_2O_3. Seemingly, the surface damage in the HSiO films under these conditions is higher than in HAlO films.

The amorphous structure of multicomponent HfO_2 based dielectrics can be obtained within a wide range of SiO_2 or Al_2O_3 incorporation. The difficulty in achieving a wider composition range is mainly related to the RFMS deposition rate (R) that is quite low for these oxides. Besides, we have practical limitations in terms of the maximum P_{rf} that can be applied and also in the minimal P_{rf} on HfO_2 target, that may reduce even further the total value of R. The means of modifying the composition in co-sputtered films is by changing

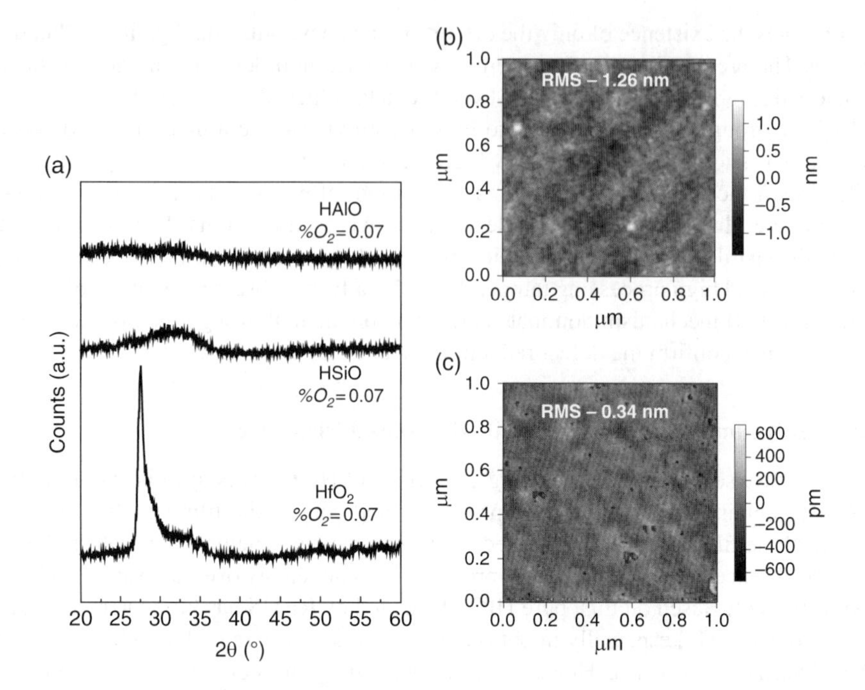

Figure 4.11 *Structure and morphology of co-sputtered HSiO and HAlO films deposited at 0.3 Pa, P_{rf} of 7.5 W cm^{-2} and a %O_2 of 0.07: a) XRD patterns, b) HSiO AFM microstructure (1×1 μm^2) and c) HAlO AFM microstructure (1×1 μm^2). Adapted with permission from ref [78] Copyright (2009) Wiley-VCH.*

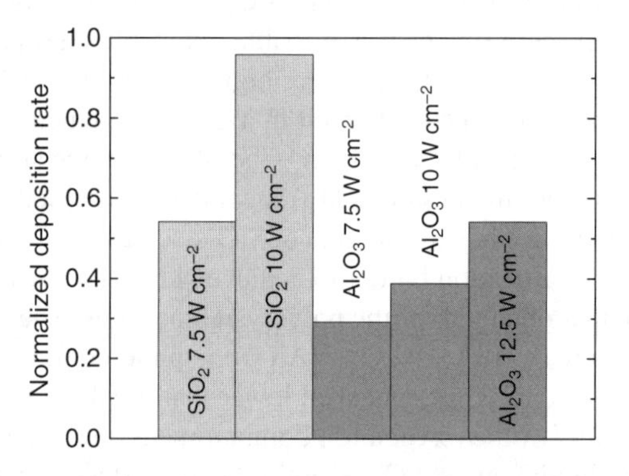

Figure 4.12 *Normalized deposition rate of SiO_2 and Al_2O_3 deposited at 0.3 Pa and with a %O_2 of 0.07 for different P_{rf}. The deposition rate for HfO_2 under these conditions is 2.4 nm min^{-1}.*

the power ratio of the targets. When keeping the P_{rf} on HfO_2 target constant, SiO_2 and Al_2O_3 incorporation can be controlled by changing the P_{rf} of these targets. Figure 4.12 shows the R for SiO_2 and Al_2O_3 normalized to the HfO_2. It is confirmed that the incorporation of SiO_2 is easier as R is higher than for Al_2O_3 (roughly two times higher for the same P_{rf}). One easy

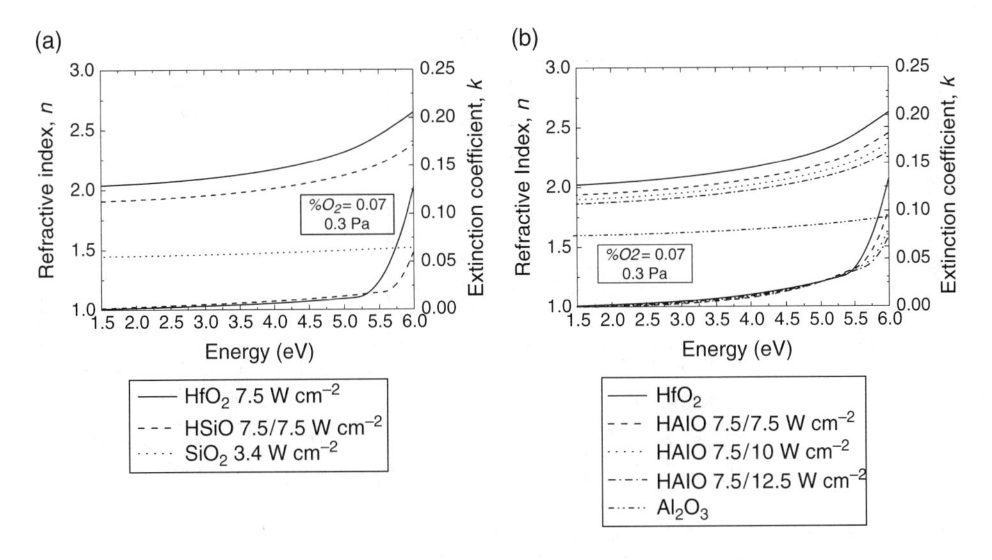

Figure 4.13 *Refractive index (n) and extinction coefficient (k) of films obtained by co-sputtering at 0.3 Pa and with a %O_2 flow ratio of 0.07: a) HSiO and b) HAlO. The P_{rf} ratio on the legend stands for the P_{rf} ratio on HfO_2 and SiO_2 (or Al_2O_3) target* [111]. *Adapted with permission from ref [111] Copyright (2001) American Institute of Physics.*

way to infer the mol % incorporation under co-sputtering is by assuming that the volume ratio within the films is equal to the *R* ratio by using the following approximation:

$$\frac{mol\,X}{mol\,Y} = \frac{\dfrac{V_X \rho_X}{M_X}}{\dfrac{V_Y \rho_Y}{M_Y}} = \frac{R_x \rho_X M_Y}{R_Y \rho_Y M_X} \tag{4.7}$$

where *V* is the volume deposited of each oxide, *M* the molar mass and ρ the density. In this case we were supposed to get compositions, within reasonable *R* values, between 33 to 50 mol % and between 20 to 30 mol % for SiO_2 and Al_2O_3 content, respectively. In this range the films are always amorphous. There are reports that even with a small incorporation amorphous structures in HfO_2 mixtures are obtained [109]. In terms of metal cations' atomic percentage the tested compositions overlap within a wide range, between 33 to 50 % for Si and 26 to 40 % for Al. For instance, in Al_2O_3 and for a normalized *R* around 0.4 the resulting film should has a theoretical mol % of 25, that is, an Al atomic % of 33. It is verified by XPS that Al atomic % is indeed around 36 in films prepared under this conditions. This means the adjustment on *R* (adjusting P_{rf} in each target) allows for the relatively precise control of the films' composition.

The spectroscopic ellipsometry data depicted in Figure 4.13a show that the refractive index of HSiO films is located between those for pure HfO_2 and SiO_2, confirming the incorporation of the latter. That is, the refractive index will move towards the SiO_2 value as its

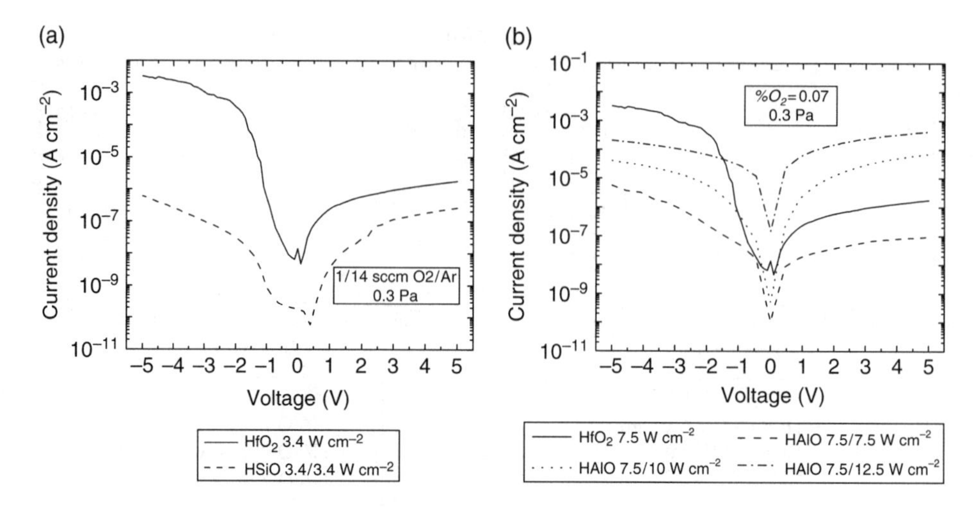

Figure 4.14 *J–V characteristics of MIS structures of co-sputtered thin films deposited on p-type c-Si at 0.3 Pa and with a %O$_2$ of 0.07: a) HSiO b) HAlO.*

content increases. Also, the band gap is higher than for HfO_2, changing from 5.31 to 5.56 eV. But why does it increase when SiO_2 is mixed to HfO_2? These two oxides are known to form silicate that has a gap higher than pure HfO_2 [112]. However, we cannot ensure that during co-sputtering they will completely react and form hafnium silicate. Beside, 100% silicate film would only exist in equilibrium conditions for a fixed HfO_2/SiO_2 ratio. In our study when 7.5 W cm^{-2} is applied to each target the *R* for HfO_2 is around twice that of the SiO_2. This means that there is an excess of HfO_2 and no 100% silicate films are obtained. So, if we had a good phase separation, our film would be composed of a mixture of silicate and HfO_2, being the band gap value determined by HfO_2 that has a lower value than HSiO. However, this does not happen as the band gap determined by an ellipsometry increase of around 0.25eV when SiO_2 is co-sputtered with HfO_2. The explanation is that HfO_2 and SiO_2 are forming an amorphous solid-solution, that can exist to a greater extent outside equilibrium conditions, and no phase separation occurs [113].

The same trends are observed for HfO_2 mixed with Al_2O_3 (Figure 4.13b). In this case the power applied to the Al_2O_3 target was changed and it is possible to see that the refractive index also moves towards the Al_2O_3 reference. The band gap also increases from 5.31 eV (for pure HfO_2) to 5.48 eV, 5.59 eV and 5.61 eV, when the P_{rf} on Al_2O_3 target is 7.5, 10 and 12.5 W cm^{-2}, respectively. Here the explanation is the same as for SiO_2. This means that a non-equilibrium amorphous solid solution exists.

Looking now to the electrical characteristics of the co-sputtered films we see the incorporation of SiO_2 or Al_2O_3 reduced the leakage current as a result of band gap enlargement and amorphous structure. However, an interesting feature is that the increments in Al_2O_3 content do not correspond to a decrease in the leakage current. As was mentioned above, the high P_{rf} needed to incorporate more Al_2O_3 degrades the deposited film due to high negative oxygen ionic bombardment. The best insulating properties are achieved for the lowest P_{rf} used in an Al_2O_3 target (see Figure 4.14).

Again, the Poole-Frenkel mechanism dominates the conduction mechanism as it fits well the leakage current variation with the electrical field. The best fitting for SiO_2 and

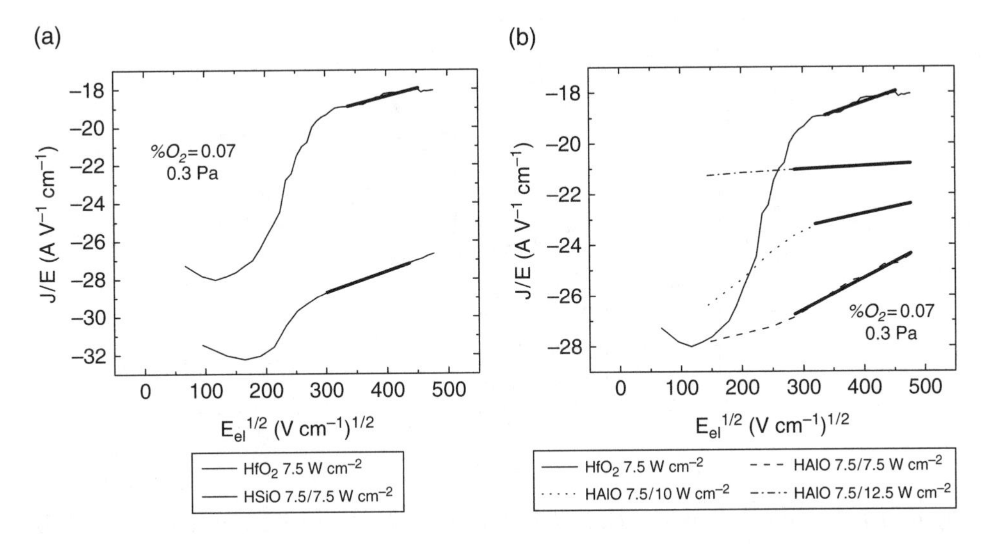

Figure 4.15 Poole-Frenkel mechanism plot and fitting in the high field regime for co-sputtered films thin films deposited at 0.3 Pa and a %O_2 of 0.07: a) HSiO and b) HAlO.

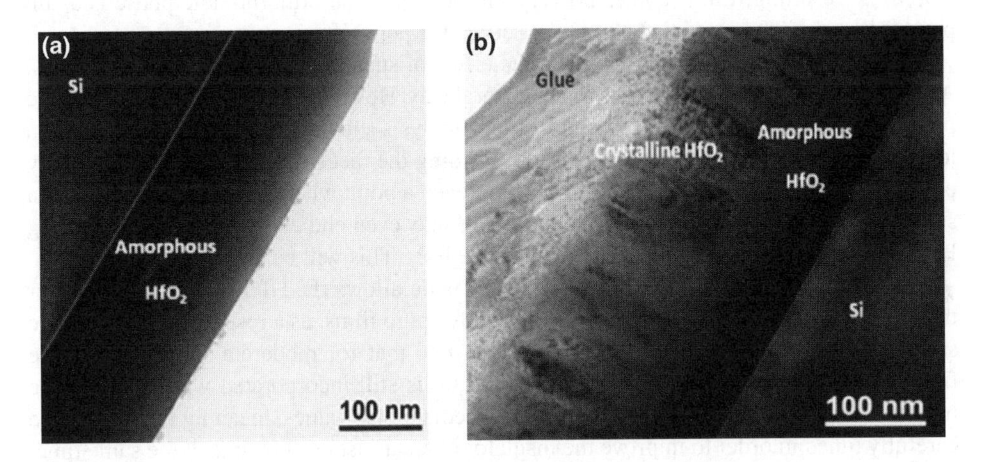

Figure 4.16 TEM cross-section images of HSiO thin films deposited a) without and b) with substrate bias.

Al_2O_3 low P_{rf} lies in the interval of theoretical β with z values between 1 and 2. However, they are closer to z=2, meaning the existence of a relative high trap density within the films. At higher P_{rf} on Al_2O_3 target the β values determined are below both β_{PF} and β_{SC}. This is ascribed again to the "anomalous" Poole-Frenkel regime (see Figure 4.15).

We have already seen for pure HfO_2 that the substrate bias may be benefical in improving the insulating properties. It is time now to see the bias effect on co-sputtered films. A typical TEM image of a HSiO thin film deposited by co-sputtering is shown in Figure 4.16. The cross-section image confirms that the films are completely amorphous, with no phase

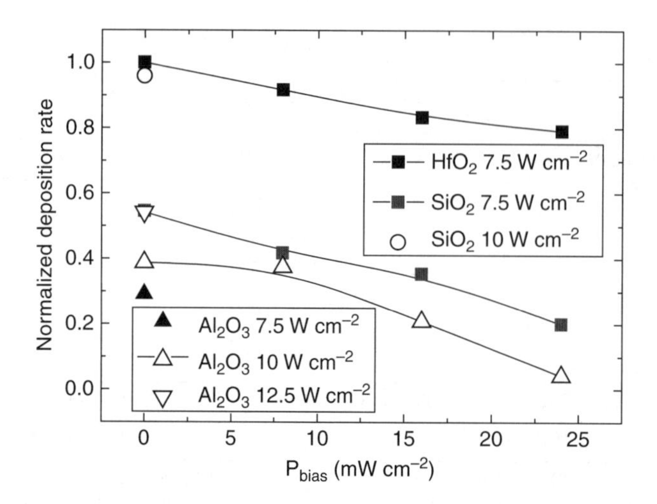

Figure 4.17 *Deposition rate of SiO_2 and Al_2O_3 deposited at 0.3 Pa and with a %O_2 of 0.07 for different P_{rf} and P_{bias}. The values were normalized to the deposition rate of HfO_2 (2.4 nm min^{-1}).*

separation being detectable. However, it is interesting to note that when the substrate is biased the resulting film becomes polycrystalline where the orthorhombic phase is again present in a similar way to that already observed for pure HfO_2.

At this point it is important to explain the effect of substrate bias on co-sputtered films. As shown in Figure 4.17, it reduces R in a general way. However, the decrease is much more pronounced for SiO_2 and Al_2O_3. Compositional analysis shows that the biased films have a low Si content. The same happens for Al_2O_3. Adding the fact that these two oxides already present lower R when compared with HfO_2 we reach a point where the films grown under a substrate bias become much more rich in Hf. We may even end close to only incorporating Hf, as in the case of co-sputtered Al_2O_3 under high P_{rf}. This will be confirmed below.

The reduced concentration of the additional oxide allows the HfO_2 to crystallize under the orthorhombic structure as already observed for pure films, as a result of the extra ionic substrate bombardment (Figure 4.18). But we note that for moderate substrate bias the films are still amorphous as some SiO_2 and Al_2O_3 is still incorporated within HfO_2. The transition region seems to happen within intermediate bias values, meaning that it must be carefully tuned in order to improve the insulator characteristics allowing, at the same time, for the incorporation of addition oxide in order to stabilize an amorphous structure.

The structural modifications are reflected in the morphology of the HSiO films that is very similar to that for pure HfO_2. For films that contain Al_2O_3 the change in the surface is more dramatic and the roughness becomes high as the substrate bias level increases. Indeed, in Figure 4.19b, c and d we can see the surface modification under substrate bias on HSiO (on the left) and HAlO films (on the right). The films grown under P_{bias} of 80 and 160 mW cm^{-2} still exhibit a smooth surface with low RMS roughness for both oxides (associated to amorphous structure). However, some differences are observed as HSiO presents a smother surface than HAlO films. When the P_{bias} is further increased the films became rough in association with high surface damage and crystallization. The RMS value is around two times higher in HAlO than in HSiO thin films.

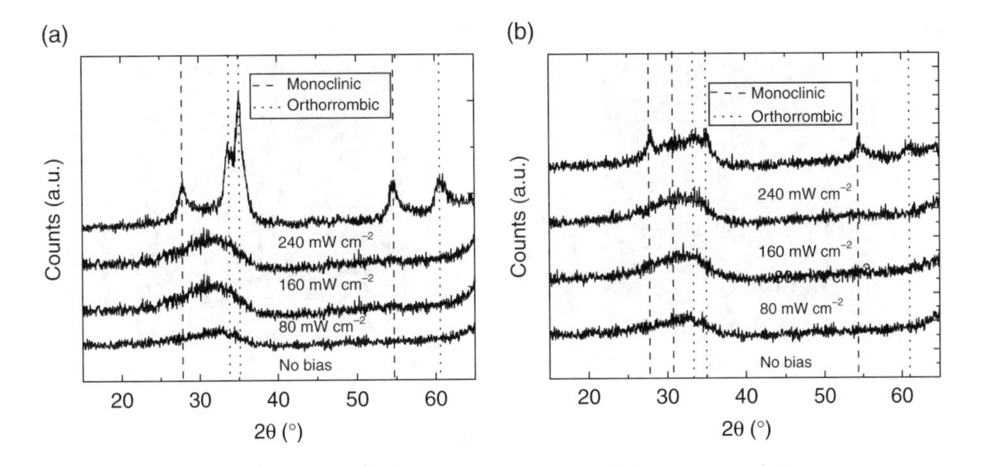

Figure 4.18 *XRD data for co-sputtered films deposited at 0.3 Pa with %O_2 of 0.07 for different P_{bias}: a) HSiO thin films deposited with a P_{rf} ratio of 7.5/7.5 W cm^{-2}; b) HAlO thin films deposited with a P_{rf} ratio of 7.5/10 W cm^{-2}.*

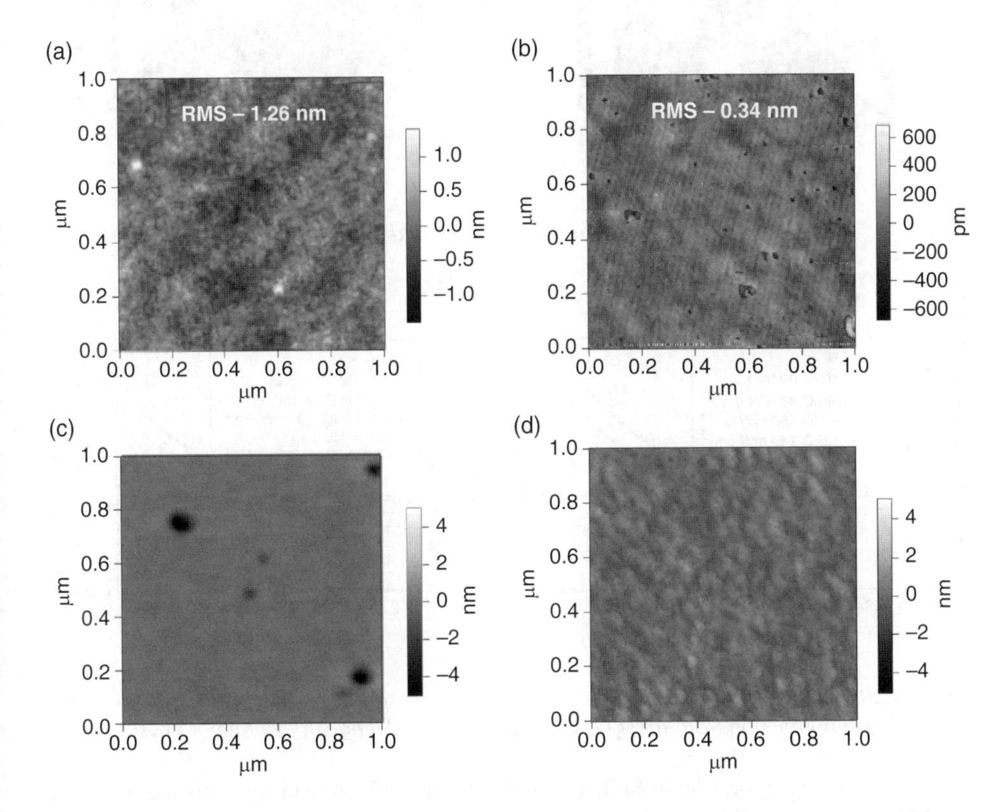

Figure 4.19 *AFM image of the HSiO and HAlO deposited using different P_{bias}: a) and b) no bias; b) and c) 80 mW cm^{-2}; d) and e) 160 mW cm^{-2}; g) and h) 240 mW cm^{-2}.*

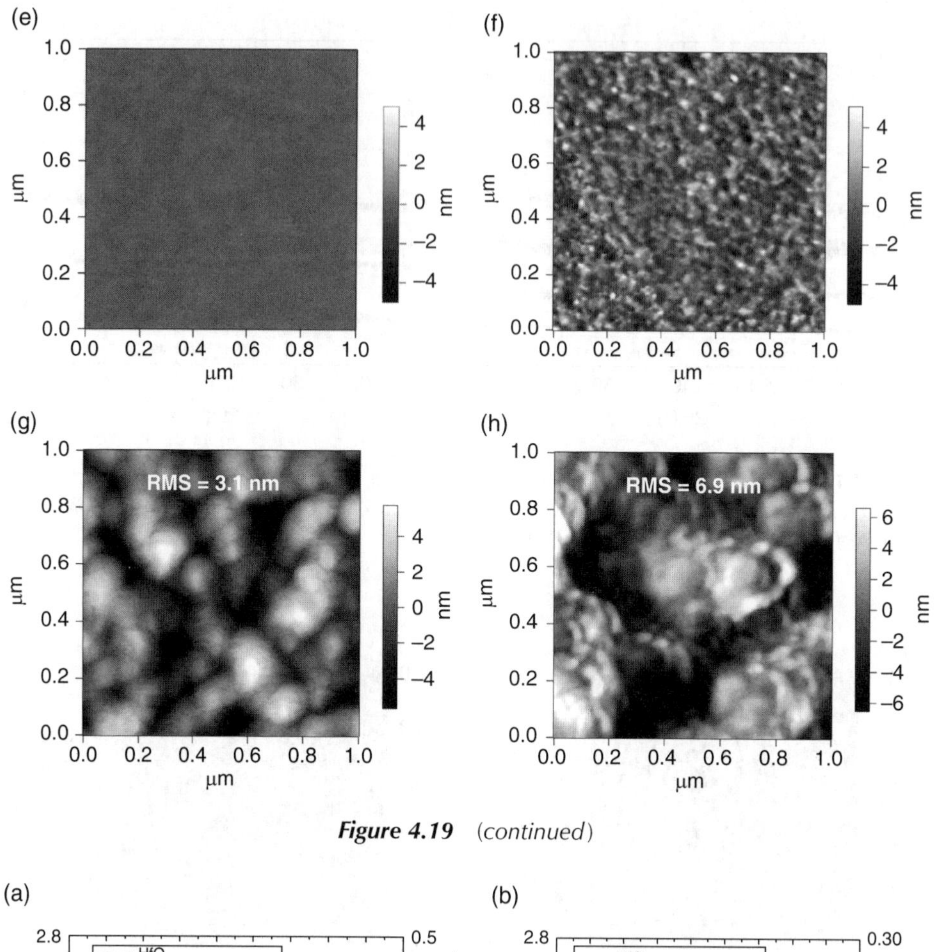

Figure 4.19 *(continued)*

Figure 4.20 *Optical properties of films obtained by co-sputtering: a) HSiO and b) HAlO. Adapted with permission from ref [111] Copyright (2001) American Institute of Physics.*

The reduced incorporation of SiO_2 and Al_2O_3 under a substrate bias is also suggested by the analysis of the n values depicted in Figure 4.20a). We have already seen that for HSiO co-sputtered films n is located between those for pure HfO_2 and SiO_2, confirming the incorporation of the latter. That is, n is moving closer to that of SiO_2 as its content increases. It

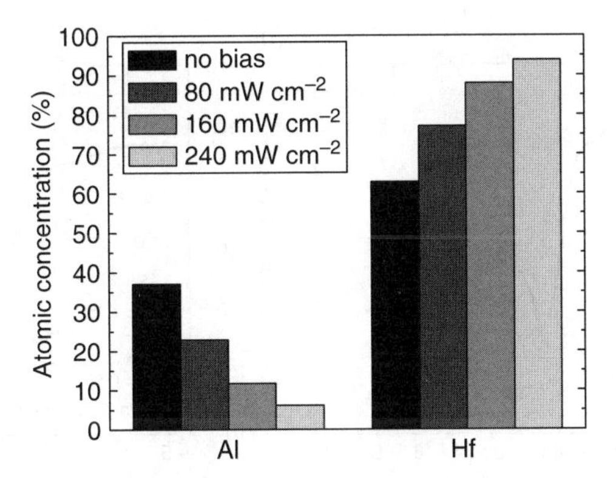

Figure 4.21 Cation atomic concentration normalized to oxygen determined by XPS for co-sputtered HAlO thin films prepared with different P_{bias}.

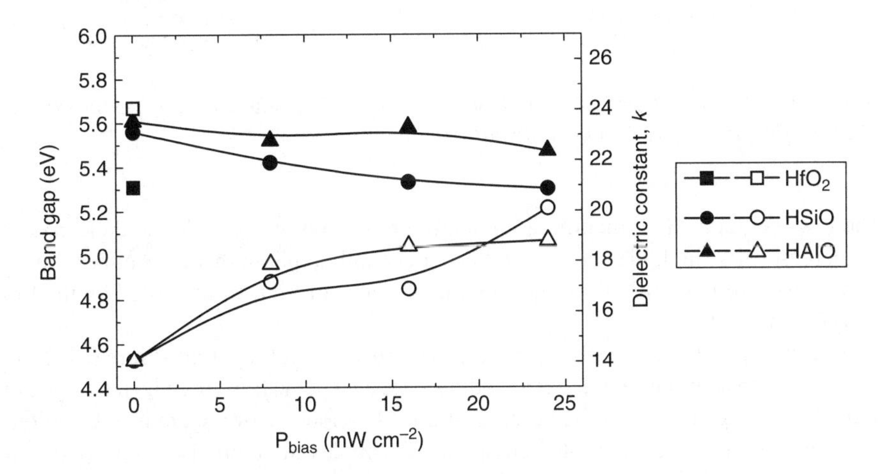

Figure 4.22 Band gap (determined by ellipsometry) and dielectric constant (determined by C–V on c-Si MIS structures at 1 MHz) of co-sputtered thin films deposited under P_{bais}. Solid symbols stand for E_g and open ones stand for κ.

is also important to note the increase on the band gap from 5.31 to 5.56 eV. However, n approaches the pure HfO_2 values as the P_{bias} is increased. We have already verified experimentally the ability of substrate bias to reduced SiO_2 incorporation (Figure 4.17). Besides, the dielectric constant value is reduced when SiO_2 is added but increases when bias is used. Despite this being also an indicator of densification, it is not believed that the values for very dense HSiO films would get close to pure HfO_2. Instead, we believe that this is related to the low SiO_2 incorporation when substrate bias is used.

The same happens for Al_2O_3. In this case, the initial R is even lower than in SiO_2 and the incorporation of HfO_2 is much more reduced under bias. It is noted that the n under a P_{bias} of 240 mW cm^{-2} is increased and exceeds the value of pure HfO_2. This is associated with the low incorporation of Al_2O_3 resulting in almost pure HfO_2 with a denser structure. The reduced Al_2O_3 incorporation is confirmed by XPS analysis where we see that under high P_{rf} the Al content is around 5 % compared with the initial 36 % without bias (Figure 4.21).

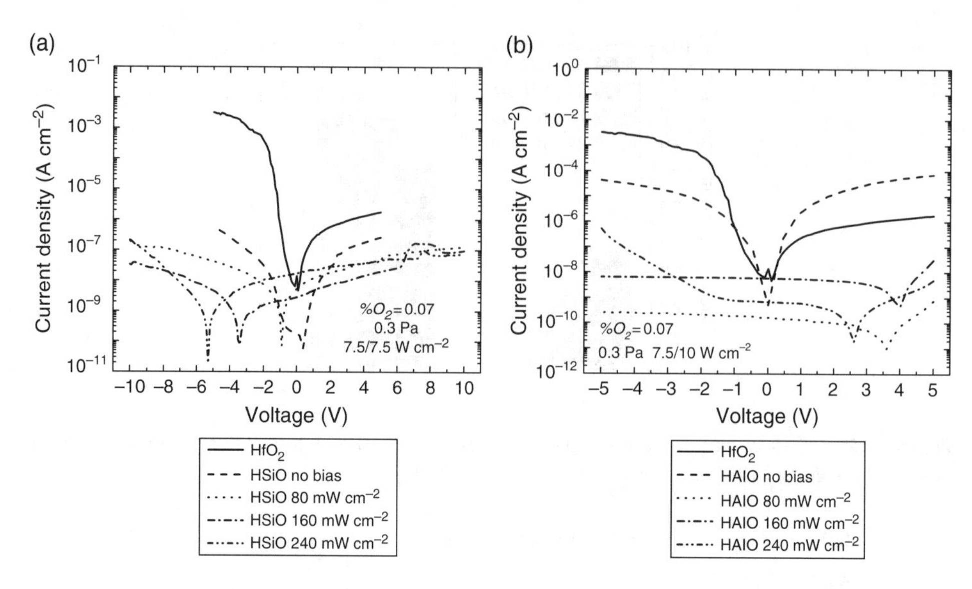

Figure 4.23 *J–V characteristics of: a) HSiO and b) HAlO. Adapted with permission from ref [111] Copyright (2001) American Institute of Physics.*

The low SiO_2 and Al_2O_3 incorporation leads a reduction on E_g while the dielectric constant increases. For high bias power we have a crystalline phase that determines the E_g to be similar to pure HfO_2 thin films, despite the presence of SiO_2 as suggested by the low κ values (Figure 4.22).

Concerning electrical properties, substrate bias must be carefully used as it may lead to an increase on the leakage current. It has already been seen that the films became polycrystalline and that the leakage may increase because of this. As shown above, there is a reduction on the leakage current for as-deposited co-sputtered samples in comparison with pure HfO_2. Under a substrate bias the leakage is further reduced. The remarkable improvement probably results from the densification of co-sputtered amorphous films under bias. However, as the high P_{bias} is achieved the leakage current starts increasing as a result of crystallization.

In summary, co-sputtering of SiO_2 and Al_2O_3 with HfO_2 has effectively suppressed the polycrystalline structure of HfO_2 films. As observed for the mixed films deposited under low P_{bias}, the additional energy rules out weak bonding and leads to densification of films, improving the J–V characteristics (Figure 4.23). When deposited under a P_{bias} of 240 mW cm^{-2}, the films present a rough surface due to crystallization and surface damage, leading to an increase in leakage current. Substrate bias, when carefully tuned, is an important parameter designed to improve the electrical properties of mixed films deposited at low temperature.

4.6.2 Multicomponent Dielectrics from Single Target

Co-sputtering may be a good alternative for a first approach study where the composition may be modified by adjusting the P_{rf} in each target. However, it was seen that this composition control sometimes becomes difficult due to different *R* that is even more critical under

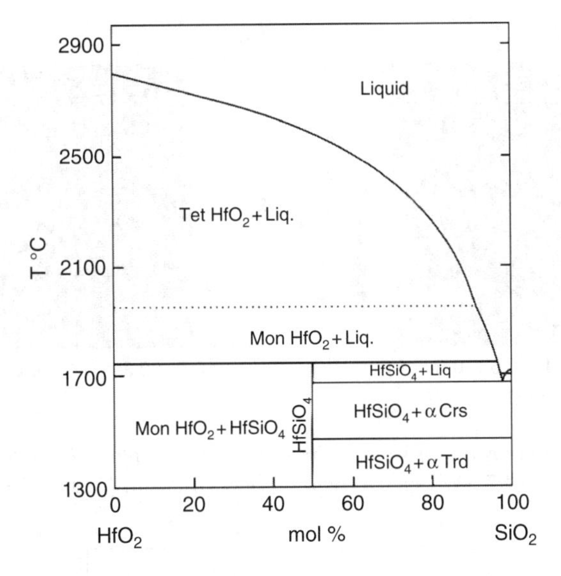

Figure 4.24 *Equilibrium phase diagram of HfO_2 and SiO_2.*

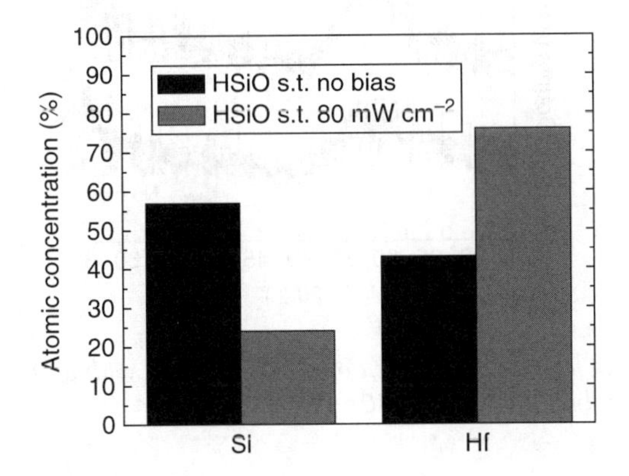

Figure 4.25 *XPS analysis of HSiO s.t. films deposited at 0.3 Pa with a P_{rf} of 7.5 W cm^{-2} and a %O_2 of 0.07. This shows the comparison between no bias and P_{bias} of 80 mW cm^{-2}.*

substrate bias. So, if a specific composition is desired for films deposited by RFMS another way is to use only one target of the aimed multicomponent dielectric. This option will be explored in the following pages and we will focus on the HfO_2 and SiO_2 example.

Figure 4.24 presents the equilibrium phase diagram of the HfO_2-SiO_2 system. One interesting composition is a molar ratio around 50/50 mol%, where we are supposed to obtain 100% $HfSiO_4$. Indeed, we verified that a target sintered at 1500°C typically results in a mixture of 98% of silicate and only 2% of HfO_2.[b]

[b] The target used for this study was developed at the Jozef Stefan Institute in Slovenia, under the framework of the Multiflexioxides European project.

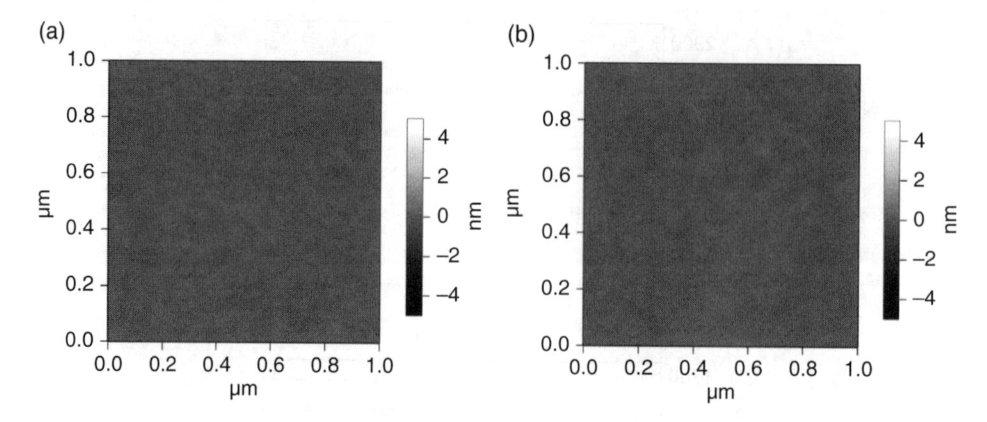

Figure 4.26 *AFM microstructure ($1 \times 1\,\mu m^2$) for HSiO s.t. films deposited at 0.3 Pa with a O_2% of 0.07 and a P_{rf} of 7.5 W cm^{-2}: a) with no bias, b) 240 mW cm^{-2} bias.*

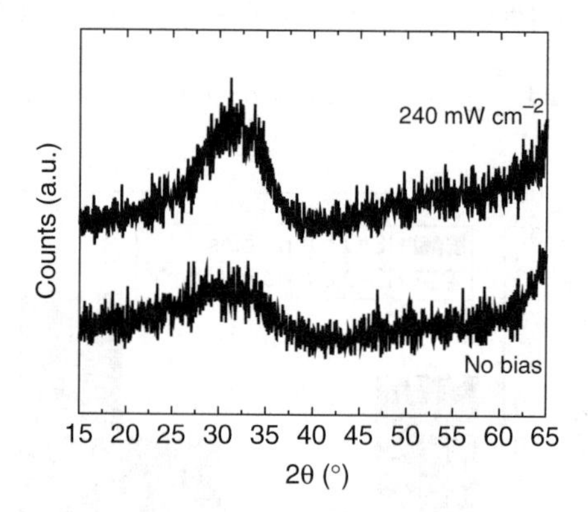

Figure 4.27 *XRD data for HSiO s.t. films deposited at 0.3 Pa with a O_2 % of 0.07 and a P_{rf} of 7.5 W cm^{-2}: a) with no bias and under 240 mW cm^{-2} bias.*

Films sputtered from this almost 100% silicate target (HSiO s.t.) are also amorphous (Figure 4.25). This is not a surprise because more SiO_2 is incorporated and it is known that $HfSiO_4$ thin films tend be amorphous [109, 110]. Still, the question remains: we cannot ensure that the deposited film has the same composition as the target. XPS results confirm that this does not happen as the atomic concentration of Si is slightly higher than 50%, but is close to the nominal value. However, under moderate bias (80 mW cm^{-2}) this relation is greatly modified where Si concentration within the films may be reduced by more than 50%. This suggests preferential etching of SiO_2 for the deposition under bias and may also explain what was already observed for co-sputtered films.

It is expected that the increase in P_{bias} leads to a further reduction in the Si content but it is not enough to reduce its value in order to allow for HfO_2 crystallization (Figure 4.26). The XRD data show that the films remain essentially amorphous, despite some short-range order seem to be present (Figure 4.27). We cannot discard the hypothesis of amorphous

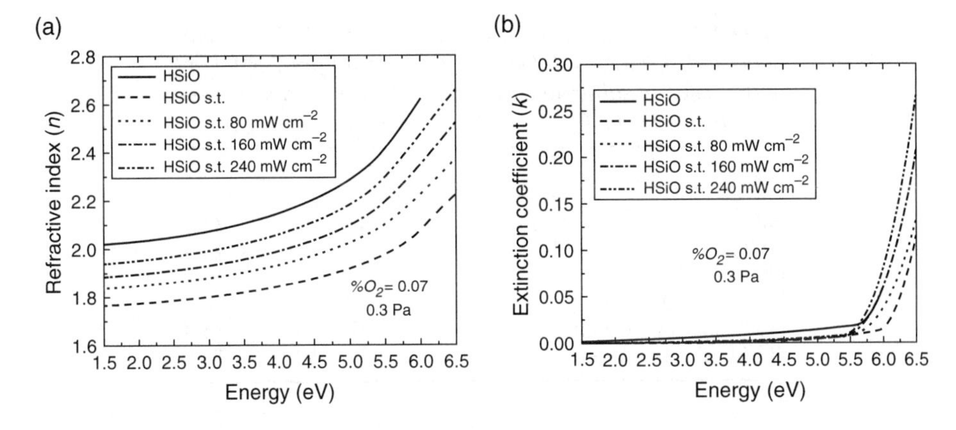

Figure 4.28 *Optical parameters for HSiO s.t. films deposited at 0.3 Pa with an O_2% of 0.07 and a P_{rf} of 7.5 W cm^{-2} under different substrate bias density a) refractive index and b) extinction coefficient. Co-sputtered films were also obtained wit a P_{rf} ratio of 7.5/7.5 W cm^{-2}.*

silicate formation on the substrate that might be more resistant to substrate bias than an amorphous solid solution. The amorphous structure results in a very smooth surface even when high P_{bias} is used. In a general way the surface roughness is lower when compared with co-sputtered films presenting a RMS value below 0.5 nm. This confirms that these films are much more resistant to ionic and electronic substrate bombardment.

In order to better understand the structural differences it is important to compare the optical and electrical properties of co-sputtered and single target films. The differences in composition are revealed by the n values that are higher for co-sputtered films due to the low SiO_2 incorporation. It is worth remembering that 7.5/7.5 W cm^{-2} ratio on HfO_2/SiO_2 targets lead approximately to a 2:1 ratio in R, corresponding to the same ratio in volume fraction. Taking into account the HfO_2 and SiO_2 molar mass and density and using equation (4.7), the volume fraction will roughly be similar to mol ratio, that is, the co-sputtered film has a composition around 67:33 HfO_2:SiO_2 mol%. The difficulty in increasing the SiO_2 content in a practical way in co-sputtered films is associated with the high power need in the target that may lead to a degradation of the deposited films. Considering this fact, sputtering from a single target is a good solution when aiming for high SiO_2 content. We thus have confirmation of the high incorporation through the n evolution. The values for single target film are lower than for co-sputtered ones. Even when bias is used n remains below that for co-sputtered films meaning that the SiO_2 content is still high (Figure 4.28).

The E_g values are also affected by the high SiO_2 incorporation increasing by about 8 % when compared with co-sputtered films. However, the selective etching of SiO_2 seems to happen as the gap value is again decreased under substrate bias. The reduction is not so pronounced as for co-sputtered films. The XPS results confirmed the Si reduction within the films.

The high SiO_2 incorporation is also verified in the κ values that are lower in single target films compared with co-sputtered ones. It tends to increase under substrate bias but is not as close to pure HfO_2 as in co-sputtered ones, confirming that these films are more resistant to selective SiO_2 etching (Table 4.4).

Table 4.4 Band gap and dielectric constant of HSiO s.t. films deposited at 0.3 Pa with $\%O_2$ of 0.07 and a P_{rf} of 7.5 W cm^{-2} under different substrate bias density. Co-sputtered films were also obtained wit a P_{rf} ratio of 7.5/7.5 W cm^{-2}.

	Eg (eV)	κ
HfO_2	5.31	24.1
HSiO	5.56	14.1
HSiO s.t.	5.92	11.1
HSiO s.t. 80 mW cm^{-2}	5.50	10.9
HSiO s.t. 160 mW cm^{-2}	5.37	15.4
HSiO s.t. 240 mW cm^{-2}	5.37	13.9

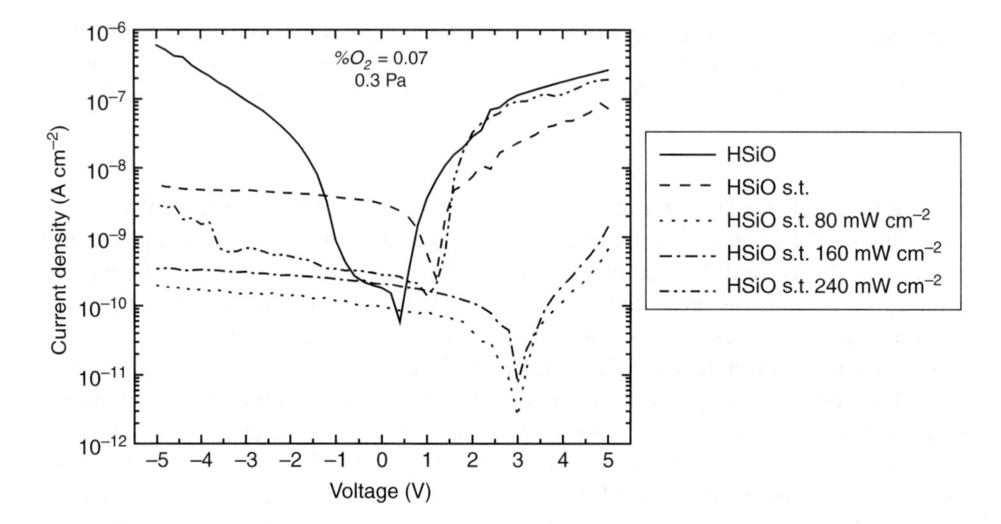

Figure 4.29 Leakage current for HSiO s.t. films deposited at 0.3 Pa with $\%O_2$ of 0.07 and a P_{rf} of 7.5 W cm^{-2} under different substrate bias density. Co-sputtered films were also obtained with a P_{rf} ratio of 7.5/7.5 W cm^{-2}.

The benefits of substrate bias in improving the insulating characteristics of sputtered films are here evident. Moderate values lead to a reduction on the leakage current in MIS structures. Apart from the selective etching of SiO_2, substrate bias is expected to break weak bonds and to promote strong ones at the same time as the films become denser (Figure 4.29).

4.7 Tantalum Oxide (Ta_2O_5)

Tantalum oxide (Ta_2O_5) is widely used as a dielectric material, especially in capacitors [114, 115]. Its application to field effect transistors was considered, taking advantage of the relatively high dielectric constant (also of around 25). Moreover, it presents an amorphous structure when processed at temperatures below 700°C, which is of interest for application

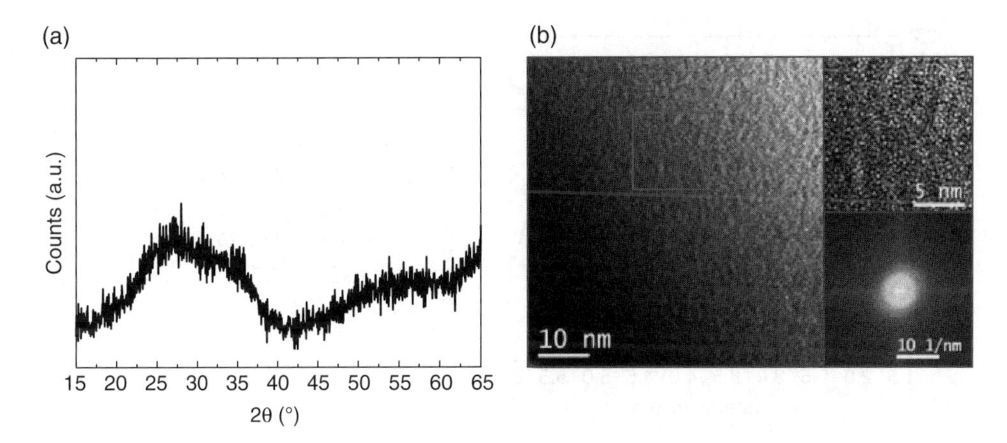

Figure 4.30 *a) XRD data on Ta_2O_5 thin films deposited at 0.3 Pa with $\%O_2$ of 0.07 and a P_{rf} of 5 W cm^{-2}. b) The TEM image and diffraction patterns show an homogenous and completely amorphous structure.*

in TFTs. Above this temperature it tends to crystallize in an orthorhombic structure [116]. Sputtered Ta_2O_5, which has not only been extensively used as a dielectric in organic TFTs [117, 118], but also in oxide TFTs [66] is selected as a starting high-κ material, due essentially to its relatively high sputtering rate even with low P_{rf}, which results in high throughput and low damage to the growing film and its interfaces. However, its application in field effect transistors has some conditions. First, the band gap of Ta_2O_5 is relatively low, on the threshold of what is acceptable to a dielectric gate. The value is around 4.5 eV but varies with different preparation and measuring techniques. In addition, misalignment of the conduction band in silicon is not favorable, but seems to be on the edge of 1 eV on oxide semiconductors. Of course those are theoretical calculations for crystalline structures. When dealing with amorphous films deviations may occur in the determined band offset values.

Because it presents some disadvantages, Ta_2O_5 oxide can be used in other dielectrics, such as mixtures with SiO_2 or Al_2O_3. The combination of Ta_2O_5 with SiO_2 or Al_2O_3 has been found to be useful for applications such as optical filters and corrosion-resistant coatings, [119–121] but in most cases the materials are grown as multilayer structures rather than as multicomponent layers. However, for a transistor's dielectric, the latter or a combination of those two is preferable due to the tendency of high-κ oxides to crystallize even when processed at low temperatures.

Sputtered Ta_2O_5 films with 300 nm tend to be amorphous within a wide range of processing conditions, as confirmed by XRD and TEM (Figure 4.30). This is quite different from what was observed for HfO_2 which tends to present a polycrystalline structure.

In the same way as has been already done for HfO_2 the deposition of Ta_2O_5 was also attempted under different conditions. But before going on to the optical properties of these layers we must go back to the SE models. The Tauc-Lorentz dispersion was also used to describe the dielectric function of Ta_2O_5 thin films. Being an amorphous material the dispersion suits well this purpose [122]. It is relevant here to mention that the extra oscillator for sub-gap absorption did not reveal any improvement of the fitting procedure.

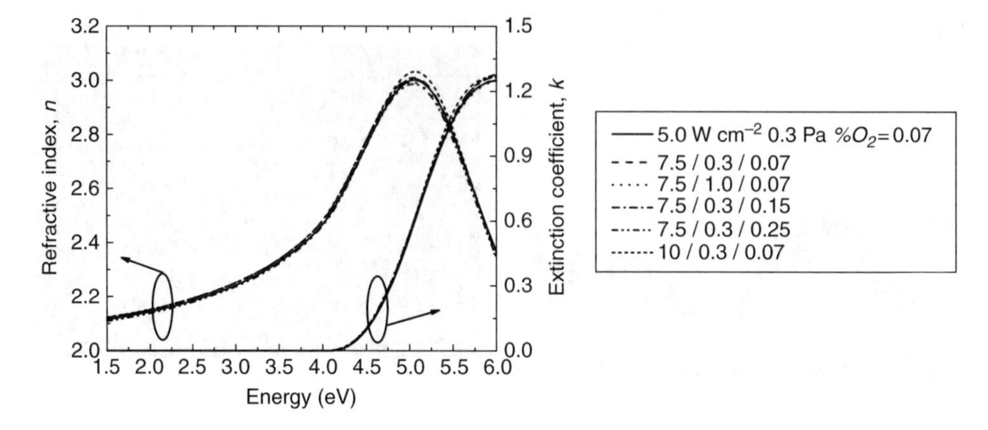

Figure 4.31 *Refractive index and extinction coefficient of Ta_2O_5 films deposited under different conditions. The P_{rf}, deposition pressure and $\%O_2$ were changed. The deposition conditions are in the following order in the legend: $P_{rf}/P_{dep}/\%O_2$.*

Table 4.5 *Tauc-Lorentz dispersion parameters for Ta_2O_5 films deposited under different conditions. In the table is also shown the dielectric constant (κ) determined from C–V characterization in p-type crystalline silicon MIS structures. The sample identification is done following the P_{rf} / P_{dep} /$\%O_2$ sequence.*

Sample	d (nm)	r (nm)	Eg (eV)	ε∞	A (eV)	E_1 (eV)	C (eV)	κ
5 / 0.3 / 0.07	305.1	0.9	4.08	0.52	177.47	5.28	1.85	30.9
7.5 / 0.3 / 0.07	316.5	0.8	4.09	0.08	196.51	5.27	1.94	31.6
7.5 / 1.0 / 0.07	288.6	1.2	4.09	0.33	198.64	5.24	1.99	28.9
7.5 / 0.3 / 0.15	293.8	0.7	4.07	0.18	187.63	5.28	1.92	25.5
7.5 / 0.3 / 0.25	277.9	0.7	4.10	0.19	195.08	5.26	1.93	28.9
10 / 0.3 / 0.07	293.8	0.9	4.04	1.34	161.78	5.31	1.73	23.8

Looking now to n and extinction coefficient obtained by SE modeling we verify that the deposition conditions, at least in the experimental range tested here, do not introduce many modifications in the films' structure and density (Figure 4.31). The broad peak in n is typical of amorphous structures and its maximum height does not present any great differences.

In order to get more information from SE characterization we must analyze the model parameters so as to better understand the influence of the deposition conditions. The E_g is around 4.1 eV for all samples. Looking at Table 4.5 we can understand the small difference verified in the n spectra. Indeed the high peak height on the graph for the sample deposited with high P_{rf} is related to an increase in the short range order and is suggested by the small value in the C parameter, that is clearly lower than for the other deposition conditions. This is a result of the high deposition rate and high energy in sputter species. The interfacial layer thickness is not shown here as it is not relevant to the discussion.

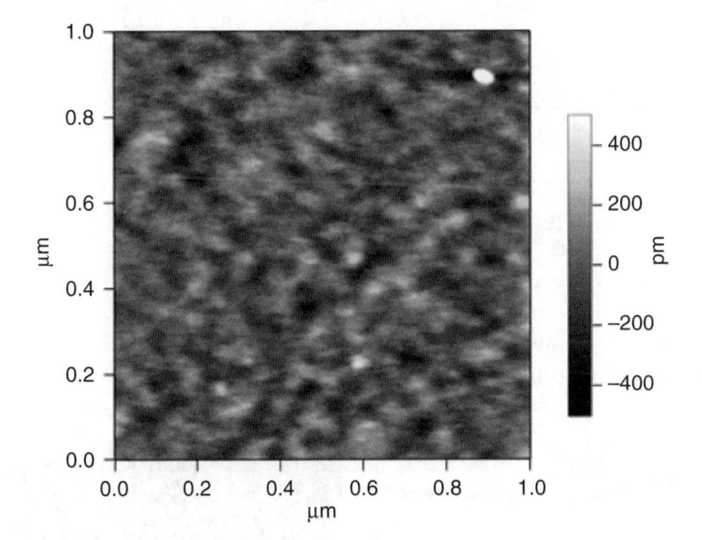

Figure 4.32 AFM microstructure ($1\times1\,\mu m^2$) of a Ta_2O_5 thin film deposited at 0.3 Pa with $\%O_2$ of 0.07 and a P_{rf} of $5\,W\,cm^{-2}$.

The amorphous characteristics of all samples is also reflected in the small value obtained for the layer that simulates the surface roughness. This value is more or less constant except for the sample deposited at high pressure. The AFM image also confirms that the films are smooth where values below 0.5 nm are obtained for RMS roughness. Of course these are very low values and the associated error is high (Figure 4.32).

The dielectric constant values determined from C–V measurements in an accumulation regime at 1 MHz in crystalline silicon MIS structures are also in the table. It is not possible to establish a trend. Just one note: the values tend to be slightly higher than those normally expected for Ta_2O_5. Relatively high values for Ta_2O_5 have already been reported [77] and may be related to leakage current [123]. Leakage current will be discussed in the next section together with multicomponent Ta_2O_5 based films.

4.7.1 Multicomponent Ta_2O_5 Based Dielectrics

As verified above the Ta_2O_5 has a relatively low E_g and so the integration of high gap oxide is a good option for enlarging it. Here an exhaustive study will not be done as for HfO_2. We will concentrate on the co-sputtering of Ta_2O_5 with SiO_2.

Figure 4.33 shows that the deposition rate for Ta_2O_5 is much higher than for SiO_2. Comparing it also with HfO_2 we see that it can double it for a P_{rf} of $7.5\,W\,cm^{-2}$. So in order to allow for more SiO_2 incorporation we decided to use a P_{rf} of 5 and $7.5\,W\,cm^{-2}$ in Ta_2O_5 and SiO_2 targets, respectively, in co-sputtered films. Under these conditions the R of Ta_2O_5 is still roughly two times that of SiO_2. Based on equation (4.7) we expect the molar ratio to be around 1 which leads to a cation atomic percentage of 67 and 33 for Ta and Si, respectively. These co-sputtered films will be called the TSiO and the P_{rf} ratio in the targets will be represented as $5/7.5\,W\,cm^{-2}$.

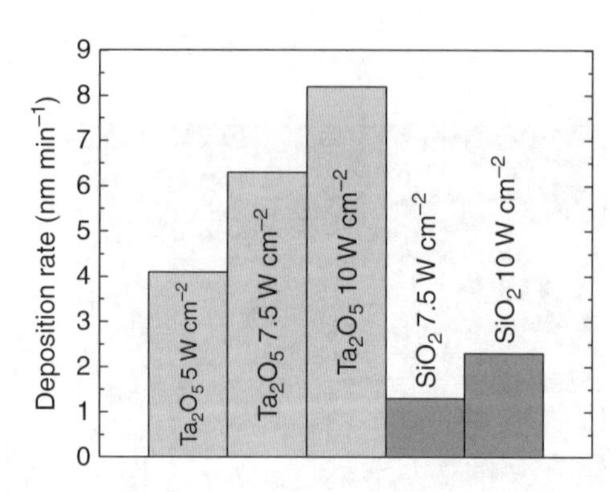

Figure 4.33 *Deposition rates of the sputtered Ta_2O_5 and SiO_2 at 0.3 Pa with %O_2 of 0.07. The values inside the bars correspond to the used P_{rf}.*

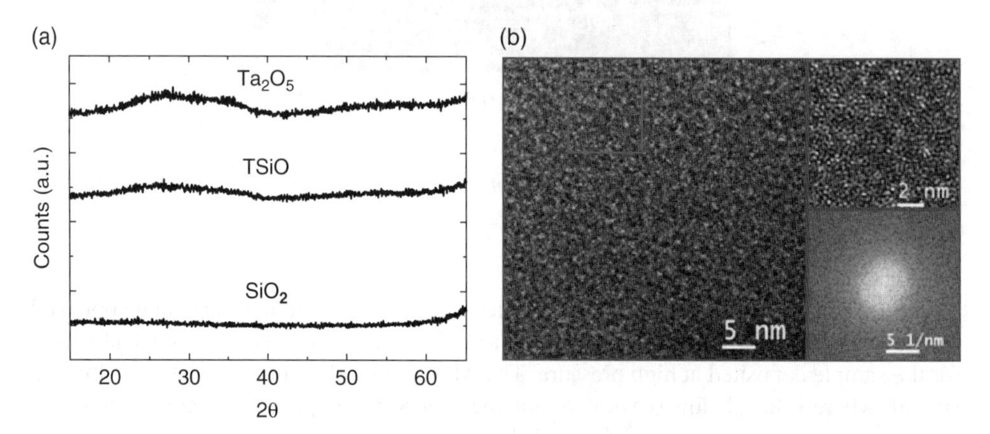

Figure 4.34 *Structural analysis: (a) XRD data for Ta_2O_5, TSiO and SiO_2 thin films deposited at 0.3 Pa with %O_2 of 0.07 (b) TEM analysis, showing an electron micrograph of TSiO film cross section. The inset shows the corresponding electron diffraction pattern. Adapted with permission from ref [66] Copyright (2010) Society for Information Display.*

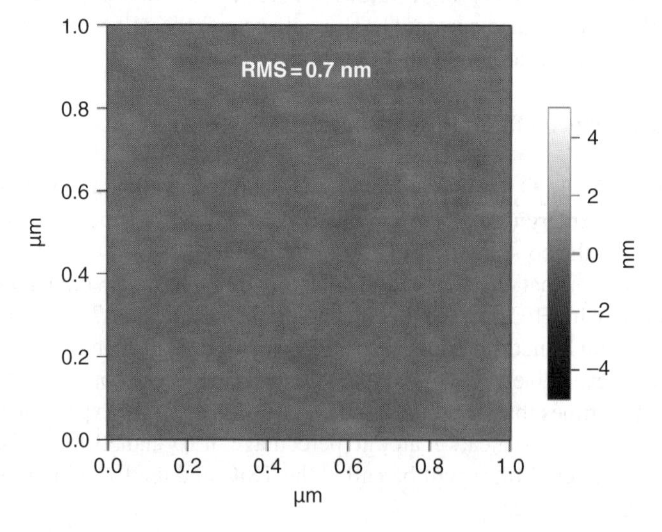

Figure 4.35 *AFM microstructure ($1 \times 1 \mu m^2$) of a TSiO thin film deposited at 0.3 Pa with %O_2 of 0.07 and a P_{rf} of 5/7.5 W cm^{-2}.*

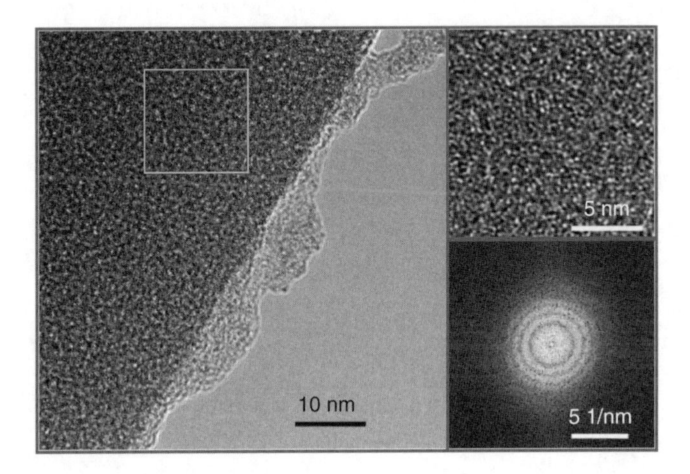

Figure 4.36 TEM analysis, showing an electron micrograph of TSiO film cross section annealed at 300°C. The inset shows the corresponding electron diffraction pattern.

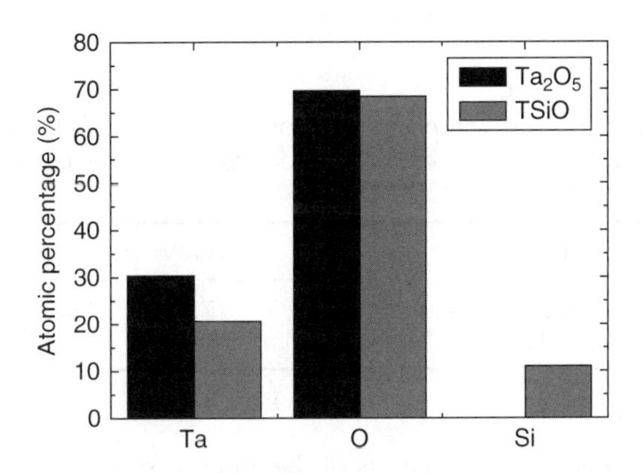

Figure 4.37 Compositional analysis obtained by XPS for the Ta_2O_5-based dielectric thin films.

Figure 4.34 and Figure 4.35 show the structural and compositional data obtained for the TSiO dielectrics. When looking at the XRD data a broad diffraction peak appears close to $2\theta=30°$, suggesting that some short-range order exists within the thin films, but crystallization (i.e., long-range ordering) is not achieved. This is expected, as Ta_2O_5 and SiO_2 are also amorphous when deposited at room temperature. TEM image shows again a perfectly homogenous film where no phase separation is detected.

Furthermore, note that these structures, even when considering only the pure Ta_2O_5 films, show a considerably lesser tendency to crystallize than other high-κ materials, such as HfO_2 or Y_2O_3, which can bring benefits in terms of electrical performance, reliability and integration [77]. Figure 4.34b shows an example of the results obtained by TEM analysis of a TSiO film, confirming that even after the annealing treatment at 300°C, the structure remains amorphous (Figure 4.36). However, the diffraction patterns suggest that the films have gained short-range order as ring features are observed.

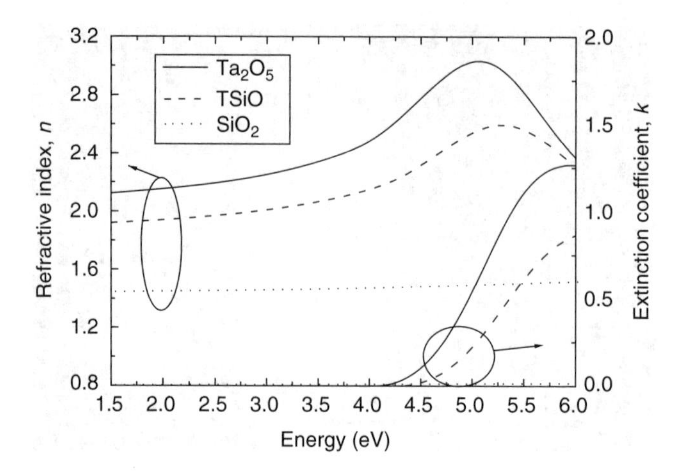

Figure 4.38 *Dependences of n and k for Ta_2O_5 and TSiO thin films deposited at 0.3 Pa with a %O_2 of 0.07. Ta_2O_5 film was deposited with a P_{rf} of 5 W cm^{-2} and TSiO film with a P_{rf} ratio of 5/7.5 mW cm^{-2}.*

Table 4.6 *Tauc-Lorentz dispersion parameters for Ta_2O_5 and TSiO films. Also shown is the dielectric constant (κ) determined from C-V characterization in p-type crystalline silicon MIS structures.*

Sample	d (nm)	r (nm)	Eg (eV)	$\varepsilon\infty$	A (eV)	E_1 (eV)	C (eV)	κ
Ta_2O_5	305.1	0.9	4.08	0.52	177.47	5.28	1.85	30.9
TSiO	292.7	1.0	4.21	1.20	131.29	5.47	1.90	17.3

Concerning the compositional analysis of the Ta_2O_5-based thin films, the results also show a good correlation between nominal (i.e., expected composition) and experimental values for the different binary compositions. It shows that Ta_2O_5 are not far from stoichiometry with an atomic % of 30.4 Ta/69.6 O close to nominal 28.6 % Ta/71.4 % O. In TSO films the cation ratio of 20.5 %/11 % is close that which is expected from the *R* ratio for Ta_2O_5 and SiO_2 (67 %/33 %) (Figure 4.37).

The spectroscopic ellipsometry data are presented in Figure 4.38 and provide good support to the structural and compositional data discussed above. Starting with the n evolution of Ta_2O_5 and SiO_2 related dielectrics (Figure 4.38), it is evident that SiO_2 is incorporated in the co-sputtered TSO film, since its n plot is located between those of pure Ta_2O_5 and SiO_2 films. Valuable information concerning the absorption rise at the onset of conduction band is given by the κ plot in Figure 4.38. An increase on E_g from 4.08 to 4.21 eV is achieved when comparing pure Ta_2O_5 and co-sputtered TSiO films, suggesting SiO_2 incorporation. Note that the E_g values presented herein are lower than those usually reported in the literature for Ta_2O_5 (around 4.5 eV), even if the analyzed films are close to the ideal stoichiometry, according to the XPS data. This is attributed to two factors: first, the simulation model used here takes into account the absorption due to tail-states (located inside the bandgap,

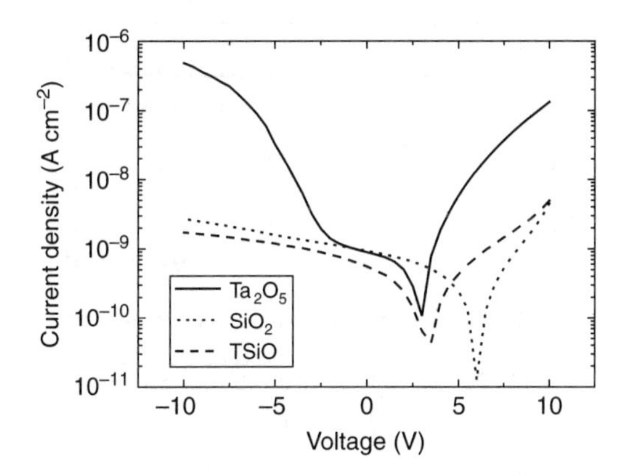

Figure 4.39 *J–V characteristics for MIS structures comprising dielectric thin films based on the tantalum-silicon oxide system. Adapted with permission from ref [66] Copyright (2010) Society for Information Display.*

close to the bands) to calculate E_g; second, the sputtering process may induce some damage to the structure, reflected in the broadening of the band-tails. Also important to note is the C parameter value that increase compared with Ta_2O_5 films deposited using the same P_{rf}. This means the short-range order is reduced by incorporation of SiO_2 (Table 4.6).

The κ values were determined from MIS capacitors and are presented in Table 4.6. They are in agreement with the XPS data, showing a trend to increase with the Ta_2O_5 content within the thin films. The binary films, especially Ta_2O_5, present κ values close to those expected for the nominal compositions, reinforcing the idea already demonstrated by XPS that their composition should be close to the theoretical stoichiometry. As expected, the multicomponent TSiO dielectrics present values between those of their constituent binary oxides.

Measurements of the current density as a function of voltage (J–V) were also performed in MIS structures, in order to evaluate the electrical reliability of these dielectrics. Low J values are a fundamental requisite, as these materials have to provide electrical insulation between the gate electrode and the semiconductor in TFT structures. Figure 4.39 shows the results obtained for some of the analyzed materials. Due to its low E_g, Ta_2O_5 presents larger J, around two orders of magnitude higher than SiO_2, even if the SiO_2 sputtering process is not entirely optimized. However, when SiO_2 and Ta_2O_5 are co-sputtered, a significant improvement is verified, with TSiO exhibiting J values similar to those of SiO_2. Additionally, TSiO has a κ more than three times larger when compared with SiO_2.

To summarize, tantalum oxide deposited by RFMS did not show a significant dependence on the processing conditions and tends to be amorphous as deposited. The films also benefit from a low surface roughness. Spectroscopic ellipsometry analysis reveals that the broadening parameter is higher for the multicomponent oxides than for the Ta_2O_5 films, suggesting a decrease of short-range order after SiO_2 incorporation. It is highly desirable to have amorphous structures, not only due to the better interface related aspects already indicated for the oxide semiconductors, but also because the grain boundaries in polycrystalline insulators potentiate high leakage currents. As expected, TSiO exhibits κ values

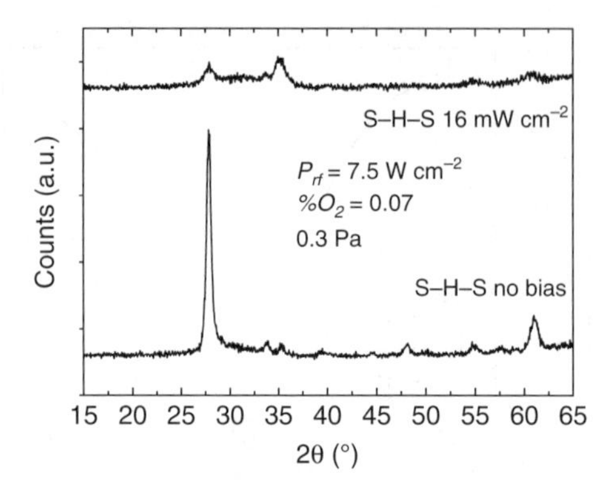

Figure 4.40 XRD data for S-H-S multilayer deposited at 0.3 Pa with %O_2 of 0.07 and a P_{rf} of 7.5 W cm^{-2}. The thickness of the multilayer is 20/180/20 nm.

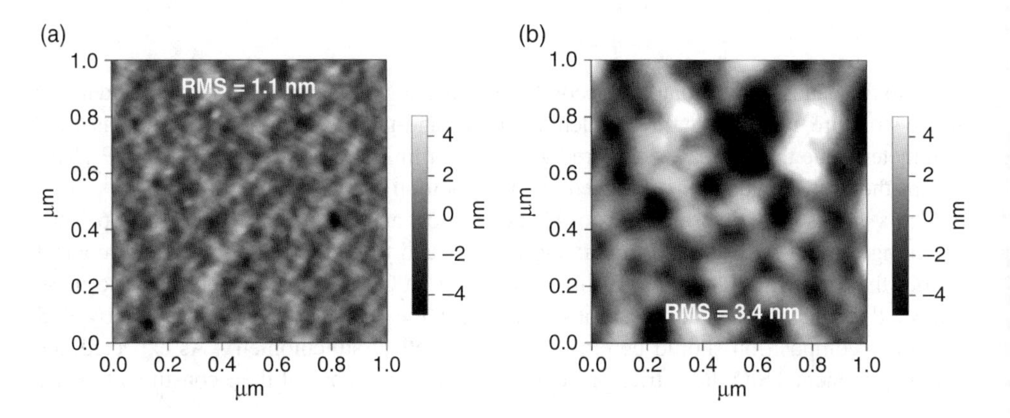

Figure 4.41 AFM microstructure (1 × 1 µm^2) of a S–H–S multilayer deposited at 0.3 Pa with %O_2 of 0.07 and a P_{rf} of 7.5 W cm^{-2}: a) without substrate bias and b) with a P_{bias} of 160 mW cm^{-2}.

intermediate to the composing binary oxides, but always closer to Ta_2O_5 (24.6), which is the predominant element. The J–V curves reveal considerable improvements for TSiO, with J being decreased by around two orders of magnitude when compared with pure Ta_2O_5, down to values close to those obtained for SiO_2.

4.8 Multilayer Dielectrics

The concept being discussed in this chapter is based on high-κ dielectrics for field effect based devices, in order to reduce the working voltage and influence of trapping states and, at the same time, increase the current in oxide based TFTs. This way SiO_2 is not considered as an option, and is only used to stabilize amorphous structures' in high-κ materials.

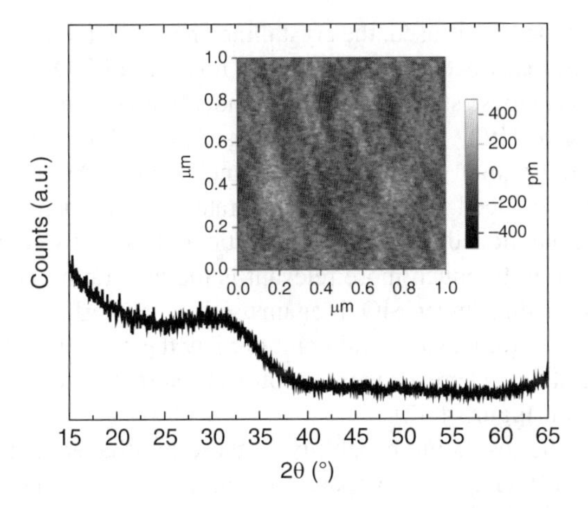

Figure 4.42 XRD pattern of a S–HS–S film deposited at 0.3 Pa with a P_{rf} of 7.5 W cm^{-2}, a %O_2 of 0.07 and a P_{bias} of 160 mW cm^{-2}. The inset shows the AFM microstructure (1×1 μm^2).

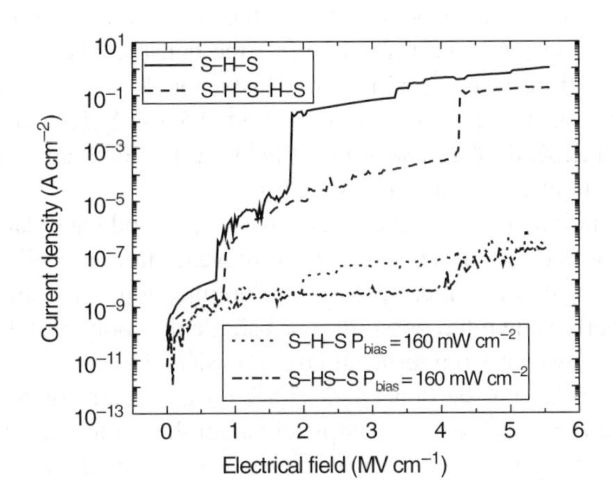

Figure 4.43 J–E_{el} characteristics for MIS structures comprising multilayer structures deposited with or without substrate bias.

However, due to its large bag gap value, SiO_2 may be of interest for multilayer layer gate dielectrics, adjusting its thickness to minimize the κ value reduction. In this section we will briefly summarize some of the most important features of these structures in terms of structural and electrical properties when deposited by RFMS.

We will start with $SiO_2/HfO_2/SiO_2$ structures (named as S–H–S) with a thickness of around 20/180/20 nm. The XRD data shows peaks associated with HfO_2 that are from the monoclinic phase (Figure 4.40). So we have amorphous double interface layers with a crystalline structure in between. An interesting point is that RMS roughness is much lower than in single HfO_2, meaning that the top SiO_2 layer may work as a buffer in getting smoother surfaces. When substrate bias is used, the crystalline phase in HfO_2 becomes predominately orthorhombic as has already been observed. Again, no peaks of SiO_2 are detected. However, the surface roughness in these stacks is modified under bias (Figure 4.41). It is not expected to be directly related to HfO_2; as we saw previously the orthorhombic phase tends to be smoother, due also to the small grain size. So, the increase on RMS roughness is due to the SiO_2 surface damage induced by the ionic substrate bombardment when bias (240 mW cm^{-2}) is used. As we mentioned before, SiO_2 is particularly sensitive to this.

One option that might be much more relevant is the integration HSiO films instead of pure HfO_2 as an intermediate layer. SiO_2 is again used at the interface layer. These multilayers are called S–HS–S. From a structural point of view the resulting film is amorphous, as was expected, since all layers are also amorphous. The surface is very smooth with a RMS roughness below 1 nm (Figure 4.42).

In order to check the insulating capability of these multilayers MIS structures in c-Si were used (Figure 4.43). Here, we compare both biased and unbiased films but also stack with two more layers (S–H–S–H–S). When bias is not used the increase in the number of layers leads to a reduction in leakage current and improvement in breakdown voltage from 2 to around 4 MV cm^{-1}. This may be related to the extra barrier the electron must overcome when crossing the insulator. However, the most effective approach is to combine the multilayer with substrate bias. Moderate P_{bias} values were used in order to minimize the surface damage, but, at the same time, improve the density. The breakdown voltage of S–H–S and S–HS–S films is above 5 MV cm^{-1}. Just a note for the dielectric constant value on these films that decreases from 14 to around 9.5 for S–H–S and S–HS–S films, respectively: it will depend, of course, on the thickness and total number of SiO_2 layers as this will reduce the total capacitance due to low κ.

To summarize, multilayer layers are considered to be good candidates for oxide TFTs dielectrics. If tuning the SiO_2 thickness so as to prevent direct tunneling it may act as an effective barrier located at the interfaces to electron injection in the dielectric conduction band. We have briefly shown the possibility of using either polycrystalline or amorphous HfO_2 based layers between two interfacial layers of SiO_2. This way the final surface morphology of these stacks will be dependent either on the intermediate layer structure or substrate bias conditions. However, a completely amorphous multilayer proved to be very promising as a gate dielectric, since leakage current can be greatly reduced.

Table 4.7 *Deposition conditions of the dielectric used in GIZO MIS structures for P_{dep}=0.3 Pa and %O_2=0.07.*

Sample	P_{rf} (W cm^{-2})	P_{bias} (mW cm^{-2})
HfO_2	7.5	–
HSiO s.t.,	7.5	–
Ta_2O_5	5.0	–
TSiO	5.0/7.5	–
S-HS-S	7.5/7.5/7.5	–
S-HS-S biased	7.5/7.5/7.5	160
S-TS-S	7.5/5–7.5/7.5	–
S-TS-S biased	7.5/5–7.5/7.5	160

4.9 High-κ Dielectrics/Oxide Semiconductors Interface

This chapter has been dedicated to the fundamental properties of high-κ dielectrics processed by RFMS. In this final section we aim to show the integration of these materials with an oxide semiconductor. GIZO has been the most studied material as a channel layer in oxide TFTs as it has been shown to be the best compromise between performance and structural and electrical stability.

As mentioned at the beginning of the chapter, so far the gate dielectric has been somehow forgotten when optimizing the TFTs' electrical characteristics. Not much is known about high κ materials and oxide semiconductor interfaces. Little exists so far on this subject, as shown by the examples of Jackson *et al.* [124] and Kimura *et al.* [125], which makes this subject a very interesting area of study.

We selected some representative dielectric layers from those analyzed in the previous sections. Table 4.7 describes the deposition conditions of dielectrics in GIZO MIS structures. We must also note that multilayers of SiO_2 with TSiO were also used (S–TS–S), despite the fact that no characterization has been shown before. We have simply followed the same approach as for the same multilayer structure with HSiO s.t.

In order to study a dielectric/semiconductor interface MIS structures are commonly used. Indeed, some were prepared in p-type c-Si so as to study the leakage current, as explained previously. In this section we will explore the MIS structure but using GIZO instead. In an ideal MIS structure without any applied voltage, the difference between the work function of metal and the semiconductor is zero, that is, the diagram of energy bands is flat (flat band condition – V_{FB} – flat band voltage) (Figure 4.44). Moreover, the only charges that exist in the structure under any state of polarization are found in the semiconductor and those in the same amount but opposite sign, are at the gate metal. Finally, under bias there is no charge transport through the insulator so it behaves ideally as if having infinite resistivity.

For an n-type semiconductor, (typically the situation in oxide TFTs) when the MIS structure is polarized directly (positive voltage on the door), the energy bands near the semiconductor bend downward, resulting in an increase in the difference between the Fermi level (E_F) and the intrinsic level (E_I). This causes the accumulation of electrons near the semiconductor-insulator interface – the accumulation regime. For a small negative voltage, the energy bands bend upward, the difference between the Fermi level decreases

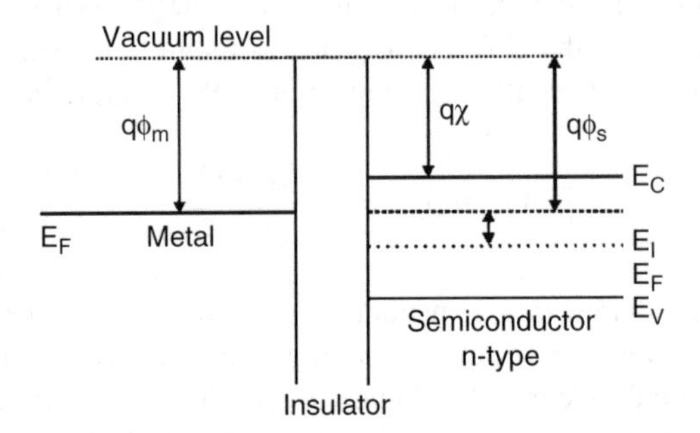

Figure 4.44 Energy band diagram for an ideal MIS structure.

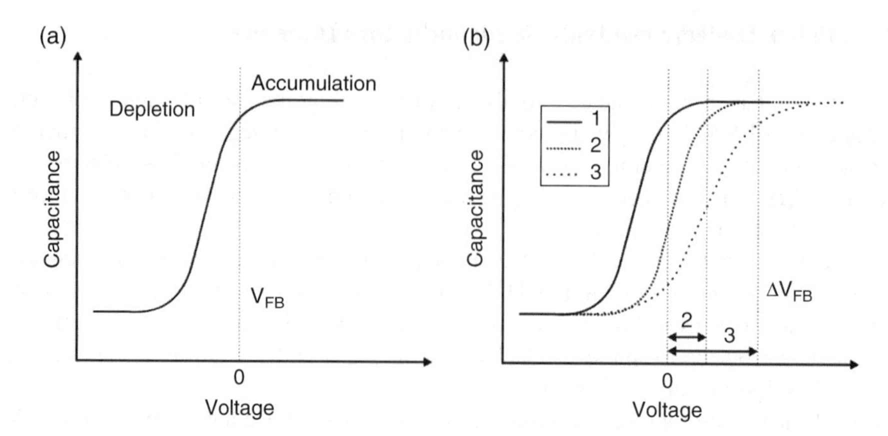

Figure 4.45 *a) CV for an ideal MIS structure, b) schematic of the possible deviations in a MIS structure real MIS structure – curve 1 represents the ideal characteristic, curve 2 the effect of* Q_f*,* Q_t *and* Q_m *and curve 3 the effect of* Q_{it}*.*

and the intrinsic level and the majority carriers (electrons) are depleted in the region of the interface – the depletion regime. These two regimes are known to occur in oxide semiconductors. However, contrary to that which normally occurs in c-Si MIS structures, as the negative voltage increases, the Fermi level hardly exceeds the intrinsic level and the concentration of holes does not exceed that of electrons near the interface. Moreover, as we are dealing with wide band gap semiconductors a high surface potential would be needed [23]. This means that the inversion regime in oxide semiconductors is not achieved.

In the accumulation regime, there is no depletion layer in the semiconductor, that is, the total capacity of the structure is approximately equal to the capacity of the insulator. Thus, this regime can be used to determine its capacitance (C_i). Indeed, the results shown previously concerning the dielectric constant were determined in this way in a p-type c-Si semiconductor.

A real MIS structure presents deviations from the ideal derived mainly by two factors. First, for the metals normally used as a gate electrode, the difference in work function for the semiconductor is not null ($\phi_{ms}=\phi_m - \phi_s \neq 0$). In addition, there are different types of charge in the insulator and at the interface, which are related to existing defects, which will, in different ways, influence the C–V curve. The most significant shift is changing the flat band voltage (V_{FB}), which is ideally zero, but a real MIS structure is given by:

$$V_{FB} = \phi_{ms} - \frac{Q_f + Q_t + Q_m}{C_i} \tag{4.8}$$

where Q_f is the fixed charge in the insulator, Q_t the trapped charge and Q_m the mobile charge. The latter two also contribute to the appearance of hysteresis. The effect on the C–V curve is represented by curve 2 in Figure 4.45b. There are still trapped charges at the interface (Q_{it}), which when present in a relevant amount will also contribute to the depletion zone being enlarged, as it influences the value of V_{FB}, distorting the C–V curve (curve

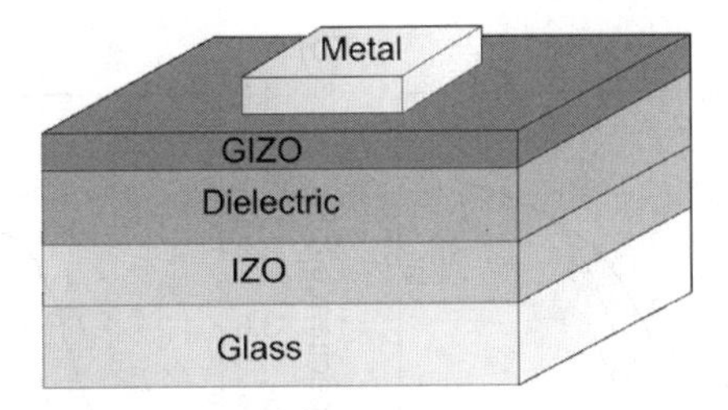

Figure 4.46 Scheme of characterized MIS structures.

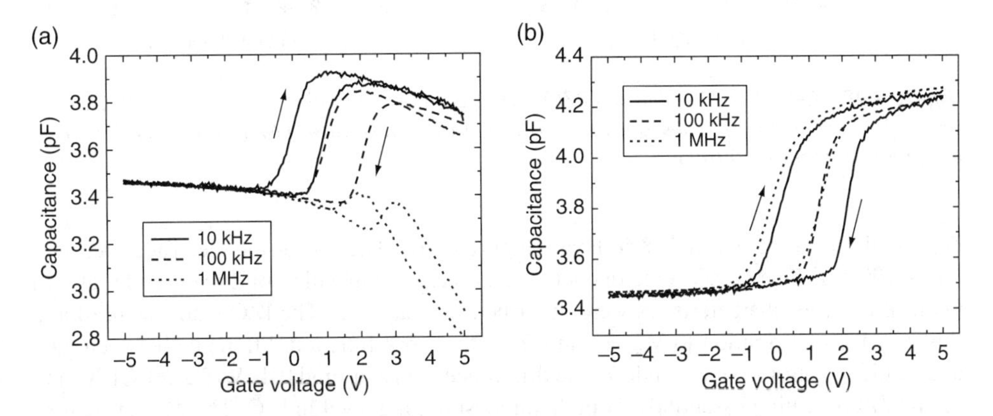

Figure 4.47 C–V characteristics of GIZO MIS structures integrating: a) HfO_2 and b) HSiO s.t. They were obtained at frequencies between 10 kHz and 1 MHz with an excitation voltage of 100 mV.

3 in Figure 4.45b). The example shown is for an n-type semiconductor, for a p-type the curve would have a symmetrical appearance.

The charges trapped at the interface (Q_{it}), usually defined as the density of states per unit area at the interface (D_{it}), are viewed as a major limitation to the application to field effect transistors, as they reduce the carrier mobility. They are a result of structural defects at the interface and cause trap states within the energy gap of the semiconductor. The fixed charges (Q_f) are located in the region near the interface (less than 5 nm), and can be active in a wide range of voltages. They are generally positive, contributing to a shift of the curve towards negative voltages, depending on the conditions in which the insulator was grown or annealed. Ionized atoms or dangling bonds in the region near the interface contribute to the appearance of this type of charge [126]. The trapped charge within the insulator (Q_t), is associated with existing defects, which can arise, for example, by high energy radiation such as X-rays or electron bombardment. They are distributed inside the dielectric and can be minimized by heat treatment. Finally, the mobile charges (Q_m) are ions in the insulator moving under the influence of temperatures or high voltages.

It is now time to look at the C–V characteristics of GIZO MIS structures. They were deposited with the same configuration as bottom gate TFTs, that is, the semiconductor is

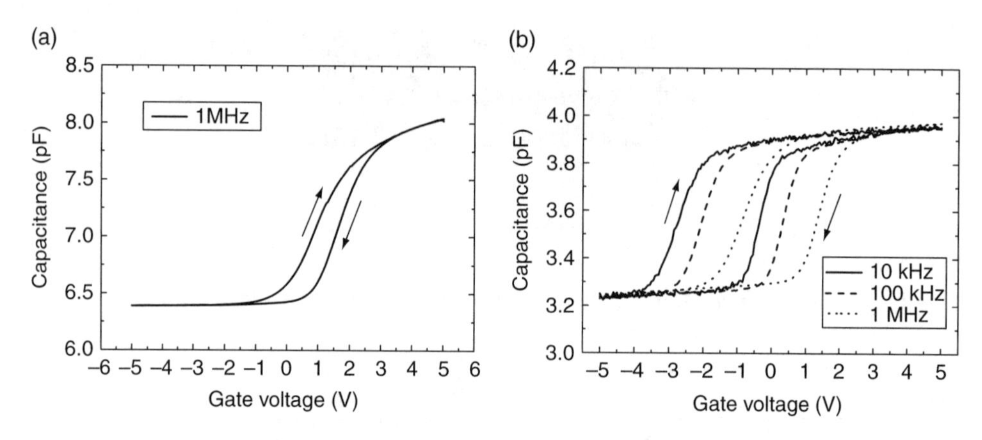

Figure 4.48 *C–V characteristics of GIZO MIS structures integrating: a) Ta_2O_5 and TSiO. They were obtained at frequencies between 10 kHz and 1 MHz (just 1 MHz for Ta_2O_5) with an excitation voltage of 100 mV.*

deposited on top of the dielectric layer (Figure 4.46). The top of the electrode area was $100 \times 100\,\mu m^2$. The GIZO semiconductor layer has a composition of 2:4:2 atomic ratio in Ga:In:Zn and its characteristics were described in Chapter 2. The IZO gate was used and this way the contribution to V_{FB} coming from the work function difference between gate and semiconductor is small. Indeed, the difference is just around 0.4 eV (5.2 for GIZO and 4.8 for IZO) coming essentially from the highest doping level in IZO. The MIS structures were then annealed at 150°C. The measuring cycle was done from accumulation to depletion and then to accumulation again (Figure 4.46).

The C–V characteristics for HfO_2 and HSiO s.t. are shown in Figure 4.47. The accurate determination of the carrier concentration of GIZO is difficult due to its low value. It becomes impossible to calculate precisely the V_{FB}. However, it is also important to see the differences in the C–V curves and how they behave for different excitation frequencies. The capacitance variation is rather small. This is supposed to be associated to the small thickness of the semiconductor layer (around 40 nm) which limits the maximum dimension of the depletion region we can achieve. Remember that we have C_i in series with the depletion region capacitance (C_{dep}) and the resulting capacitance is dominated by the smallest one. However, as the depletion layer increases the C_{dep} decreases and becomes comparable with C_i and so, the overall capacitance series association decreases. In theory, the maximum depletion width is limited by the semiconductor thickness (around 40 nm), and so the minimum capacitance of the system won't be much lower than the total capacitance.

Another factor that may limit the capacitance variation is the interface states. This may create a parallel capacitance (C_{ss}) with a depletion region capacitance (C_{dep}) that leads to overall capacitance enhancement. Depending on the number of interface states, the C_{ss} can be large and if higher than C_{dep} it will fully dominate the final capacitance value ($C_iC_{dep}/(C_i+C_{dep})+C_{ss}$), making the total capacitance even less sensitive to variation in C_{dep}.

When HfO_2 is used in GIZO MIS structures we observe a large capacitance dispersion and flat band shift (ΔV_{FB}) as the frequency increases (Figure 4.47). On the other hand, amorphous HSiO thin films do not show such dispersion. The flat band shift (around 1 V)

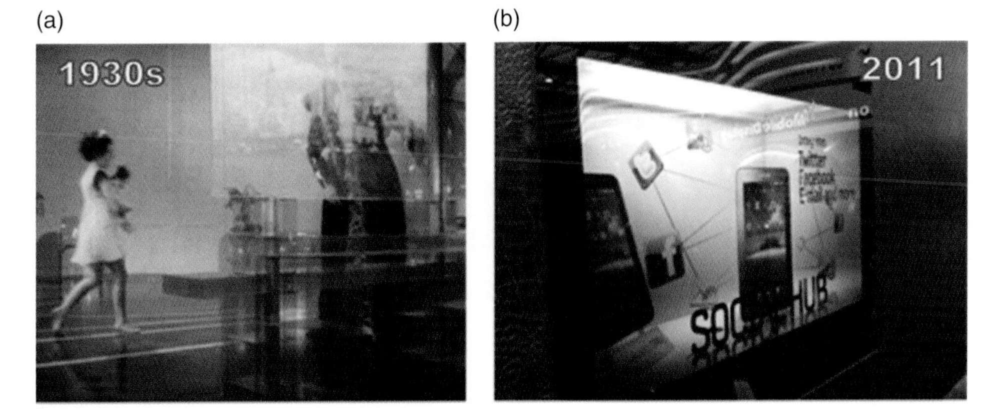

Plate 1.1 *Transparent displays: a) early vision, in H.G. Wells' 1930s novel* The Shape of Things to Come *[1]; b) Samsung's 22" transparent LCD panel now being mass-produced in 2011 [2]. Reproduced from [2] Copyright (2011) Samsung Corp.*

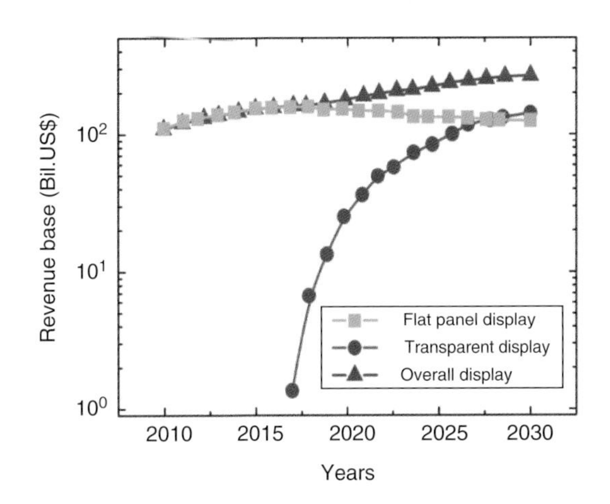

Plate 1.2 *Transparent display technology evolution and global display market (adapted from [3]). Adapted with permission from [3] Copyright (2011) DisplayBank.*

Transparent Oxide Electronics: From Materials to Devices, First Edition. Pedro Barquinha, Rodrigo Martins, Luis Pereira and Elvira Fortunato.

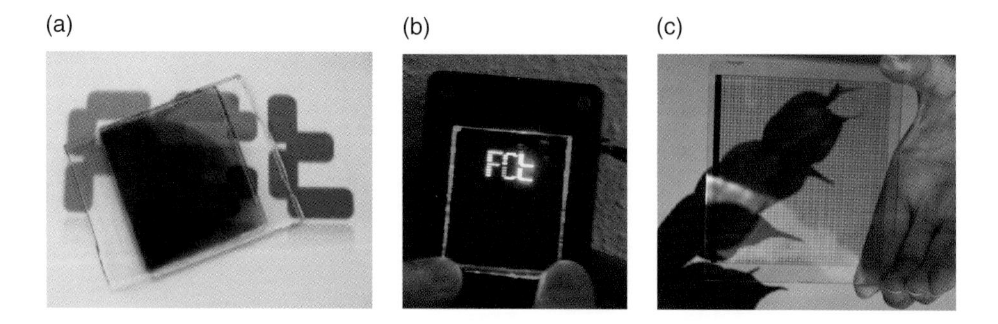

Plate 1.3 *Some applications of TCOs at CENIMAT: a) electrochromic windows; b) passive matrix LED display; c) see-through solar cell.*

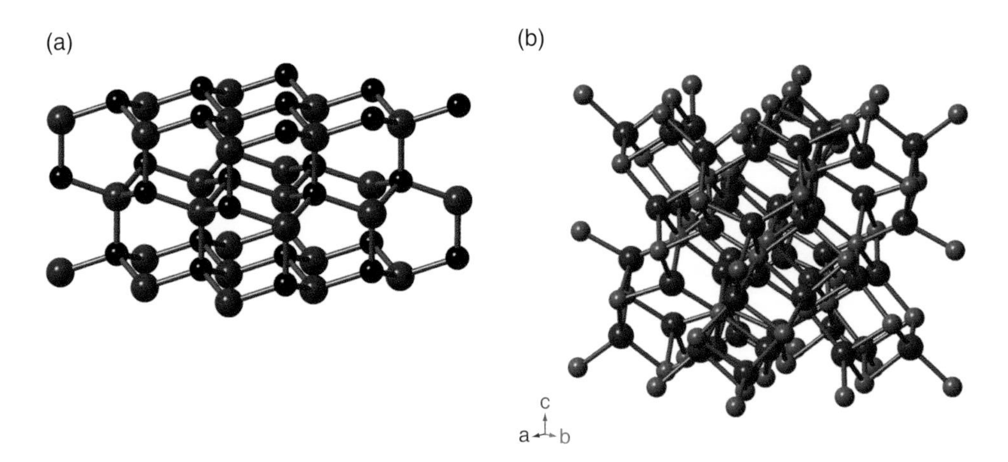

Plate 2.1 *Crystalline structure commonly adopted by a) ZnO, hexagonal (wurtzite) and b) In_2O_3, cubic (bixbyite). The small spheres represent the metallic cations, while the large spheres represent the oxygen anions.*

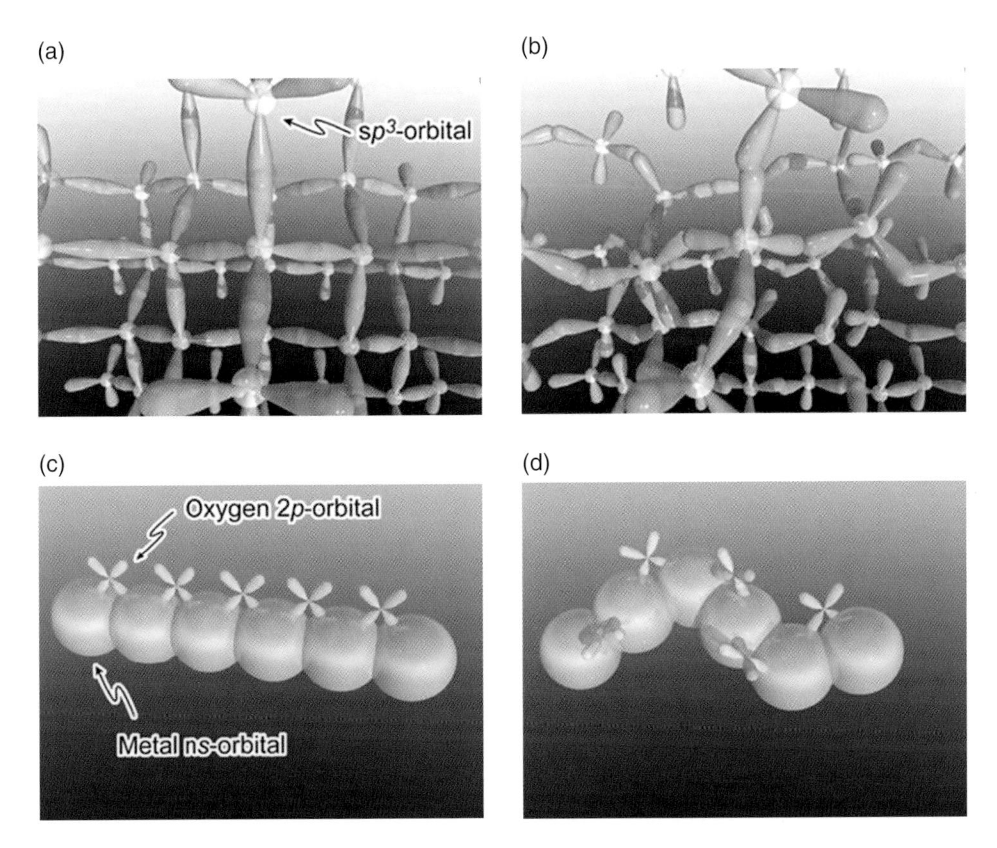

Plate 2.2 *Schematics proposed by Nomura et al. of the orbitals composing the CBM on covalent semiconductors with sp^3 orbitals and ionic semiconductors with ns orbitals ($n \geq 4$): a) covalent crystalline; b) covalent amorphous; c) ionic crystalline; d) ionic amorphous [52]. Reproduced with permission from [52] Copyright (2004) Macmillan Publishing Ltd.*

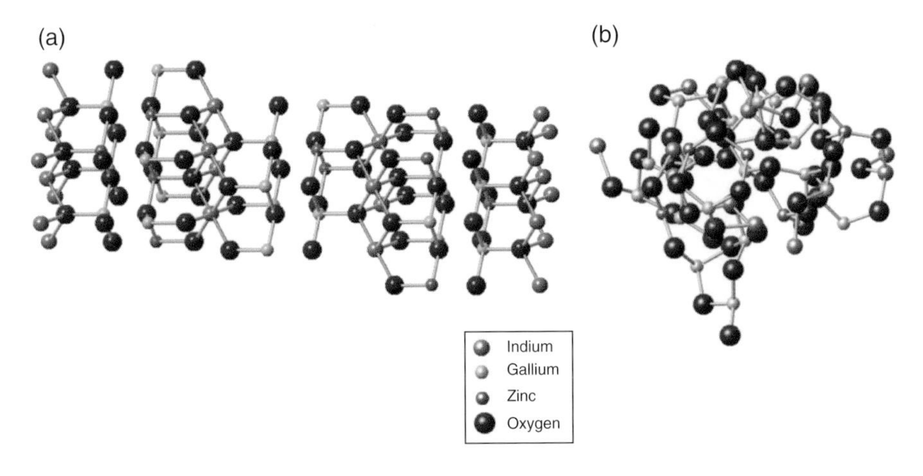

Plate 2.3 *Schematic structures of a) crystalline and b) amorphous GIZO.*

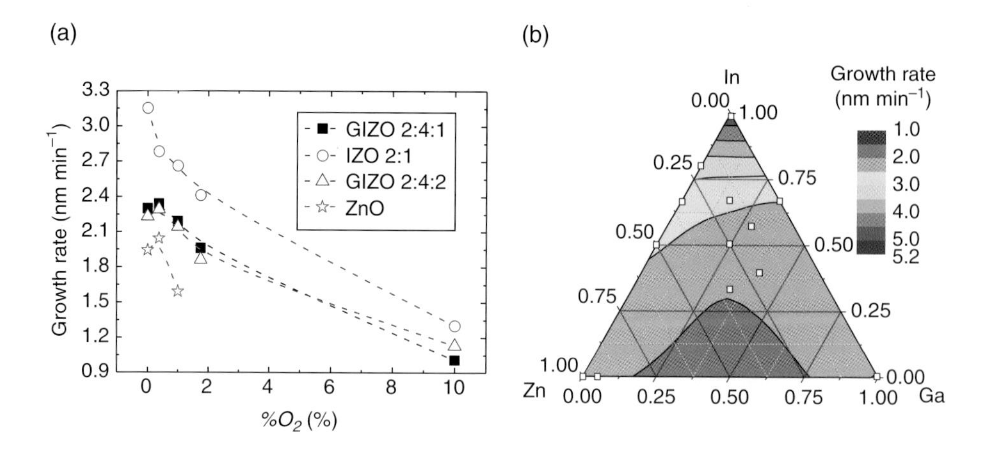

Plate 2.4 *Dependence of the growth rate on a) %O_2, and b) target composition.*

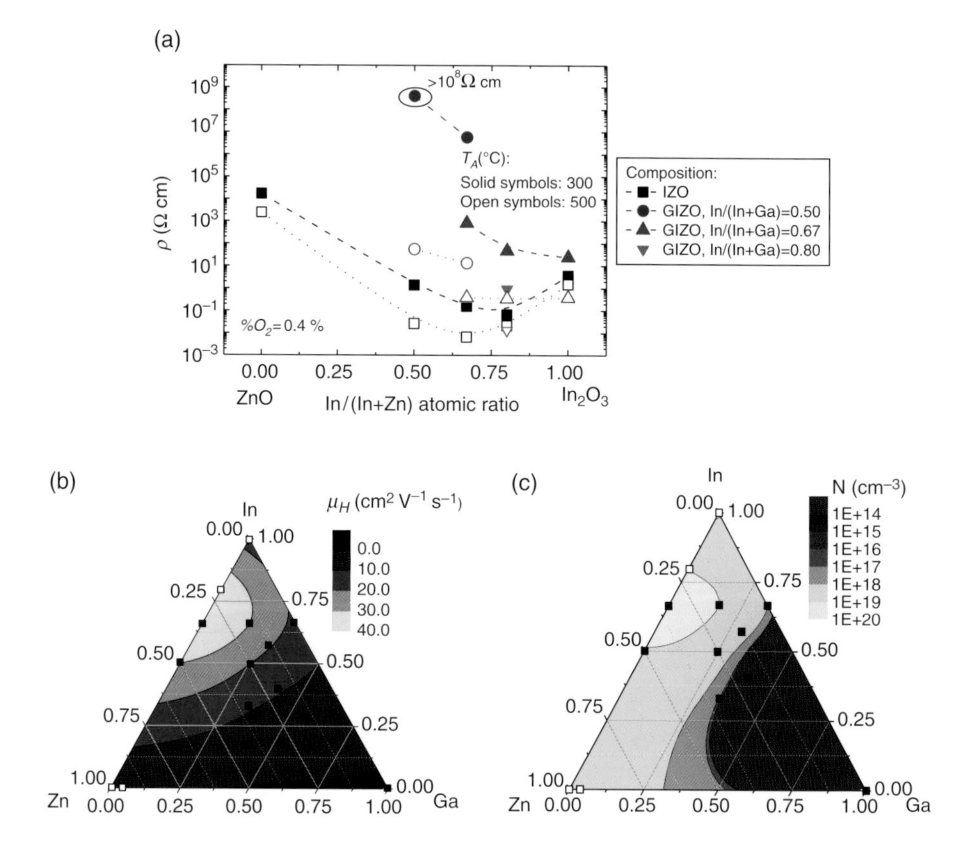

Plate 2.10 *Effect of target composition on the electrical properties of oxide semiconductor thin films produced under the same deposition conditions: a) Effect of In/(In+Zn) and In/(In+Ga) atomic ratios on ρ, for TA=300 and 500°C; b) and c) are ternary diagrams for gallium-indium-zinc oxide system, for films annealed at 500°C, showing μ and N, respectively. The filled and empty squares denote amorphous and polycrystalline films, respectively.*

Plate 2.29 *a) SEM, b) AFM and c) XRD data for ZTO thin films deposited by spray-pyrolysis at 400°C.*

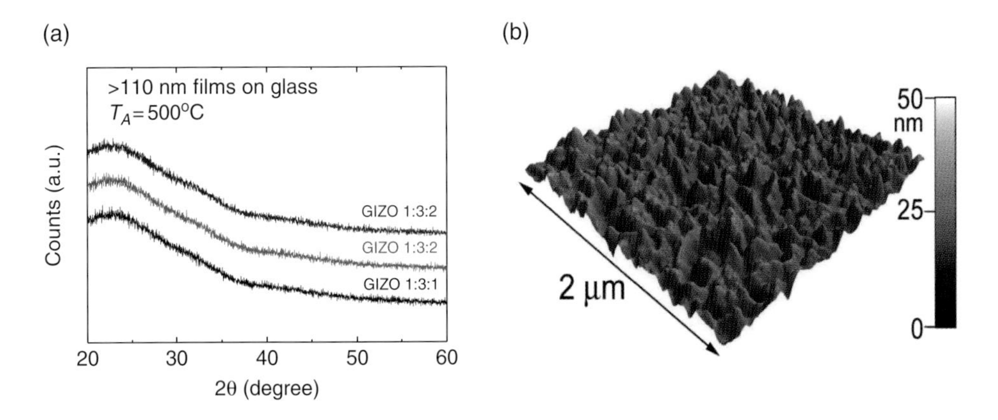

Plate 2.31 *a) XRD and b) AFM data for GIZO thin films deposited by spin-coating, with T_A=500°C. AFM data refers to GIZO 1:3:1 films. Adapted with permission from ref [129] Copyright (2010) American Institute of Physics.*

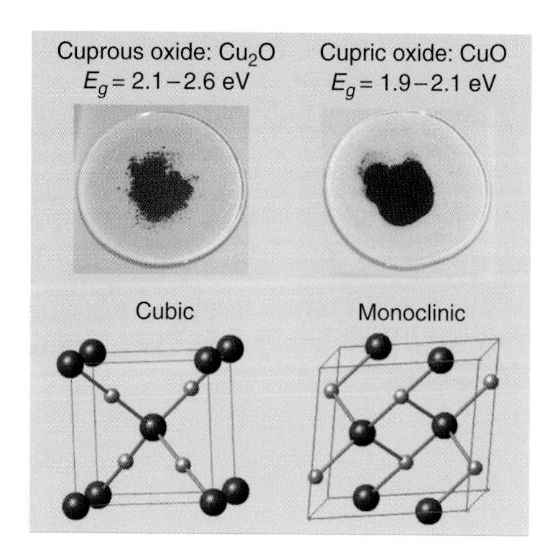

Plate 3.3 *Copper (I) oxide or cuprous oxide is one of the principal oxides of copper. The compound can appear either yellow or red, depending on the size of the particles. Cu_2O crystallizes in a cubic structure. Copper (II) oxide or cupric oxide belongs to the monoclinic crystal system. It is a black solid with an ionic structure.*

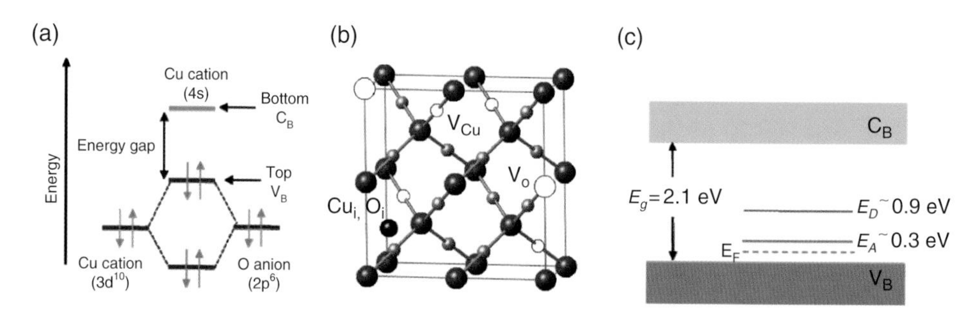

Plate 3.4 *a) Chemical bond between an oxide ion and a cation that has a closed-shell electronic configuration; b) representation of the more important defects in Cu_2O; c) a simple electronic model proposed by Brattain, with a compensated semiconductor with one acceptor level at 0.3 eV and a deep donor level at 0.9 eV from V_B. Adapted with permission from ref [87] Copyright (2010) American Institute of Physics.*

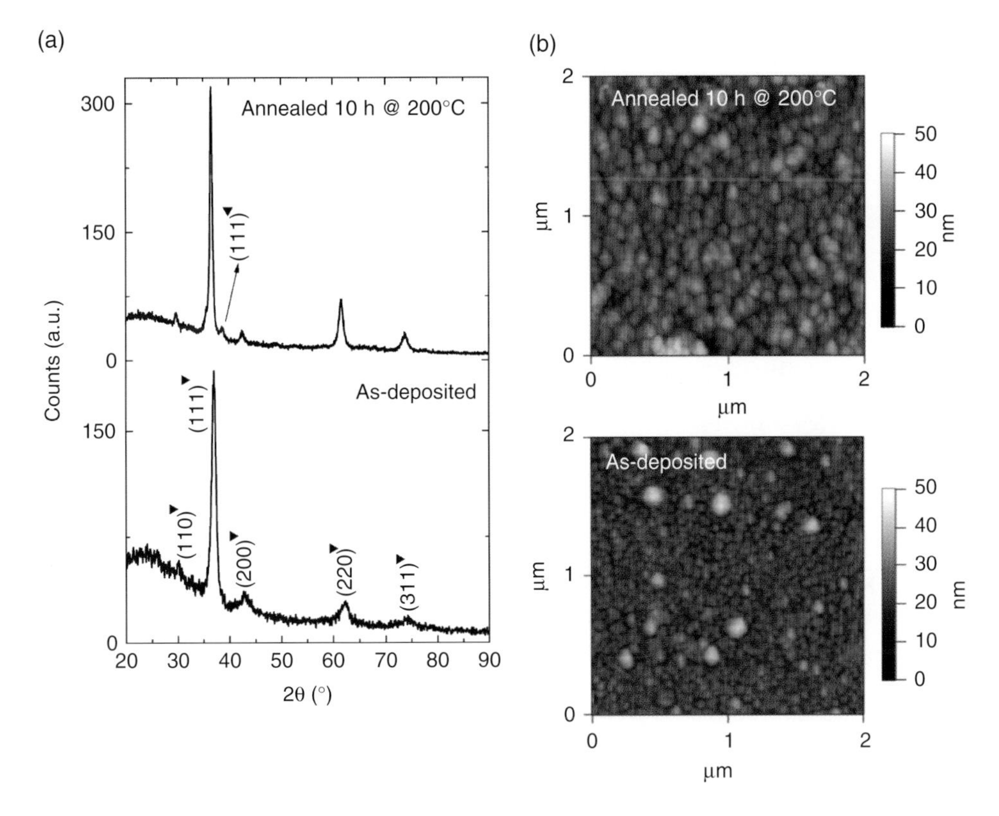

Plate 3.6 *a) X-ray pattern of the Cu_2O film annealed at 200°C in air for 10 hours in comparison with the as-deposited film. (Symbol representations: ▼-Cu_2O phase; ◄-CuO phase); b) AFM images of the corresponding films. Adapted with permission from ref [87] Copyright (2010) American Institute of Physics.*

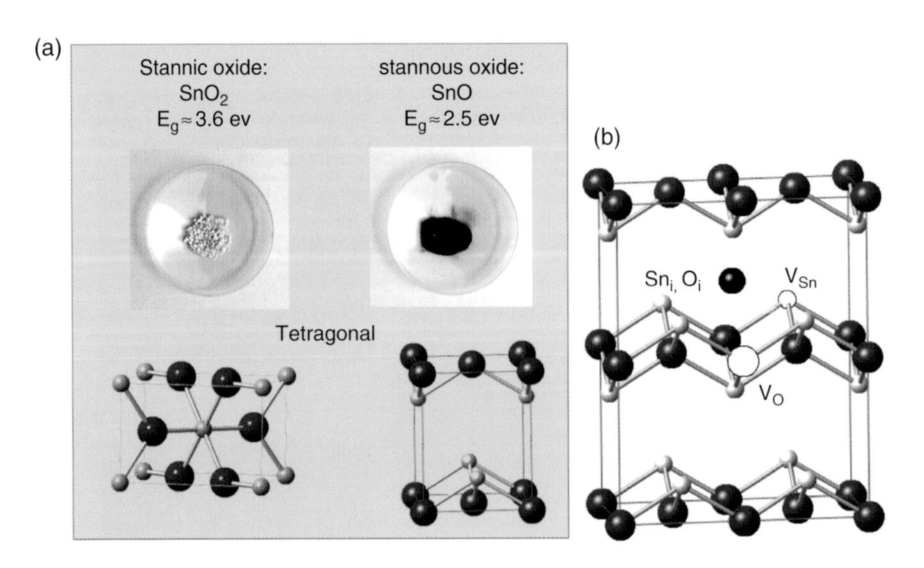

Plate 3.11 *a) Unit cell of the crystal structure of SnO_2 and SnO; b) Sketch of the unit cell of tin oxide showing the presence of interstitial atoms and vacants of oxygen and tin in the structure.*

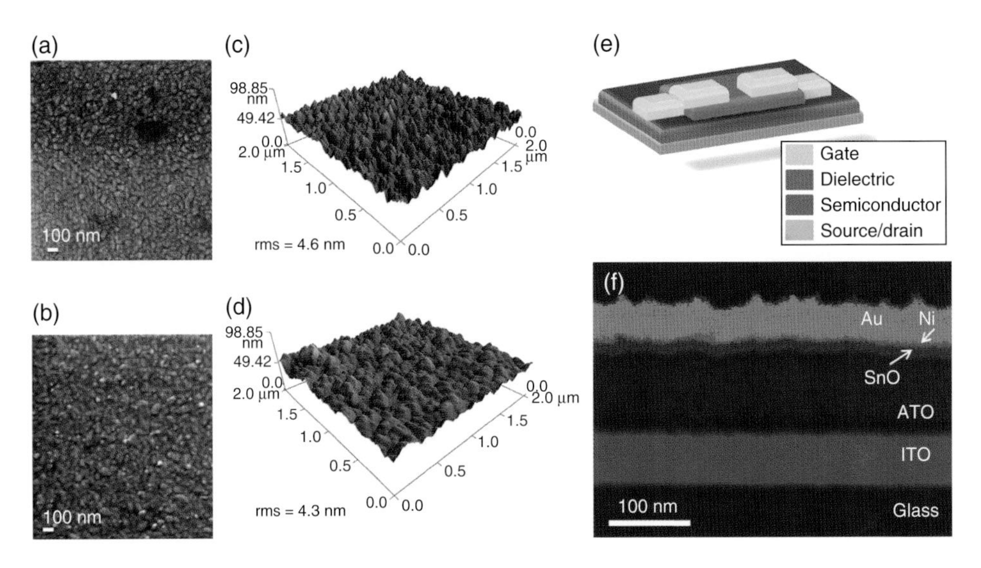

Plate 3.15 *SEM images of post-annealed tinoxide used as channel layers prepared at O_{pp} of a) 3.0 % and b) 3.6 %, for TFT applications. Corresponding AFM images of the TFT channel layers are shown in c) and d). e) Schematic illustration of SnO_x TFT structure employed for the present study. f) Cross-sectional SEM image of the SnO_x channel layer.*

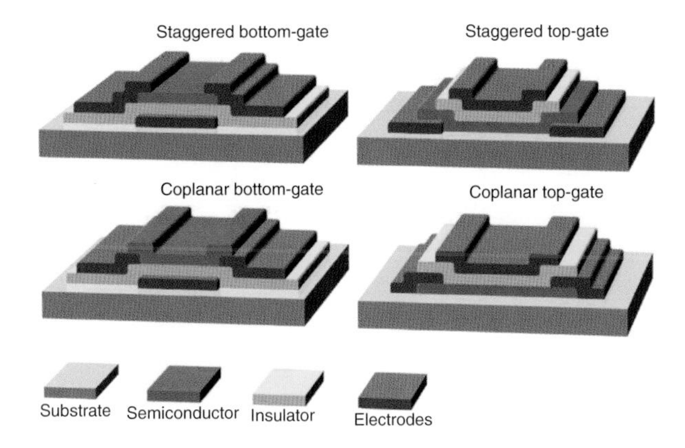

Plate 5.1 *Schematics showing some of the most conventional TFT structures, according to the position of the gate electrode and the distribution of the electrodes relative to the semiconductor.*

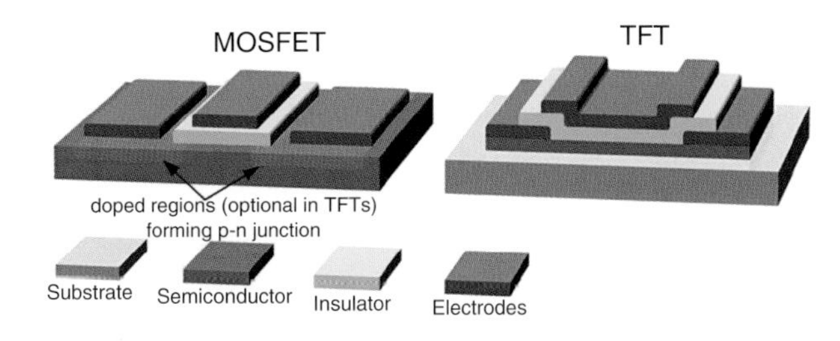

Plate 5.2 *Comparison between the typical structures of MOSFETs and TFTs.*

Plate 5.7 *Example of integration of organic TFTs in an OLED display by Sony [38]. Reprinted with permission from [38] Copyright (2007) Sony Corporation.*

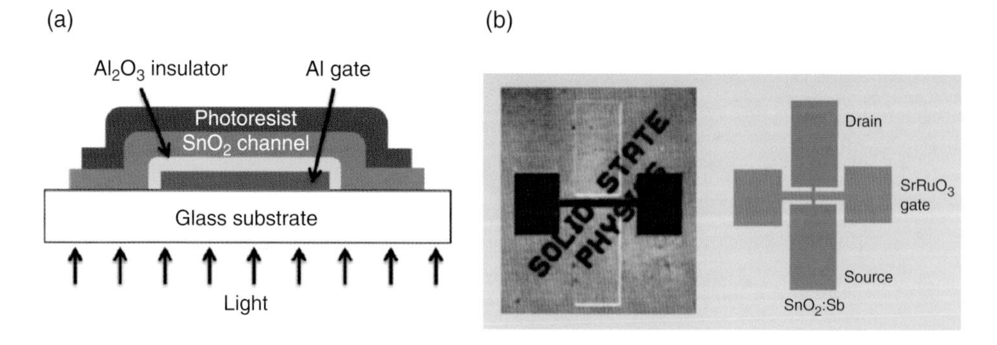

Plate 5.8 *Initial work on oxide TFTs: a) Schematic of the SnO_2 TFT reported by Klasens and Koelmans in 1964. Reprinted with permission from [44] Copyright (2008) Springer Science + Business Media; b) top-view and layout of the ferroelectric Sb:SnO_2 TFT reported by Prins et al. in 1996. Reprinted with permission from [42] Copyright (1996) American Institute of Physics.*

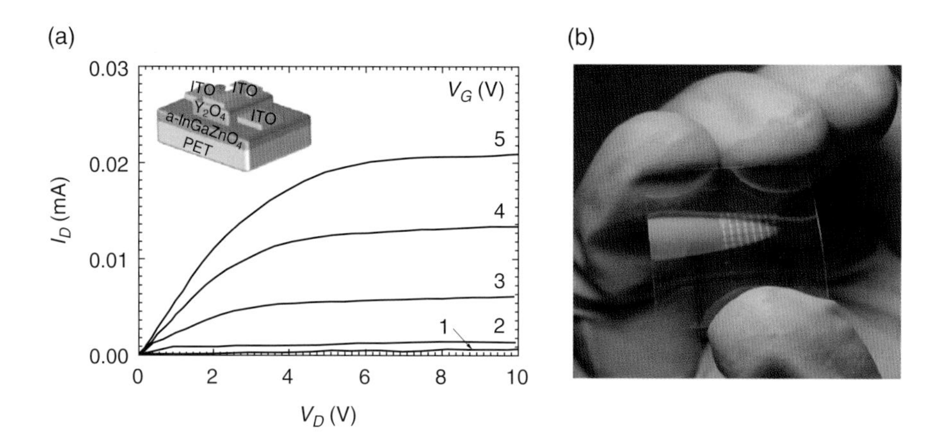

Plate 5.9 *GIZO TFTs on flexible substrates presented by Nomura et al. in 2004: a) output characteristics; Reprinted with permission from [60] Copyright (2004) Macmillan Publishing; b) photograph of the flexible TFT sheet bent at R=30 mm (adapted from [60]).*

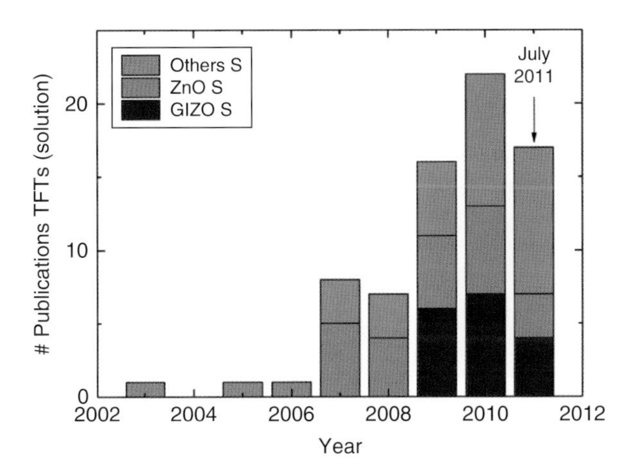

Plate 5.10 *Number of publications related to solution-processed oxide TFTs over the years.*

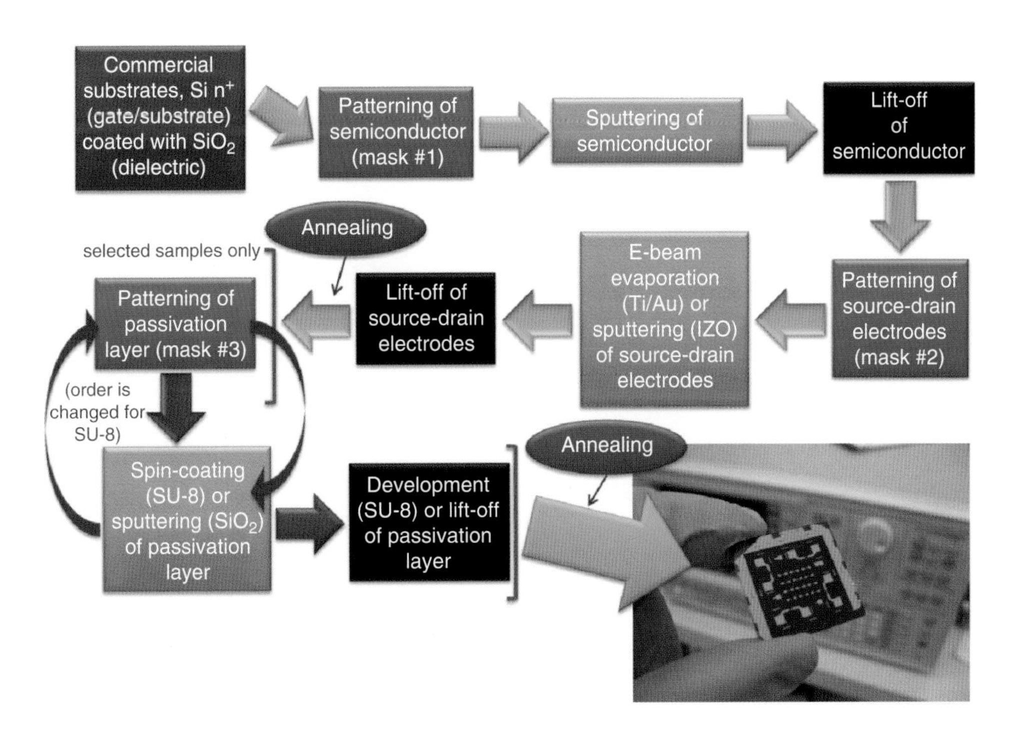

Plate 5.11 *Process flow used to produce oxide TFTs employing commercial Si/SiO_2 substrates.*

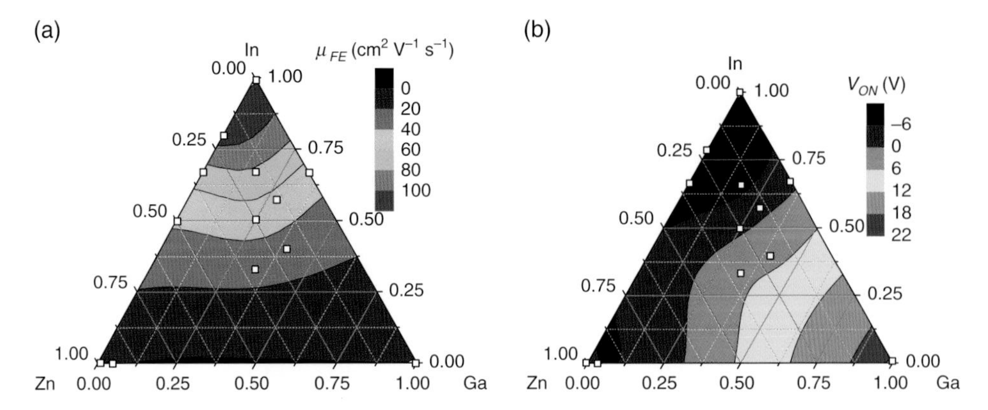

Plate 5.15 *a) μ_{FE} and b) V_{ON} obtained for TFTs with different oxide semiconductor compositions, in the gallium-indium-zinc oxide system. Devices annealed at 150°C, with $\%O_2 = 0.4\ \%$. The red and yellow symbols denote amorphous and polycrystalline semiconductor films, respectively.*

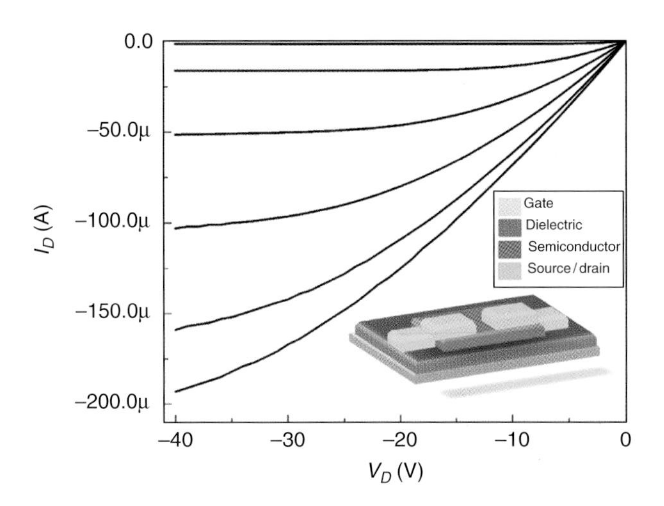

Plate 5.33 *Output characteristics of p-type SnO TFT, annealed at 200°C. Gate voltage is varied from 0 to −50V in −10V steps. The inset shows a schematic of the device structure. Adapted with permission from [147] Copyright (2011) Wiley-VCH.*

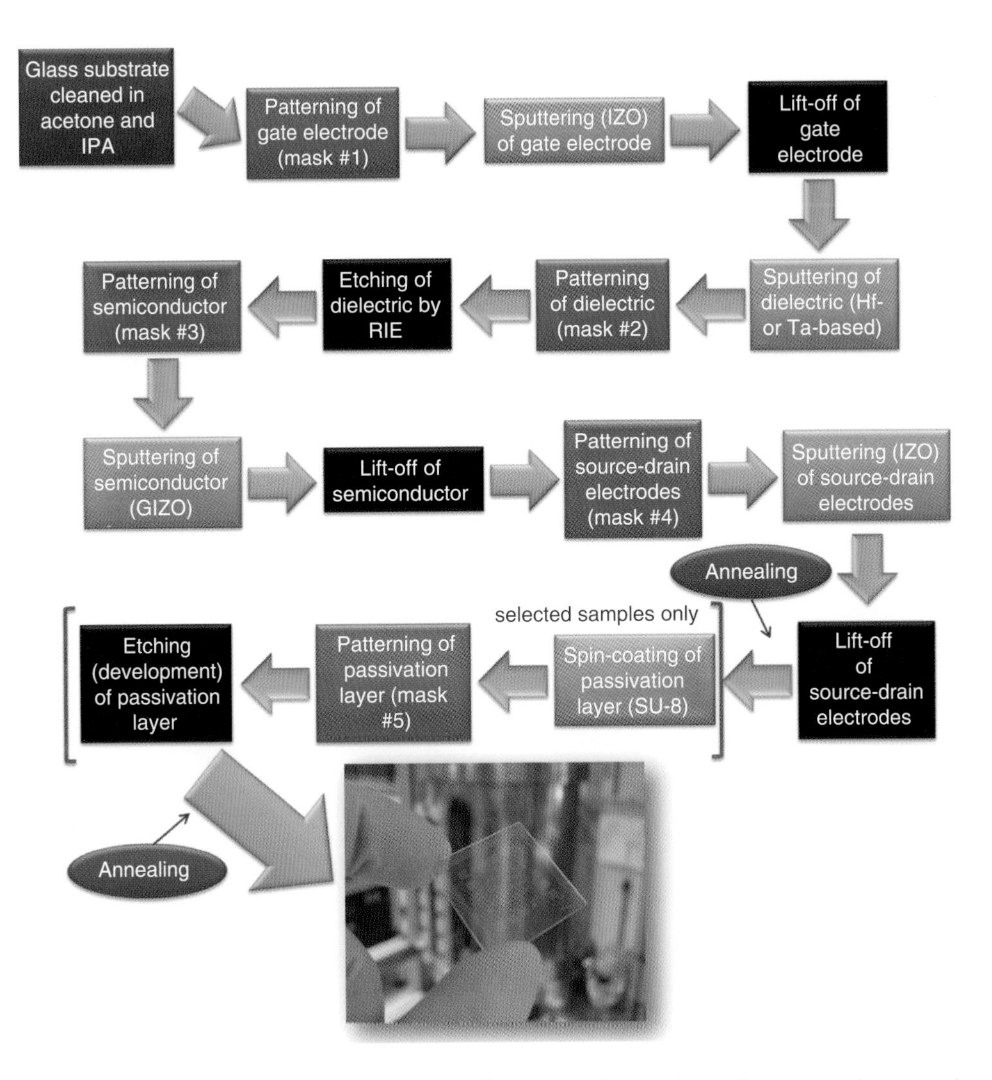

Plate 5.35 *Process flow used to produce oxide TFTs employing glass substrates and sputtered dielectrics.*

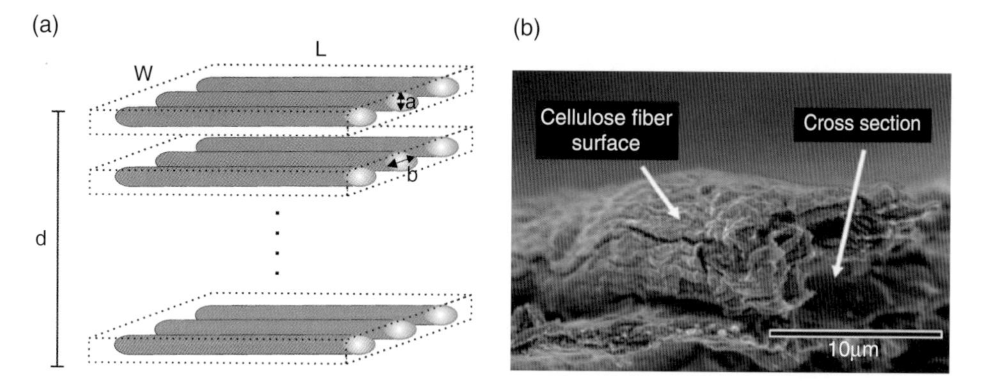

Plate 6.7 *(a) Sketch of the distribution of the fibers along the planes, and of how such planes are associated along the dielectric width and its thickness. (b) Cross-section SEM image of a fiber [30]. Reprinted with permission from [30] Copyright (2011) Wiley-VCH.*

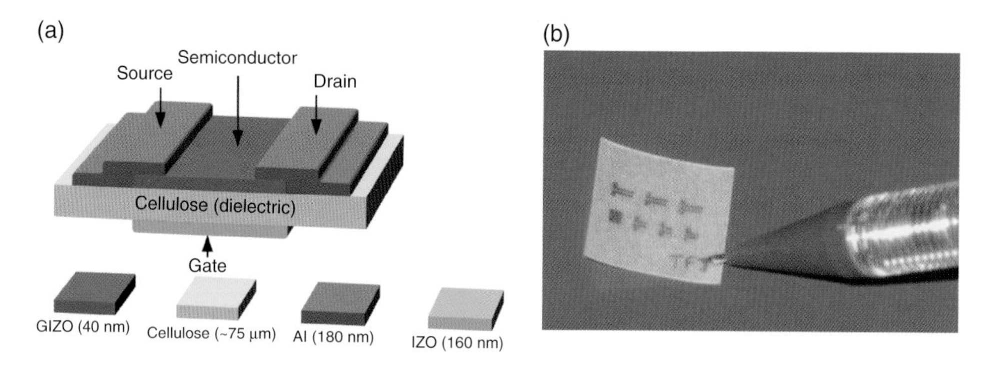

Plate 6.9 *a) Schematic representation of the field effect transistors structure using the cellulose sheet as the gate dielectric, where are also seen the other constituents of the final paper transistor; b) picture of the set of paper transistors.*

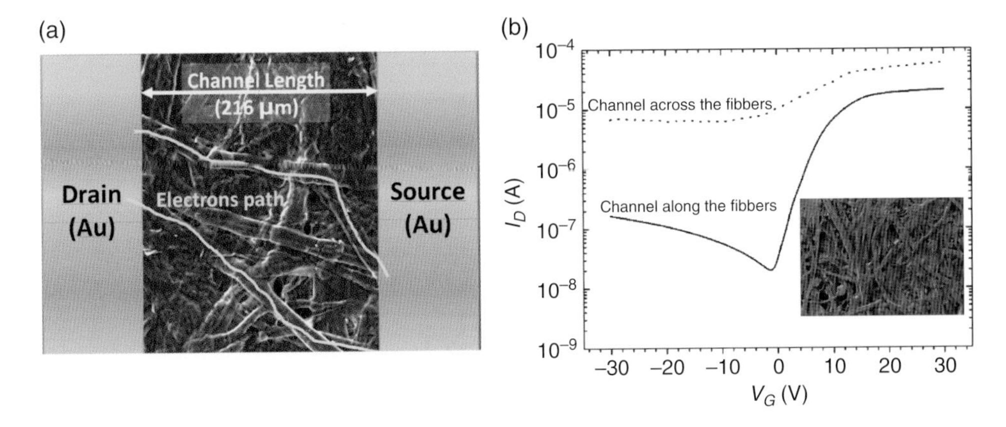

Plate 6.11 *a) A real image of the 2D surface of the paper TFT showing the discrete channel region limited by the continue drain and source metal regions. Within the fibers are sketched two possible carrier pathways between drain and source under a certain V_D, for a fixed V_G applied to the back gate electrode; b) transfer characteristics of paper FETs where the channel was processed along the fibers and across the fibers, in the paper structure shown in the inset, respectively.*

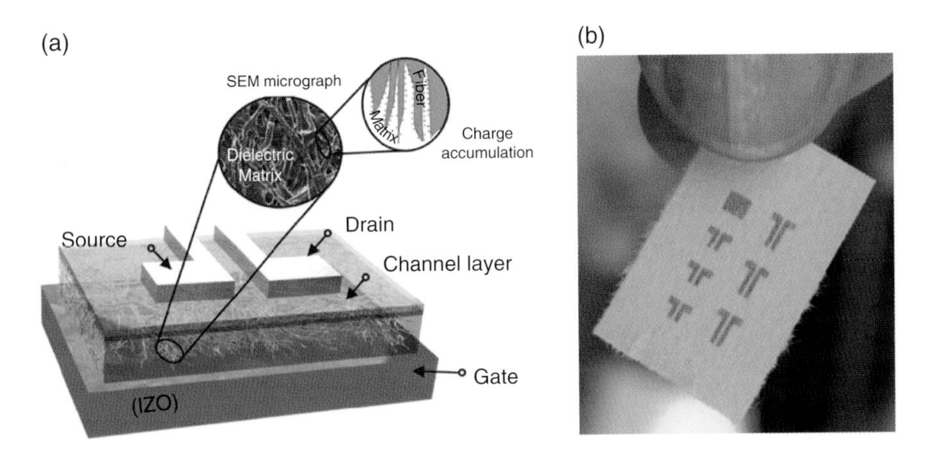

Plate 6.12 *a) Sketch of the distribution of the fibers along the planes, and of how such planes are associated along the dielectric width and its thickness. Reprinted with permission from [27] Copyright (2008) American Institute of Physics.*

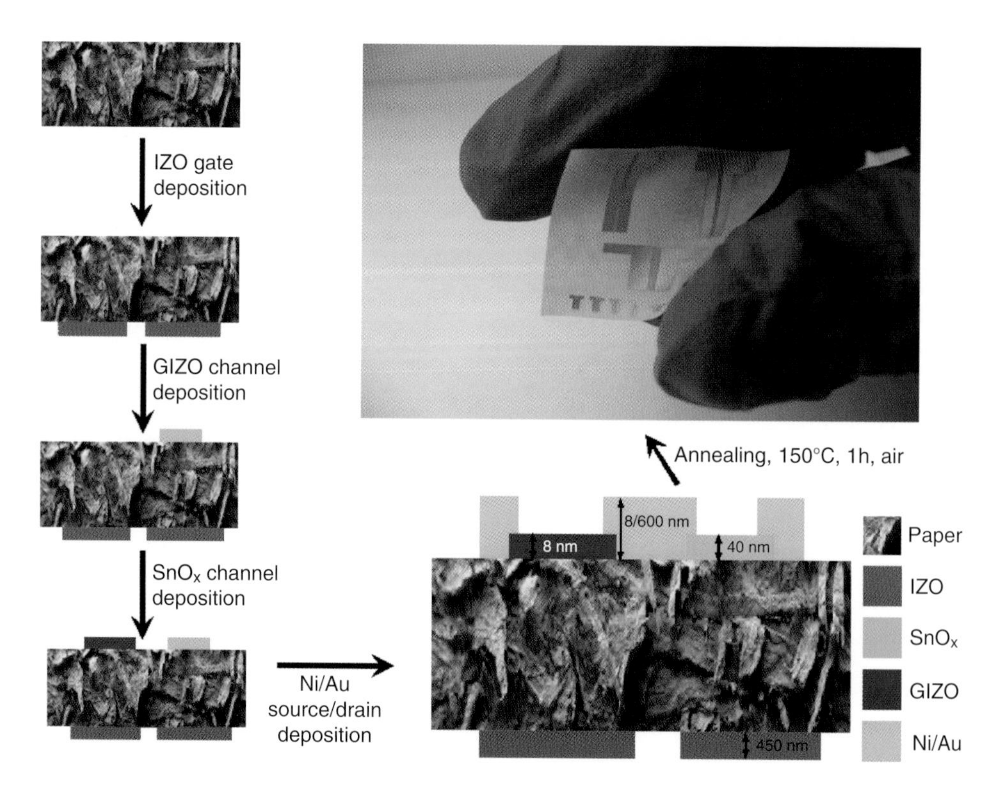

Plate 6.18 *Cross sectional schematic of fabrication sequence of the paper-CMOS showing all layers that constitute the final device and how they are interconnect as well as image of the real device [181]. Reprinted with permission from [181] Copyright (2011) Wiley-VCH.*

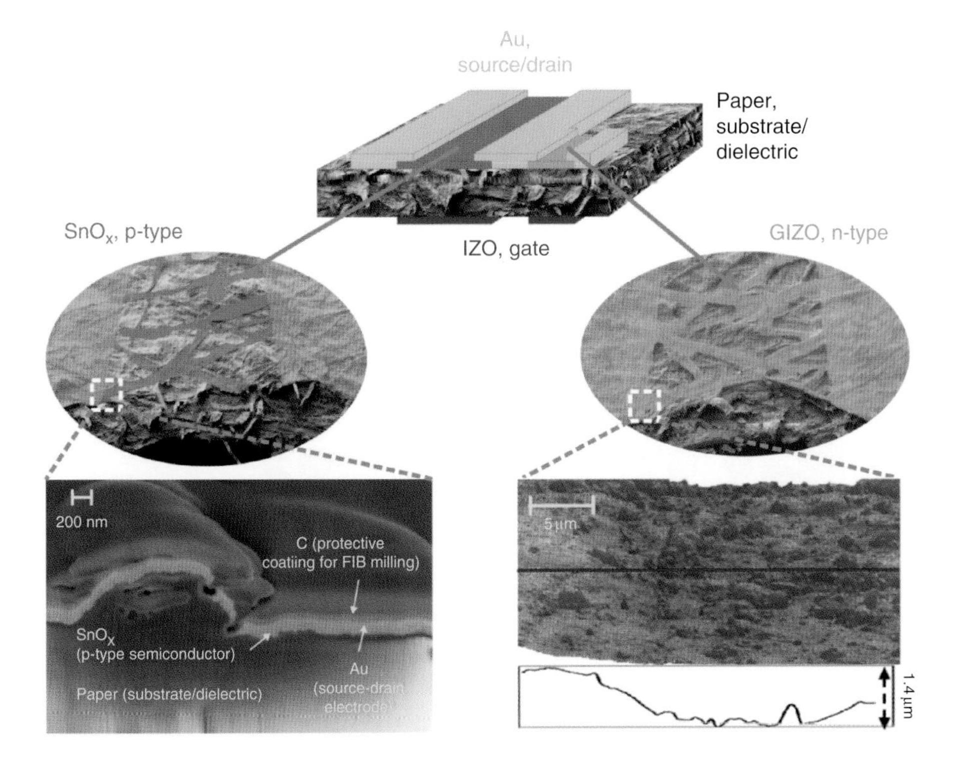

Plate 6.19 *Schematic of the paper-CMOS inverter layered on, and integrated with, paper showing the top and back view of the n-and p-channel FETs whose channels are based on GIZO (40 nm) and SnO_x (8 nm), respectively. The drain and source contacts are based on Ni/Au (8/120 nm thick) films and gate electrodes on the backside are based on highly conductive a-IZO films, with thickness around 450 nm. The cross sectional SEM image of the p-FET along one fiber, shows the carbon protective coating on the device before milling the paper by a focus ion beam, along with the Au on the drain and the SnO_x film. The AFM image on the lower right shows the GIZO surface on the paper substrate.*

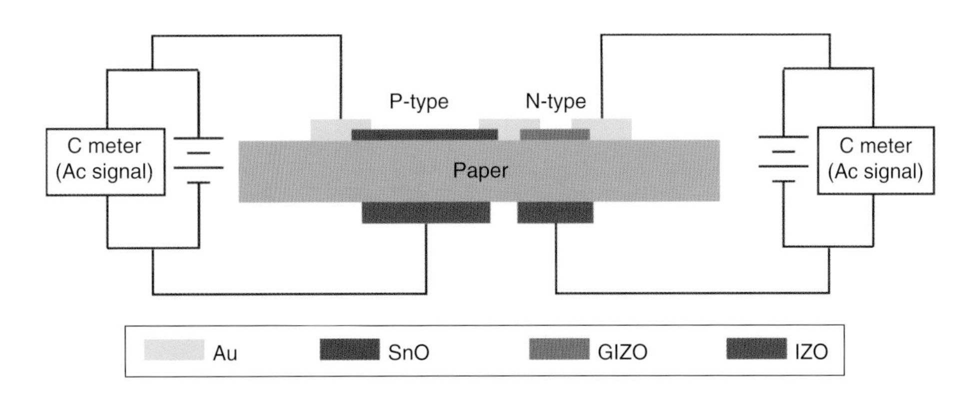

Plate 6.20 *Schematic of the capacitance-voltage measurement.*

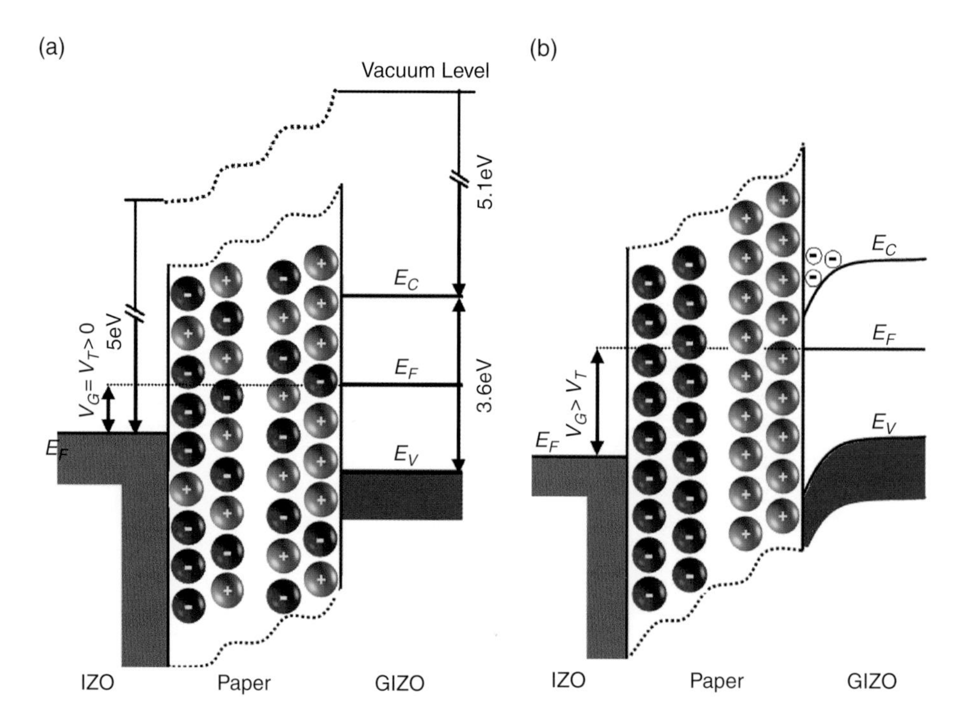

Plate 6.23 *Simplified band diagram of an n-channel FET for (a) $V_G = V_T > 0$ and (b) $V_G > V_T$. Reprinted with permission from [181] Copyright (2011) Wiley-VCH.*

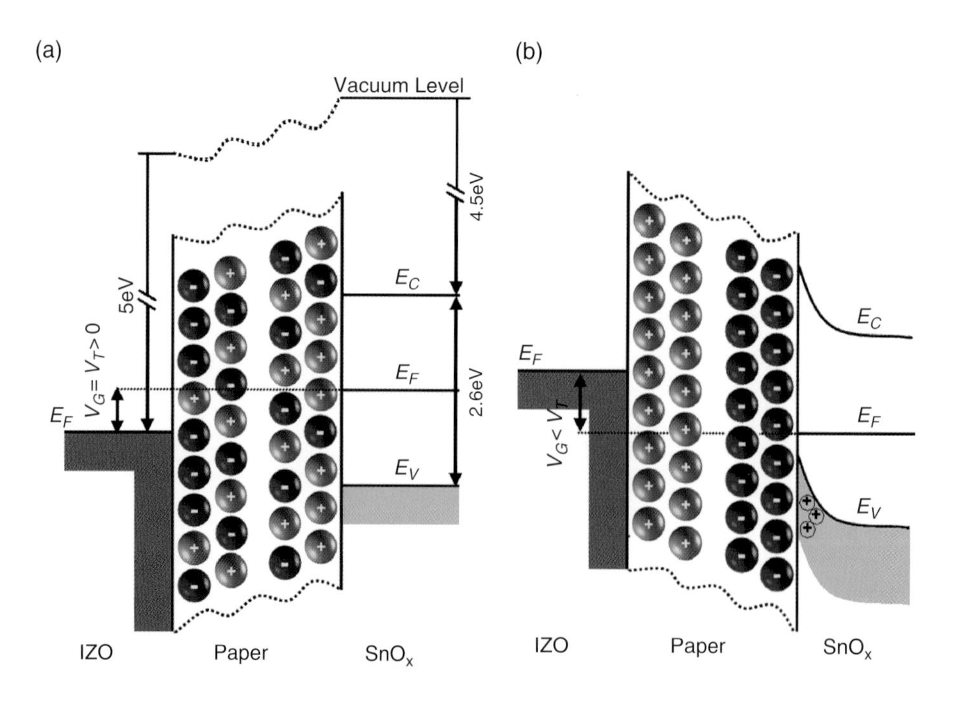

Plate 6.24 *Simplified band diagram of an p-channel FET for (a) $V_G = V_T > 0$ (b) and $|V_G| > V_T$. Reprinted with permission from [181] Copyright (2011) Wiley-VCH.*

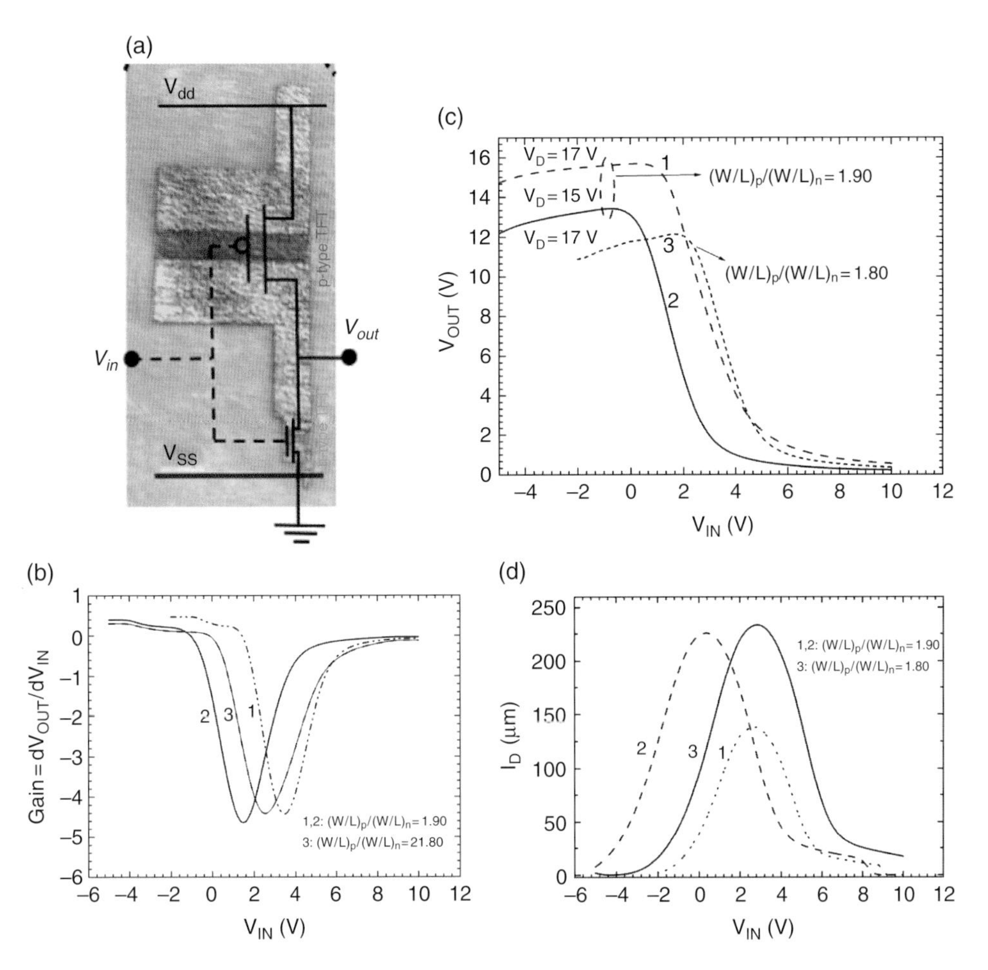

Plate 6.26 *(a) Image of the CMOS on paper where the large $(W/L)_p \approx 19.0$ and small $(W/L)_n \approx 10.0$ correspond to the p-FET and n-FET, respectively; (b) VTC of the CMOS inverter for the different configurations as numbered in Table 6.4. This is used to extract the high and low states, associated with the input and output voltages (V_{IL}, V_{OH} and V_{IH}, V_{OL}). (c) Gain and circuit leakage current, I_D, for different configurations as indicated in Table 6.4. Reprinted with permission from [181] Copyright (2011) Wiley-VCH.*

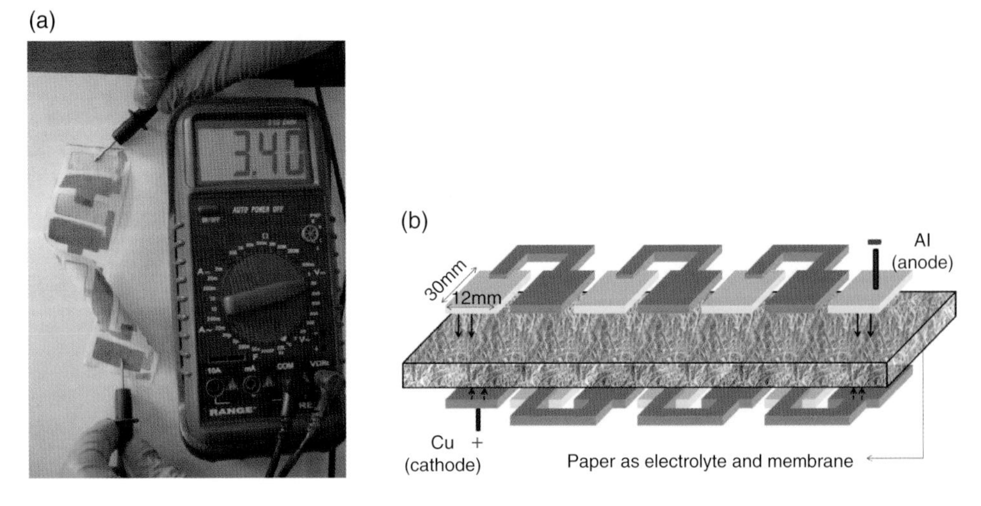

Plate 6.27 *Paper battery (sketch and structure): a) photograph of the paper battery, with seven elements integrated in series, during the measurement of open circuit voltage with a multimeter; b) Schematic of the integrated (in series) paper battery, revealing its constituents, as depicted in the sketch.*

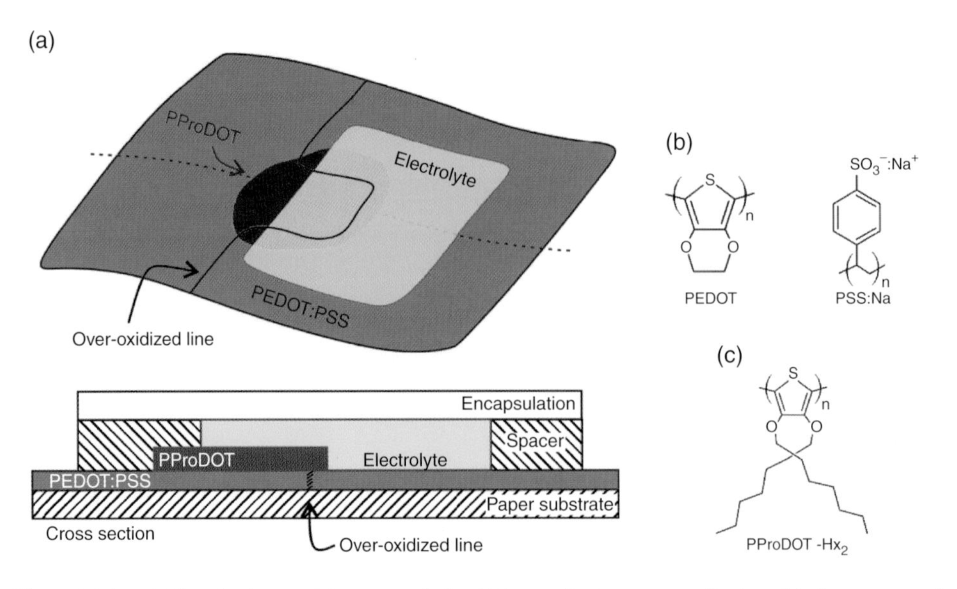

Plate 6.31 *a) The device architecture of the bi-layer device paper display; b) the repeated unit structure; c) PProDOT-Hx2, as adapted from [195]. Reprinted with permission from [195] Copyright (2007) Wiley-VCH.*

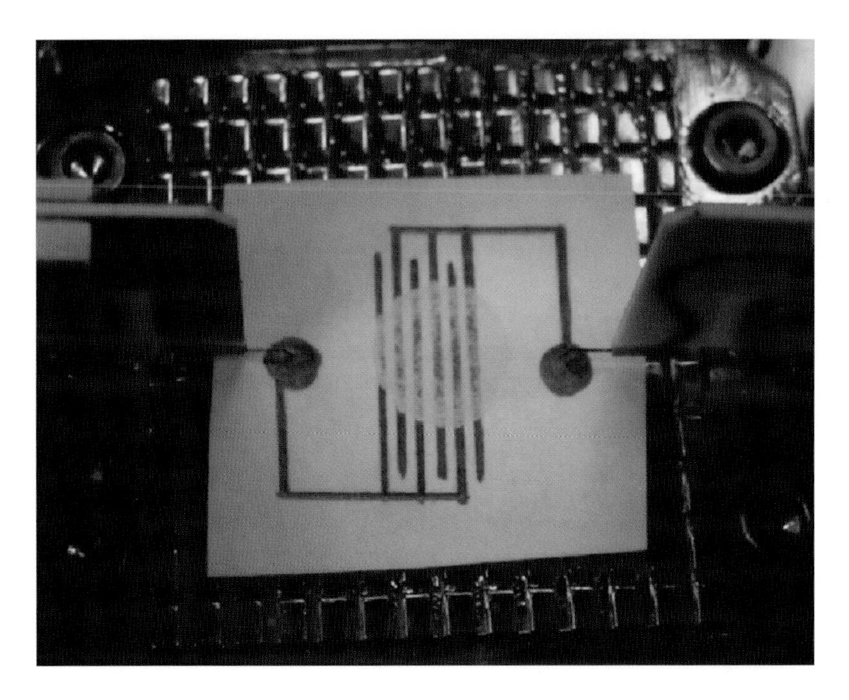

Plate 6.33 *Interdigitized sensor in a paper substrate connected to a probe station [198]. Reprinted with permission from [198] Copyright (2011) American Chemical Society.*

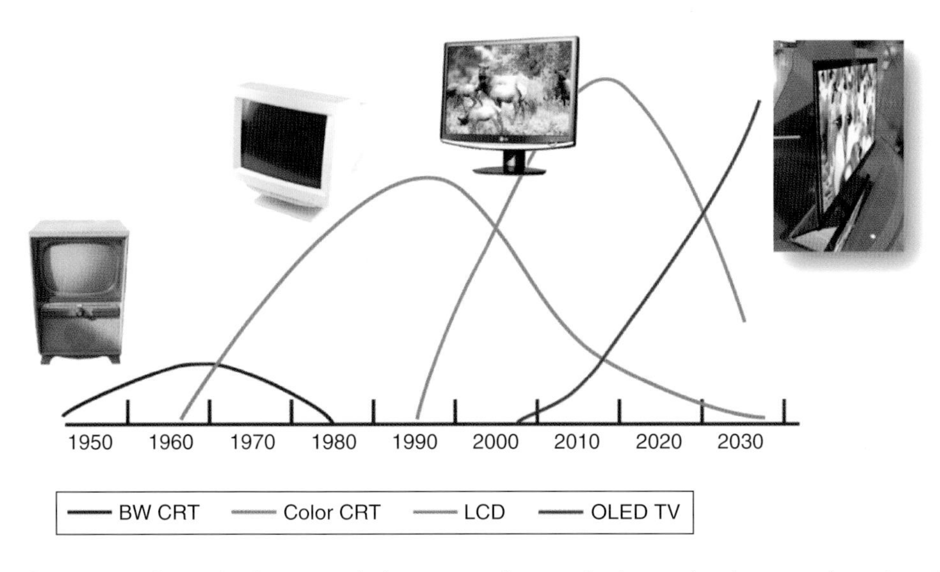

Plate 7.1 *Life cycle forecast of the most relevant display technologies. Adapted with permission from [7] Copyright (2011) DisplayBank.*

Plate 7.3 *Transparent PM chipLED display using IZO electrodes.*

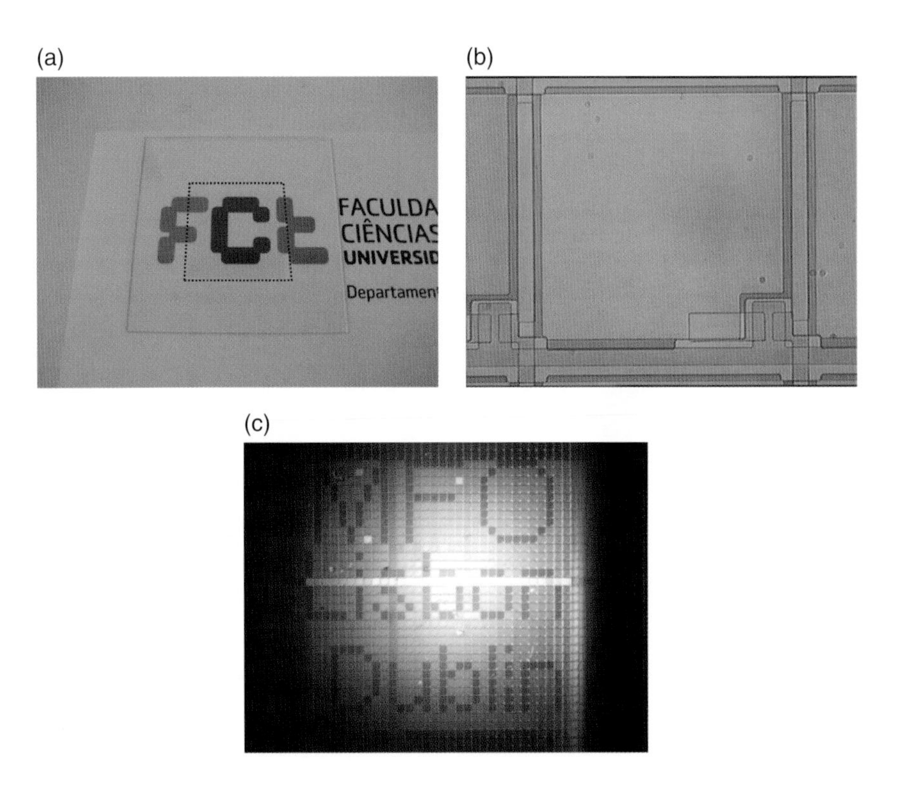

Plate 7.4 *AM backplane with GIZO TFTs produced at CENIMAT for a LCD display: a) photograph of a transparent AM on glass; b) micrograph of pixel area (350×350 μm); c) prototype after integration with LCD frontplane [16]. Reprinted with permission from [16] Copyright (2009) Electrochemical Society.*

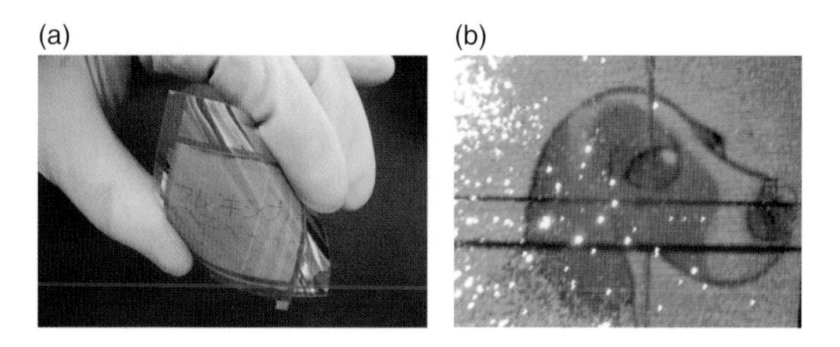

Plate 7.5 *Prototypes of displays with oxide TFT backplanes: a) flexible electrophoretic display. Reprinted with permission from [19] Copyright (2008) Wiley-VCH; b) OLED display driven by solution-processed IZO TFTs [23]. Reprinted from [23] Copyright (2009) Society for Information Display.*

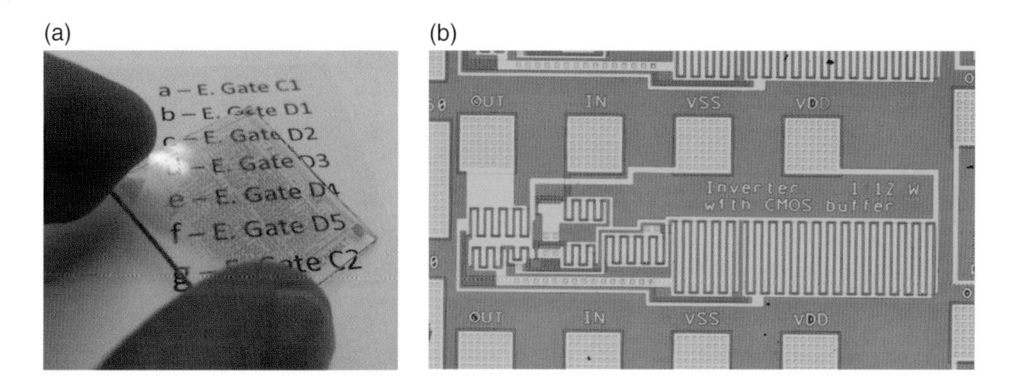

Plate 7.8 *a) Transparent CMOS inverters on glass; b) optical micrograph of an oxide CMOS inverter.*

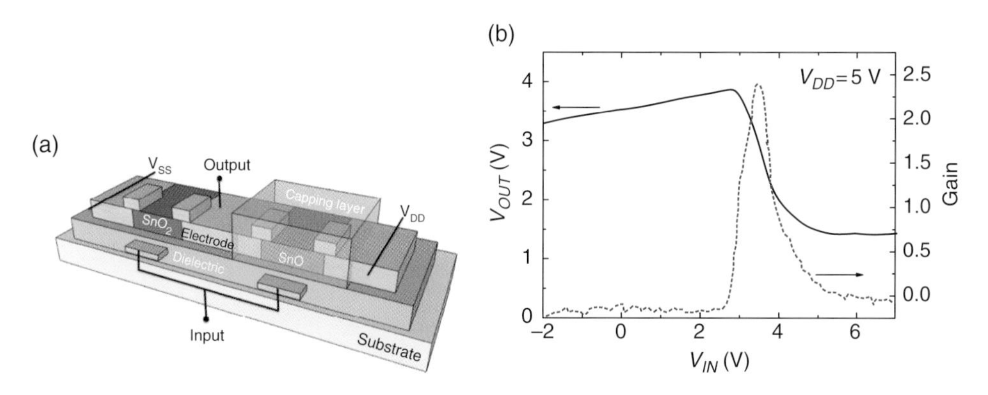

Plate 7.11 *Schematic of a conceptual CMOS inverter using n-type SnO_2 and p-type SnO TFTs (adapted from [34]); b)VTC for an ambipolar SnO CMOS inverter (adapted from [35]). Adapted from [35] Copyright (2011) Wiley-VCH.*

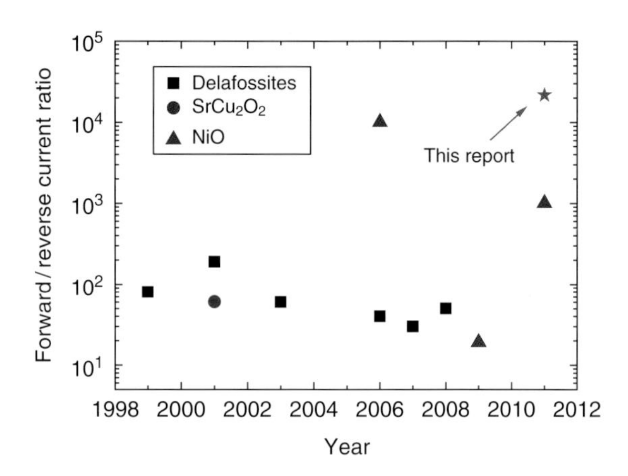

Plate 7.12 *Evolution of the forward/reverse current ratio on thin film all-oxide heterojuntions produced on glass substrates [37–41, 44–48].*

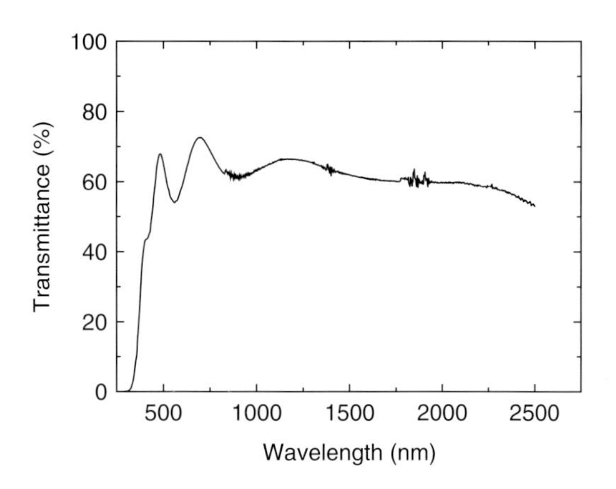

Plate 7.15 *Total transmittance of the stack glass/IZO/GIZO/NiO.*

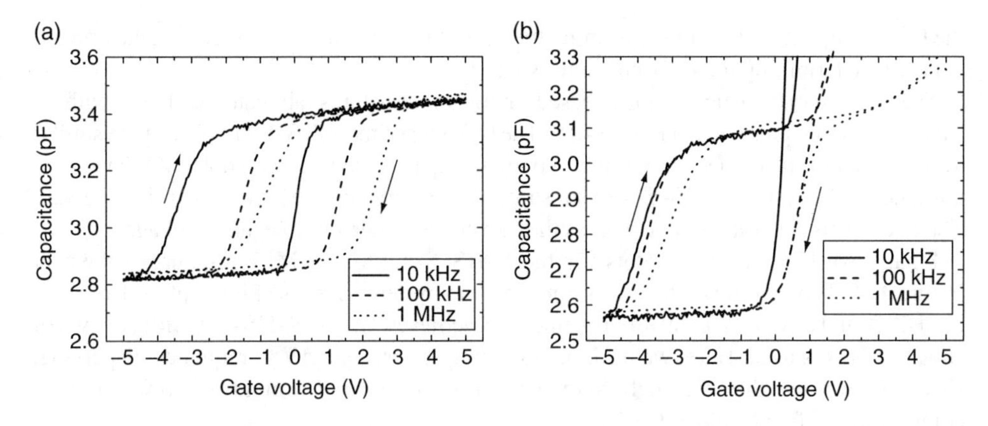

Figure 4.49 *C–V characteristics of GIZO MIS structures integrating: a) S–TS–S multilayer and b) S–TS–S multilayer with P_{bias} of 80 mW cm^{-2}. They were obtained at frequencies between 10 kHz and 1 MHz with an excitation voltage of 100 mV.*

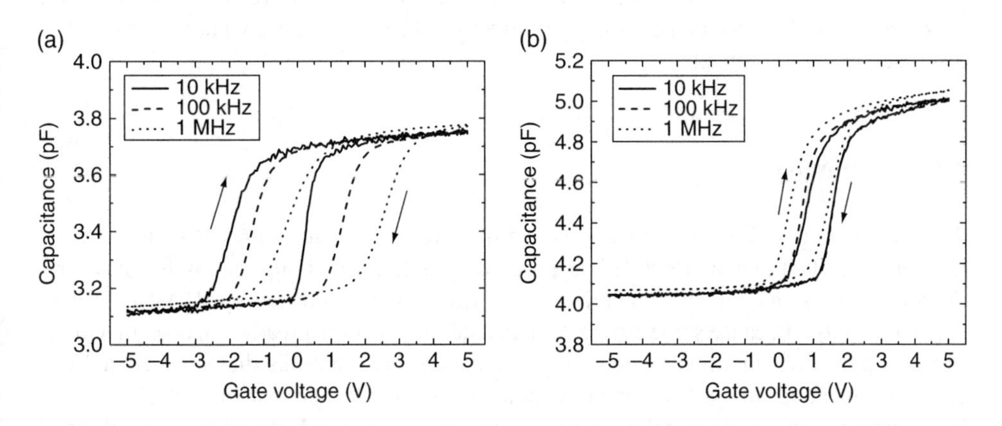

Figure 4.50 *C–V characteristics of GIZO MIS structures integrating: a) S–HS–S multilayer and b) S–HS–S multilayer with P_{bias} of 80 mW cm^{-2}. They were obtained at frequencies between 10 kHz and 1 MHz with an excitation voltage of 100 mV.*

and hysteresis amplitude are reduced for high frequencies meaning that slow states are no longer responding. The hysteresis is clockwise which means that we have electron trapping at or close to the semiconductor/dielectric interface [23].

Looking now at Ta_2O_5 based dielectrics in GIZO MIS structures we observe anomalous features in Ta_2O_5 MIS structures' C–V characteristics for frequencies, below 100 kHz (not shown). It is not possible to compare this with other samples but we must note that the hysteresis amplitude at 1 MHz is lower than on TSiO, being indeed the smallest one observed among the samples analyzed. In TSiO no variation on the accumulation capacitance with frequency was detected but the ΔV_{FB} reaches almost 2 V and moves to more negative voltages at low frequencies. The high frequency curve is centered at around 0 V which means the fixed charge density is low, assuming the small Φ_{ms} as already discussed. So, the trapped charge is determining the C–V characteristics.

The C–V characteristics of Ta_2O_5 based multilayer structures fabricated with and without substrate bias are presented in Figure 4.49. The behavior of the S–TS–S sample is quite similar to that of TSiO. Just a decrease in the maximum capacitance is observed as SiO_2 layers are introduced. Concerning S–TS–S with substrate bias an anomalous behavior is again observed. The fixed charge density will be high as the curves are centered for negative voltages.

Finally we have the C–V plots for the S–HS–S multilayer MIS structures shown in Figure 4.50. When a substrate bias is not used, the hysteresis is also high, not far from that for HSiO and TSiO. A great improvement is achieved in the S–HS–S multilayer when substrate bias is used. Here, we have the lowest hysteresis and ΔV_{FB} frequency dispersion. However, we see that the high frequency curve is centred around 1 V revealing the contribution of fixed charge density.

To complete this brief analysis we can conclude that amorphous dielectrics present better C–V characteristics than polycrystalline ones with low frequency dispersion of capacitance. Despite the good insulating characteristics revealed by multilayer stacks deposited under substrate bias the C–V curves at high frequency are shifted showing the presence of fixed charge density that will affect the TFTs' threshold voltage. This will be discussed in the next chapter.

4.10 Summary

At the beginning of this chapter we proposed sputtered high-κ dielectrics as a valid alternative for oxide TFTs technology. This is particularly interesting in up scaling for mass production as the semiconductor materials are normally also deposited by RFMS. We have considered HfO_2 because some strong background information already exists on this material for application in silicon technology. We have shown that this dielectric crystallizes easily when deposited by RFMS even without any intentional substrate heating. This is not beneficial in terms of electrical properties as leakage current in MIS structures tends to be high. However, we have shown that if substrate bias is used denser films can be obtained and leakage current is reduced. This has proven to be more effective than thermal annealing in polycrystalline films.

In order to achieve amorphous dielectric layers HfO_2 was mixed with SiO_2 and Al_2O_3 by co-sputtering in what we call multicomponent dielectrics. This way we managed also to increase the E_g value, which is relevant as, theoretically, HfO_2 is not much above the edge of 1 eV for band offset with some oxide semiconductors. The composition can be tuned in a limited range (in a practical sense) by adjusting the P_{rf} in each target. When combining substrate bias with co-sputtering good insulating characteristics are achieved. However, the substrate bias must be carefully tuned in order to prevent week SiO_2 or Al_2O_3 incorporation in the growing films as this will result in HfO_2 crystallization.

One other option for depositing multicomponent dielectrics consists in using a single target with a fixed composition. We have also used this approach to deposit films close to a 50/50 mol%, that is, the stoichiometric composition of hafnium silicate ($HfSiO_4$). It was verified experimentally that the resulting sputtered films are slightly Si rich but close to nominal composition. Bias can also be used to improve the density and insulating characteristics of the dielectrics produced. These films remain amorphous after increasing P_{bias}.

In this chapter we have also centered our attention on Ta_2O_5 based dielectrics. Despite the theoretically unfavorable band offset with oxide semiconductors, this material may be interesting since thin films can preserve a stable amorphous structure up to high temperatures. The surface is also very smooth, an interesting characteristic in bottom gate TFTs. We have also tried the approach of "alloying" Ta_2O_5 with SiO_2. This way we were able to increase slightly the gap and reduce the leakage current on MIS structures.

The multilayer approach was also shown. In this particular case we have studied the structural and electrical characteristics of SiO_2 and HfO_2 based dielectrics. The most promising approach is the deposition of a stack of SiO_2/HSiO/SiO_2, that is, amorphous HSiO is placed between two thin SiO_2 layers. This innovative approach of combining multilayers with multicomponent dielectrics for oxides for TFTs allows for low leakage current density (the lowest obtained in the results here presented) with a break down field above 5 MV cm^{-1}.

At the end of this chapter we have shown the C–V characteristics of GIZO MIS structures integrating some of dielectric layers developed. This is of particular interest as it represents the real dielectric/semiconductor interface. We have seen that multilayer dielectrics based on SiO_2 and HfO_2 deposited under substrate bias present better interface characteristics in GIZO MIS structures but may be affected by a fixed charge within the dielectric.

Finally, we must highlight the importance of substrate bias in improving the dielectrics' characteristics. This has proven to be a valid solution for films deposited without any intentional substrate heating and is an effective option for the production of high performance devices on low cost substrates.

References

[1] K. Nomura, H. Ohta, A. Takagi, T. Kamiya, M. Hirano, and H. Hosono (2004) Room-temperature fabrication of transparent flexible thin-film transistors using amorphous oxide semiconductors, *Nature* **432**, 488–92.

[2] R.L. Hoffman, B.J. Norris, and J.F. Wager (2003) ZnO-based transparent thin-film transistors, *Applied Physics Letters* **82**, 733.

[3] E.M.C. Fortunato, P.M.C. Barquinha, A.C.M.B.G. Pimentel, A.M.F. Gonçalves, A.J.S. Marques, R.F.P. Martins, and L.M.N. Pereira (2004) Wide-bandgap high-mobility ZnO thin-film transistors produced at room temperature, *Applied Physics Letters* **85**, 2541.

[4] P.F. Carcia, R.S. McLean, M.H. Reilly, and G. Nunes (2003) Transparent ZnO thin-film transistor fabricated by rf magnetron sputtering, *Applied Physics Letters* **82**, 1117.

[5] J.-S. Park, J.K. Jeong, Y.-G. Mo, H.D. Kim, and S.-I. Kim (2007) Improvements in the device characteristics of amorphous indium gallium zinc oxide thin-film transistors by Ar plasma treatment, *Applied Physics Letters* **90**, 262106.

[6] P. Barquinha, a Pimentel, a Marques, L. Pereira, R. Martins, and E. Fortunato (2006) Influence of the semiconductor thickness on the electrical properties of transparent TFTs based on indium zinc oxide, *Journal of Non-Crystalline Solids* **352**, 1749–1752.

[7] R. Hoffman (2006) Effects of channel stoichiometry and processing temperature on the electrical characteristics of zinc tin oxide thin-film transistors, *Solid-State Electronics* **50**, 784–7.

[8] P. Barquinha, L. Pereira, G. Gonçalves, R. Martins, and E. Fortunato (2008) The Effect of Deposition Conditions and Annealing on the performance of high-mobility GIZO TFTs, *Electrochemical and Solid-State Letters* **11**, H248.

[9] A.R. Merticaru, A.J. Mouthaan, and F.G. Kuper (2006) Current degradation of a-Si:H/SiN TFTs at room temperature and low voltages, *IEEE Transactions on Electron Devices* **53**, 2273–9.

[10] K. Long, A.Z. Kattamis, I.-C. Cheng, H. Gleskova, S. Wagner, and J.C. Sturm (2006) Stability of amorphous-silicon TFTs deposited on clear plastic substrates at 250°C to 280°C, *IEEE Electron Device Letters* **27**, 111–13.

[11] L. Mariucci, G. Fortunato, A. Bonfiglietti, M. Cuscuna, A. Pecora, and A. Valletta (2004) Polysilicon TFT structures for kink-effect suppression, *IEEE Transactions on Electron Devices* **51**, 1135–42.

[12] D.W.M. Zhang (2006) Metal-replaced junction for reducing the junction parasitic resistance of a TFT, *IEEE Electron Device Letters* **27**, 269–71.

[13] P.F. Carcia, R.S. McLean, M.H. Reilly, M.K. Crawford, E.N. Blanchard, A.Z. Kattamis, and S. Wagner (2007) A comparison of zinc oxide thin-film transistors on silicon oxide and silicon nitride gate dielectrics, *Journal of Applied Physics* **102**, 074512.

[14] Y.-L. Wang, L.N. Covert, T.J. Anderson, W. Lim, J. Lin, S.J. Pearton, D.P. Norton, J.M. Zavada, and F. Ren (2008) RF characteristics of room-temperature-deposited, small gate dimension indium zinc oxide TFTs, *Electrochemical and Solid-State Letters* **11**, H60.

[15] W. Lim, S.-H. Kim, Y.-L. Wang, J.W. Lee, D.P. Norton, S.J. Pearton, F. Ren, and I.I. Kravchenko (2008) Stable room temperature deposited amorphous $InGaZnO_4$ thin film transistors, *Journal of Vacuum Science & Technology B: Microelectronics and Nanometer Structures* **26**, 959.

[16] R.M. Wallace and G. Wilk (2002) Alternative gate dielectrics for microelectronics, *MRS Bulletin* **27**, 186–91.

[17] M. Hirose, M. Koh, W. Mizubayashi, H. Murakami, K. Shibahara, and S. Miyazaki (2000) Fundamental limit of gate oxide thickness scaling in advanced MOSFETs, *Semiconductor Science and Technology*, **15**, 485–90.

[18] G.D. Wilk, R.M. Wallace, and J.M. Anthony (2001) High-κ gate dielectrics: Current status and materials properties considerations, *Journal of Applied Physics* **89**, 5243.

[19] N.R. Mohapatra, M.P. Desai, S.G. Narendra, and V.R. Rao (2002) The effect of high-κ gate dielectrics on deep submicrometer CMOS device and circuit performance, *IEEE Transactions on Electron Devices* **49**, 826–31.

[20] J. Robertson (2002) Band offsets of high dielectric constant gate oxides on silicon, *Journal of Non-Crystalline Solids* **303**, 94–100.

[21] R. Degraeve, E. Cartier, T. Kauerauf, R. Carter, L. Pantisano, A. Kerber, and G. Groeseneken (2002) On the electrical characterization of high- κ dielectrics, *MRS Bulletin* **27**, 222–5.

[22] J. Robertson (2002) Electronic structure and band offsets of high-dielectric-constant gate oxides, *MRS Bulletin* **27**, 217–21.

[23] J.F. Wager, D.A. Keszler, and R.E. Presley (2008) *Transparent Electronics*, Boston, MA, Springer US.

[24] J. Robertson, and B. Falabretti (2006) Band offsets of high-κ gate oxides on high mobility semiconductors, *Materials Science and Engineering: B* **135**, 267–71.

[25] J. Robertson, K. Xiong, and S. Clark (2006) Band gaps and defect levels in functional oxides, *Thin Solid Films* **496**, 1–7.

[26] P. King, R. Lichti, Y. Celebi, J. Gil, R. Vilão, H. Alberto, J. Piroto Duarte, D. Payne, R. Egdell, I. McKenzie, C. McConville, S. Cox and T. Veal (2009) Shallow donor state of hydrogen in In_2O_3 and SnO_2: Implications for conductivity in transparent conducting oxides, *Physical Review B* **80**, 1–4.

[27] Y. Ogo, H. Hiramatsu, K. Nomura, H. Yanagi, T. Kamiya, M. Hirano, and H. Hosono (2008) p-channel thin-film transistor using p-type oxide semiconductor, SnO, *Applied Physics Letters* **93**, 032113.

[28] E. Fortunato, R. Barros, P. Barquinha, V. Figueiredo, S.-H.K. Park, C.-S. Hwang, and R. Martins (2010) Transparent p-type SnO thin film transistors produced by reactive rf magnetron sputtering followed by low temperature annealing, *Applied Physics Letters* **97**, 052105.

[29] E. Fortunato, V. Figueiredo, P. Barquinha, E. Elamurugu, R. Barros, G. Gonçalves, S.-H.K. Park, C.-S. Hwang, and R. Martins (2010) Thin-film transistors based on p-type Cu_2O thin films produced at room temperature, *Applied Physics Letters* **96**, 192102.

[30] X.Z. X. Zou, G. Fang, L. Yuan, M. Li, and W. Guan (2010) Top-gate low-threshold voltage p-Cu2O thin-film transistor grown SiO2/Si substrate using a high-κ HfON gate dielectric, *IEEE Electron Device Letters* **31**, 827–9.

[31] H.J.H. Chen, B.B. Yeh, and W.-Y. Chou (2008) Low temperature post-annealing of ZnO thin-film transistors with high-k gate dielectrics, *ECS Transactions, ECS*, 315–22.
[32] J. Kwo (2003) Advances in high κ gate dielectrics for Si and III–V semiconductors, *Journal of Crystal Growth* **251**, 645–50.
[33] Z. Wang, T. Kumagai, H. Kokawa, J. Tsuaur, M. Ichiki, and R. Maeda (2005) Crystalline phases, microstructures and electrical properties of hafnium oxide films deposited by sol–gel method, *Journal of Crystal Growth* **281**, 452–7.
[34] Y.-S. Lin and R. Puthenkovilakam, J.P. Chang (2002) Dielectric property and thermal stability of HfO2 on silicon, *Applied Physics Letters* **81**, 2041–3.
[35] M. Cho, J. Park, H.B. Park, C.S. Hwang, J. Jeong, and K.S. Hyun (2002) Chemical interaction between atomic-layer-deposited HfO2 thin films and the Si substrate, *Applied Physics Letters* **81**, 334.
[36] J.-B. Cheng, A.-D. Li, Q.-Y. Shao, H.-Q. Ling, D. Wu, Y. Wang, Y.-J. Bao, M. Wang, Z.-G. Liu, N.-B. Ming (2004) Growth and characteristics of La_2O_3 gate dielectric prepared by low pressure metalorganic chemical vapor deposition, *Applied Surface Science* **233**, 91–8.
[37] K. Frohlich, R. Luptak, E. Dobrocka, K. Husekova, K. Cico, A. Rosova, M. Lukosius, A. Abrutis, P. Pisecny, and J.P. Espinos (2006) Characterization of rare earth oxides based MOSFET gate stacks prepared by metal-organic chemical vapour deposition, *Materials Science in Semiconductor Processing* **9**, 1065–72.
[38] Y. Ohshita, A. Ogura, M. Ishikawa, T. Kada, A. Hoshino, T. Suzuki, H. Machida, and K. Soai (2006) HfO_2 and $Hf_{1-x}Si_xO_2$ thin films grown by metal-organic CVD using Tetrakis(diethylamido) hafnium, *Chemical Vapor Deposition*, **12**, 130–5.
[39] J.-S.M. Lehn, S. Javed, and D.M. Hoffman (2006) New precursors for the CVD of zirconium and hafnium oxide films, *Chemical Vapor Deposition* **12**, 280–4.
[40] S. Abermann, G. Pozzovivo, J. Kuzmik, G. Strasser, D. Pogany, J.-F. Carlin, N. Grandjean, and E. Bertagnolli (2007) MOCVD of HfO_2 and ZrO_2 high- k gate dielectrics for InAlN/AlN/GaN MOS-HEMTs, *Semiconductor Science and Technology* **22**, 1272–5.
[41] P.F. Carcia, R.S. McLean, and M.H. Reilly (2006) High-performance ZnO thin-film transistors on gate dielectrics grown by atomic layer deposition, *Applied Physics Letters* **88**, 123509.
[42] A.C. Jones, H.C. Aspinall, P.R. Chalker, R.J. Potter, K. Kukli, A. Rahtu, M. Ritala, and M. Leskel (2004) Some recent developments in the MOCVD and ALD of high-k dielectric oxides, *Journal of Materials Chemistry* **14**, 3101.
[43] A. Callegari, E. Cartier, M. Gribelyuk, H.F. Okorn-Schmidt, and T. Zabel (2001) Physical and electrical characterization of Hafnium oxide and Hafnium silicate sputtered films, *Journal of Applied Physics* **90**, 6466.
[44] E. Atanassova and A. Paskaleva (2002) Breakdown fields and conduction mechanisms in thin Ta_2O_5 layers on Si for high density DRAMs, *Microelectronics Reliability* **42**, 157–73.
[45] S. Xing (2003) Preparation of hafnium oxide thin film by electron beam evaporation of hafnium incorporating a post thermal process, *Microelectronic Engineering* **66**, 451–6.
[46] N.A. Chowdhury, R. Garg, and D. Misra (2004) Charge trapping and interface characteristics of thermally evaporated HfO_2, *Applied Physics Letters* **85**, 3289.
[47] Y.Y. Lebedinskii, A. Zenkevich, E.P. Gusev, and M. Gribelyuk (2005) In situ investigation of growth and thermal stability of ultrathin Si layers on the HfO_2/Si (100) high-κ dielectric system, *Applied Physics Letters* **86**, 191904.
[48] K. Honda, A. Sakai, M. Sakashita, H. Ikeda, S. Zaima, and Y. Yasuda (2004) Pulsed Laser Deposition and Analysis for Structural and Electrical Properties of HfO_2–TiO_2 Composite Films, *Japanese Journal of Applied Physics* **43**, 1571–6.
[49] P.D. Kirsch, C.S. Kang, J. Lozano, J.C. Lee, and J.G. Ekerdt (2002) Electrical and spectroscopic comparison of HfO_2/Si interfaces on nitrided and un-nitrided Si(100), *Journal of Applied Physics* **91**, 4353.
[50] B.-Y. Tsui and H.-W. Chang (2003) Formation of interfacial layer during reactive sputtering of hafnium oxide, *Journal of Applied Physics* **93**, 10119.
[51] L. Pereira, A. Marques, H. Aguas, N. Nedev, S. Georgiev, E. Fortunato, and R. Martins (2004) Performances of hafnium oxide produced by radio frequency sputtering for gate dielectric application, *Materials Science and Engineering B* **109**, 89–93.

[52] H. Grüger (2004) High quality r.f. sputtered metal oxides (Ta_2O_5, HfO_2) and their properties after annealing, *Thin Solid Films* **447–448**, 509–15.
[53] C. Chaneliere, J.L. Autran, R.A.B. Devine, and B. Balland (1998) Tantalum pentoxide (Ta_2O_5) thin films for advanced dielectric applications, *Materials Science and Engineering: R: Reports* **22**, 269–322.
[54] W.-J. Qi, R. Nieh, B.H. Lee, L. Kang, Y. Jeon, and J.C. Lee (2000) Electrical and reliability characteristics of ZrO_2 deposited directly on Si for gate dielectric application, *Applied Physics Letters* **77**, 3269.
[55] W.K. Chim, T.H. Ng, B.H. Koh, W.K. Choi, J.X. Zheng, C.H. Tung, A.Y. Du (2003) Interfacial and bulk properties of zirconium dioxide as a gate dielectric in metal–insulator–semiconductor structures and current transport mechanisms, *Journal of Applied Physics* **93**, 4788.
[56] T.-M. Pan, C.-L. Chen, W.-W. Yeh, and W.-J. Lai (2007) Physical and electrical properties of lanthanum oxide dielectrics with Al and Al/TaN metal gates, *Electrochemical and Solid-State Letters* **10**, H101.
[57] E.K. Evangelou, C. Wiemer, M. Fanciulli, M. Sethu, and W. Cranton (2003) Electrical and structural characteristics of yttrium oxide films deposited by rf-magnetron sputtering on n-Si, *Journal of Applied Physics* **94**, 318.
[58] T.-M. Pan and J.-D. Lee (2007) Physical and electrical properties of yttrium oxide gate dielectrics on si substrate with NH_3 plasma treatment, *Journal of The Electrochemical Society* **154**, H698.
[59] H. Yabuta, M. Sano, K. Abe, T. Aiba, T. Den, H. Kumomi, K. Nomura, T. Kamiya, and H. Hosono (2006) High-mobility thin-film transistor with amorphous $InGaZnO_4$ channel fabricated by room temperature rf-magnetron sputtering, *Applied Physics Letters* **89**, 112123.
[60] H. Kumomi, K. Nomura, T. Kamiya, and H. Hosono (2008) Amorphous oxide channel TFTs, *Thin Solid Films* **516**, 1516–22.
[61] J.S. Lee, S. Chang, S.-M. Koo, and S.Y. Lee (2010) High-performance a-IGZO TFT with ZrO_2 gate dielectric fabricated at room temperature, *IEEE Electron Device Letters* **31**, 225–7.
[62] W. Lim, S. Kim, Y.-L. Wang, J.W. Lee, D.P. Norton, S.J. Pearton, F. Ren, and I.I. Kravchenko (2008) High-performance indium gallium zinc oxide transparent thin-film transistors fabricated by radio-frequency sputtering, *Journal of The Electrochemical Society*, **155**, H383.
[63] H.J.H. Chen, B.B.L. Yeh, H.-C. Pan, and J.-S. Chen (2008) ZnO transparent thin-film transistors with HfO_2/Ta_2O_5 stacking gate dielectrics, *Electronics Letters* **44**, 186.
[64] L. Yuan, X. Zou, G. Fang, J. Wan, H. Zhou, and X. Zhao (2011) High-performance amorphous indium gallium zinc oxide thin-film transistors $HfO_xN_y/HfO_2/HfO_xN_y$ tristack gate dielectrics, *IEEE Electron Device Letters* **32**, 42–4.
[65] Q. Yao and D. Li (2005) Fabrication and property study of thin film transistor using rf sputtered ZnO as channel layer, *Journal of Non-Crystalline Solids* **351**, 3191–4.
[66] P. Barquinha, L. Pereira, G. Gonçalves, D. Kuscer, M. Kosec, A. Vilà, A. Olziersky, J.R. Morante, R. Martins and E. Fortunato (2010) Low-temperature sputtered mixtures of high-κ and high bandgap dielectrics for GIZO TFTs, *Journal of the Society for Information Display* **18**, 762.
[67] J. Kim, B. Ahn, C. Lee, K. Jeon, H. Kang, and S. Lee (2008) Characteristics of transparent ZnO based thin film transistors with amorphous HfO_2 gate insulators and Ga doped ZnO electrodes, *Thin Solid Films* **516**, 1529–32.
[68] C.J. Chiu, S.P. Chang, and S.J. Chang (2010) High-performance a-IGZO thin-film transistor usingTa_2O_5 gate dielectric, *IEEE Electron Device Letters* **31**, 1245–7.
[69] Y.T. Hou, M.F. Li, H.Y. Yu, and D.-L. Kwong (2003) Modeling of tunneling currents through HfO_2 and $(HfO_2)_x(Al_2O_3)_{1-x}$ gate stacks, *IEEE Electron Device Letters* **24**, 96–8.
[70] H.D. Kim, Y. Roh, Y. Lee, J.E. Lee, D. Jung, and N.-E. Lee (2004) Effects of annealing temperature on the characteristics of $HfSi_xO_y/HfO_2$ high-k gate oxides, *Journal of Vacuum Science & Technology A: Vacuum, Surfaces, and Films* **22**, 1347.
[71] D. a Neumayer and E. Cartier (2001) Materials characterization of ZrO_2–SiO_2 and HfO_2–SiO_2 binary oxides deposited by chemical solution deposition, *Journal of Applied Physics* **90**, 1801.
[72] K.J. Hubbard and D.G. Schlom (1996) Thermodynamic stability of binary oxides in contact with silicon, *Journal of Materials Research* **11**, 2757–76.
[73] D.G. Schlom and J.H. Haeni (2002) A thermodynamic approach to selecting alternative gate dielectrics, *MRS Bulletin* **27**, 198–204.

[74] V. Misra, G. Lucovsky, and G. Parsons (2002) Issues in high-κ gate stack interfaces, *MRS Bulletin*, **27** 212–16.
[75] B.K. Park, J. Park, M. Cho, C.S. Hwang, K. Oh, Y. Han, and D.Y. Yang (2002) Interfacial reaction between chemically vapor-deposited HfO_2 thin films and a HF-cleaned Si substrate during film growth and postannealing, *Applied Physics Letters* **80**, 2368.
[76] S.-G. Lim, S. Kriventsov, T.N. Jackson, J.H. Haeni, D.G. Schlom, A.M. Balbashov, R. Uecker, P. Reiche, J.L. Freeour, and G. Lucovsky (2002) Dielectric functions and optical bandgaps of high-K dielectrics for metal-oxide-semiconductor field-effect transistors by far ultraviolet spectroscopic ellipsometry, *Journal of Applied Physics* **91**, 4500.
[77] L. Pereira, P. Barquinha, E. Fortunato, R. Martins, D. Kang, C. Kim, H. Lim, I. Song, and Y. Park (2008) High k dielectrics for low temperature electronics, *Thin Solid Films* **516**, 1544–8.
[78] L. Pereira, P. Barquinha, G. Gonçalves, A. Vilà, A. Olziersky, J. Morante, E. Fortunato, and R. Martins (2009) Sputtered multicomponent amorphous dielectrics for transparent electronics, *Physica Status Solidi (a)* **206**, 2149–54.
[79] J.J. Gand (2007) Optical properties and structure of HfO_2 thin films grown by high pressure reactive sputtering, *Journal of Physics D: Applied Physics* **40**, 5256–65.
[80] M. Frank, S. Sayan, S. Dormann, T. Emge, L. Wielunski, E. Garfunkel, Y. Chabal (2004) Hafnium oxide gate dielectrics grown from an alkoxide precursor: structure and defects, *Materials Science and Engineering B* **109**, 6–10.
[81] H. Kim, P.C. McIntyre, and K.C. Saraswat (2003) Effects of crystallization on the electrical properties of ultrathin HfO_2 dielectrics grown by atomic layer deposition, *Applied Physics Letters* **82**, 106.
[82] J. Sundqvist (2003) Atomic layer deposition of polycrystalline HfO2 films by the HfI_4–O_2 precursor combination, *Thin Solid Films* **427**, 147–51.
[83] N.V. Nguyen, A.V. Davydov, D. Chandler-Horowitz, and M.M. Frank (2005) Sub-bandgap defect states in polycrystalline hafnium oxide and their suppression by admixture of silicon, *Applied Physics Letters* **87**, 192903.
[84] L. Pereira, P. Barquinha, E. Fortunato, and R. Martins (2005) Influence of the oxygen/argon ratio on the properties of sputtered hafnium oxide, *Materials Science and Engineering B* **118**, 210–13.
[85] F.L. Martínez, M. Toledano-Luque, J.J. Gandía, J. Cárabe, W. Bohne, J. Röhrich, *et al.* (2007) Optical properties and structure of HfO_2 thin films grown by high pressure reactive sputtering, *Journal of Physics D: Applied Physics* **40**, 5256–65.
[86] J. Leger, a Atouf, P. Tomaszewski, and A. Pereira (1993) Pressure-induced phase transitions and volume changes in HfO_2 up to 50 GPa., *Physical Review. B, Condensed Matter* **48**, 93–8.
[87] A. Jayaraman, S. Wang, S. Sharma, and L. Ming (1993) Pressure-induced phase transformations in HfO_2 to 50 GPa studied by Raman spectroscopy, *Physical Review. B, Condensed Matter* **48**, 9205–11.
[88] H. Wang, Y. Wang, J. Feng, C. Ye, B.Y. Wang, H.B. Wang, Q. Li, Y. Jiang, A.P. Huang, and Z.S. Xiao (2008) Structure and electrical properties of HfO_2 high-k films prepared by pulsed laser deposition on Si (100), *Applied Physics A* **93**, 681–4.
[89] S. Nam (2002) Influence of annealing condition on the properties of sputtered hafnium oxide, *Journal of Non-Crystalline Solids* **303**, 139–43.
[90] J.C. Moreno-Marín, I. Abril, and R. Garcia-Molina (1999) Radial profile of energetic particles bombarding the substrate in a glow discharge, *Journal of Vacuum Science & Technology A: Vacuum, Surfaces, and Films* **17**, 528.
[91] V. Tvarozek, I. Novotny, P. Sutta, S. Flickyngerova, K. Schtereva, and E. Varinsky (2007) Influence of sputtering parameters on crystalline structure of ZnO thin films, *Thin Solid Films* **515**, 8756–60.
[92] M.C. Cheynet, S. Pokrant, F.D. Tichelaar, and J.-L. Rouvière (2007) Crystal structure and band gap determination of HfO_2 thin films, *Journal of Applied Physics* **101**, 054101.
[93] J. Price, P.Y. Hung, T. Rhoad, B. Foran, and A.C. Diebold (2004) Spectroscopic ellipsometry characterization of $Hf_xSi_yO_z$ films using the Cody–Lorentz parameterized model, *Applied Physics Letters* **85**, 1701.
[94] Y.J. Cho, N.V. Nguyen, C. a Richter, J.R. Ehrstein, B.H. Lee, and J.C. Lee (2002) Spectroscopic ellipsometry characterization of high-k dielectric HfO_2 thin films and the high-temperature annealing effects on their optical properties, *Applied Physics Letters* **80**, 1249.

[95] G. He, L.D. Zhang, G.H. Li, M. Liu, L.Q. Zhu, S.S. Pan, and Q. Fang (2005) Spectroscopic ellipsometry characterization of nitrogen-incorporated HfO_2 gate dielectrics grown by radio-frequency reactive sputtering, *Applied Physics Letters* **86**, 232901.
[96] D. Bhattacharyya and A. Biswas (2005) Spectroscopic ellipsometric study on dispersion of optical constants of Gd_2O_3 films, *Journal of Applied Physics* **97**, 053501.
[97] D. Bhattacharyya (2000) Spectroscopic ellipsometry of TiO_2 layers prepared by ion-assisted electron-beam evaporation, *Thin Solid Films* **360**, 96–102.
[98] K. Postava, M. Aoyama, T. Yamaguchi, and H. Oda (2001) Spectroellipsometric characterization of materials for multilayer coatings, *Applied Surface Science* **175–176**, 276–80.
[99] G.E. Jellison and F.A. Modine (1996) Parameterization of the optical functions of amorphous materials in the interband region, *Applied Physics Letters* **69**, 371.
[100] H. Chen and W.Z. Shen (2005) Perspectives in the characteristics and applications of Tauc-Lorentz dielectric function model, *The European Physical Journal B* **43**, 503–7.
[101] H. Takeuchi and D. Ha, T.-J. King (2004) Observation of bulk HfO_2 defects by spectroscopic ellipsometry, *Journal of Vacuum Science & Technology A: Vacuum, Surfaces, and Films* **22**, 1337.
[102] L. Pereira, P. Barquinha, E. Fortunato, and R. Martins (2006) Low temperature processed hafnium oxide: structural and electrical properties, *Materials Science in Semiconductor Processing* **9**, 1125–32.
[103] J.R. Yeargan (1968) The Poole-Frenkel effect with compensation present, *Journal of Applied Physics* **39**, 5600.
[104] C. Chaneliere, S. Four, J.L. Autran, R.A.B. Devine, and N.P. Sandler (1998) Properties of amorphous and crystalline Ta_2O_5 thin films deposited on Si from a $Ta(OC_2H_5)_5$ precursor, *Journal of Applied Physics* **83**, 4823.
[105] H. Hirose and Y. Wada (1964) Dielectric properties and DC conductivity of vacuum-deposited SiO films, *Japanese Journal of Applied Physics* **3**, 179–90.
[106] T.E. Hartman (1966) Electrical conduction through SiO films, *Journal of Applied Physics* **37**, 2468.
[107] J. Simmons (1967) Poole-Frenkel effect and Schottky effect in metal-insulator-metal systems, *Physical Review*, **155**, 657–60.
[108] Y. Yang (2004) High-temperature phase stability of hafnium aluminate films for alternative gate dielectrics, *Journal of Applied Physics* **95**, 3772.
[109] L. Feng, Z. Liu, and Y. Shen (2009) Compositional, structural and electronic characteristics of HfO_2 and HfSiO dielectrics prepared by radio frequency magnetron sputtering, *Vacuum* **83**, 902–5.
[110] S.V. Ushakov, A. Navrotsky, Y. Yang, S. Stemmer, K. Kukli, M. Ritala, M. A. Leskelä, P. Fejes, A. Demkov, C. Wang, B.-Y. Nguyen, D. Triyoso, and P. Tobin (2004) Crystallization in hafnia- and zirconia-based systems, *Physical Status Solidi (B)* **241**, 2268–78.
[111] Z.L. Pei, L. Pereira, G. Goncalves, P. Barquinha, N. Franco, E. Alves, A.M.B. Rego, R. Martins, and E. Fortunato (2009) Room-Temperature Cosputtered HfO_2-Al_2O_3 Multicomponent Gate Dielectrics, *Electrochem. Solid State Lett.* **12**, G65–G68.
[112] a Callegari, E. Cartier, M. Gribelyuk, H.F. Okorn-Schmidt, and T. Zabel (2001) Physical and electrical characterization of Hafnium oxide and Hafnium silicate sputtered films, *Journal of Applied Physics* **90**, 6466.
[113] A. Navrotsky and S.V. Ushakov (2005) Thermodynamics of oxide systems relevant to alternative gate dielectrics, in: A.A.D. and A. Navrotsky (eds), *Materials Fundamentals of Gate Dielectrics*, Springer, pp. 57–108.
[114] T.Y. Chang, X. Wang, D.A. Evans, S.L. Robinson, and J.P. Zheng (2002) Tantalum oxide–ruthenium oxide hybrid capacitors, *Journal of Power Sources* **110**, 138–43.
[115] C. Liu, S. Chang, J. Chen, S. Chen, J. Lee, and U. Liaw (2004) High-quality ultrathin chemical-vapor-deposited Ta_2O_5 capacitors prepared by high-density plasma annealing, *Materials Science and Engineering B* **106**, 234–41.
[116] T. Dimitrova (2001) Crystallization effects in oxygen annealed Ta_2O_5 thin films on Si, *Thin Solid Films*, **381**, 31–8.
[117] C. Bartic, H. Jansen, A. Campitelli, and S. Borghs (2002) Ta_2O_5 as gate dielectric material for low-voltage organic thin-film transistors, *Organic Electronics* **3**, 65–72.

[118] M. Mizukami, N. Hirohata, T. Iseki, K. Ohtawara, T. Tada, S. Yagyu, T. Abe, T. Suzuki, Y. Fujisaki, Y. Inoue, S. Tokito, and T. Kurita (2006) Flexible AM OLED panel driven by bottom-contact OTFTs, *IEEE Electron Device Letters* **27**, 249–51.
[119] C. Tang, C. Jaing, K. Wu, and C. Lee (2009) Residual stress of graded-index-like films deposited by radio frequency ion-beam sputtering, *Thin Solid Films* **517**, 1746–9.
[120] C. Chaneliere, J.L. Autran, R.A.B. Devine, and B. Balland (1998) Tantalum pentoxide (Ta_2O_5) thin films for advanced dielectric applications, *Materials Science and Engineering: R: Reports* **22**, 269–322.
[121] T. Kaneko, N. Akao, N. Hara, and K. Sugimoto (2005) In situ ellipsometry analysis on formation process of Al_2O_3-Ta_2O_5 films in ion beam sputter deposition, *Journal of The Electrochemical Society* **152**, B133.
[122] N.V. Nguyen, C.A. Richter, Y.J. Cho, G.B. Alers, and L.A. Stirling (2000) Effects of high-temperature annealing on the dielectric function of Ta_2O_5 films observed by spectroscopic ellipsometry, *Applied Physics Letters* **77**, 3012.
[123] I. Ciofi, M.R. Baklanov, Z. Tőkei, and G.P. Beyer (2010) Capacitance measurements and k-value extractions of low-k films, *Microelectronic Engineering* **87**, 2391–2406.
[124] W.B. Jackson, R.L. Hoffman, and G.S. Herman (2005) High-performance flexible zinc tin oxide field-effect transistors, *Applied Physics Letters* **87**, 193503.
[125] M. Kimura and S. Imai (2010) Degradation evaluation of a-IGZO TFTs for application to AM-OLEDs, *IEEE Electron Device Letters* **31**, 963–5.
[126] S.M. Sze (1985) *Semiconductor Devices – Physics and Technology*, John Wiley & Sons Ltd.

5

The (R)evolution of Thin-Film Transistors (TFTs)

5.1 Introduction: Device Operation, History and Main Semiconductor Technologies

Since the early years of research into thin-film transistors (TFTs), these devices have been seen as potential candidates for low cost and high density electronic switches, their main application being in the active matrix backplanes of liquid crystal displays (LCDs). The materials and fabrication techniques of TFTs were (and continue to be) studied exhaustively from the 1960s, and nowadays they permit a multitude of applications, ranging from small size, flexible and low cost displays to those that have large size, high resolution and high speed.

Rather than covering in detail all the technological aspects related to TFTs (there are several excellent books on these topics, for instance [1] and [2]), this chapter intends to provide the fundamental background to these devices, regarding their structure, physics and history. Particular relevance is given to the current and upcoming semiconductor material technologies for TFTs, including the emerging oxide semiconductors.

5.1.1 Device Structure and Operation

TFTs are three terminal field-effect devices, whose working principle relies on the modulation of the current flowing in a semiconductor placed between two electrodes (source and drain). A dielectric layer is inserted between the semiconductor and a transversal electrode (gate), being the current modulation achieved by the capacitive injection of carriers close to the dielectric/semiconductor interface, known as field effect [1].

Transparent Oxide Electronics: From Materials to Devices, First Edition. Pedro Barquinha, Rodrigo Martins, Luis Pereira and Elvira Fortunato.

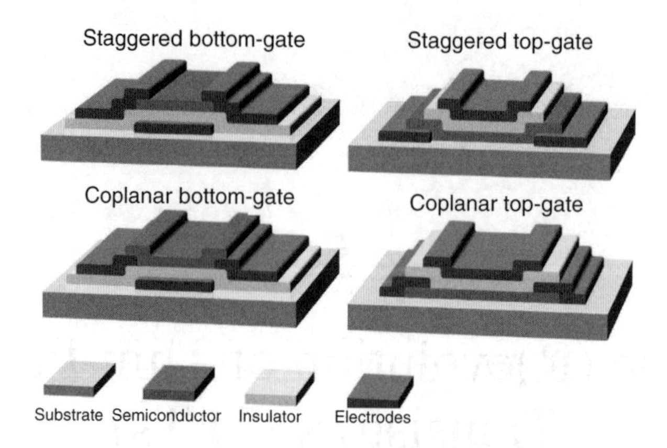

Figure 5.1 *Schematics showing some of the most conventional TFT structures, according to the position of the gate electrode and the distribution of the electrodes relative to the semiconductor.*

Figure 5.1 (see color plate section) shows the most common TFT structures. Depending on whether the source-drains are on opposite sides or on the same side of the semiconductor, the structures are identified as staggered and coplanar. Inside these, top- and bottom-gate structures exist, depending on whether the gate electrode is at the top or bottom of the structure [2, 3].

Each of these structures presents advantages or disadvantages, largely dictated by the materials used to fabricate the TFTs. For instance, the staggered bottom-gate configuration is widely used for the fabrication of a-Si:H TFTs, due to the easier processing and enhanced electrical properties. As a-Si:H is light sensitive, the usage of this configuration is advantageous for the application of these TFTs in LCDs, since the metal gate electrode shields the semiconductor material from the effect of the backlight present in these displays [3]. On the other hand, a coplanar top-gate structure is normally preferred for poly-Si TFTs. This is mostly attributed to the fact that the crystallization process of the semiconductor material generally requires high temperatures that could degrade the properties of other materials (and their interfaces) previously deposited, being favored if the semiconductor is a flat and continuous film without any layers beneath it [3]. Furthermore, the fact that in bottom-gate structures (both staggered and coplanar) the semiconductor surface is exposed to air can bring undesirable instability effects but can also be explored as an easy way to modify its properties, for instance by the facilitated adsorption of impurities during annealing or plasma treatments in a suitable atmosphere [1].

The schematics in Figure 5.1 only show the fundamental layers of a TFT, but other layers can also be introduced for different purposes. As an example, in silicon technology it is common to use highly doped semiconductor layers at the source-drain regions, in order to form low-resistance contacts. Also, an insulating film is often deposited on top of the semiconductor layer in staggered bottom-gate structures. This can allow for more accurate etching of the source-drain electrodes, without damaging the semiconductor surface, i.e. acting as an etch-stopper [2, 4]. This insulating layer can also have interesting effects in

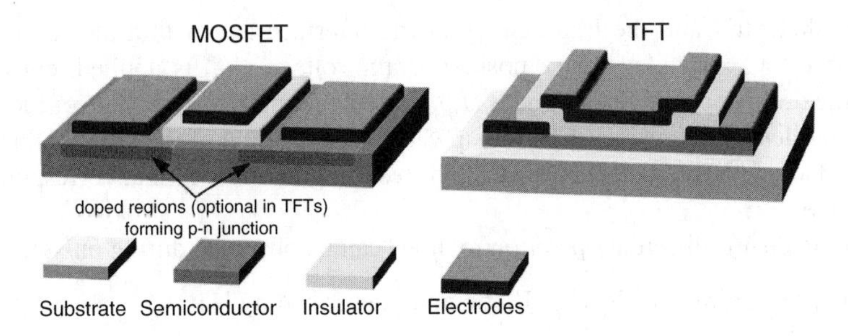

Figure 5.2 *Comparison between the typical structures of MOSFETs and TFTs.*

semiconductor films that strongly interact with environmental species such as oxygen or moisture, which is reflected in the large variations of the electrical properties exhibited by the TFTs [1]. This is particularly relevant for some of the first semiconductors used in TFTs, such as CdSe, and also for oxide semiconductors, as will be shown in this chapter. Additionally, the insulating layer on top of the TFT structure can work as an effective mechanical and chemical protection from subsequent processes for the devices, such as their integration with liquid crystal cells [5].

TFTs are quite similar to other field-effect devices in terms of operation and composing layers, such as the well known MOSFETs used in high performance applications such as microprocessors or memories. However, important differences exist between these devices, some of which can be seen by inspecting their typical structures (Figure 5.2, see color plate section). First, while TFTs use an insulating substrate, normally glass, in MOSFETs the silicon wafer acts both as the substrate and the semiconductor. Thus, higher performance arises naturally for MOSFETs, given that the electrons flow in a single crystalline semiconductor, rather than in a polycrystalline or amorphous one in TFTs. Also, the temperatures involved in the fabrication of both devices are quite different: while processing temperatures exceeding 1000°C are common for MOSFETs, for instance to create the dielectric layer, in TFTs they are limited by parameters such as the softening point of the substrate, which for most common glass substrates does not exceed 600–650°C [2]. In addition, MOSFETs have p-n junctions at the source-drain regions, which are absent in TFTs. This is related to another important difference in device operation: even if both TFTs and MOSFETs rely on the field effect to modulate the conductance of the semiconductor close to its interface with the dielectric, in TFTs this is achieved by an accumulation layer, while in MOSFETs an inversion region has to be formed close to that interface, i.e., a n-type conductive layer is created in a p-type silicon substrate.

The ideal operation of an n-type TFT depends on the existence of an electron accumulation layer at the dielectric/semiconductor interface. This is achieved for a gate voltage (V_G)[a] higher than a certain threshold voltage (V_T), corresponding to downward band-bending of the semiconductor close to its interface with the dielectric. In a real device, V_T deviates from 0 V, being a function of the gate electrode-semiconductor work function difference, the background carrier concentration of the semiconductor (N), the charge density residing

[a] Throughout this analysis, the source electrode is always assumed to be grounded.

within the dielectric and the trap density at the interface and within the semiconductor [1, 6]. For $V_G>V_T$, provided that a positive drain voltage (V_D) is applied, current flows between the drain and source electrodes (I_D), corresponding to the *on*-state of the TFT. For $V_G<V_T$, regardless of the value of V_D the upward band-bending of the semiconductor close to the interface with the dielectric is verified, resulting in a low I_D that corresponds to the TFT *off*-state.

Depending on V_D, different operation regimes can be observed during on-state:

- Pre-pinch-off regime, for $V_D<V_G-V_T$, being I_D described by [1]:

$$I_D = C_i \mu_{FE} \frac{W}{L}\left[(V_G-V_T)V_D - \frac{1}{2}V_D^2\right] \tag{5.1}$$

where C_i is the gate capacitance per unit area, μ_{FE} is the field-effect mobility, W is the channel width and L is the channel length. For very low V_D, the quadratic term can be neglected, yielding a linear relation between I_D and V_D, often designated by linear regime.

- Post-pinch-off or saturation regime, for $V_D>V_G-V_T$, being I_D described by:

$$I_D = C_i \mu_{sat} \frac{W}{2L}(V_G-V_T)^2 \tag{5.2}$$

where μ_{sat} is the saturation mobility. In this regime, the semiconductor close to the drain region becomes depleted, a phenomenon designated by pinch-off that leads to the saturation of I_D.

The equations describing the ideal operation of TFTs are based on the "gradual channel approximation" initially proposed by Shockley [7]. Even if this describes fairly well most of the TFT behavior, assumptions such as V_G independent μ_{FE} and μ_{sat} are generally not valid, particularly for oxide TFTs, where $\mu_{FE}(V_G)$ and $\mu_{sat}(V_G)$ should be considered instead [8].

The static characteristics of TFTs are accessed by their output and transfer characteristics, shown in Figures 5.3a and b, respectively.

In output characteristics, V_D is swept for different V_G values, allowing one to observe clearly the pre- and post-pinch-off regimes described above. Different qualitative information can be assessed by output characteristics: for instance, a decreasing separation between I_D–V_D curves for increasing V_G is indicative of μ degradation for that V_G range; the flatness of the I_D–V_D curves at the post-pinch-off regime permits one to evaluate if the channel layer can be fully depleted close to the drain electrode for the range of V_D and V_G used. Good saturation is an important requisite in electronic circuits, especially if the TFT is used as a current limiter [1]; the low V_D region (linear) also provides useful information regarding contact resistance, as demonstrated in section 5.2.1.4.

On the other hand, transfer characteristics, where V_G is swept for a constant V_D, permit one to extract a large number of quantitative electrical parameters, which are summarized below:

- *On/off* – this is simply defined as the ratio of the maximum to the minimum I_D. The minimum I_D is generally given by the noise level of the measurement equipment or by the gate leakage current (I_G), while the maximum I_D depends on the semiconductor material itself and on the effectiveness of capacitive injection by the field effect. *On/off* above

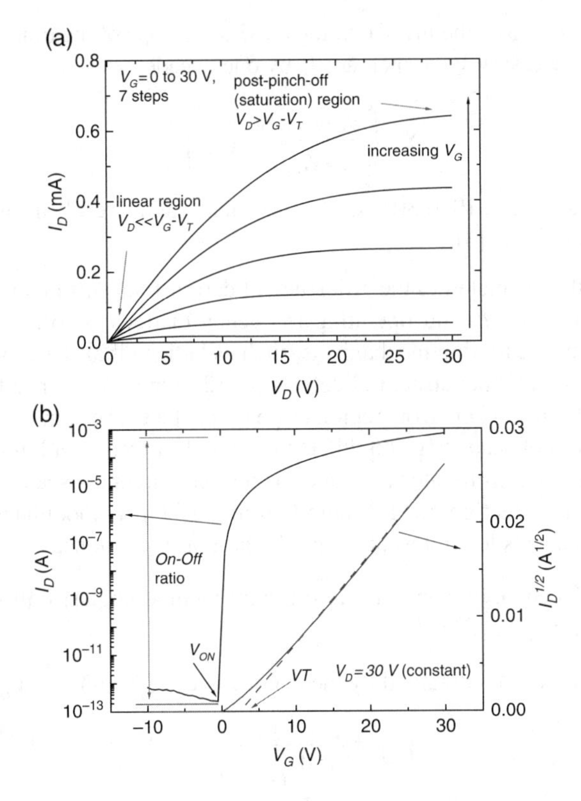

Figure 5.3 Typical a) output and b) transfer characteristics of a n-type oxide TFT.

10^6 are typically obtained in TFTs and a large value is required for their successful usage as electronic switches [3];

- *V_T and turn-on voltage (V_{ON})* – as previously defined, V_T corresponds to the V_G for which an accumulation layer or conductive channel is formed close to the dielectric/semiconductor interface, between the source and drain electrodes. For an n-type TFT, depending upon whether V_T is positive or negative, the devices are designated as enhancement or depletion mode, respectively (conversely, for p-type TFTs the opposite applies). Both types are useful for circuit fabrication, but generally the enhancement mode is preferable, because no V_G is required to turn off the transistor, making circuit design easier and minimizing power dissipation [9]. V_T can be determined using different methodologies, such as linear extrapolation of the I_D–V_G plot (for low V_D) or of the $I_D^{1/2}$–V_G plot (for high V_D), V_G corresponding to a specific I_D, ratio of conductance and transconductance, among others [10]. Even if considering only one methodology, great ambiguity can arise in the determination of V_T (for instance, by using different fitting parameters on the linear regressions). The concept of V_{ON} is widely used in the literature, simply corresponding to the V_G at which I_D starts to increase as seen in a *log* I_D–V_G plot, or in other words, the V_G necessary to fully turn-off the transistor [11].

- *Subthreshold swing (S)* – the inverse of the maximum slope of the transfer characteristic, it indicates the necessary V_G to increase I_D by one decade:

$$S = \left(\frac{dlog(I_D)}{dV_G} \bigg|_{\max\ I_D} \right)^{-1} \tag{5.3}$$

Typically, S<<1, around 0.10–0.30 V dec^{-1} and small values result in higher speeds and lower power consumption [3].

- *Mobility (μ)* – this is related to the efficiency of carrier transport in a material, affecting directly the maximum I_D and operating frequency of devices [6]. In a material, μ is affected by several scattering mechanisms, such as lattice vibrations, ionized impurities, grain boundaries and other structural defects [6, 12]. On a TFT, since the movement of carriers is constrained to a narrow region close to the dielectric/semiconductor interface, additional sources of scattering should be considered, such as Coulomb scattering from dielectric charges and from interface states or surface roughness scattering [12]. Still, as seen for instance in section 5.2.1.2, note that in a TFT μ is modulated by V_G, turning scattering mechanisms less relevant for particular bias conditions.

Mobility of a TFT can be extracted using different methodologies. Following Schroder's nomenclature, one may have [12]:

- Effective mobility (μ_{eff}) – obtained by the conductance (g_d) with low V_D:

$$\mu_{eff} = \frac{g_d}{C_i \frac{W}{L}(V_G - V_T)} \tag{5.4}$$

- Field-effect mobility (μ_{FE}) – obtained by the transconductance (g_m) with low V_D:

$$\mu_{FE} = \frac{g_m}{C_i \frac{W}{L} V_D} \tag{5.5}$$

- Saturation mobility (μ_{sat}) – obtained by the transconductance with high V_D:

$$\mu_{sat} = \frac{\left(\frac{d\sqrt{I_D}}{dV_G} \right)^2}{\frac{1}{2} C_i \frac{W}{L}} \tag{5.6}$$

Each methodology has its advantages and drawbacks. Even if μ_{eff} includes the important effect of V_G, it requires the determination of V_T and is more sensitive to contact resistance (low V_D). This last issue is also verified for μ_{FE}, but μ_{FE} does not require V_T and is easily calculated by the derivative of transfer characteristics, turning it a widely used parameter. Finally, μ_{SAT} does not require V_T and is less sensitive to contact reisistance. However, it describes a situation where the channel is pinched-off, i.e., its effective length is smaller than L, which is intrinsically not assumed by equation 5.2.

Other methodologies can be found in the literature, such as the average (μ_{avg}) and incremental (μ_{inc}) mobility proposed by Hoffman [11]. While the former provides an average value of all the carriers induced in the channel, the latter probes the mobility of carriers as they are incrementally added to the channel, providing valuable insights into carrier transport.

5.1.2 Brief History of TFTs

The first reports on TFTs date from almost 80 years ago and are attributed to Lilienfeld and Heil [13–16]. At that time, little was known about semiconductor materials and vacuum techniques for producing thin films were far from being established. Hence, these first reports are actually concept patents and no evidence exists about the production of working devices. Still, in these patents, the idea of controlling the current flow in a material by the influence of a transversal electrical field was already present. One of Lilienfeld first patents, published in 1930, describes the basic principle of what is known today as the metal-semiconductor field-effect transistor (MESFET, Figure 5.4a), while other, published three years later, shows the concept of a device where an insulating material (aluminum oxide) is introduced between the semiconductor (cooper sulfide) and the field-effect (or gate) electrode (aluminum), anticipating the so-called metal-insulator-semiconductor field-effect transistor (MISFET, Figure 5.4b) [17]. It is also noteworthy that in the MISFET patent, the thickness specified for the insulating layer, 100 nm, is very close to that used nowadays in TFTs [1]. The first reports of current modulation by field effect are attributed to Shockley and co-workers at Bell Laboratories in the 1940s, by using a germanium thin film separated from a gate electrode by a thin film of mica [1]. The very small change in the conductivity of germanium, considerably lower than that expected theoretically, was attributed to charge trapping in surface states. Even if the result was disappointing regarding the application of the structure as a practical device, it enabled great advances to be made in the study of the surface defects of semiconductors. In these studies, in late 1947, Bardeen and Brattain [18] discovered the point-contact transistor (Figure 5.5) and later on, in 1948, Shockley [19] proposed the p-n junction transistor. But the need for a solid-state amplifier with a higher input impedance than the junction transistor persisted. This was achieved with the junction field-effect transistor (JFET) proposed by Shockley in 1952 [20]. In this case, a reverse biased p-n junction separated the conductive channel from the gate electrode. This work was extremely important for the definition of the theoretical operation principles of field-effect devices, including the well known "gradual channel approximation", being that most of these principles are still used in the analysis of modern TFTs and other field-effect devices. The work of Shockley, Bardeen and Brattain related to semiconductors and transistors resulted in their receiving the Noble Prize for physics in 1956.

It took one more decade for the first TFT to be produced. This was achieved by Weimer at the RCA Laboratories in 1962 [22]. Weimer used a vacuum technique (evaporation) to deposit gold electrodes, a polycrystalline cadmium sulfide (CdS) n-type semiconductor and a silicon monoxide (SiO) insulator, using shadow masks to define the patterns of these layers. Other important work from Borkan and Weimer also provided an analysis of the characteristics of these devices, based on the models initially proposed by Shockley [23]. During the 1960s further work on TFTs appeared, such as that by Shallcross and Weimer [24, 25], using semiconductors such as CdSe or tellurium, respectively. In 1967 Sihvonen *et al.* demonstrated the possibility of printing a TFT (actually, his nomenclature was *Graphic Active Device (GAD)*), using a semiconductor based on CdS:CdSe inks, dielectrics based on silicate cements and electrodes based on a Hg:In paste. Being that these preliminary

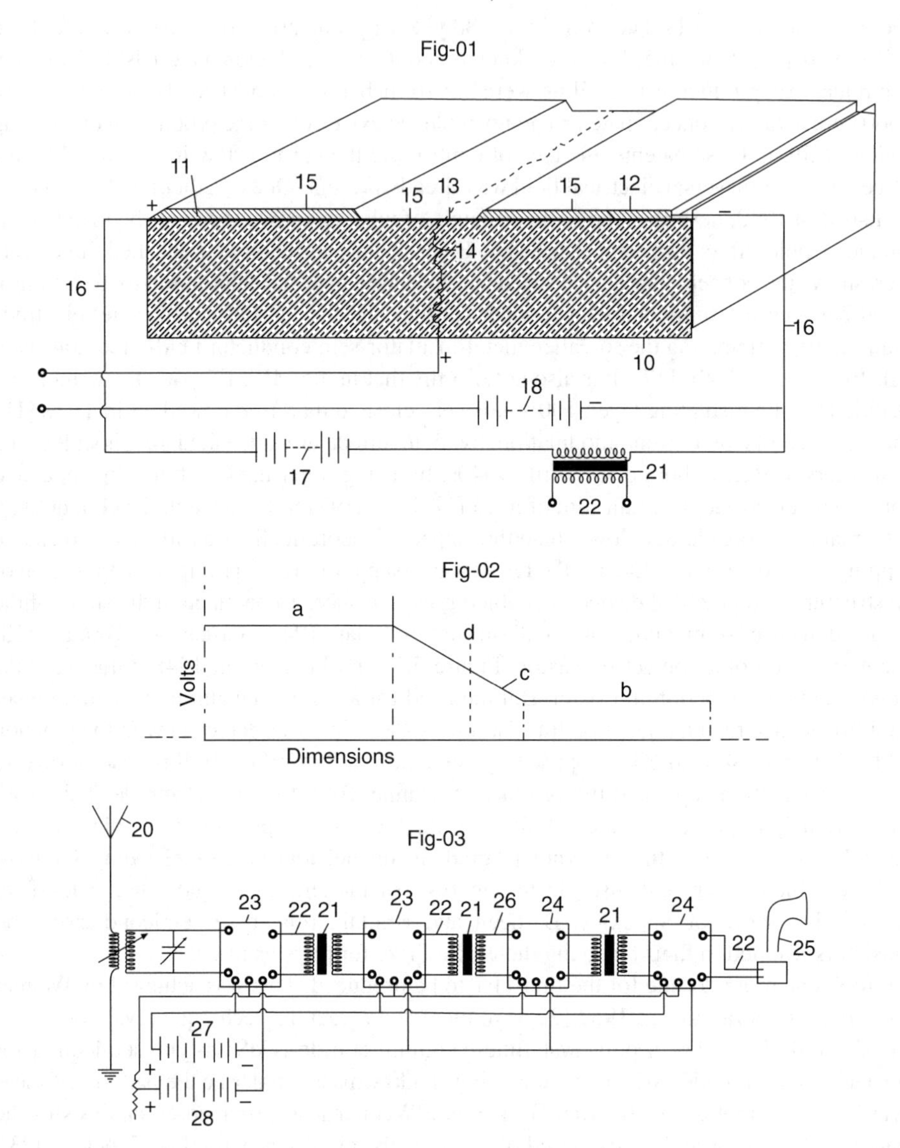

Figure 5.4 *Initial patents of field-effect devices submitted by Lilienfeld: a) MESFET; Reprinted from [17] US Patent 1745175 Copyright (1930) J E Lilienfeld. b) MISFET. Reprinted from [17] US Patent 1900018 Copyright (1933) J E Lilienfeld.*

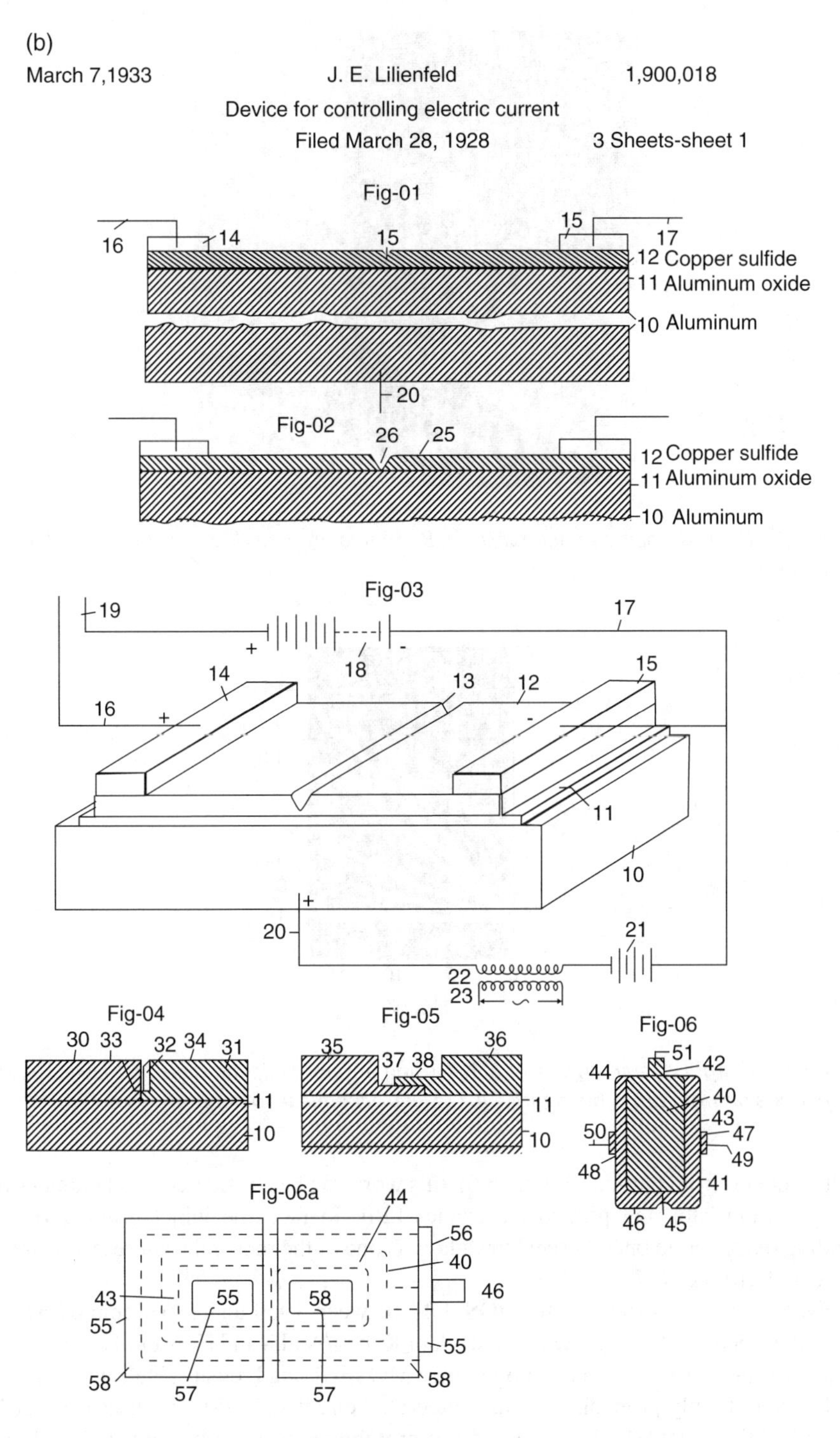

Figure 5.4 (continued)

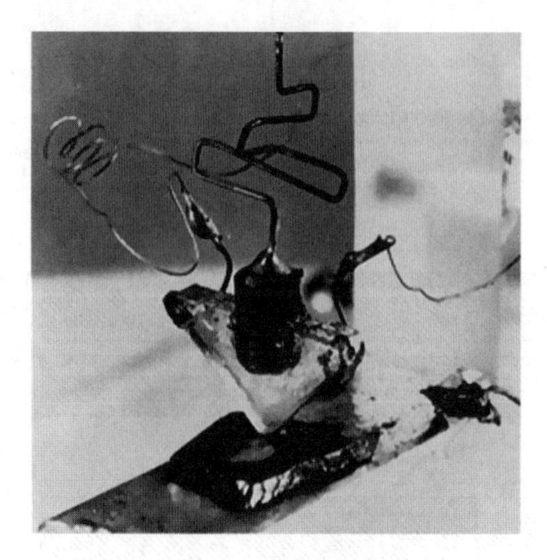

Figure 5.5 The first point-contact transistor. Reprinted from [21] Copyright (2011) University of Maryland.

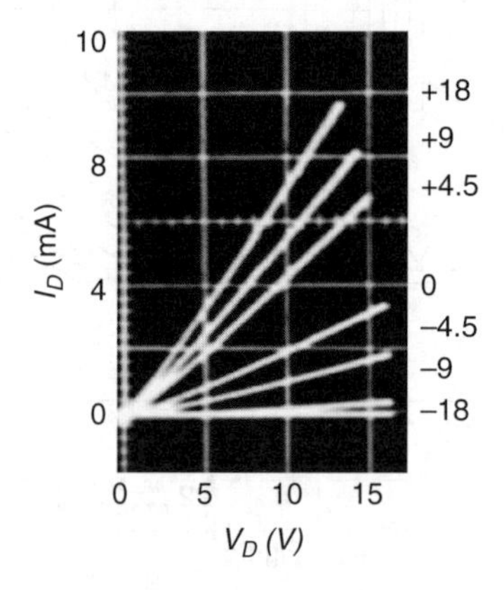

Figure 5.6 Output characteristics of a TFT using an inorganic-ink dielectric layer. Reprinted with permission from [26] Copyright (1967) Electrochemical Society.

results were far from optimal (Figure 5.6), this work can be considered as a landmark in the history of transistor and printed electronics [26]. In fact, the inkjet printing of TFTs, including oxide-based ones, is seen nowadays as one of the most relevant research areas for low-cost electronics [27].

However, the emergence of the MOSFETs between 1960 and 1963, employing single crystalline silicon technology, with the work of Kahng/Atalla and Hofstein/Heiman, pushed researchers away from TFT technology, since MOSFETs had remarkable electrical properties that were highly promising for fast integrated circuits [1, 28]. Nevertheless, for large area applications, MOSFETs represented a prohibitive cost when compared with TFTs. The turning point that once again motivated the research on TFTs was the work presented by Lecnher *et al.* [29] in 1971, where the authors proposed the use of TFTs to control each pixel of a LCD, obtaining considerably less crosstalk, lower response time and higher contrast ratios than that achieved when controlling liquid crystals with the more conventional x–y disposed electrodes. A prototype using an array of CdSe TFTs integrated in a LCD was demonstrated by Brody and co-workers in 1973 [30].

However, the greatest innovation in terms of TFT development, that dictated the success of this technology for so many decades, was the introduction of a-Si:H as a semiconductor material in TFTs. This was reported by LeComber, Spear and Ghaith [31, 32] in 1979, as a natural consequence of their work on the analysis of the density of states in disordered structures. In spite of exhibiting considerably lower μ than polycrystalline materials such as CdSe (about 1 against 150–200 $cm^2 V^{-1} s^{-1}$), a-Si:H was perfectly suitable for the application of TFTs as switching elements in LCDs, since it allowed for low cost, good reproducibility and uniformity in large areas and *on/off* exceeding 10^6.

The study of a-Si:H TFTs continued during the 1980s, with new improvements being achieved regarding fabrication processes, structures and material combinations. Also, knowledge of the physics of devices employing this semiconductor material was considerably broadened, for instance by the work of Powell and co-workers [33, 34]. Although CdSe TFTs were still being studied with the main propose of obtaining a viable alternative for a-Si:H TFTs when high μ was needed, this objective was mostly achieved with the appearance of poly-Si as a semiconductor material in TFTs. A poly-Si-based TFT was initially reported by Depp *et al.* [2] in 1980 and over the next few years it became possible to achieve $\mu_{FE} \approx 400\,cm^2 V^{-1} s^{-1}$, which begins to approach the values typically obtained in MOSFETs. The large μ_{FE} of these transistors allowed for their application not only as switching elements but also as driver circuitry devices, theoretically decreasing the costs and complexity of LCD fabrication. However, poly-Si TFTs had a large cost, mostly because they required high temperature fabrication processes, which were only compatible with quartz substrates, not with normal glass. A considerable decrease in the processing temperature of poly-Si TFTs to around 550°C was only reported in 1991, by Little *et al.* from Seiko-Epson [35]. However, these low-temperature poly-Si (LTPS) TFTs could not easily penetrate in the LCD market, which was already widely dominated by a-Si:H TFTs. Also, the intrinsic limitation of the polycrystalline structure of the semiconductor material imposed several restrictions on large area processing, as happened previously with CdSe TFTs.

The application of organic materials as semiconductors in TFTs was another interesting technology introduced in the late 1980s. In 1986, Tsumura *et al.* [36] reported on TFTs using an electrochemically grown polythiophene organic semiconductor, resulting in an enhancement mode p-type device with *on/off*$\approx 10^3$ and a low $\mu_{SAT} \approx 10^{-5}\,cm^2 V^{-1} s^{-1}$. The low μ_{SAT} of this initial work was attributed to the amorphous structure of the organic layer and to its low doping level, inhibiting the mediation of carriers between polymer chains by dopants. At the early 1990s, Garnier [37] reported TFTs using evaporated hexathiophene as a semiconductor material, resulting in devices with comparable performance to a-Si:H TFTs. Organic semiconductors have, however, a great advantage over a-Si:H, which is their extremely low processing temperature. This is the reason why organic semiconductor devices in general and organic TFTs in particular are pointed to as one of the most promising technologies for flexible electronics (Figure 5.7, see color plate section).

To the author's knowledge, the first reports on TFTs employing transparent semiconducting oxides (TSOs) as channel layers are almost coincidental with the initial

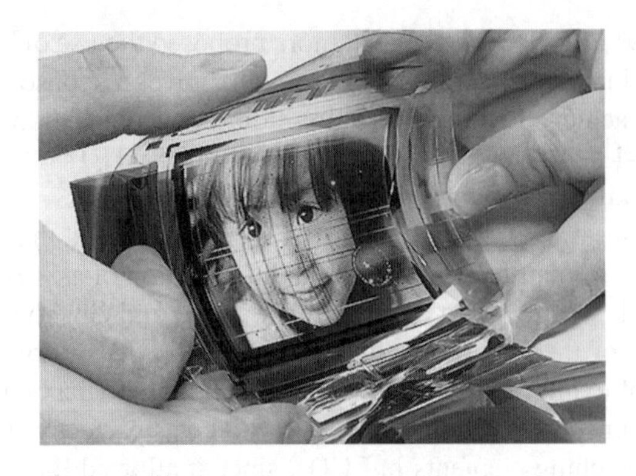

Figure 5.7 Example of integration of organic TFTs in an OLED display by Sony. Reprinted with permission from [38] Copyright (2007) Sony Corporation.

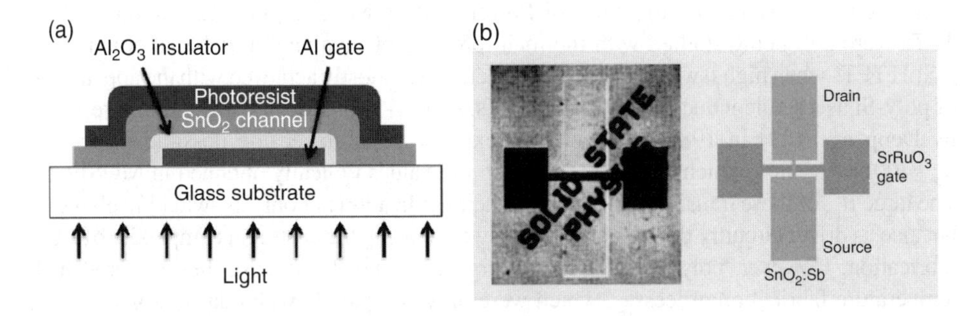

Figure 5.8 Initial work on oxide TFTs: a) Schematic of the SnO_2 TFT reported by Klasens and Koelmans in 1964. Reprinted with permission from [42] Copyright (1996) American Institute of Physics; b) top-view and layout of the ferroelectric Sb:SnO_2 TFT reported by Prins et al. in 1996. Reprinted with permission from [44] Copyright (2008) Springer Science + Business Media.

CdS TFTs reported by Weimer. In fact, back in 1964, Klasens and Koelmans [39] proposed a TFT comprising an evaporated SnO_2 semiconductor on glass, with aluminum source-drain and gate electrodes and an anodized Al_2O_3 gate dielectric. The usage of a TSO serves essentially to demonstrate a new self-lift-off process intended to define the pattern of source-drain electrodes (Figure 5.8a, see color plate section), with few details being given regarding electrical properties. A lithium-doped ZnO single crystal was also tested as a channel layer material by Boesen and Jacobs in 1968, but a very small I_D modulation by V_G and no I_D saturation was observed in their devices [40]. Similar properties were obtained by Aoki and Sasakura in SnO_2 TFTs in 1970 [41]. In 1996, ferroelectric field-effect devices with SnO_2:Sb and In_2O_3 TSOs, by Prins *et al.* and Seager [42, 43], respectively, were reported. The main intent was to demonstrate hysteresis associated with ferroelectric behavior, with little information being provided regarding device performance. Still, it is noteworthy that full transparency is achieved in a TFT in [42] and it is only severely affected

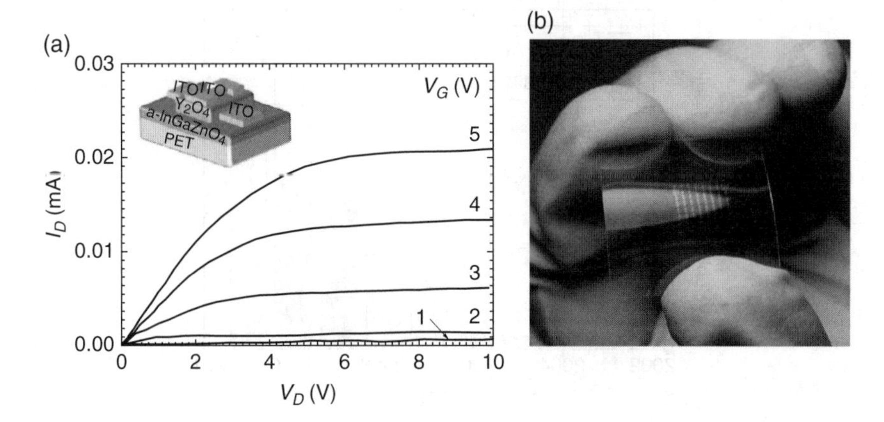

Figure 5.9 GIZO TFTs on flexible substrates presented by Nomura et al. in 2004: a) output characteristics; Reprinted with permission from [60] Copyright (2004) Macmillan Publishing. b) photograph of the flexible TFT sheet bent at R=30 mm.

by the $SrRuO_3$ gate electrode (Figure 5.8b), although the authors mention that fully transparent TFTs were fabricated by using a heavily doped SnO_2 gate electrode.

With the new millennium renewed interest in oxide TFTs emerged, with the works of Masuda *et al.*, Hoffman *et al.* and Carcia *et al.* on ZnO TFTs [9, 45, 46]. Even if in most of the cases high (post-)processing temperatures (above 450°C) were required to obtain good performance devices, electrical properties began to become comparable or even superior to those typically exhibited by a-Si:H and organic TFTs, mostly in terms of μ_{FE}, which could already be as high as 2.5 cm^2 V^{-1} s^{-1}. However, the work from Carcia *et al.* showed for the first time that room-temperature sputtered ZnO layers could also provide similar electrical properties. Low- or even room-temperature processing of ZnO and simultaneously improved device performance were in fact some of the main targets of different reports on oxide TFTs over the following years [47–49]. Other highlights in 2003–2004 were non-vacuum processes to produce the ZnO layers [50]; the first simulations of ZnO TFTs assuming that their properties were largely dictated by the polycrystalline structure of the TSO [51]; new methods for mobility extraction in ZnO TFTs [11]; the application of ZnO TFTs as UV photodetectors [52, 53]; the fabrication of SnO_2 TFTs [54]; and the use of In_2O_3 or ZnO nanowires as channel layers [55–58].

Multicomponent TSOs for the channel layer of TFTs were initially suggested by Nomura *et al.* [59] in 2003. In Nomura's work, a $InGaO_3(ZnO)_5$ (or GIZO) single-crystalline semiconductor layer was epitaxially grown in an yttria-stabilized zirconia substrate, resulting in devices with an impressive μ_{eff}≈80 cm^2 V^{-1} s^{-1}, V_{ON}≈−0.5 V and *on/off*≈10^6. Despite the very high processing temperature (1400°C) required to obtain the single-crystalline film, this work demonstrated the great potential of oxide TFTs. This potential was reinforced by work from the same author in the next year, where an amorphous GIZO layer was deposited at room-temperature by pulsed laser deposition (PLD) and used to fabricate transparent TFTs on flexible substrates (Figure 5.9, see color plate section) [60]. Despite the lower performance when compared with the single-crystalline TFTs, these

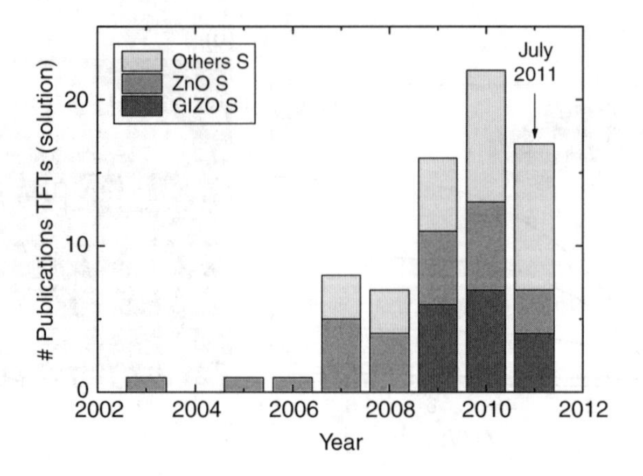

Figure 5.10 *Number of publications related to solution-processed oxide TFTs over the years.*

flexible GIZO TFTs exhibited respectable μ_{sat}≈9 cm^2 V^{-1} s^{-1}, V_T≈1–2 V and *on/off*≈10^3, proving the low sensitivity of multicomponent oxides to structural disorder.

In the following years binary compounds such as ZnO or In_2O_3 continued to be investigated as channel layers of oxide TFTs, but most attention was undoubtedly drawn to multicomponent oxides. Several combinations of cations with $(n-1)d^{10}ns^0$ ($n \geq 4$) electronic configuration began to be used to this end, with ZTO [8, 61–63], IZO [64–67], and GIZO [68–73] being those most widely explored. Nowadays, after the numerous studies regarding the compositional and (post-)processing conditions dependence of these TSOs, oxide TFTs are able to provide remarkable electrical properties, considerably superior to a-Si:H or organic TFTs, such as μ_{FE} above 10 cm^2 V^{-1} s^{-1}, close to 0 V_{ON}, *on/off* exceeding 10^7 and S≈0.20 V dec^{-1}, with the indium-based semiconductors having the added advantage of allowing for very low or even room temperature processing. In addition, solution processing of oxide TFTs is an area of growing interest (Figure 5.10, see color plate section), with device performance already starting to approach that of vacuum-processed oxide TFTs, even with maximum processing temperatures close to 200°C [27, 74]. Solution processing of oxide TFTs brings important advantages regarding the easy and low-cost fabrication of devices in large areas.

5.1.3 Comparative Overview of Dominant TFT Technologies

Several properties of the different TFT technologies regarding semiconductor materials were indicated in the previous section. Here, a more detailed comparison between the dominant TFT technologies is made, based on performance, stability, cost and processing of the devices. Table 5.1 provides a summary of this for oxide, a-Si:H, poly-Si and organic TFTs.

a-Si:H TFTs are unquestionably the most mature technology. These devices have been studied for almost 40 years, and their strengths and limitations are well known. Moreover, there is a huge industrial implementation of a-Si:H TFTs in active matrices for LCDs. Oxide TFTs, even if far from the maturity of a-Si:H TFTs, are probably responsible for one of the fastest research-to-production migration processes seen in recent years in

Table 5.1 *Comparison between dominant TFT technologies regarding semiconductor materials [75–79].*

TFT properties	Oxides	a-Si:H	poly-Si (LTPS)	Organics
μ ($cm^2 V^{-1} s^{-1}$)	1 to 100	1 max	50 to 100	0.1–1
S ($V dec^{-1}$)$^+$	0.1 to 0.6	0.4 to 0.5	0.2 to 0.3	0.1 to 1.0
Leakage current (A)$^+$	10^{-13}	~10^{-12}	~10^{-12}	~10^{-12}
Reproducibility	High	High	Low	Low
TFT for AMOLEDs	4 to 5 masks	4 to 5 masks	5 to 9 masks	Poor applicability*
Manufacturing cost	Low	Low	High	Low
Long term TFT reliability	High (forecast)	Low	High	Low in air
Yield	High	High	Low	High
Process temperature (°C)	RT to 350	~250	<550	RT
CMOS fabrication	Yes and large areas	Very low performance	Yes, not for large areas	Low performance

+Highly dependent on the dielectric layer.
*Most organic semiconductor layers are patterned using shadow masks.

microelectronics. This is a consequence not only of the great advantages of this TFT technology, but also of the possibility of using the existing processing and lithographic tools of the LCD industry, for instance, the sputtering systems used to fabricate TCO films used as top LCD electrodes. If physical techniques such as sputtering are considered, another advantage arises with regard to oxide semiconductors in comparison with silicon-based technologies, since only argon and oxygen need to be used, rather than explosive gases such as silane, phosphine or diborane.

Due to the polycrystalline structure of poly-Si, this material represents a great disadvantage regarding large area deposition, being the lack of uniformity and reproducibility in large areas, one of the greatest drawbacks of this technology. On the contrary, a-Si:H, organic and oxide semiconductors are suitable for large area deposition, since they can have amorphous structures. Processing temperatures are also normally higher for poly-Si TFTs, in order to obtain larger grain sizes, hence higher performance. But even a-Si:H generally requires temperatures exceeding 250°C in order to achieve good performance levels. Oxides and mainly organic semiconductors require the lowest processing temperatures, making them compatible with inexpensive glass or even plastic substrates, making them suitable for flexible electronics. The lower temperatures also have the advantage of resulting in lower costs for these technologies and as mentioned in the previous section, the emergence of solution processed oxide TFTs allows one to anticipate further cost reductions. However, benefiting from a well established industry, a-Si:H TFTs also represent nowadays low production costs. Regarding poly-Si, the fact that it requires high temperature or complex/expensive crystallization methods, such as excimer laser annealing (ELA), and the fact that complex pixel architectures are needed in displays to compensate for the non-uniform device properties (increases in the number of process masks), results in higher-cost TFTs.

Concerning electrical performance and stability, poly-Si TFTs have the advantage of exhibiting the highest μ_{FE}, which even for LTPS TFTs can surpass 100 cm^2 V^{-1} s^{-1} [80]. As pointed out earlier, this enables one to use them also for the drivers' circuitry in LCDs. a-Si:H and organic semiconductor TFTs exhibit the smallest μ_{FE}, typically less than 1 cm^2 V^{-1} s^{-1}, while oxide TFTs exhibit intermediate μ_{FE} values between poly-Si and a:Si:H TFTs. Even if little is known yet about the long-term stability of oxide TFTs, the reports presented over recent years show that reversible V_T shift under constant bias or current stress is the predominant instability mechanism (several results are presented for this topic in section 5.2). Furthermore, given the large E_G of oxide semiconductors, their properties are not dramatically changed when exposed to visible light, in contrast to a-Si:H (and also to poly-Si, although to a considerably less extent), where degradation can arise due to the creation of dangling bonds according to the Staebler-Wronski effect, with the initial properties only being reestablished after an annealing treatment [81]. This requires the usage of light shields in silicon-based TFTs, increasing process complexity, cost, and contributing to a decrease in the aperture ratio when these TFTs are used in displays, which directly affects the brightness levels [3]. Even if UV filters are used for oxide TFTs to block wavelengths lower than 400–450 nm, these filters do not compromise the overall appearance of the devices [82]. Organic materials also represent a well-known sensitivity to environmental species, such as water and oxygen, which can even permanently affect their properties, hence a meticulous passivation step is required for these devices.

The feasibility of CMOS devices is also an important aspect to consider if fast and low-power consumption integrated circuits are envisaged. From the semiconductor technologies presented in Table 5.1, poly-Si seems to be the most adequate for this, given the reasonably good performance of both n- and p-type TFTs. However, the reliable large area processing of CMOS is naturally an issue for poly-Si technology. Organic TFTs are mostly p-type, with n-type ones exhibiting worse properties, although organic CMOS inverters can be fabricated [83]. For a-Si:H and oxide semiconductors n-type conduction is far more frequent. However, p-type TSOs are rapidly gaining in relevance, and even if their performance level is far from its n-type counterparts, preliminary CMOS inverters are already a reality, as demonstrated in Chapter 7.

Naturally, all the TFT technologies have their advantages and drawbacks, but oxide TFTs seem to be a very attractive technology, enabling transparent, low cost, low temperature and high performance transistors and applications.

5.2 Fabrication and Characterization of Oxide TFTs

Oxide TFTs have been extensively explored at CENIMAT. Research started with sputtered ZnO TFTs in 2003, followed by multicomponent n-type oxides such as IZO, GIZO and GZTO. From 2009, intense activity also began on p-type oxide TFTs, with sputtered SnO and Cu_2O. Integration of TSOs with sputtered dielectrics has also been achieved, establishing as one of the main requirements the deposition of all the layers at low temperature (or even without any intentional substrate heating), in order to make the overall process compatible with low-cost and flexible substrates. More recently, from 2010, solution processed TSOs have also started to be integrated in TFT structures. This section summarizes our work on oxide TFTs, showing our main results and their dependence on

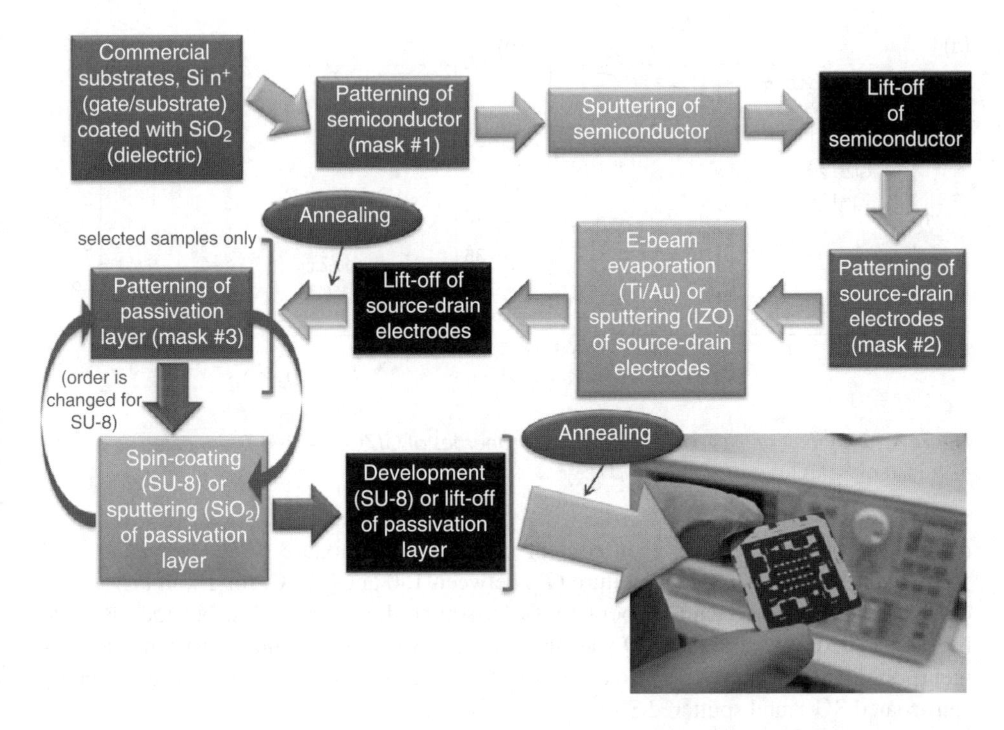

Figure 5.11 Process flow used to produce oxide TFTs employing commercial Si/SiO_2 substrates.

the (post-) processing conditions. Besides typical static characterization (transfer/output characteristics), special focus is also given to electrical stress measurements.

5.2.1 N-type GIZO TFTs by Physical Vapor Deposition

Among all the available TSOs, sputtered GIZO is the most widely studied material for application as a channel layer of oxide TFTs. Hence, a more detailed analysis is provided in this chapter regarding devices whose semiconductor layer falls inside the Ga-In-Zn oxide system. In order to analyze how the different (post-) processing parameters and compositions inside this material system affect device performance and stability, a structure employing 100 nm thick SiO_2 dielectric deposited by a well established plasma-enhanced chemical vapor deposition (PECVD) process at 400°C in heavily doped silicon wafers is used to fabricate the transistors. Si/SiO_2 provides a reproducible and reliable base structure for the devices, with PECVD SiO_2 being preferred over thermally grown SiO_2 due to its lower processing temperature, allowing for a more accurate comparison between the oxide semiconductor-based devices produced here and "real-world" TFTs.

The process flow is depicted in Figure 5.11 (see color plate section). The semiconductor layer (≈40 nm thick) was sputtered using a home-made rf magnetron sputtering system, without intentional substrate heating. Several ceramic target compositions within the Ga-In-Zn oxide system were tested, as well as different (post-) processing parameters,

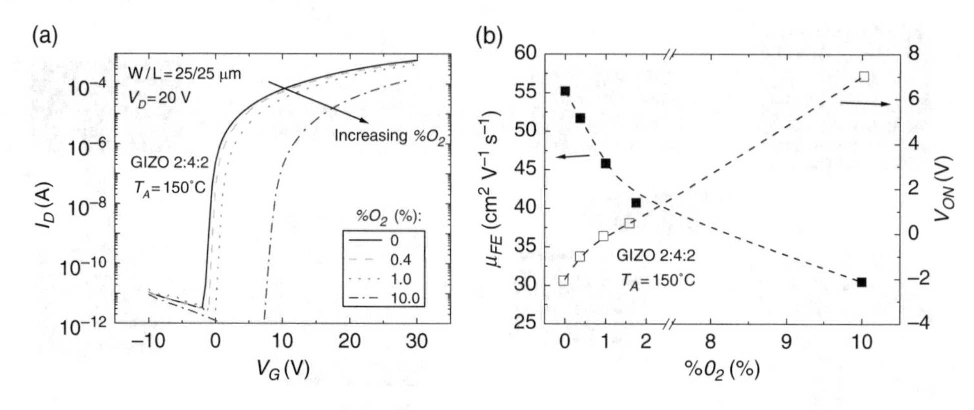

Figure 5.12 *Effect of %O_2 on the electrical properties of GIZO 2:4:2 TFTs annealed at 150°C: a) transfer characteristics; b) μ_{FE} and V_{ON}.*

namely the percentage of oxygen content in the Ar+O_2 mixture (%O_2, between 0 and 10.0%) and the annealing temperature (T_A, between 150 and 300°C, for 1h, in air).[b] Most of the devices have e-beam evaporated Ti/Au source-drain electrodes, but the effect of replacing Ti/Au by sputtered IZO was also analyzed (including contact resistance assessment). Similar structures were also used to evaluate the effect of passivation layers, namely spin-coated SU-8 and sputtered SiO_2.

5.2.1.1 *Effect of oxygen content in the Ar+O_2 mixture*

As shown in Chapter 2, %O_2 is one of the key processing parameters dictating the properties of oxide semiconductors. Thus, the electrical performance of TFTs is also expected to be significantly affected by the %O_2 used to produce its oxide semiconductor layer. Figure 5.12 shows the effect of %O_2 of the electrical properties of GIZO TFTs, produced using a GIZO 2:4:2 target composition and T_A=150°C.

In Figure 5.12 it is seen that V_{ON} increases with %O_2. As seen in Chapter 2, given that oxygen vacancies are the main source of free electrons in oxide semiconductors, higher %O_2 results in lower N and thus in a shift of E_F towards the midgap. Hence, for higher %O_2 a larger number of unfilled traps have to be filled with the charges induced by V_G before E_F reaches tail and extended states, where the induced carriers start to contribute to increase I_D. In addition, the more severe bombardment phenomena by energetic ions for higher %O_2 can also yield a less compact film and generate more defects, both in GIZO's bulk and at its interface with SiO_2, contributing also to the increase of V_{ON}. The trends of V_{ON} and μ_{FE} with %O_2 are related, with lower V_{ON} devices exhibiting larger μ_{FE} (Figure 5.12b), because both parameters depend on N. This would be expected, given that for higher background N films a larger fraction of the charges induced by V_G is readily available to flow between source and drain electrodes. As carrier transport in oxide semiconductors is enhanced with larger N (as long as N values typical of ionized impurity scattering mechanisms are not achieved, see section 2.2.3.4), μ_{FE} is increased. Similar relations were found by other authors [64, 85, 86].

[b] Although not explored in this chapter, other processing parameters were naturally explored, such as the deposition pressure (p_{dep}), the rf power density (P_{rf}) and the oxide semiconductor thickness (d_s). Ref. [84] provides a more detailed overview on this, specifically oriented for GIZO TFTs.

All of the devices in Figure 5.12 present *on/off* in the range of 10^8–10^9, which are comparable or even superior to the best performing a-Si:H TFTs. The *off*-current is always defined by the I_G level, meaning that the GIZO layer can be fully depleted regardless $\%O_2$. S increases from ≈0.25 V dec^{-1} for $\%O_2$=0–1.0% to ≈0.35 V dec^{-1} for $\%O_2$=10.0%, agreeing with the suggested larger defect density in GIZO bulk and/or in its interface with SiO_2 for large $\%O_2$ [87]. Concomitant with this, the V_{ON} shift in consecutive measurements of transfer characteristics (ΔV_{ON}) and the hysteresis (which are always positive and clockwise, respectively, consistent with electron trapping at or near the GIZO/SiO_2 interface) are almost negligible for $\%O_2$=0–1.0% but significant (≈6 V) for $\%O_2$=10.0% [84].

Based on the analysis presented above, the $\%O_2$ conditions yielding best performance are $\%O_2$=0 and 0.4%. Nevertheless, for non-annealed devices or T_A<150°C, $\%O_2$=0% often results in highly conductive GIZO layers, leading to always-on devices without appreciable channel conductivity modulation. Hence, $\%O_2$ between 0.4 and 1.0% are normally used to achieve better control and reproducibility of the transistors electrical characteristics. The dependence of IZO 2:1 and GIZO 2:4:1 TFTs on $\%O_2$ was also investigated. Similar dependences of $\%O_2$ are observed, but the effects mentioned above regarding films produced without oxygen are even more severe, since these compositions yield films with higher N than GIZO 2:4:2. The effect of $\%O_2$ was also studied for polycrystalline ZnO TFTs. For this binary compound similar effects are verified, although the properties are even more dependent on $\%O_2$, i.e the processing window required to obtain good performance devices is considerably narrower than for IZO or GIZO. For ZnO, with T_A=150°C, devices only show significant field-effect for $\%O_2$=0 and 0.4%. This is in agreement with that which was verified in Chapter 2, where ZnO was seen to be highly sensitive to $\%O_2$, much more than (G)IZO, which arises mainly due to the existence of grain boundaries in ZnO that act as additional paths for oxygen absorption/desorption processes. The higher sensitivity to substrate bombardment of ZnO thin films should also play an important role for this result. In fact, other authors also found ZnO TFT performance to be highly dependent on substrate bombardment: for instance, Yao *et al.* showed that by putting a grounded mesh between the substrate and the target, the *on/off* was increased by one order of magnitude, due to the less severe substrate bombardment [88].

The overall control of properties and the large process window regarding $\%O_2$ variation achieved for TFTs based on (G)IZO constitutes a great advantage for multicomponent amorphous oxide systems over binary polycrystalline systems.

5.2.1.2 Effect of composition (binary, ternary and quaternary compounds)

The composition of oxide semiconductors has a fundamental role in defining their electrical properties, as demonstrated in Chapter 2. To evaluate the effect in terms of TFT performance, devices were fabricated using several target compositions inside the Ga-In-Zn oxide system, encompassing binary, ternary and quaternary compounds. Semiconductor layers were produced with $\%O_2$=0.4%, being the final devices annealed at T_A=150°C. Transfer characteristics are presented in Figures 5.13 and 5.14 for binary/ternary and ternary/quaternary compounds, respectively, being the trends of μ_{FE} and V_{ON} with composition shown in Figure 5.15 (see color plate section).

The analysis of the data allows one to observe significant differences and trends. Starting with the TFTs comprising binary compounds, large differences are obtained for In_2O_3, ZnO and Ga_2O_3 devices. This is naturally related to the different electrical properties of the respective thin films. For these deposition conditions, $N \approx 10^{18}$ cm^{-3} for In_2O_3, which renders E_F to be very close to CBM, making it impossible to deplete the semiconductor with reasonable V_G values. Hence, even if a very large μ_{FE} is achieved for the In_2O_3 TFTs, due to the E_F pinning above CBM as V_G is increased, the devices are not usable as transistors, since they cannot be switched off. On the other hand, Ga_2O_3 films have non-measurable ρ (i.e., $>10^8$ Ω cm), presumably due to a very low N and large density of empty traps, in addition to the large bandgap, above 4 eV. This results in very poor device performance, with $V_{ON} > 20$ V, $\mu_{FE} \approx 0.02$ cm^2 V^{-1} s^{-1} and large ΔV_{ON}, above 6 V, which is not improved even for $T_A = 500$°C. Chiang *et al.* also reported similar characteristics for Ga_2O_3 TFTs, with $\mu_{inc} \approx 0.05$ cm^2 V^{-1} s^{-1} and $V_{ON} > 10$ V for $T_A = 800$°C [85]. ZnO seems to be the best binary compound for oxide TFT application, at least considering the range of deposition conditions used herein. In fact, close to 0 V_{ON}, *on/off* exceeding 10^6 and $\Delta V_{ON} \approx 1$ V are achieved on these ZnO TFTs. Still, the small slope of the transfer characteristics is synonym of a relatively high S (0.90 V dec^{-1}) and the low maximum I_D is indicative of a low μ_{FE} (1.6 cm^2 V^{-1} s^{-1}) as compared with the other devices depicted in Figure 5.13. Two reasons can be given as the main justification for the moderate performance of the ZnO TFTs: first, $\rho \approx 10^7$ Ω cm, which is indicative of low N and/or large unfilled trap densities;[c] second, the existence of grain boundaries, which are associated with depletion regions with high potential barriers that greatly affect the movement of free carriers.

Multicomponent amorphous oxides, in general, allow for considerably better TFT performance than binary compounds. Rather than having carrier transport limited by grain boundaries, in amorphous oxides carrier transport is mostly dictated by the potential barriers located around the CBM, associated with the structural randomness, which can easily be surpassed in properly processed films by increasing V_G. Furthermore, the relative proportions of the cations directly controls the background N within a broad range, which has direct implications for transistor performance. For IZO TFTs with In/(In+Zn) atomic ratios between 0.50 and 0.80 it is verified that the properties tend to be closer to the predominant binary compound, i.e., for In/(In+Zn)=0.80 both I_D and μ_{FE} are similar to In_2O_3 TFTs,[d] while for In/(In+Zn)=0.50 the properties start to move away from those of In_2O_3 TFTs toward those of ZnO TFTs. Intermediate properties are obtained for In/(In+Zn)=0.67. This means that N and μ_{FE} are higher for increased indium content, but only the IZO 1:1 TFTs are able to present clear *on*- and *off*-states within the V_G range used here. Considerably better device performance is achieved using IGO instead of IZO. This arises as a direct consequence of the strong bonds that gallium forms with oxygen [89], suppressing the generation of free carriers and raising ρ relative to IZO films. In fact, besides other important drawbacks of IZO such as increased light sensitivity, the difficulty in controlling the background N down to low values (at least $<10^{16}$ cm^{-3}) in IZO makes researchers (and industry) move for materials such as GIZO in order to fabricate the active

[c] In fact, it was verified that the properties of ZnO TFTs are improved by growing the ZnO films under a pure argon atmosphere and keeping all the other deposition parameters unchanged. Although this seems to contradict what is generally found in the literature for ZnO TFTs, where most authors have to use $Ar+O_2$ atmospheres to control N to proper levels, one has to bear in mind that the overall deposition conditions (and not only $\%O_2$) dictate the final properties exhibited by films and devices and ZnO is particularly sensitive to small changes on the processing conditions.

[d] Note that even if the In_2O_3 thin films deposited under the conditions used to produce these TFTs are polycrystalline, their high N makes them almost unaffected by grain boundary scattering effects, so they can exhibit similar μ_{FE} to indium-rich multicomponent amorphous oxides.

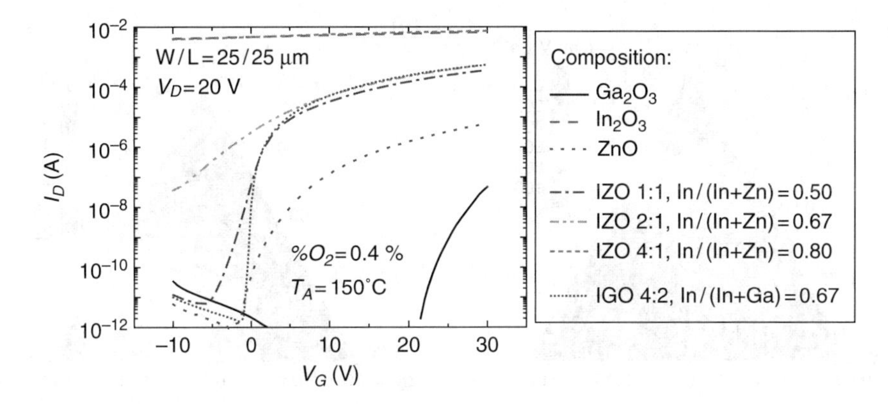

Figure 5.13 *Effect of oxide semiconductor target composition on the transfer characteristics of TFTs annealed at 150°C: binary and ternary compounds.*

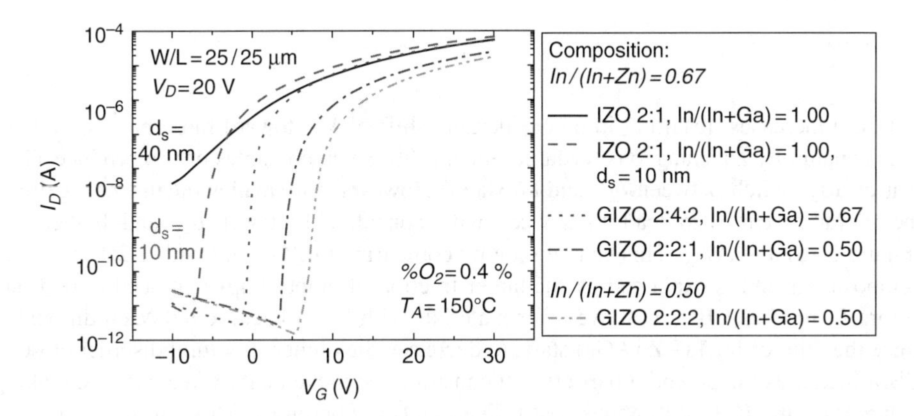

Figure 5.14 *Effect of oxide semiconductor target composition on the transfer characteristics of TFTs annealed at 150°C: ternary and quaternary compounds. The effect of decreasing d_s from 40 to 10 nm for IZO 2:1 is also shown.*

layers of TFTs. However, note that the problem of high background N can be attenuated by decreasing d_s: for instance, by using for IZO 2:1 $d_s = 10$ nm rather than the "standard" 40 nm used for the oxide TFTs presented throughout this chapter, the transistors can be fully switched-off for $V_G < -7$ V (Figure 5.13). This is justified by the fact that the depletion region created at the back (i.e., air exposed) surface of the active layer due to the interaction with environmental oxygen can be extended to the semiconductor/dielectric interface when a low d_s is used.[e]

The effect of adding different amounts of gallium to IZO is shown in Figure 5.14. Given the stronger bonds of gallium with oxygen, background N decreases as the gallium content

[e] A detailed analysis of the effect of d_s can be found in ref. [84].

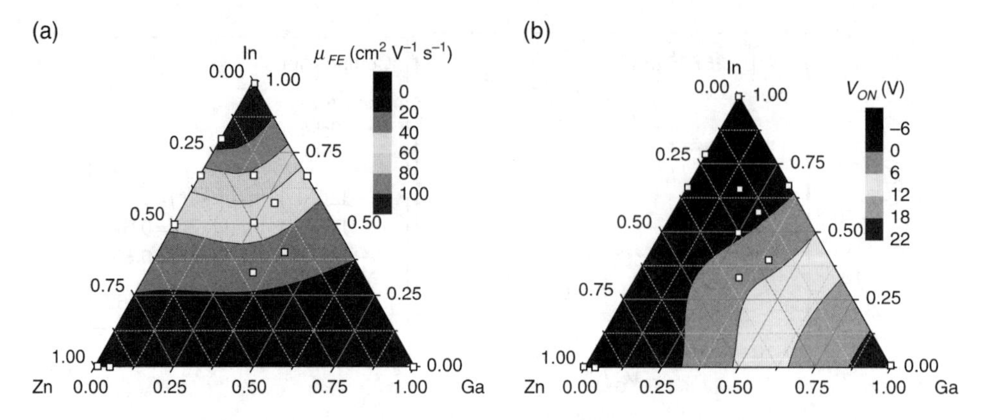

Figure 5.15 *a) μ_{FE} and b) V_{ON} obtained for TFTs with different oxide semiconductor compositions, in the gallium-indium-zinc oxide system. Devices annealed at 150°C, with $\%O_2$=0.4 %. The red and yellow symbols denote amorphous and polycrystalline semiconductor films, respectively.*

in GIZO increases, resulting in a considerable shift of V_{ON} toward more positive values. This is naturally advantageous for large In/(In+Zn), since it enables one to produce TFTs that clearly switch between *off*- and *on*-states. However, for smaller In/(In+Zn), gallium incorporation can also lead to devices with considerably lower μ_{FE} and higher V_{ON} (Figure 5.15) and ΔV_{ON}. This can be seen by comparing GIZO 2:4:2 with 2:2:1 and 2:2:2 compositions and is explained by the larger fraction of empty traps that need to be filled before E_F reaches CBM or above it for materials with lower background N. Additionally, since the ratio of In/(In+Zn+Ga) starts to decrease, the higher structural disorder close to CBM increases the potential barriers that constrain the movement of free carriers, making it harder for the E_F to shift above CBM. The trends verified for In/(In+Zn) in Figure 5.13 can also be observed in Figure 5.14, for fixed In/(In+Ga) atomic ratios (compare GIZO 2:2:1 with 2:2:2), although to a smaller extent than in gallium-free materials, since the properties start to be dominated by the gallium content when this element is present.

Naturally, all of the compositional effects also depend on the remaining processing parameters and so the specific results presented here for a particular composition should not be taken as an entirely strict rule: in fact, a large number of reports regarding stable TFTs with close to 0 V_{ON} exist in the literature, for compositions around 2:2:2 and 2:2:1 [4, 87, 90]. Nevertheless, the overall trends verified here for IZO [91] and GIZO [70, 72, 92–94] devices were also verified by other authors.

Despite the increase of V_G seen above V_{ON} leads invariably to the enhancement of μ_{FE}, as predicted by conventional field-effect theory, the transconductance (hence μ_{FE}) is changed in different ways, depending on the composition and structure of the oxide semiconductor material. Figure 5.16 illustrates this for GIZO 2:4:2 (amorphous) and ZnO (polycrystalline) TFTs, both having semiconductor layers deposited with $\%O_2$=0.4 % and subjected to T_A=150°C. Given that the PECVD SiO_2 has a high breakdown voltage, the V_G range is extended for both cases in order to see all of the μ_{FE}-V_G regimes.

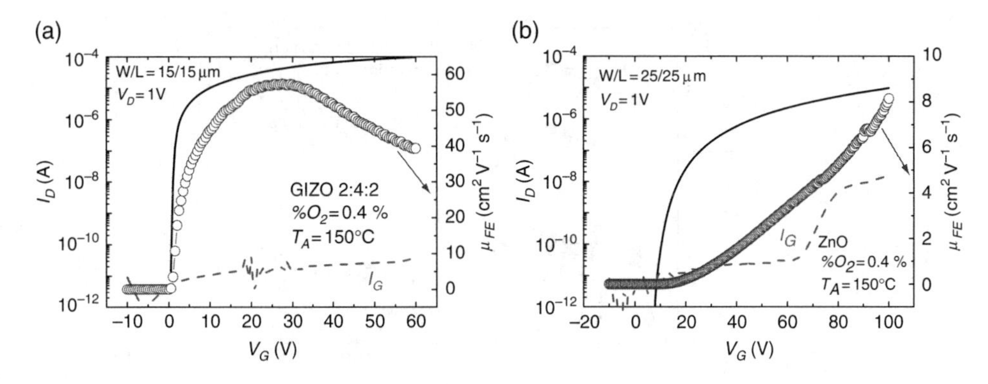

Figure 5.16 *Transfer characteristics and μ_{FE}-V_G plots measured for high V_G for TFTs annealed at 150°C, based on a) GIZO 2:4:2 and b) ZnO.*

For the GIZO TFT (Figure 5.16a) an almost abrupt increase of μ_{FE} is verified after V_{ON}, given the small S. Physically, E_F is raised very quickly above CBM by V_G, since the trap density is very low and very large μ_{FE} can be achieved when the small potential barriers associated with structural disorder are surpassed, which happens for $V_G \approx 20$–30 V, where μ_{FE} is maximum. As V_G gets higher than these values, the conductive channel is drawn closer to the GIZO/SiO_2 interface, which contributes to increased scattering effects of the large density of induced charges, resulting in a decrease of μ_{FE} [1]. An electron injection barrier at the source electrode may also contribute to this drop in μ_{FE}, as proposed by Dehuff *et al.* [64]. A considerably different μ_{FE}–V_G trend is observed for the ZnO TFT (Figure 5.16b). In this case the increase of μ_{FE} with V_G is much more gradual, due essentially to the polycrystalline structure of ZnO with small grain sizes, hence the large density of grain boundaries. Some models describe quite well the behavior of polycrystalline TFTs, namely that proposed by Levinson in 1982 [95] and more recently that of Hossain, specifically designed for ZnO TFTs [51]. From these models it can be seen that the barrier height associated with the depletion regions at the grain boundaries is modulated by the total N, which has contributions both by the background N and by the charges induced by V_G. Furthermore, for smaller grain sizes the width of the depletion regions can extend deep inside the crystallites and even overlap with adjacent depletion regions, resulting in very high ρ. For the ZnO TFTs presented in Figure 5.16b the modulation of these effects by V_G should be dominant and overshadow ZnO/SiO_2 interface scattering, even if V_G is increased up to 100 V. Although measured in a smaller V_G range, Nishii *et al.* also obtained a similar μ_{FE}-V_G trend in ZnO TFTs [47]. For a large V_G range, saturation (and even decreasing) of mobility is observed by Hoffman for V_G>70 V, on ZnO TFTs produced on thermal SiO_2 [11]. However, note that mobility is extracted by Hoffman using a different methodology, designated by incremental mobility (μ_{inc}), which probes the mobility of carriers as they are added incrementally to the channel, rather than by averaging the mobility of all the carriers present in the channel for a given V_G [11]. In fact, by using the more conventional "average mobility" (μ_{avg}), physically similar to μ_{FE}, mobility saturation is not achieved in Hoffman's work. By plotting μ_{inc}–V_G for the data depicted in Figure 5.16b, saturation is still not obtained, which can be due to the different ZnO processing conditions (plausible given the considerably higher μ_{FE}

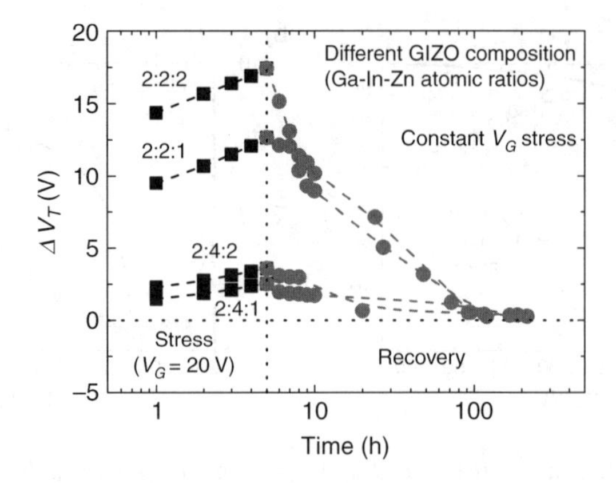

Figure 5.17 *ΔV_T obtained on gate-bias stress measurements for 5h, for oxide TFTs with different GIZO compositions, annealed at 150°C. Reprinted with permission from [96] Copyright (2011) Elsevier.*

achieved by Hoffman) and also to the effect of I_G that for the PECVD SiO_2 starts to increase considerably for V_G>70V.

Some observations should also be made regarding the other polycrystalline binary compound, In_2O_3. If the devices employing this semiconductor are annealed at a higher T_A to allow for complete crystallization, such as T_A=300°C, intermediate properties to those observed in Figure 5.16a and b are observed. Although grain boundaries can still affect the movement of the free carriers, their effect should be considerably smaller, given the larger grain size and larger N of this material. Hence, in the unbiased state, most of the traps associated with grain boundaries can be compensated by the background N, leaving more of the V_G induced charges available to increase the transconductance.

Stress measurements are crucial in order to understand instability mechanisms in devices as well as to predict whether the technology is suitable to be used in integrated circuits. Hence, the dependence of GIZO composition on the electrical stability of oxide TFTs under gate-bias stress measurements was also studied, for GIZO layers with $\%O_2$=0.4% and T_A=150°C (Figure 5.17). Stress measurements were carried out by applying a constant V_G=20V for 5h, while keeping the source and drain electrodes grounded.

As happens with other semiconductor technologies [97–99], reversible charge trapping at or close to the dielectric/semiconductor interface, resulting in a shift of V_T (ΔV_T), is the most frequent instability mechanism [82, 100–104]. The same is verified here for our devices, with 2:2:2 and 2:2:1 GIZO compositions resulting in the most unstable devices, as would be predictable given the larger ΔV_{ON} and hysteresis verified for these compositions. Nevertheless, regardless of composition, the initial properties can be recovered some hours after stress without any subsequent annealing treatment, which is consistent with conventional charge trapping mechanisms rather than defect state creation or ionic drift [44, 82, 105, 106]. Despite GIZO 2:4:1 and 2:4:2 compositions provide the best performing and more stable transistors, it will be seen in section 5.2.1.5 that even for optimized devices stability can be greatly improved using a passivation layer on top of GIZO.

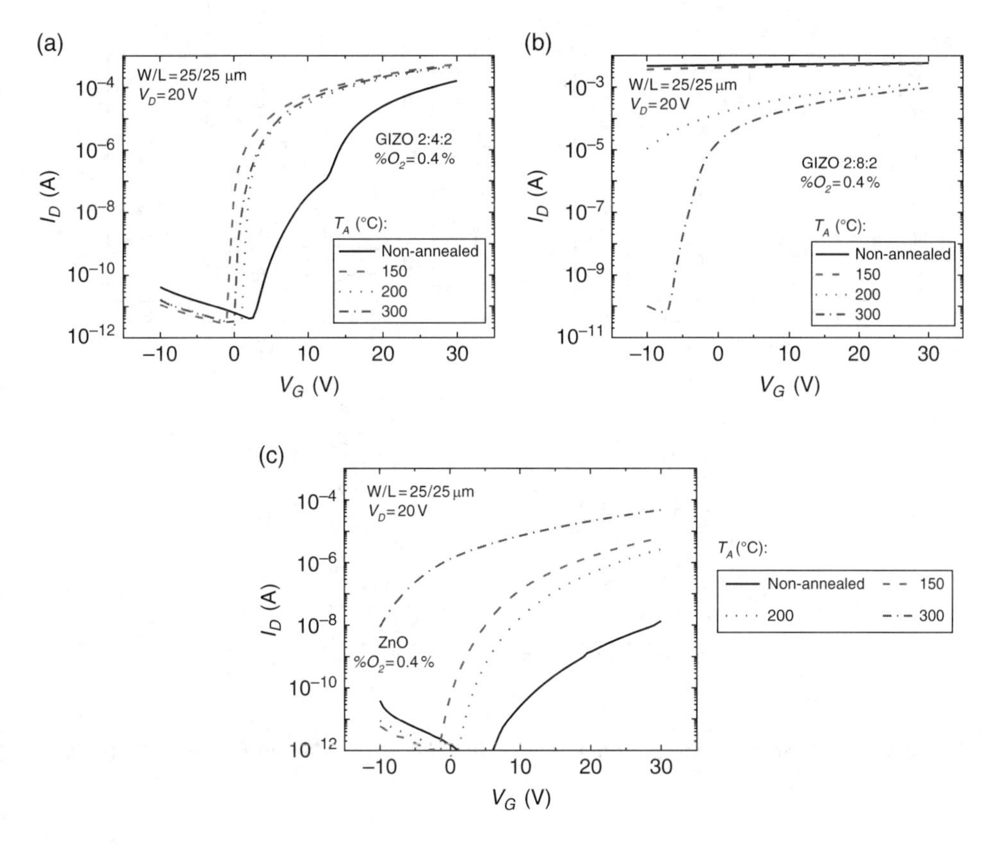

Figure 5.18 *Effect of T_A on the transfer characteristics exhibited by TFTs with different semiconductor compositions, with $\%O_2$=0.4 %: a) GIZO 2:4:2; b) GIZO 2:8:2; c) ZnO.*

5.2.1.3 *Effect of annealing temperature*

Even if flexible and low-cost electronics demands the use of very low T_A, typically below 200°C, depending on the composition and processing conditions of oxide semiconductors higher T_A might be required to achieve proper performance and stability levels. Furthermore, it was verified during our research that reproducibility is greatly improved after a low temperature annealing treatment, typically below 300°C [107]. Figures 5.18 and 5.19 show the effect of T_A on oxide TFTs with different semiconductor compositions, produced with $\%O_2$=0.4 %.

As stated in the previous section, for the specific processing conditions used here, GIZO 2:4:2 is one of the compositions providing better transistor performance for low T_A (150°C). Interestingly, this composition also results in the less significant changes of properties with T_A, with the largest evolution being verified between non-annealed and 150°C annealed TFTs (Figures 5.18a and 5.19a). This could be associated with the removal by the low-temperature annealing treatment of residual contaminations due to the complex processing steps, namely the lithographic processes. These residuals can be one of the causes for the non-idealities frequently observed in the transfer characteristics of non-annealed devices, such as in Figure 5.18a, which can be associated with acceptor trap levels [44, 108]. Despite

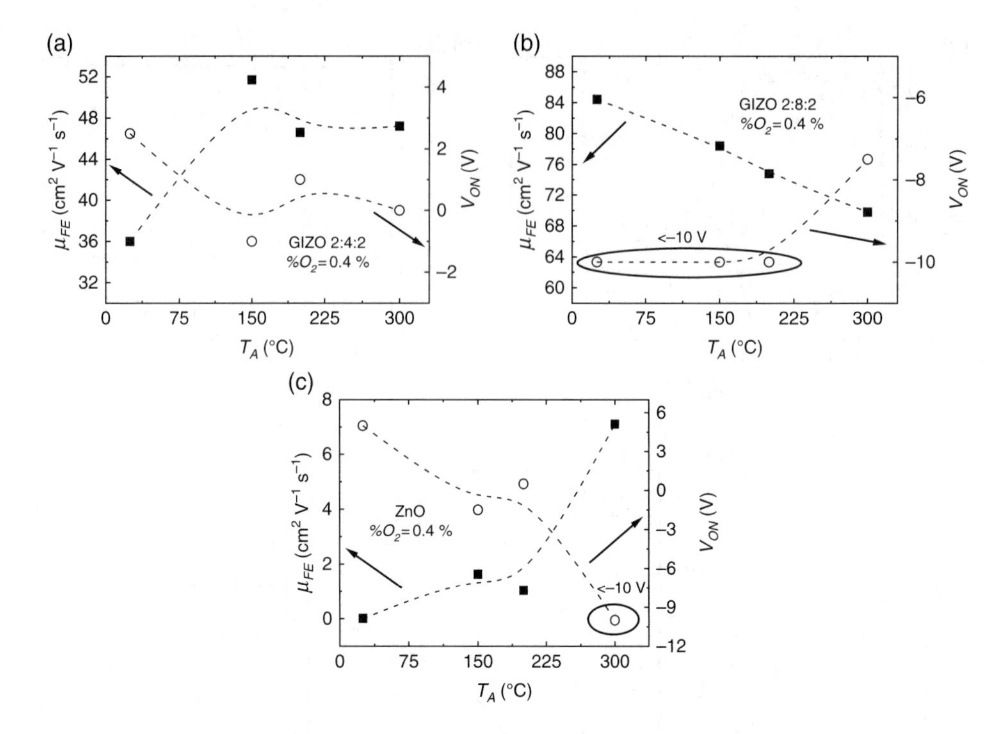

Figure 5.19 *Electrical properties obtained for the devices depicted in Figure 5.18: a) GIZO 2:4:2; b) GIZO 2:8:2; c) ZnO.*

the optimized processing conditions and composition used in this case, and even if the density of defect states in the semiconductor and/or at the GIZO/SiO_2 interface could be decreased by increasing T_A above 150°C, the combined effect of this with the slight source-drain electrodes degradation at 300°C (see section 5.2.1.4) turns the overall transistor performance barely changed between $T_A = 150$ and 300°C.

The effect of T_A is much more evident for compositions that are far from ideal, for instance GIZO 2:8:2, which for the processing conditions used here results in highly conductive thin films. In this case, only after $T_A = 200$°C is it possible to have significant channel conductivity modulation by V_G, although the devices still remain in the *on*-state for all the V_G range used here (Figures 5.18b and b). After annealing at 300°C, *off*-state can be achieved but the devices still work in depletion mode ($V_{ON} = -7.5$ V). Since the as-deposited films have a very high N ($\approx 10^{20}$ cm^{-3}), the evolution of properties with increasing T_A is being controlled essentially by the N variation, which only decreases to the 10^{17} cm^{-3} range at $T_A =$ 300°C, with the improvement of semiconductor/dielectric interface and decrease of the trap states density as T_A increases being less relevant for this case.

ZnO TFTs are also analyzed regarding T_A dependence in Figures 5.18c and 5.19c. The results show that the electrical properties are significantly modified with T_A. The non-annealed devices exhibit rather poor performance, concomitant with a large defect density and low N of the ZnO films. This is especially important given the polycrystalline nature of this semiconductor, making the transport mechanism controlled essentially by the depletion layers of the grain boundaries [44]. But given that grain boundaries should behave as preferential paths for interdiffusion processes, the effects of T_A are enhanced in this material. Hence, ρ is significantly decreased for T_A=300°C, resulting in a large enhancement of μ_{FE}. However, this is accompanied by a severe decrease of V_{ON}, rendering the device to be in the *on*-state even for V_G=−10V.

These results show, once again, the very narrow process window required to obtain good performance devices based on ZnO, being that the process window is highly amplified when using amorphous multicomponent oxides.

5.2.1.4 *Influence of source-drain electrodes material*

In order to fabricate TFTs with good electrical properties and to use them as elements of circuits such as active matrix backplanes, one of the essential requirements is to form good electrical contacts between the semiconductor and the source-drain regions [5]. In fact, several reports in the literature concerning oxide TFTs show important non-idealities in devices that are attributed to contact effects, such as current crowding in the output characteristics for low V_D [109–112], μ_{FE} degradation and increase of pinch-off voltage [107].

A dual layer comprising titanium and gold is often used for source-drain electrodes in oxide TFTs (e.g. [90, 111, 113]). This combination generally allows one to obtain reliable Ohmic contacts with low specific contact resistance to ZnO related materials, at least for T_A up to 300°C [114]. But for the fabrication of fully transparent devices, alternative materials have to be studied. At CENIMAT, highly transparent and conductive IZO, deposited by sputtering without intentional substrate heating, is explored as an electrode material, not only for TFTs, but also for other devices such as OLEDs [115, 116]. Figure 5.20 compares the electrical performance of GIZO TFTs produced with Ti/Au and IZO source-drain electrodes. The GIZO layer was produced with a 2:4:2 target composition and $\%O_2$=0.4%, being the final devices annealed at 150 and 300°C.

The properties at T_A=150°C are quite similar for both source-drain materials, being that the most notorious difference in the μ_{FE} degradation occurs at high V_G for devices with IZO electrodes. When annealed at 300°C, both devices show a decrease on μ_{FE}, but the effect is much more significant for IZO, which starts presenting noisy characteristics, specially for large V_G. This is further evidenced by the output characteristics of the T_A=300°C TFTs, represented in Figure 5.20b and c for Ti/Au and IZO electrodes, respectively. The insets in these figures show that for low V_D (i.e., linear region) Ti/Au electrodes allow for larger I_D and less current crowding effects than IZO.

Non-idealities at the source-drain/semiconductor contact region in TFTs can generally arise due to the creation of Schottky barriers and large contact resistance, although sometimes both are simply defined as "contact effects" [117]. When a metal and a semiconductor are joined there is a charge transfer between both materials until their E_F levels are aligned.[f] The direction of charge flow is dictated by the work function differences of the metal (Φ_M) and semiconductor (Φ_S), being the barrier height (Φ_B) formed after contact given by $\Phi_B=\Phi_M-\chi$, where χ is the electron affinity of the semiconductor [6, 12]. Accumulation, neutral and depletion contacts can be obtained, for $\Phi_M<\Phi_S$, $\Phi_M=\Phi_S$ or $\Phi_M>\Phi_S$, respectively. Although accumulation-type contacts would naturally be preferred to

[f]For simplicity, no interfacial layers are considered in this exposition.

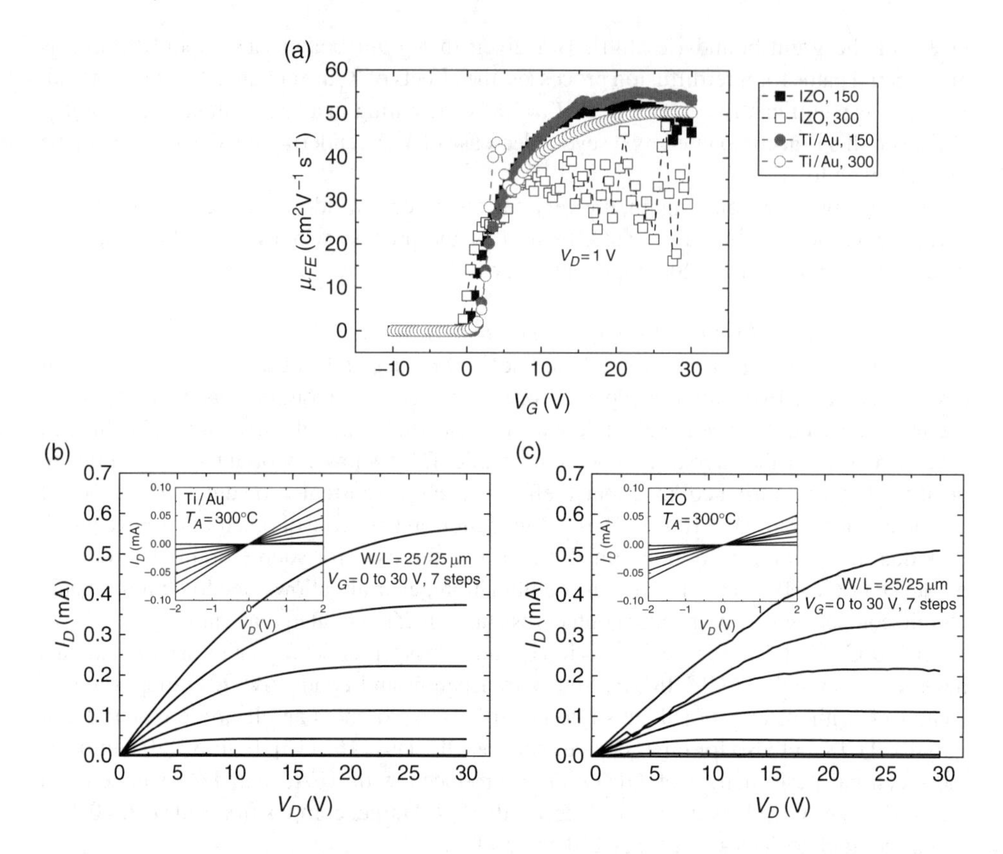

Figure 5.20 *Electrical properties of GIZO 2:4:2 TFTs (%O_2=0.4 %) with Ti/Au and IZO source-drain electrodes, for T_A = 150 and 300°C: a) μ_{FE}-V_G plots; b) output characteristics for devices with Ti/Au electrodes (T_A = 300°C); c) output characteristics for devices with IZO electrodes (T_A = 300°C).*

achieve good ohmic behavior, in practice, depletion-type contacts are obtained with most of the materials. Hence, to have an ohmic contact, most of the work is focused on decreasing Φ_B as well as the depletion layer width (x_d), which depends on $N^{-1/2}$. In fact, for very thin depletion regions, even if Φ_B is relatively high, carriers can tunnel through instead of going over the potential barrier (field emission rather than thermionic emission) [6, 12].

Considering the same χ for both IZO and Ti/Au devices, Φ_B is higher when IZO is used for source-drain electrodes, because Φ_{IZO}≈4.8 eV and Φ_{Ti}=4.3 eV (Φ_{IZO} and Φ_{Ti} were measured by Kelvin probe).[g] In addition, the GIZO layer can be slightly oxidized during the IZO electrodes sputtering process, decreasing N, as suggested by Shimura *et al.* [118]. If this would be the case, it would result in the widening of x_d, but given the linear *I-V* characteristics obtained for IZO electrodes at T_A=150°C, it seems that this effect is not particularly relevant here, at least to the extent of significantly inhibiting the tunneling of electrons through the potential barrier of the contact. The interface quality between IZO and GIZO is expected to be good, with low

[g] IZO is considered here as a metal, given its high degeneracy.

density of defects, given the similar amorphous structures and smooth surfaces of both materials. On the other hand, it is reported that very thin TiO_x layers are created when Ti is deposited on top of oxide semiconductors, such as ZnO, even before any annealing treatment. This is attributed to the very high affinity of titanium with oxygen [119, 120]. Although the formation of this TiO_x interfacial layer could be seen as disadvantageous for good contact properties, the fact is that it increases N near the semiconductor surface (via oxygen vacancies creation), resulting in easy tunneling of carriers through the thin oxide barrier. The Au film on top of the very thin Ti layers prevents oxidation of the bulk electrode when exposed to air. Due to the combination of these phenomena, the properties of GIZO TFTs employing whether IZO or Ti/Au source-drain electrodes are similar when $T_A = 150°C$.

When T_A is increased to 300°C, several effects are observed at the contacts that might justify the large differences obtained in terms of device performance. Regarding IZO, its Φ_M is increased due to the shifting of E_F towards the midgap, in response to the decreased N. This results in a significant increase of IZO's ρ by almost one order of magnitude relative to $T_A = 150°C$ and in a larger Φ_B, degrading contact properties (note that the optimized GIZO films used herein do not present significant electrical performance variation between $T_A = 150$ and 300°C). For Ti/Au electrodes, it is verified by TOF-SIMS analysis that for $T_A \geq 250°C$ interdiffusion of elements occurs, similar to that which was verified in Ti/Al/Pt/Au contacts on ZnO [114, 116]. Besides the increased oxidation of Ti that could arise as a consequence of a higher T_A, (this could not be confirmed by the current TOF-SIMS setup, given the low signal obtained for oxygen) the interdiffusion effects should even allow some Au to reach the Ti/GIZO contact layers, changing significantly the overall contact properties. For Au-richer interfaces, a higher Φ_M (thus Φ_B) is expected, given that $\Phi_{Au} > \Phi_{Ti}$. In agreement with this, note that pure Au source-drain electrodes (i.e., without the thin Ti layer) are reported to result in worse performing GIZO TFTs, namely in terms of μ_{FE} and V_T [118].

Important information about contact resistance on TFTs can be extracted by measuring several devices with different channel lengths at low V_D and assuming that the total TFT on-resistance (R_T) is given by [2]:

$$R_T = \frac{V_D}{I_D} = r_{ch} L + 2R_{SD} \tag{5.7}$$

where r_{ch} is the channel resistance per channel length unit and $2R_{SD}$ is the total (source + drain) series resistance. R_{SD} includes the contributions both of the contact itself as well as the semiconductor regions between the contact and the channel. The width-normalized $2R_{SD}$ and $r_{ch} \times L$ values can be plotted against V_G so as to infer about their relative contributions to R_T. This is shown in Figure 5.21a for IZO and Ti/Au devices, with $T_A = 150°C$. As expected for this T_A, the series resistance contribution using either IZO or Ti/Au is essentially the same, only marginally larger for the former. This figure also illustrates other important aspects:

- Both r_{ch} and $2R_{SD}$ decrease with increasing V_G. The large r_{ch} dependence on V_G would be expected having in mind field-effect theory, since more carriers are being induced in the channel as V_G increases. Regarding the $2R_{SD}$ dependence, one should account for the fact that the source-drain to gate overlap is full for the present devices, since the silicon wafer works as the gate electrode. Hence, carriers are also induced in the GIZO regions outside

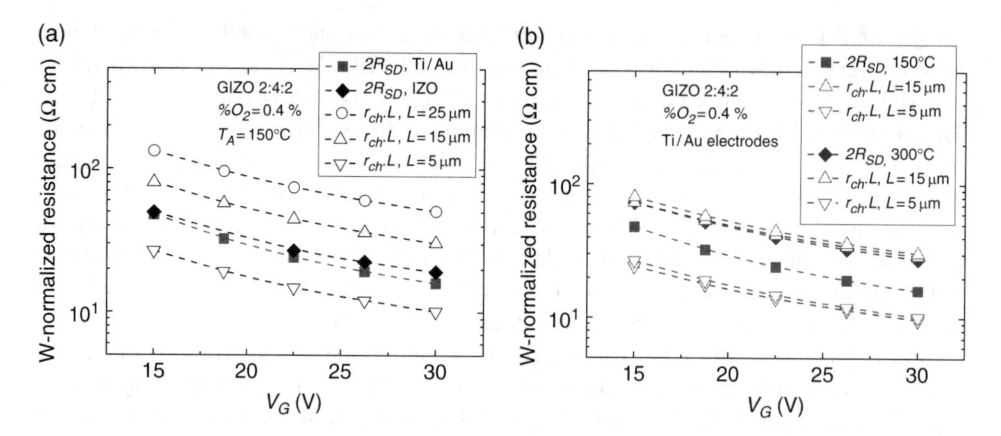

Figure 5.21 *Width-normalized $2R_{SD}$-V_G and r_{ch}.L-V_G plots for different GIZO 2:4:2 TFTs (%O_2=0.4 %): a) devices with Ti/Au and IZO source-drain electrodes, annealed at 150°C; b) devices with Ti/Au source-drain electrodes annealed at 150 and 300°C. Given the low T_A in a), r_{ch}.L is assumed to be the same for both Ti/Au and IZO cases.*

the channel length defined by L, reducing the overall series resistance as V_G increases. In a-Si:H TFTs similar relations were found, being that the R_{SD}-V_G dependence is considerably lower for reduced source-drain to gate overlaps [121].

- For L between 15 and 5 µm R_{SD} starts to be similar or even higher to r_{ch}. This is extremely relevant for designing TFTs based on the current processes and materials, since it indicates that for this L range the device properties begin to be dominated by contact effects rather than intrinsic semiconductor and dielectric/semiconductor interface characteristics.

The same principle can be used to compare devices annealed at different temperatures, as presented in Figure 5.21b for TFTs employing Ti/Au source-drain electrodes. As predicted by the previous analysis, R_{SD} increases after T_A=300°C, revealing degraded contact properties. Additionally, r_{ch} barely decreases for T_A=300°C, confirming that ρ of GIZO thin films produced under these conditions is not significantly affected by this T_A range. Furthermore, while TFTs annealed at 150°C only show $R_{SD}>r_{ch}$ for L=5–10 µm, the R_{SD} of TFTs annealed at 300°C is already comparable to r_{ch} for L=15 µm, meaning that contact resistance starts to be critical for larger L devices as T_A increases.

In reports related to GIZO TFTs using MoW source-drain electrodes, the authors found the devices to be contact limited at considerably larger L, around 30 µm [4, 69]. Although the source-drain material selection seems to be able to justify the difference by itself, the GIZO deposition conditions and composition can also contribute to this. Width normalized $2R_{SD}$ and $r_{ch}\times L$ plots are presented in Figure 5.22 for TFTs having non-ideal GIZO layers, produced with 2:2:1 composition and %O_2=10.0 %, being the final devices annealed at 150°C.

The results show that the semiconductor layer can greatly affect both $2R_{SD}$ and r_{ch} and devices begin to be contact limited for L above 15 µm, even if the TFTs were only annealed at 150°C. A large increase of r_{ch} when compared with the values presented in Figure 5.21 would be expected, since both composition and %O_2 greatly affect intrinsic semiconductor properties, as seen above. Regarding the increase of $2R_{SD}$, it should be ascribed not only to

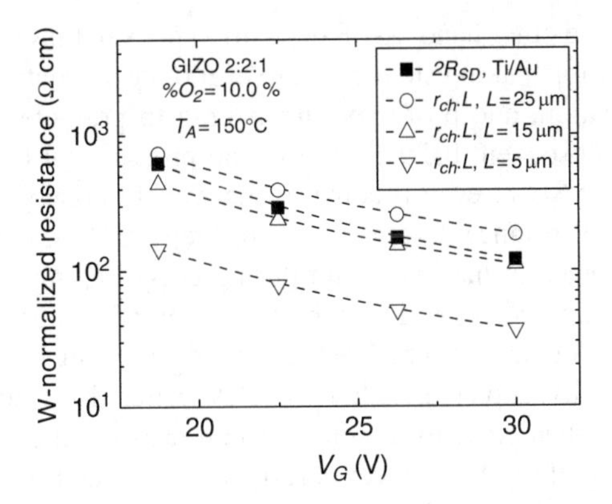

Figure 5.22 *Width-normalized $2R_{SD}$-V_G and $r_{ch}.L$-V_G plots for TFTs employing a non-ideal GIZO process (2:2:1, $\%O_2$=10.0 %), with Ti/Au source-drain electrodes and T_A=150°C.*

the higher ρ of the access regions inside the GIZO layer but also to degraded metal/semiconductor contact properties. In fact, both Φ_B and x_d are expected to be larger for the present case: regarding Φ_B, it should be increased due to the higher gallium content of GIZO, which makes its χ become lower, since Ga_2O_3 has lower χ than any of the other binary compounds present in GIZO (ZnO and In_2O_3) [70]. Given the lower N of GIZO, x_d is also increased, as x_d scales with $N^{-1/2}$.

5.2.1.5 *Influence of passivation layer*

A passivation layer is required in a TFT in order to protect it from environmental or subsequent processing conditions that could change its electrical properties [4, 80, 122] or to prepare it for integration processes in LCDs or OLED displays [5]. In addition, passivation layers seem to improve the uniformity of oxide TFTs over large areas [123].

Figure 5.23 shows the effect of two different passivation layers, sputtered SiO_2 and spin-coated SU-8 [124–126], on the performance and stability of GIZO TFTs. A GIZO 2:4:2 composition and $\%O_2$=0.4 % were used for these devices, which were annealed at 200°C before and after the passivation layer deposition. Despite the slight shift of V_{ON} towards negative V_G, the overall properties and stability of SU-8 passivated devices after annealing are greatly improved over non-passivated transistors, while the sputtered SiO_2 passivation results in always-on devices [127].

The results can be understood based on the passivation material and on its deposition technique. During the pump down time to sputter the SiO_2 layer some of the oxygen lying at GIZO's back surface can be removed, decreasing or even removing the depletion layer close to that back surface. Additionally, during sputtering of an insulator material such as SiO_2, which requires deposition with high P_{rf} and low p_{dep} in order to ensure reasonable growth rates, intense substrate bombardment can arise, breaking weak metal cation-oxygen bonds, resulting in oxygen vacancy generation, and thus higher N. Then, if the deposited passivation layer has a large density of positive charges at its bulk and/or close to its interface with the semiconductor, a significant electrons accumulation layer is created close to GIZO's back surface, making channel conductivity modulation hard to achieve, even for a large negative V_G. On the other hand, SU-8 deposition does not involve vacuum processes, only spin-coating, multiple baking steps taken to evaporate solvents, UV exposure to initiate chemical amplification, and post-exposure baking to promote cross-linking reaction and development. A subsequent baking process can promote further cross-linking of the polymeric network. For SU-8, even if some of the weakly bonded oxygen might lost its bonding with GIZO, for instance during the baking steps, UV exposure or due to the rearrangements that take place in the SU-8 film during cross-linking, there is no vacuum or substrate bombardment processes to favor the formation of a large accumulation layer at the GIZO back surface. In fact, even before the second annealing process, SU-8 passivated TFTs can be almost fully depleted with $V_G = -10\,V$. After the second annealing step at 200°C, electrical properties close to the ones of non-passivated devices are obtained. At this stage, cross-linking of the SU-8 layer is fully achieved (confirmed by FTIR analysis) and the remaining H^+ ions available at the passivation layer, close to its interface with GIZO, can probably capture some electrons from GIZO, re-establishing a depletion layer close to the back surface of GIZO. A second annealing step was also used by other authors to re-establish good device properties after passivation [103, 128].

Besides the negligible sensitivity of SU-8 passivated GIZO TFTs to vacuum or high moisture environments [84], the effectiveness of SU-8 passivation is also reinforced by the improvement of the stability under constant I_D stress measurements (Figure 5.23b). In fact, a ΔV_T as small as 0.46 V was obtained for SU-8 passivated TFTs after 24 h $I_D = 10\,\mu A$ stress, against 2.28 V for non-passivated ones. This stability enhancement is essentially attributed to the barrier formed by the passivation layer, inhibiting oxygen adsorption/desorption processes that are known to be potentiated when ZnO-based materials are subjected to electric fields [129]. On the other hand, for non-passivated devices, oxygen can be adsorbed as the stress measurement progresses, widening the depletion layer close to the back surface and increasing V_T. Improvements in the stability of passivated oxide TFTs are also reported in the literature, for instance by Cho *et al.* and Levy *et al.* [103, 130].

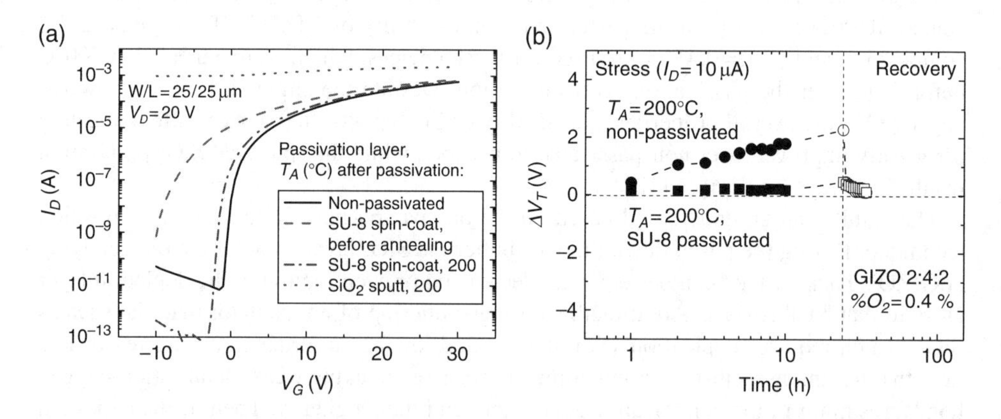

Figure 5.23 *Electrical properties of GIZO 2:4:2 TFTs ($\%O_2 = 0.4\,\%$, $T_A = 200°C$) with SU-8 and SiO_2 passivation layers: a) transfer characteristics; b) ΔV_T vs stress/recovery time obtained from constant I_D stress measurements with and without SU-8 passivation.*

5.2.2 N-type GZTO TFTs by Physical Vapor Deposition

Even if from the analysis presented above for GIZO TFTs it may be observed that high indium contents on the active layer lead to the best performing devices, indium-free alternatives are also important to consider, given the high cost and relative scarcity of indium, for instance when compared with zinc [131]. GZTO TFTs were fabricated using a similar process flow to that presented in Figure 5.11, being the GZTO layer (≈50 nm thick) obtained by co-sputtering from Ga-doped ZnO and metallic tin targets [132]. GZTO sputtering was performed both without intentional substrate heating and at 150°C, while annealing was carried out in nitrogen atmosphere, during 1h, between 150 and 300°C. Figure 5.24 and Table 5.2 show the electrical properties obtained for these devices (non-passivated).

Even if the processing conditions and composition of GZTO are not yet fully optimized, the TFTs already present good electrical properties, particularly when GZTO thin films are produced at 150°C (Figure 5.24b). As with GIZO TFTs, the overall properties are improved as T_A increases, but for GZTO the effect of T_A is even more significant. Furthermore, at least for the present deposition conditions and composition of GZTO, higher T_A than for GIZO

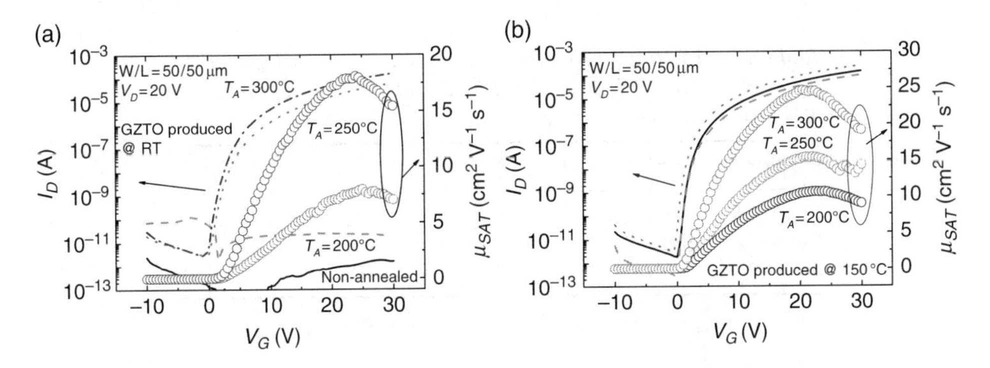

Figure 5.24 *Transfer characteristics and μ_{FE}-V_G plots for GZTO TFTs annealed at different T_A, with GZTO layer co-sputtered a) without intentional substrate heating and b) at 150°C.*

Table 5.2 *Electrical properties for the GZTO TFTs depicted in Figure 5.24b (GZTO deposited at 150°C). For comparison, results for a GZTO TFT where the GZTO layer was co-sputtered without intentional substrate heating are also presented.*

T_A (°C)	*on/off*	μ_{SAT} (cm^2 V^{-1} s^{-1})	V_{ON} (V)	S (V dec^{-1})
200	3.4×10^8	10.6	0	0.45
250	8.0×10^7	15.4	0	0.46
300	8.2×10^7	24.6	−0.5	0.38
300 (GZTO deposited @ RT)	5.8×10^7	18.1	−0.5	0.62

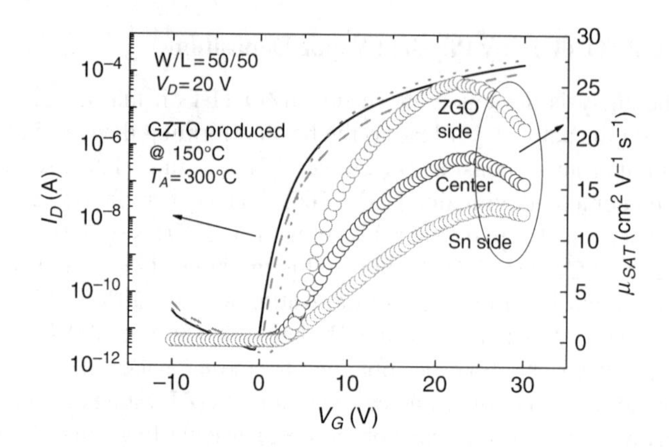

Figure 5.25 Transfer characteristics and μ_{FE}-V_G plots for GZTO TFTs where the GZTO layer was produced in different positions of the substrate holder (at 150°C), yielding films with different compositions. Devices annealed at 300°C.

Table 5.3 Electrical properties for the GZTO TFTs depicted in Figure 5.25. GZTO deposited at 150°C and final devices annealed at 300°C.

Position	on/off	μ_{SAT} (cm^2 V^{-1} s^{-1})	V_{ON} (V)	S (V dec^{-1})
ZGO side	1.1×10^8	25.4	1.5	0.54
Center	5.8×10^7	18.1	−0.5	0.62
Sn side	3.4×10^7	13.0	0	0.64

is required in order to achieve good electrical performance. As proposed by Hosono *et al.* [71], this should be related to the small difference in the thermodynamic stability between tin 2+ and 4+ valence states. As T_A increases, atomic rearrangements should result in the conversion of Sn^{2+} into Sn^{4+}, and given that Sn^{2+} has absorption for lower energies than Sn^{4+}, this results in larger absorption tail state intensity for low T_A, which is concomitant with degraded transistor performance. In fact, for as-deposited films, the absorption tail state intensity of GZTO is reported to be two orders of magnitude larger than GIZO, with significant improvements obtained only for T_A>300°C [71].

To evaluate the effect of different GZTO compositions, films with similar thicknesses were simultaneously sputtered at 150°C into three Si/SiO_2 substrates placed in a parallel arrangement with respect to the two magnetrons (Figure 5.25). The TFTs were then annealed at 300°C (Table 5.3). Similar to that which was verified for GIZO, the films having a higher concentration of the most abundant cation (while ensuring that an amorphous structure is preserved) present improved carrier transport properties, namely higher μ. Given that zinc is the predominant cation for all the GZTO compositions studied herein, the best transistor performance is achieved for the "ZGO side" devices, whose active layer has a higher zinc content, and thus fewer potential barriers close to CBM associated with the structural disorder generated by the other composing cations. Note that contrary to the indium-rich compositions studied in the Ga-In-Zn oxide system, which despite the high μ_{FE} resulted in negative V_{ON} (or even always-on) devices, for the GZTO compositions studied herein $N < 10^{16}$ cm^{-3}, assuring that a clear off-state is always achieved.

5.2.3 N-type Oxide TFTs by Solution Processing

The solution processing of TSOs to be used as active layers of oxide TFTs is nowadays a research topic of great interest, promising a significant cost-reduction when compared with conventional vacuum processing. Both ZTO (spray-pyrolysis and sol-gel spin-coating) and GIZO (sol-gel spin-coating) are being studied to that end at CENIMAT. The devices presented in the following pages were produced using a simple structure, comprising glass/ITO/ATO substrates where the oxide semiconductor is deposited and then subjected to annealing treatments in air, between 200 and 500°C, depending on the material and deposition technique. Then, aluminum source-drain electrodes were e-beam evaporated, their pattern being defined using a shadow mask during deposition.

Although the main intent here is to demonstrate the feasibility of good performance transistors even at this early stage of solution processing of oxide TFTs, it should be pointed out that the semiconductor layers of the devices presented here are not patterned, which should result in some overestimation for the μ values due to fringing currents [108]. However, relatively large W/L ratios (>10) are chosen for these devices to minimize that source of error [133].

5.2.3.1 ZTO TFTs by spray-pyrolysis

Figure 5.26 shows the transfer and output characteristics of ZTO TFTs, for ZTO deposited by spray-pyrolysis (20 nm thick) and subjected to $T_A = 200°C$ for 30 min. prior to aluminum source-drain electrode deposition.

The devices still exhibit a large *off*-current and no hard saturation is obtained, suggesting that the background N should be reduced for proper transistor operation. However, good channel conductivity modulation with V_G can be observed and promising electrical properties such as $\mu_{FE} \approx 10$ cm^2 V^{-1} s^{-1}, *on/off* $\approx 4 \times 10^4$, $V_T = 0.6$ V and $S = 0.4$ V dec^{-1} are already obtained at this early stage of research.

5.2.3.2 ZTO TFTs by sol-gel spin-coating

20 nm thick ZTO layers prepared by spin-coating were annealed at 500°C for 1 h for the production of ZTO TFTs. Fresh devices exhibit $\mu_{FE} = 5.2$ cm^2 V^{-1} s^{-1}, *on/off* $\approx 10^8$, $V_{ON} = -0.9$ V and $S = 0.15$ V dec^{-1} [134]. The stability of ZTO TFTs under illumination with different wavelengths was studied by measuring them under red ($\lambda = 630$ nm), green ($\lambda = 540$ nm) and blue ($\lambda = 410$ nm) light illuminations (Figure 5.27). The duration of light illumination was varied by up to 30 minutes with light intensity fixed at 50 μW cm^{-2} in all cases. The only significant variations arose under blue light exposure, with the *off*-current increasing by one order of magnitude and V_{ON} shifting ≈1 V towards negative V_G values. This could be attributed to photo generated carriers that increase N [135, 136] and to light-induced oxygen desorption [137].

The stability of the ZTO TFTs under gate bias stress was also evaluated, both for positive (10 V) and negative (−10 V) V_G, during 5000 s, keeping the source and drain electrodes at 0 V. The only visible effect is a V_{ON} (or V_T) shift in the positive and negative directions

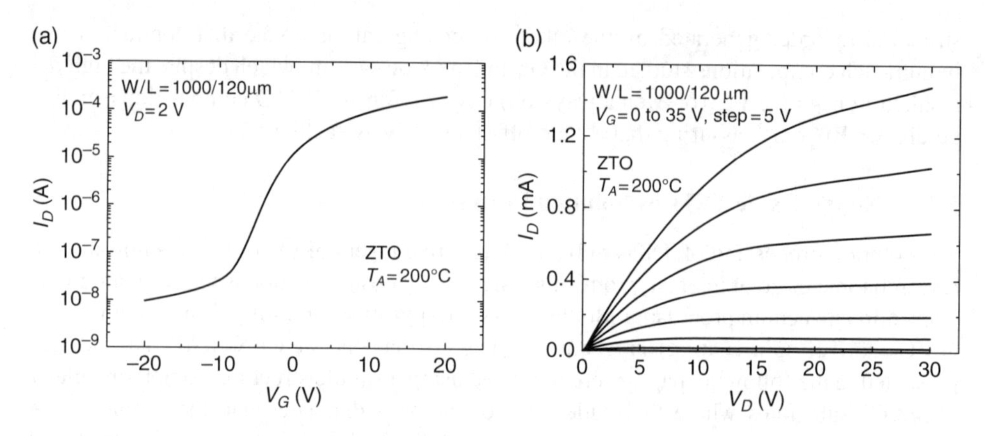

Figure 5.26 *a) Transfer and b) output characteristics of ZTO TFTs with ZTO deposited by spray-pyrolysis. Devices annealed at 200°C.*

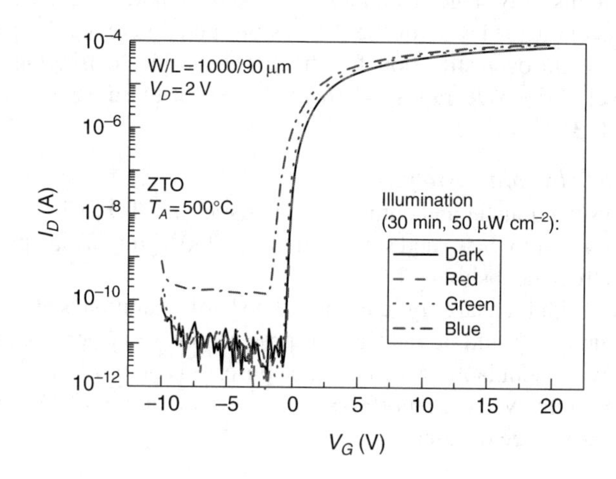

Figure 5.27 *Effect of illumination with different wavelengths on the transfer characteristics of spin-coated ZTO TFTs. Devices annealed at 500°C. Reprinted with permission from [134] Copyright (2011) IEEE.*

(Figure 5.28a and b, for V_G=+10 and −10 V, respectively), which is fully recoverable after stress, being consistent with charge trapping at or close to the ZTO/ATO interface. The shift of V_T due to a charge trapping mechanism can usually be fitted with a stretched-exponential equation [101, 104]:

$$\Delta V_T = V_0 \left\{ 1 - exp\left[-\left(\frac{t}{\tau} \right)^{\beta} \right] \right\} \tag{5.8}$$

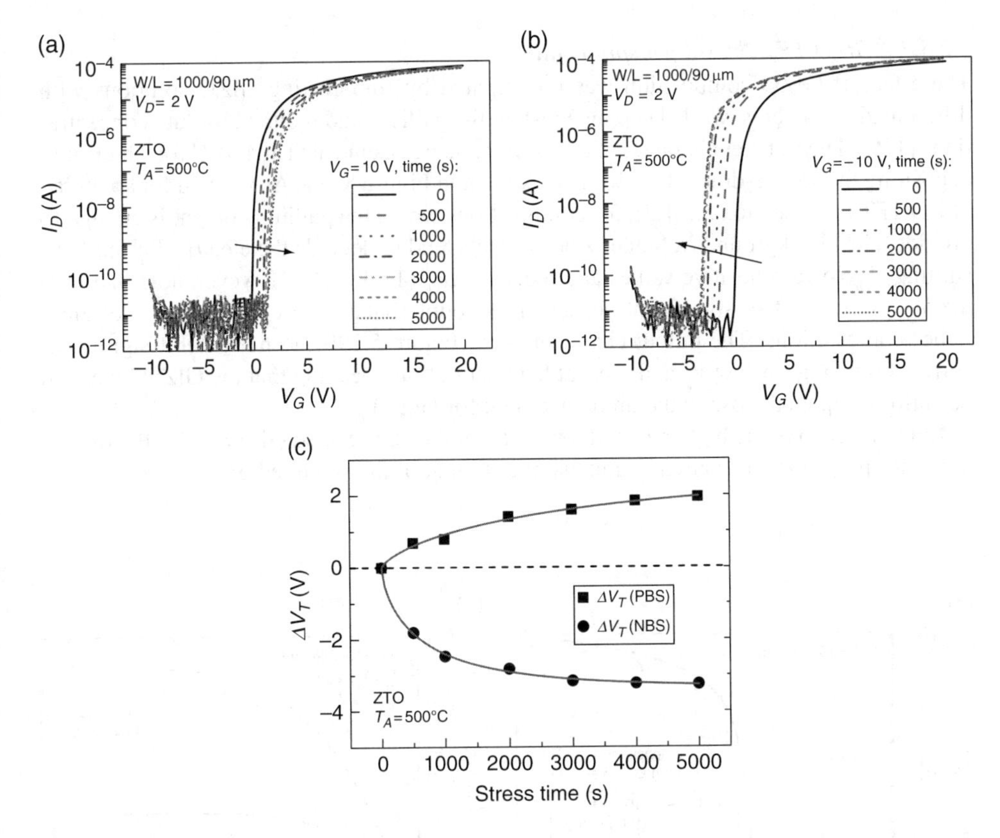

Figure 5.28 *Constant gate bias stress measurements in spin-coated ZTO TFTs: a) transfer characteristics evolution for V_G=10V; b) transfer characteristics evolution for V_G=−10V; c) stretched-exponential fitting to experimental ΔV_T data. Reprinted with permission from [134] Copyright (2011) IEEE.*

where V_0 is the ΔV_T at infinite time, β is the stretched-exponential exponent, and τ is the constant characteristic trapping time for stress phase or de-trapping time for recovery phase. The fitting is in good agreement with experimental data (Figure 5.28c), suggesting that the positive (or negative) shifting of V_T may be attributed to the temporal trapping of electrons (or holes) in the dielectric or at the semiconductor/dielectric interface [101, 104]. In addition, similar to that which was exposed to sputtered GIZO TFTs (see section 5.2.1.5), the effect of enhanced oxygen adsorption/desorption processes when ZnO-based materials are subjected to electric fields should also contribute to the V_T shifts verified under stress in these unpassivated ZTO TFTs.

Even if these solution-processed ZTO TFTs already exhibit comparable bias stress stability with regard to the sputtered GIZO TFTs presented before, further improvement should be possible by adjusting the ZTO layer composition and processing conditions [82, 138].

5.2.3.3 GIZO TFTs by sol-gel spin-coating

The effect of GIZO composition was investigated by spin-coating GIZO solutions with different zinc contents, while keeping constant the gallium and indium absolute concentrations [139]. The transfer characteristics for these devices, annealed at 400°C for 1 h prior to depositing the source-drain electrodes, are shown in Figure 5.29a. As verified for sputtered GIZO TFTs (see section 5.2.1.2), for compositions where the gallium content is enough to provide a low background N, higher zinc contents tend to degrade the electrical properties, shifting V_{ON} toward positive values and reducing μ_{FE} (Table 5.4). However, note that here the V_{ON} shift is not so significant because it is counterbalanced by the higher d_s of zinc-richer compositions. The output characteristics (Figure 5.29b) do not exhibit significant current crowding for low V_D and present hard saturation, meaning that the GIZO layer can be entirely depleted close to the drain electrode for large V_D.

Note that despite the higher T_A and lower μ_{FE} of the spin-coated GIZO TFTs, the overall transistor performance is already quite similar to that of the sputtered ones.

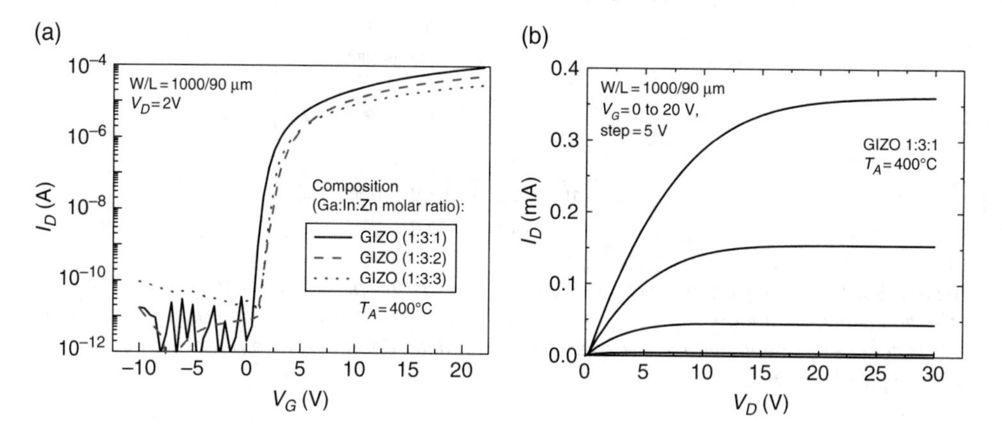

Figure 5.29 *a) Transfer characteristics of spin-coated GIZO TFTs with different GIZO compositions; b) output characteristics of spin-coated GIZO 1:3:1 TFTs. Devices annealed at 400°C. Adapted with permission from [139] Copyright (2010) American Institute of Physics.*

Table 5.4 *Electrical properties for the devices depicted in Figure 5.29a.*

Ga:In:Zn molar ratio	GIZO thickness (nm)	μ_{FE} (cm^2 V^{-1} s^{-1})	*on/off*	V_{ON} (V)	S (V dec^{-1})
1:3:1	23	5.8	6×10^7	0.5	0.28
1:3:2	28	3.1	1×10^7	1.0	0.47
1:3:3	32	1.6	4×10^6	1.2	0.42

Adapted with permission from ref [139] Copyright (2010)

5.2.4 P-type Oxide TFTs by Physical Vapor Deposition

Despite the large interest and intense research activity in oxide TFTs, at present almost all of the existing literature deals with devices employing n-type TSOs as active layer materials. However, fast and low power consumption CMOS transparent circuits require that p-type TSOs are investigated. For p-type oxides, the carrier conduction path (valence band) is mainly formed by oxygen *p* asymmetric orbitals, which severely limit μ. Hence, p-type oxides have considerably lower μ when compared to their n-type counterparts, whose carrier conduction paths are mainly derived from the cations' large and spherical *s*-orbitals. Recently, much attention has been given to copper-based semiconductors, of which the delafossite family $CuMO_2$ (M= Al, Ga, In, Y, Sc, La, etc.) is the most important. Besides these materials, simple binary compounds such as Cu_2O and SnO can also exhibit p-type conductivity. In this section the initial results achieved at CENIMAT in p-type oxide TFTs employing sputtered Cu_2O and SnO as active layers are presented.

P-type oxide TFTs were fabricated by a process flow similar to that presented in Figure 5.11 for GIZO TFTs. In this case, glass/ITO/ATO substrates were used rather than Si/SiO_2, with the high-κ dielectric (ATO) having the advantage of inducing a larger charge density by capacitive effect than SiO_2, which is crucial for semiconductor layers and interfaces with a large trap state density, as expected for p-type TSOs. Both Cu_2O (40 nm thick) and SnO (30 nm thick) were sputtered from metallic targets in an $Ar+O_2$ atmosphere. The source-drain electrodes of Ni/Au (e-beam) and IZO (sputtering) were tested. The final devices were annealed at 200°C for 1 h in air [140, 141].

5.2.4.1 *Cu_2O TFTs by sputtering*

Figure 5.30 shows the transfer and output characteristics for Cu_2O TFTs with IZO source-drain electrodes. The average optical transmission in the visible range of the entire device structure is ≈80%, allowing one to classify these p-type TFTs as transparent devices. Judging by the low V_D region of the output characteristics, IZO electrodes seem to provide reasonably good contact properties for Cu_2O. The transistors exhibit a low $\mu_{FE} \approx 1.2 \times 10^{-3}$

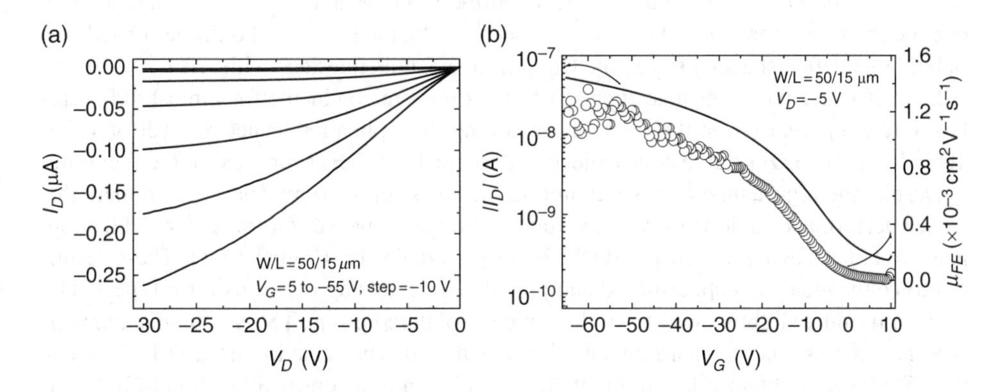

Figure 5.30 *a) Output and b) transfer characteristics of p-type sputtered Cu_2O TFTs. Devices annealed at 200°C. Adapted with permission from [140] Copyright (2010) American Institute of Physics.*

Table 5.5 *State of the art concerning p-type Cu_xO TFTs for the last three years. μ column contains data extracted by different methodologies (μ_{FE}, μ_{SAT}...).*

Deposition technique	T_{dep}/T_A (°C)	Substrate - Dielectric	μ ($cm^2 V^{-1} s^{-1}$)	*on/off*	Reference
PLD	700	$MgO – Al_2O_x$	0.26	6	Matsuzaki *et al.* [143, 144]
rf mag. sputtering	RT/200	$Si-SiO_2$	0.4	10^4	Sung *et al.* [145]
PLD	500	$Si-SiO_2$ - HfON	4.3	3×10^6	Zou *et al.*[146]
rf mag. sputtering	RT/200	Glass-ATO*	1.2×10^{-3}	2×10^2	Fortunato *et al.*[140]

*ATO – superlattice of Al_2O_3 and TiO_2.

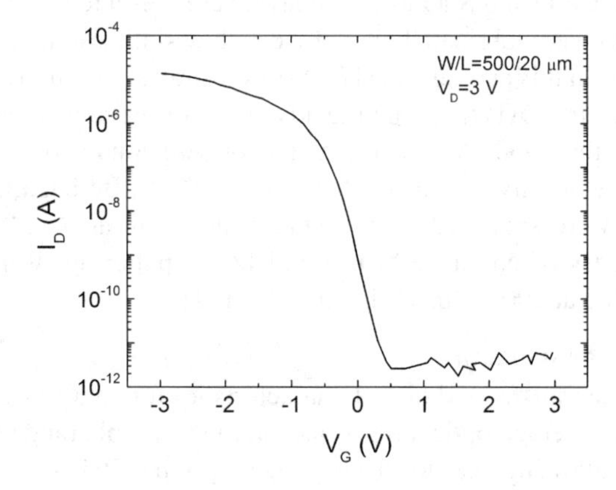

Figure 5.31 *Transfer characteristics of p-type Cu_xO TFTs where Cu_2O is deposited by PLD at 500°C (adapted from [146]).*

$cm^2 V^{-1} s^{-1}$, *on/off*$=2\times10^2$ and $V_T=-12$V. Furthermore, a high V_G range is necessary to operate the transistors and a large $S\approx14$V dec^{-1} is obtained, which should be related to a rather large defect state density at the Cu_2O film and at its interface with ATO.

Even if Cu_2O has been investigated as an oxide semiconductor for almost 100 years [142], few reports exist in the literature regarding its application as active layer of TFTs. Table 5.5 summarizes the results achieved on p-type Cu_2O TFTs reported in the literature.

Despite the high temperatures required in the work reported by Zou *et al.*, remarkable device performance at low operation voltages could be achieved, for transistors with a top-gate structure, having a high-κ HfON film as gate dielectric (Figure 5.31). These results reinforce the idea that copper-based compounds are promising materials for p-type TSOs.

Even if going slightly away from the topic of sputtered p-type TSOs, a final example is presented in this section regarding the achievement of p-type behavior in Cu_2O TFTs where the TSO layer is obtained by simple thermal oxidation of a copper thin film [147]. It was found that for T_A between 200 and 300°C a single cubic Cu_2O phase and p-type conductivity is obtained. As proof of the concept behind this methodology, a p-type TFT was produced, being the preliminary results presented in Figure 5.32.

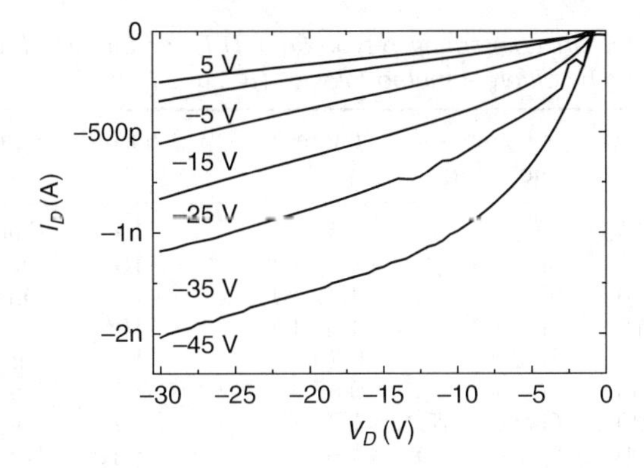

Figure 5.32 *Output characteristics for a p-type Cu_2O TFT where Cu_2O is obtained by thermal oxidation in air at 200°C of a copper thin film deposited by e-beam evaporation.*

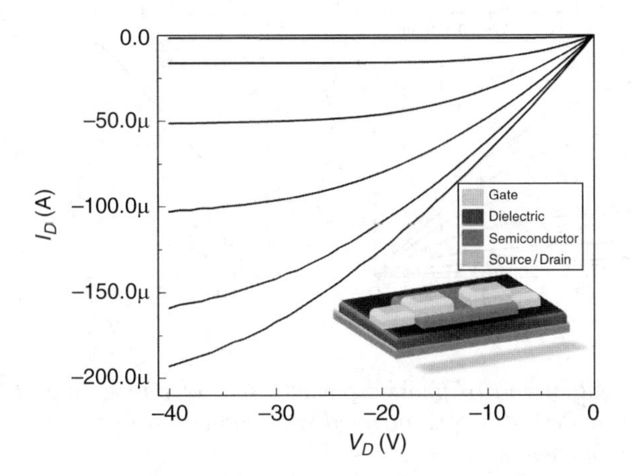

Figure 5.33 *Output characteristics of p-type SnO TFT, annealed at 200°C. Gate voltage is varied from 0 to −50V in −10V steps. The inset shows a schematic of the device structure. Adapted with permission from [148] Copyright (2011) Wiley-VCH.*

5.2.4.2 SnO TFTs by sputtering

SnO TFTs are also being explored as p-type oxide TFTs. At CENIMAT, devices exhibiting $\mu_{FE}\approx 1.1\,cm^2\,V^{-1}\,s^{-1}$, *on/off*$=10^3$ and $V_T=-5\,V$ were obtained [141]. More recently, by controlling the SnO oxidation state it was possible to obtain TFTs with improved performance, such as $\mu_{SAT}=4.6\,cm^2\,V^{-1}\,s^{-1}$ and *on/off*$>7\times10^4$, which are the highest values reported so far for p-type oxide TFTs [148]. Figure 5.33 (see color plate section) shows the output characteristics of these SnO TFTs with Ni/Au source-drain electrodes.

Reports on p-type SnO TFTs in the literature are more numerous than for Cu_2O TFTs (Table 5.6). To the author's knowledge, the first report on SnO TFTs was made by Ogo *et al.*, using epitaxially grown SnO films by PLD at 575°C [149]. Devices exhibiting clear p-type operation were obtained, with $\mu_{FE}=1.3\,cm^2\,V^{-1}\,s^{-1}$, *on/off*$\approx10^2$, $V_T=4.8\,V$ and $S\approx7\,V\,dec^{-1}$

Table 5.6 *State of the art concerning p-type SnO TFTs for the last three years. μ column contains data extracted by different methodologies (μ_{FE}, μ_{SAT}...).*

Deposition technique	T_{dep}/T_A (°C)	Substrate - Dielectric	μ (cm² V⁻¹ s⁻¹)	*on/off*	Reference
PLD	575/200	YSZ - Al_2O_x	1.3	~10^2	Ogo *et al.* [149]
Evaporation	RT/100	Si - SiO_2	1.1×10^{-2}	~10^3	Ou *et al.* [150]
Evaporation	RT/100	Si - SiO_2	4.7×10^{-3}	~10^2	Dhananjay *et al.* [151]
Evaporation	RT/310	Si - SiO_2	4.0×10^{-5}	~10^2	Lee *et al.* [152]
rf sputtering	RT/400	Si- SiN_x	0.24	~10^2	Yabuta *et al.* [153]
Evaporation	RT/400	Si - SiO_2	0.87	~10^2	Liang *et al.* [154]
rf sputtering	RT/200	Glass – ATO*	1.2	~10^3	Fortunato *et al.* [141]
rf sputtering	RT/200	Glass – ATO*	4.6	7×10^4	Fortunato *et al.* [148]
PLD	RT/250	Si - SiO_2	0.75	~10^3	Nomura *et al.* [155]

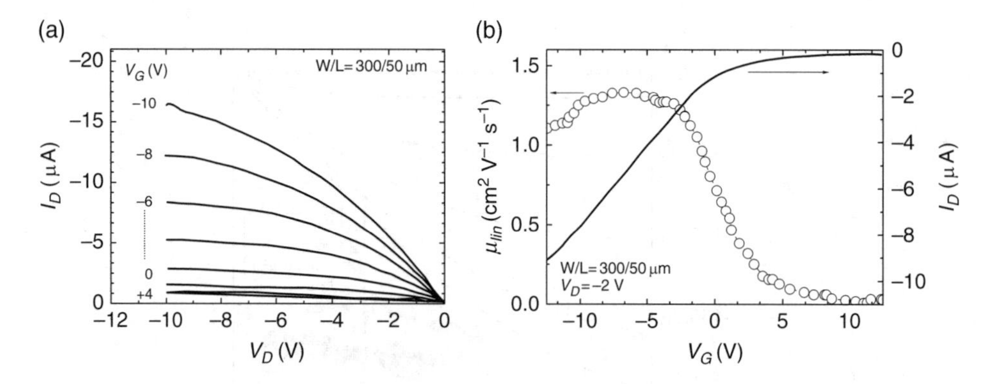

Figure 5.34 *a) Output and b) transfer characteristics of p-type SnO TFTs where SnO is epitaxially grown by PLD at 575°C. Adapted with permission from [149] Copyright (2008) American Institute of Physics.*

(Figure 5.34). The TFTs operate in the depletion mode and the large *off*-current is attributed to the large hole density (>10^{17} cm^{-3}).

By an analysis of Table 5.6, it is noteworthy that even if the electrical properties are far from optimal, (post-)processing temperatures as low as 100°C already allow one to fabricate p-type SnO TFTs. Furthermore, in a recent publication, Nomura *et al.* showed that SnO can be used as an ambipolar semiconductor, which enables the fabrication of complementary-like inverters using two SnO-based ambipolar TFTs [155].

5.2.5 N-type GIZO TFTs with Sputtered Dielectrics

Given that most of the working principles of a TFT rely on the phenomena taking place at or close to the semiconductor/dielectric interface, the performance and stability of these devices greatly depends on the dielectric layer, as already shown by different authors for oxide TFT [59, 68, 156]. All of the devices fabricated at CENIMAT presented until now in this chapter make use of a commercial dielectric processed at temperatures above 400°C, either PECVD SiO_2 or atomic layer deposition (ALD) ATO. But low-cost and flexible

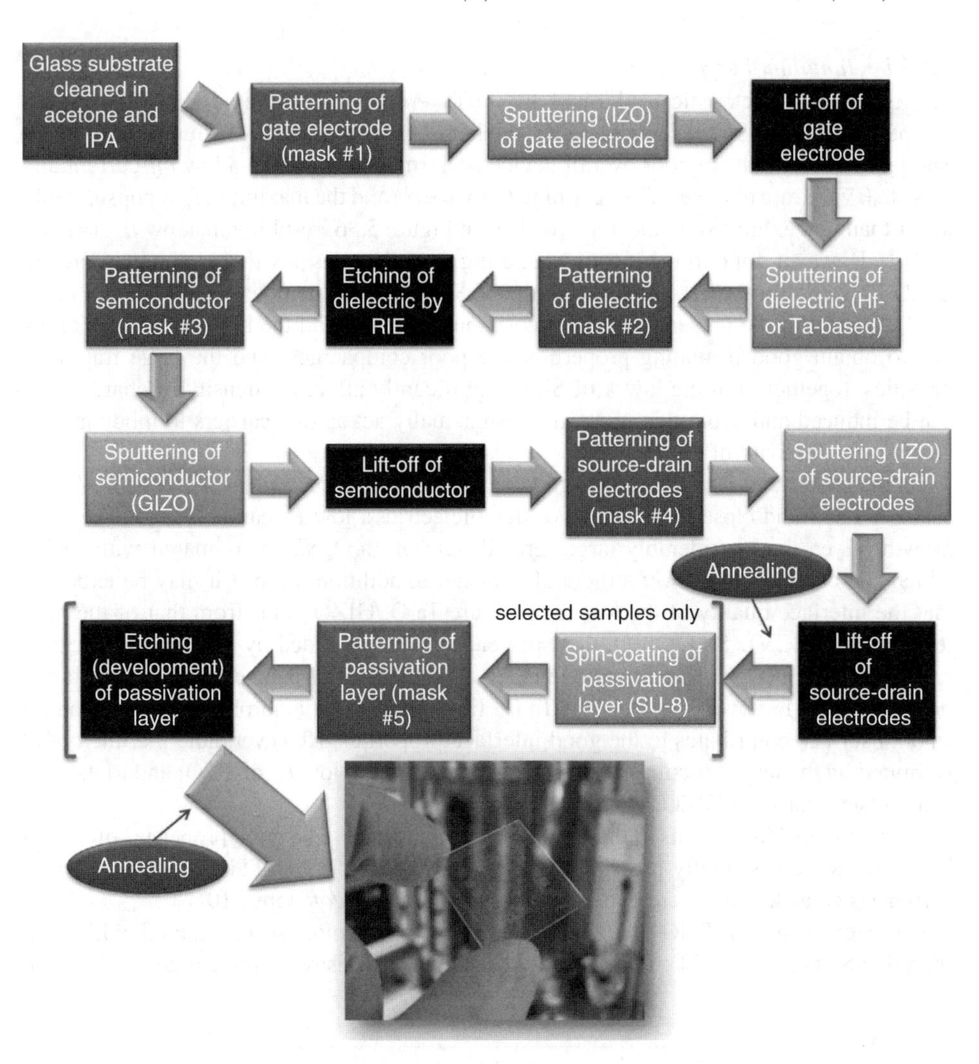

Figure 5.35 *Process flow used to produce oxide TFTs employing glass substrates and sputtered dielectrics.*

electronics requires lower processing temperatures, typically below 200°C. Furthermore, the commercial substrates/dielectrics used in the previous sections do not allow for the patterning of the gate electrode, which is mandatory if the application of the transistors in integrated circuits is envisaged. Hence, the integration of the previously optimized TSO and TCO materials and processes with the sputtered dielectrics presented in Chapter 4 is required. To evaluate the performance and stability of the different dielectric layers studied, based on tantalum and hafnium oxides, sputtered GIZO with a 2:4:2 composition and $\%O_2=0.4\%$ was selected as the channel layer, while sputtered IZO was used to fabricate the gate, source and drain electrodes of staggered bottom-gate transparent TFTs. The thickness of the dielectric layer was ≈350 nm, being the final devices annealed at 150°C. The process flow is presented in Figure 5.35 (see color plate section).

5.2.5.1 Tantalum-based dielectrics

The transfer characteristics exhibited by TFTs employing sputtered SiO_2, Ta_2O_5 and co-sputtered Ta_2O_5-SiO_2 (denoted TSiO) dielectrics are presented in Figure 5.36 [157]. Sputtered SiO_2 results in poor overall device performance: although a low *off*-current and close to 0 V V_{ON} are obtained, S is very high (≈1 V dec^{-1}) and the maximum I_D is considerably lower than that exhibited by the other devices in Figure 5.36, resulting in a low μ_{FE} (≈1 cm^2 V^{-1} s^{-1}). However, for optimized SiO_2 sputtering processes it is possible to obtain improved GIZO TFTs performance, mainly in terms of S and μ_{FE}, as demonstrated by Ofuji *et al.* [158]. Even if in the present case the amorphous structure and the high-E_G of SiO_2 allow one to obtain good insulating properties, the poor compactness and the large trap state densities, together with the low κ of SiO_2, significantly affect the density of charges that can be induced and, from this, those that can actually act as free carriers to modulate the channel conductivity of the TFT.

On the other hand, Ta_2O_5 provides good device performance, such as μ_{FE}>30 cm^2 V^{-1} s^{-1}, S=0.3 V dec^{-1} and close to 0 V V_{ON}. For this dielectric, a low P_{rf} can be used during film growth (given the considerably larger growth rate of Ta_2O_5 when compared with SiO_2), which can contribute to lower structural damage. In addition, even if it may be expected that the interface quality of the low-temperature Ta_2O_5/GIZO is far from that of the high temperature PECVD SiO_2/GIZO, the extra capacitance provided by the high-κ dielectric enables easier interface trap filling due to the large density of induced charges that it can provide [44]. Moreover, the fact that Ta_2O_5 films present an amorphous structure and a smooth surface contributes to the good interface properties. However, note that the *on/off* is limited by the large *off*-current, which is attributed to the low E_G of Ta_2O_5 and to its poor band offset relative to GIZO.

Co-sputtered TSiO results in the best performing devices. The good properties of Ta_2O_5-based TFTs are essentially preserved, but given the larger E_G of TSiO, I_G (thus, the *off*-current) is considerably decreased, allowing one to raise *on/off* above 10^6.

However, even with TSiO (with Ta_2O_5 this issue is even more severe) the reliability and reproducibility of the TFTs is far from what would be desirable for circuit application.

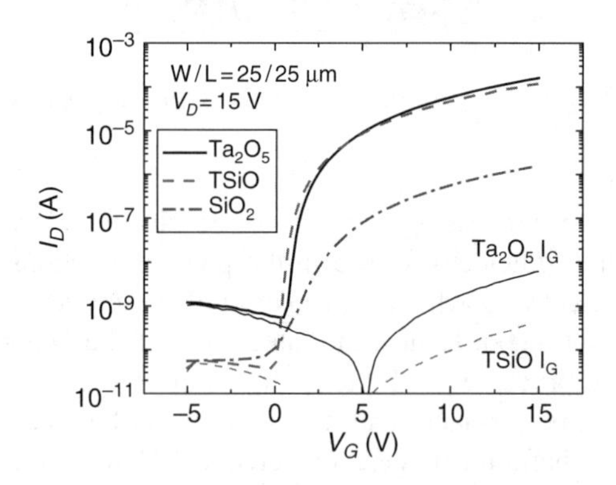

Figure 5.36 *Transfer characteristics of GIZO TFTs using sputtered SiO_2, Ta_2O_5 and co-sputtered TSiO dielectrics. Devices annealed at 150°C. Adapted with permission from [157] Copyright (2009) The Electrochemical Society.*

In fact, due to the narrow band offset between TSiO (and Ta_2O_5) and GIZO, it is often found in a substrate several devices with shorted-gates, or in less extreme cases, devices with large variations on their electrical properties, namely on their *off*-current. This effect is shown in Figure 5.37a for TSiO. Even if some TSiO-based TFTs can sustain repeated and severe stress tests, such as constant $I_D = 10\,\mu A$ for 24h [157], alternative materials or dielectric structures should be investigated so as to improve yield.

To this end we investigated multilayer dielectrics composed by TSiO stacked between two thin layers (≈20 nm) of SiO_2, denoted by S–TS–S. These dielectric layers were produced with and without r.f. substrate bias (SB) during film growth. It is known that biasing the substrate can increase the purity of the deposited thin film by resputtering poorly bonded surface atoms, but it can also lead to damaged structures, depending on the materials and the bias magnitude [159, 160]. Even if hysteresis is slightly increased for the multilayer structures when compared with TSiO (mostly due to the non-optimized SiO_2/GIZO interface), the reproducibility, reliability and overall transistor performance are significantly

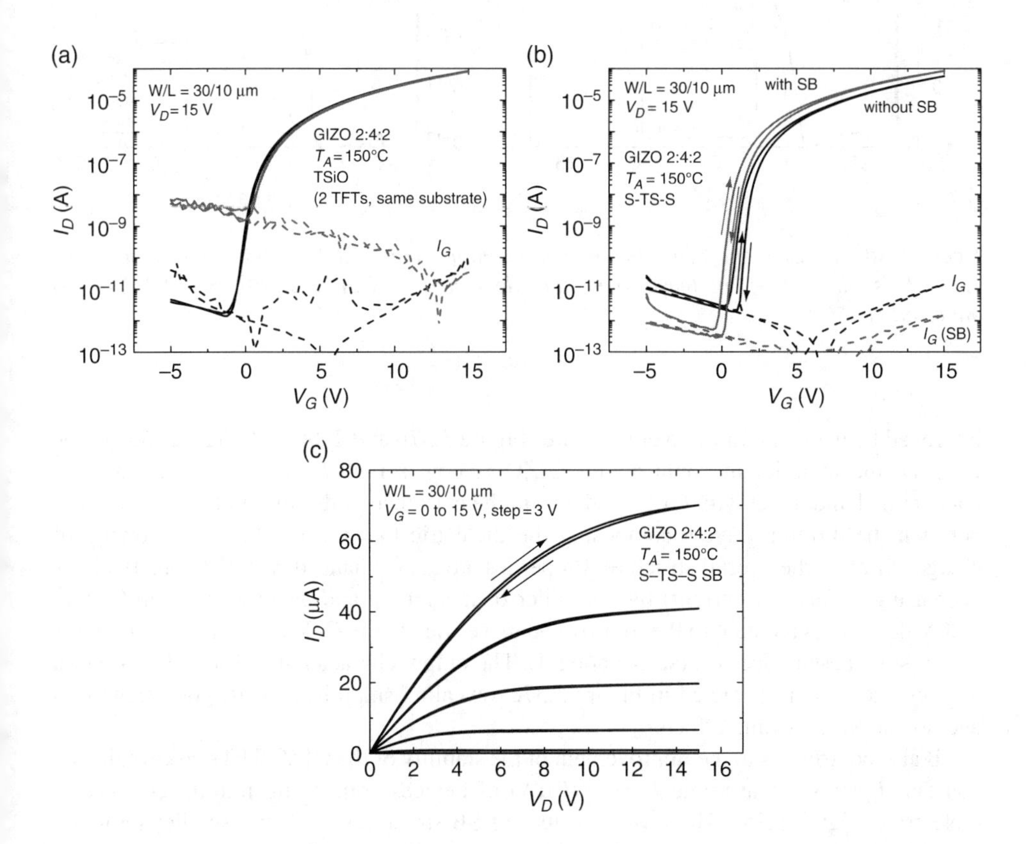

Figure 5.37 *Electrical properties of GIZO TFTs with tantalum-based dielectrics: a) transfer characteristics for two TFTs in the same substrate with TSiO dielectric; b) transfer characteristics for TFTs with multilayer S–TS–S dielectric deposited with and without SB; c) output characteristics for a TFT with multilayer S–TS–S SB dielectric. Devices annealed at 150°C.*

Table 5.7 *Electrical properties for the GIZO TFTs depicted in Figure 5.37.*

Dielectric	μ_{FE} (cm² V⁻¹ s⁻¹)	V_{ON} (V)	On/off	S (V dec⁻¹)
TSiO	12.0	−1.2	5.8×10^7	0.32
S-TS-S	11.3	0.4	3.0×10^7	0.25
S-TS-S SB	12.7	−0.4	3.0×10^8	0.20

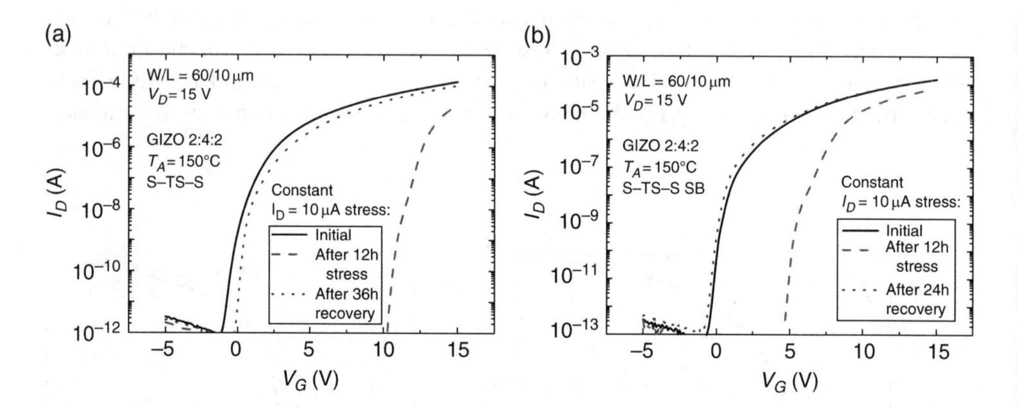

Figure 5.38 *Transfer characteristics evolution during constant I_D stress measurements for GIZO TFTs with S–TS–S multilayer dielectrics deposited a) without and b) with SB. Devices annealed at 150°C.*

improved by using the multilayer structure (Figure 5.37b and Table 5.7). This is due to the fact that the material in contact with GIZO is now a high-E_G dielectric with a large conduction band offset with GIZO and also to the existence of discontinuities or interfaces between the different layers composing the dielectric that prevent the easy flowing of charges through the overall structure. Properties are also enhanced with SB, which allows decreasing I_G (and *off*-current) by one order of magnitude and decrease S, from 0.25 to 0.20 V dec⁻¹, in agreement to the improvements verified in the C–V plots of MIS structures comprising these dielectrics (see Chapter 4). The output characteristics for S–TS–S-based devices with SB are depicted in Figure 5.37c, revealing small hysteresis, good saturation and no current crowding effects.

SB also contributes to the improvement of the stability of the GIZO TFTs, as revealed by constant I_D stress measurements (I_D=10 μA) of devices comprising multilayer S–TS–S dielectrics (Figure 5.38). The lower S values of SB structures suggest a smaller semiconductor/dielectric interface trap density, which also contributes to the improvement of stability under stress, given that the dominant instability mechanism on these devices is reversible charge trapping at or close to the dielectric/semiconductor interface, resulting in a ΔV_T towards higher V_G as the stress measurement progresses.

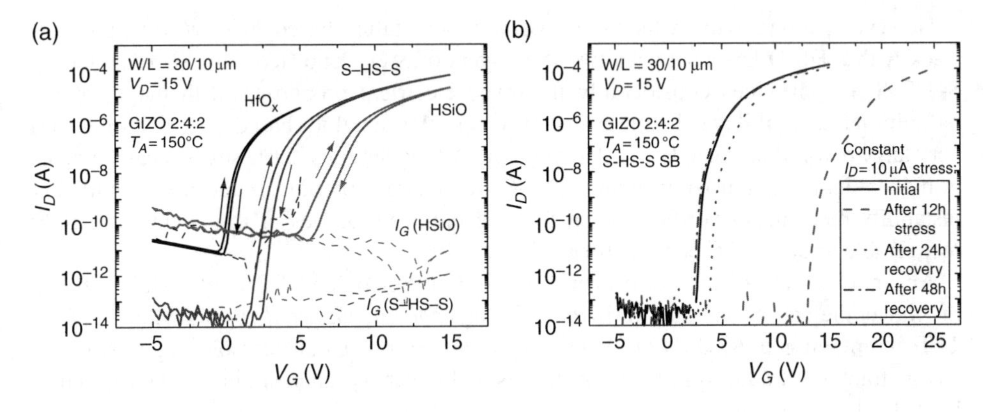

Figure 5.39 *Electrical properties of GIZO TFTs with hafnium-based dielectrics: a) transfer characteristics for TFTs with HfO_2, HSiO and S–HS–S SB dielectrics; b) transfer characteristics evolution during constant I_D stress measurements for a TFT with S–HS–S SB multilayer dielectric.*

Table 5.8 *Electrical properties for the GIZO TFTs depicted in Figure 5.39.*

Dielectric	μ_{FE} ($cm^2 V^{-1} s^{-1}$)	V_{ON} (V)	On/off	S ($V dec^{-1}$)
HfO_2	5.7	−0.6	5.0×10^5	0.25
HSiO	7.0	5.0	9.0×10^5	0.76
S-HS-S SB	17.0	1.2	6.4×10^9	0.22

5.2.5.2 *Hafnium-based dielectrics*

Hafnium-based dielectrics have also been investigated in GIZO TFTs. Given the higher E_G of HfO_2 compared with Ta_2O_5, and despite the higher P_{rf} required for HfO_2 film growth, the yield of working GIZO TFTs employing this dielectric is slightly higher, but I_G starts to sharply increase for V_G>5 V (Figure 5.39a). The small range of usable V_G and the lack of reliability/reproducibility of HfO_2-based TFTs might be associated with the higher bombardment effects during film growth due to the higher P_{rf}, which can broaden the band-tails and decrease the band-offsets with GIZO, and also with the polycrystalline structure of sputtered HfO_2 films. It is well known that grain boundaries act as preferential paths for impurity diffusion and leakage current, resulting in inferior dielectric reliability. Besides that, amorphous materials generally present smoother surfaces, resulting in improved interface properties [44, 161]. In fact, it was found that the present HfO_2 films have rough surfaces (RMS roughness≈4.5 nm), associated with the dominance of the monoclinic crystalline phase [162].

Even if amorphous films are obtained when sputtering from a multicomponent HfO_2-SiO_2 (denoted as HSiO) target and devices with this dielectric can withstand measurements up to higher V_G than HfO_2, overall transistor performance is actually worse than for HfO_2 (Figure 5.39a and Table 5.8). In addition, poor stability under constant I_D stress is verified for these HSiO-based devices, with a recoverable ΔV_T>15 V being obtained after 10h of stress. This is accompanied by an irreversible increase of the *off*-current up to ≈10 nA.

However, similarl to that which was observed for tantalum-based dielectrics, a multilayer stack with a HSiO layer between two thin layers of SiO_2 deposited with SB (denoted by S–HS–S SB) provides considerably improved electrical properties, with the I_G of the hafnium-based multilayer being even lower than the tantalum-based one, resulting in a very large *on/off* that exceeds 10^9. Although the ΔV_T under constant I_D stress measurements is higher than for transistors with S–TS–S SB, charge trapping is the only significant instability mechanism verified, being that the initial properties of the devices are fully recoverable after 48h of recovery time (Figure 5.39b).

An improvement of the performance and stability of oxide TFTs with hafnium-based multilayer dielectrics was also verified by Chang *et al.* and Lee *et al.*, by showing that ZnO TFTs employing a $Al_2O_3/HfO_2/Al_2O_3$ dielectric exhibit a considerable suppression of charge trapping, traduced in lower hysteresis and lower ΔV_T than similar TFTs with HfO_2 [163, 164].

References

[1] A.C. Tickle (1969) *Thin-Film Transistors – A New Approach to Microelectronics*. New York: John Wiley & Sons, Inc.

[2] C.R. Kagan and P. Andry (2003) *Thin-film transistors*. New York: Marcel Dekker, Inc.

[3] J.-H. Lee, D.N. Liu, and S.-T. Wu (2008) *Introduction to Flat Panel Displays*. West Sussex, UK: John Wiley & Sons, Ltd.

[4] M. Kim, J.H. Jeong, H.J. Lee, T.K. Ahn, H.S. Shin, J.S. Park, J.K. Jeong, Y.G. Mo, and H.D. Kim (2007) High mobility bottom gate InGaZnO thin film transistors with SiO_x etch stopper, *Applied Physics Letters* **90**, 212114-1–212114-3.

[5] A. Sato, K. Abe, R. Hayashi, H. Kumomi, K. Nomura, T. Kamiya, M. Hirano, and H. Hosono (2009) Amorphous In-Ga-Zn-O coplanar homojunction thin-film transistor, *Applied Physics Letters* **94**, 133502-1–133502-3.

[6] E.S. Yang (1988) *Microelectronic Devices*. Singapore: McGraw-Hill.

[7] S.M. Sze (1981) *Physics of Semiconductor Devices*, 2nd edn New York: John Wiley & Sons, Inc.

[8] R.L. Hoffman (2006) Effects of channel stoichiometry and processing temperature on the electrical characteristics of zinc tin oxide thin-film transistors, *Solid-State Electron.* **50**, 784–7.

[9] R.L. Hoffman, B.J. Norris, and J.F. Wager (2003) ZnO-based transparent thin-film transistors, *Applied Physics Letters* **82**, 733–5.

[10] A. Ortiz-Conde, F.J.G. Sanchez, J.J. Liou, A. Cerdeira, M. Estrada, and Y. Yue (2002) A review of recent MOSFET threshold voltage extraction methods, *Microelectron. Reliab.* **42**, 583–96.

[11] R.L. Hoffman (2004) ZnO-channel thin-film transistors: Channel mobility, *Journal of Applied Physics* **95**, 5813–19.

[12] D.K. Schroder (2006) *Semiconductor Material and Device Characterization*, 3rd edn New Jersey: John Wiley & Sons, Inc.

[13] J.E. Lilienfield (1930) "Method and apparatus for controlling electric currents," U.S., patent number 1745175.

[14] J.E. Lilienfield (1932) "Amplifier for electric currents," U.S., patent number 1877140.

[15] J.E. Lilienfield (1933) "Device for controlling electric currents," U.S., patent number 19000140.

[16] O. Heil (1935) "Improvements in or relating to electrical amplifiers or other control arrangements," U.K., patent number 439457.

[17] http://chem.ch.huji.ac.il/history/lilienfeld.htm.

[18] J. Bardeen and W.H. Brattain (1948) The transistor, a semi-conductor triode, *Physical Review* **74**, 230–1.

[19] W. Shockley (1949) The theory of p-n junctions in semiconductors and p-n junction transistors, *Bell System Technical Journal* **28**, 435–89.

[20] W. Shockley (1952) A unipolar field-effect transistor, *Proceedings of the Institute of Radio Engineers* **40**, 1365–76.
[21] http://www.ece.umd.edu/class/enee312-2.S2005/.
[22] P.K. Weimer (1962) TFT - new thin-film transistor, *Proceedings of the Institute of Radio Engineers* **50**, 1462–9.
[23] H. Borkan and P.K. Weimer (1963) An analysis of the characteristics of insulated-gate thin-film transistors, *RCA Review* **24**, 153–65.
[24] P.K. Weimer (1964) P-type tellurium thin-film transistor, *Proc. IEEE* **52**, 608–9.
[25] F.V. Shallcross (1963) Cadmium selenide thin-film transistors, *Proc. IEEE*, **51**, 851.
[26] Y.T. Sihvonen, S.G. Parker, and D.R. Boyd (1967) Printable insulated gate field effect transistors, *Journal of the Electrochemical Society* **114** (1967), 96–102.
[27] M.G. Kim, M.G. Kanatzidis, A. Facchetti, and T.J. Marks (2011) Low-temperature fabrication of high-performance metal oxide thin-film electronics via combustion processing, *Nature Materials* **10**, 382–8.
[28] S.R. Hofstein and F.P. Heiman (1963) Silicon insulated-gate field-effect transistor, *Proc. IEEE*, **51**, 1190–1202.
[29] B.J. Lechner, F.J. Marlowe, E.O. Nester, and J. Tults (1971) Liquid crystal matrix displays, *Proceedings of the Institute of Electrical and Electronics Engineers* **59**, 1566–79.
[30] T.P. Brody, J.A. Asars, and G.D. Dixon (1973) 6×6 inch 20 lines-per-inch liquid-crystal display panel, *IEEE Trans. Electron Devices* **ED20**, 995–1001.
[31] P.G. Lecomber, W.E. Spear, and A. Ghaith (1979) Amorphous-silicon field-effect device and possible application, *Electron. Lett.* **15**, 179–81.
[32] A. Madan and R. Martins (2009) From materials science to applications of amorphous, microcrystalline and nanocrystalline silicon and other semiconductors PREFACE, *Philos. Mag.* **89**, 2431–4.
[33] M.J. Powell and J. Pritchard (1983) The effect of surface-states and fixed charge on the field-effect conductance of amorphous-silicon, *Journal of Applied Physics* **54**, 3244–8.
[34] M.J. Powell (1989) The physics of amorphous-silicon thin-film transistors, *IEEE Trans. Electron Devices* **36**, 2753–63.
[35] T.W. Little, H. Koike, K. Takahara, T. Nakazawa, and H. Ohshima (1991) "A 9.5 inch, 1.3 mega-pixel low-temperature poly-Si TFT-LCD fabricated by SPC of very thin-films and an ECR-CVD gate insulator," in *Conference Record of the 1991 International Display Research Conference* Playa Del Rey: Soc Information Display, 219–22.
[36] A. Tsumura, H. Koezuka, and T. Ando (1986) *Macromolecular electronic device: Field-effect transistor with a polythiophene thin film* Applied Physics Letters 49, 1210–2.
[37] F. Garnier, G. Horowitz, X. H. Peng, and D. Fichou (1990) An all-organic soft thin-film transistor with very high carrier mobility, *Adv. Mater.* **2**, 592–4.
[38] http://www.sony.net/SonyInfo/csr/environment/technology/.
[39] H.A. Klasens and H. Koelmans (1964) A tin oxide field-effect transistor, *Solid-State Electron.* **7**, 701–2.
[40] G.F. Boesen and J.E. Jacobs (1968) ZnO field-effect transistor, *Proceedings of the Institute of Electrical and Electronics Engineers* **56**, 2094–5.
[41] A. Aoki and H. Sasakura (1970) Tin oxide thin film transistors, *Jpn. Journal of Applied Physics* **9**, 582.
[42] M.W.J. Prins, K.O. GrosseHolz, G. Muller, J.F.M. Cillessen, J.B. Giesbers, R.P. Weening, and R.M. Wolf (1996) A ferroelectric transparent thin-film transistor, *Applied Physics Letters* **68**, 3650–2.
[43] C.H. Seager, D.C. McIntyre, W.L. Warren, and B.A. Tuttle (1996) Charge trapping and device behavior in ferroelectric memories, *Applied Physics Letters* **68**, 2660–2.
[44] J.F. Wager, D.A. Keszler, and R.E. Presley (2008) *Transparent Electronics*. New York: Springer.
[45] S. Masuda, K. Kitamura, Y. Okumura, S. Miyatake, H. Tabata, and T. Kawai (2003) Transparent thin film transistors using ZnO as an active channel layer and their electrical properties, *Journal of Applied Physics* **93**, 1624–30.
[46] P.F. Carcia, R.S. McLean, M.H. Reilly, and G. Nunes (2003) Transparent ZnO thin-film transistor fabricated by rf magnetron sputtering, *Applied Physics Letters* **82**, 1117–19.

[47] J. Nishii, F. M. Hossain, S. Takagi, T. Aita, K. Saikusa, Y. Ohmaki, I. Ohkubo, S. Kishimoto, A. Ohtomo, T. Fukumura, F. Matsukura, Y. Ohno, H. Koinuma, H. Ohno, and M. Kawasaki (2003) High mobility thin film transistors with transparent ZnO channels, *Jpn. Journal of Applied Physics Part 2 - Lett.* **42**, L347–L349.

[48] E.M.C. Fortunato, P.M.C. Barquinha, A. Pimentel, A.M.F. Goncalves, A.J.S. Marques, R.F.P. Martins, and L.M.N. Pereira (2004) Wide-bandgap high-mobility ZnO thin-film transistors produced at room temperature, *Applied Physics Letters* **85**, 2541–3.

[49] E. Fortunato, P. Barquinha, A. Pimentel, A. Goncalves, A. Marques, L. Pereira, and R. Martins (2004) "Zinc oxide thin-film transistors," in *NATO Advanced Research Workshop on Zinc Oxide as a Material for Micro- and Optoelectronic Applications*, St Petersburg, 225–38.

[50] B.J. Norris, J. Anderson, J.F. Wager, and D.A. Keszler (2003) Spin-coated zinc oxide transparent transistors, J. *Phys. D-Appl. Phys.* **36**, L105–L107.

[51] F.M. Hossain, J. Nishii, S. Takagi, A. Ohtomo, T. Fukumura, H. Fujioka, H. Ohno, H. Koinuma, and M. Kawasaki (2003) Modeling and simulation of polycrystalline ZnO thin-film transistors, *Journal of Applied Physics* **94**, 7768–77.

[52] H.S. Bae, M.H. Yoon, J.H. Kim, and S. Im (2003) Photodetecting properties of ZnO-based thin-film transistors, *Applied Physics Letters* **83**, 5313–15.

[53] H.S. Bae and S. Im (2004) Ultraviolet detecting properties of ZnO-based thin film transistors, *Thin Solid Films* **469–70**, 75–9.

[54] R.E. Presley, C.L. Munsee, C.H. Park, D. Hong, J.F. Wager, and D.A. Keszler (2004) Tin oxide transparent thin-film transistors, J. *Phys. D-Appl. Phys.* **37**, 2810–13.

[55] D. Zhang, C. Li, S. Han, X. Liu, T. Tang, W. Jin, and C. Zhou (2003) Electronic transport studies of single-crystalline In_2O_3 nanowires, *Applied Physics Letters* **82**, 112–5.

[56] Y.W. Heo, L.C. Tien, Y. Kwon, D.P. Norton, S.J. Pearton, B.S. Kang, and F. Ren (2004) Depletion-mode ZnO nanowire field-effect transistor, *Applied Physics Letters* **85**, 2274–6.

[57] Q.H. Li, Q. Wan, Y.X. Liang, and T.H. Wang (2004) Electronic transport through individual ZnO nanowires, *Applied Physics Letters* **84**, 4556–8.

[58] Z. Fan, D. Wang, P.-C. Chang, W.-Y. Tseng, and J.G. Lu (2004) ZnO nanowire field-effect transistor and oxygen sensing property, *Applied Physics Letters* **85**, 5923–5.

[59] K. Nomura, H. Ohta, K. Ueda, T. Kamiya, M. Hirano, and H. Hosono (2003) Thin-film transistor fabricated in single-crystalline transparent oxide semiconductor, *Science* **300**, 1269–72.

[60] K. Nomura, H. Ohta, A. Takagi, T. Kamiya, M. Hirano, and H. Hosono (2004) Room-temperature fabrication of transparent flexible thin-film transistors using amorphous oxide semiconductors, *Nature* **432**, 488–92.

[61] H.Q. Chiang, J.F. Wager, R.L. Hoffman, J. Jeong, and D.A. Keszler (2005) High mobility transparent thin-film transistors with amorphous zinc tin oxide channel layer, *Applied Physics Letters* **86**, 013503-1–013503-3.

[62] P. Gorrn, M. Sander, J. Meyer, M. Kroger, E. Becker, H.H. Johannes, W. Kowalsky, and T. Riedl (2006) Towards see-through displays: Fully transparent thin-film transistors driving transparent organic light-emitting diodes, *Advanced Materials* **18**, 738–41.

[63] W.B. Jackson, R.L. Hoffman, and G.S. Herman (2005) High-performance flexible zinc tin oxide field-effect transistors, *Applied Physics Letters* **87**, 193503-1–193503-3.

[64] N.L. Dehuff, E.S. Kettenring, D. Hong, H.Q. Chiang, J.F. Wager, R.L. Hoffman, C.H. Park, and D.A. Keszler (2005) Transparent thin-film transistors with zinc indium oxide channel layer, *Journal of Applied Physics* **97**, 064505-1–064505-5.

[65] P. Barquinha, A. Pimentel, A. Marques, L. Pereira, R. Martins, and E. Fortunato (2006) Influence of the semiconductor thickness on the electrical properties of transparent TFTs based on indium zinc oxide, J. *Non-Cryst. Solids* **352**, 1749–52.

[66] B. Yaglioglu, H.Y. Yeom, R. Beresford, and D.C. Paine (2006) High-mobility amorphous In_2O_3-10 wt % ZnO thin film transistors, *Applied Physics Letters* **89**, 062103-1–062103-3.

[67] D.C. Paine, B. Yaglioglu, Z. Beiley, and S. Lee (2008) Amorphous IZO-based transparent thin film transistors, *Thin Solid Films* **516**, 5894–8.

[68] A. Suresh, P. Wellenius, A. Dhawan, and J. Muth (2007) Room temperature pulsed laser deposited indium gallium zinc oxide channel based transparent thin film transistors, *Applied Physics Letters* **90**, 123512-1–123512-3.

[69] J.S. Park, J.K. Jeong, Y.G. Mo, H.D. Kim, and S.I. Kim (2007) Improvements in the device characteristics of amorphous indium gallium zinc oxide thin-film transistors by Ar plasma treatment, *Applied Physics Letters* **90**, 262106-1–262106-3.
[70] D. Kang, I. Song, C. Kim, Y. Park, T.D. Kang, H.S. Lee, J.W. Park, S.H. Baek, S.H. Choi, and H. Lee (2007) Effect of Ga/In ratio on the optical and electrical properties of GaInZnO thin films grown on SiO_2/Si substrates, *Applied Physics Letters* **91**, 091910-1–091910-3.
[71] H. Hosono, K. Nomura, Y. Ogo, T. Uruga, and T. Kamiya (2008) Factors controlling electron transport properties in transparent amorphous oxide semiconductors, *J. Non-Cryst. Solids* **354**, 2796–2800.
[72] T. Iwasaki, N. Itagaki, T. Den, H. Kumomi, K. Nomura, T. Kamiya, and H. Hosono (2007) Combinatorial approach to thin-film transistors using multicomponent semiconductor channels: An application to amorphous oxide semiconductors in In-Ga-Zn-O system, *Applied Physics Letters* **90**, 242114-1–242114-3.
[73] P. Barquinha, L. Pereira, G. Goncalves, R. Martins, and E. Fortunato (2009) Toward high-performance amorphous GIZO TFTs, *J. Electrochem. Soc.* **156**, H161–H168.
[74] K.K. Banger, Y. Yamashita, K. Mori, R. L. Peterson, T. Leedham, J. Rickard, and H. Sirringhaus (2011) Low-temperature, high-performance solution-processed metal oxide thin-film transistors formed by a 'sol-gel on chip' process, *Nature Materials* **10**, 45–50.
[75] R.A. Street (2009) Thin-film transistors, *Advanced Materials* **21**, 2007–22.
[76] S.D. Brotherton (1995) Poly crystalline silicon thjin-film transistors, *Semiconductor Science and Technology* **10**, 721–38.
[77] J.E. Anthony, A. Facchetti, M. Heeney, S.R. Marder, and X.W. Zhan (2010) n-Type organic semiconductors in organic electronics, *Advanced Materials* **22**, 3876–92.
[78] H. Sirringhaus (2009) Reliability of organic field-effect transistors, *Advanced Materials* **21**, 3859–73.
[79] T. Kamiya, K. Nomura, and H. Hosono (2010) Present status of amorphous In-Ga-Zn-O thin-film transistors, *Science and Technology of Advanced Materials* **11**, 044305-1 – 044305-23.
[80] H.D. Kim, J.K. Jeong, H.J. Chung, and Y.G. Mo (2008) Invited paper: Technological challenges for large-size AMOLED display, in *International Symposium of the Society-for-Information-Display (SID 2008)*, Los Angeles, CA, 291–4.
[81] D.L. Staebler and C.R. Wronski (1977) Reversible conductivity changes in discharge-produced amorphous Si, *Applied Physics Letters* **31**, 292–4.
[82] P. Gorrn, P. Holzer, T. Riedl, W. Kowalsky, J. Wang, T. Weimann, P. Hinze, and S. Kipp (2007) Stability of transparent zinc tin oxide transistors under bias stress, *Applied Physics Letters* **90**, 063502-1–063502-3.
[83] A. Dzwilewski, P. Matyba, and L. Edman (2010) Facile fabrication of efficient organic CMOS circuits, *J. Phys. Chem. B* **114** 135–140.
[84] P. Barquinha, R. Martins, and E. Fortunato (2011) N-type Oxide Semiconductor Thin-Film Transistors, in *Advances in GaN and ZnO-based Thin Film, Bulk and Nanostructured Materials and Devices*. Under publication, S.J. Pearton, Ed. New York: Springer.
[85] H.Q. Chiang, D. Hong, C.M. Hung, R.E. Presley, J.F. Wager, C.H. Park, D.A. Keszler, and G.S. Herman (2006) Thin-film transistors with amorphous indium gallium oxide channel layers, *J. Vac. Sci. Technol. B* **24**, 2702–5.
[86] H.Q. Chiang, B.R. McFarlane, D. Hong, R.E. Presley, and J.F. Wager (2008) Processing effects on the stability of amorphous indium gallium zinc oxide thin-film transistors, *J. Non-Cryst. Solids* **354**, 2826–30.
[87] J.H. Jeong, H.W. Yang, J.S. Park, J.K. Jeong, Y.G. Mo, H.D. Kim, J. Song, and C.S. Hwang (2008) Origin of subthreshold swing improvement in amorphous indium gallium zinc oxide transistors, *Electrochem. Solid State Lett.* **11**, H157–H159.
[88] Q.J. Yao and D.J. Li (2005) Fabrication and property study of thin film transistor using rf sputtered ZnO as channel layer, *J. Non-Cryst. Solids* **351**, 3191–4.
[89] H. Hosono (2006) Ionic amorphous oxide semiconductors: Material design, carrier transport, and device application, *J. Non-Cryst. Solids* **352**, 851–8.
[90] K. Nomura, T. Kamiya, H. Ohta, M. Hirano, and H. Hosono (2008) Defect passivation and homogenization of amorphous oxide thin-film transistor by wet O_2 annealing, *Applied Physics Letters* **93**, 192107-1–192107-3.

[91] M.G. McDowell and I.G. Hill (2009) Influence of Channel Stoichiometry on Zinc Indium Oxide Thin-Film Transistor Performance, *IEEE Trans. Electron Devices* **56**, 343–7.
[92] G.H. Kim, B.D. Ahn, H.S. Shin, W.H. Jeong, H.J. Kim, and H.J. Kim (2009) Effect of indium composition ratio on solution-processed nanocrystalline InGaZnO thin film transistors, *Applied Physics Letters* **94**, 233501-1–233501-3.
[93] H. Kumomi, K. Nomura, T. Kamiya, and H. Hosono (2008) Amorphous oxide channel TFTs, *Thin Solid Films* **516**, 1516–22.
[94] J.K. Jeong, J.H. Jeong, H.W. Yang, J.S. Park, Y.G. Mo, and H.D. Kim (2007) High performance thin film transistors with cosputtered amorphous indium gallium zinc oxide channel, *Applied Physics Letters* **91**, 113505-1–113505-3.
[95] J. Levinson, F.R. Shepherd, P.J. Scanlon, W.D. Westwood, G. Este, and M. Rider (1982) Conductivity behavior in polycrystalline semiconductor thin film transistors, *Journal of Applied Physics* **53**, 1193–1202.
[96] A. Olziersky, P. Barquinha, A. Vilà, C. Magaña, E. Fortunato, J.R. Morante, R. Martins, Role of Ga2O3-In2O3-ZnO channel composition on the electrical performance of thin-film transistors, *Materials Chemistry and Physics*, in press.
[97] M.J. Powell, C. Vanberkel, I.D. French, and D.H. Nicholls (1987) Bias dependence of instability mechanisms in amorphous-silicon thin-film transistor, *Applied Physics Letters* **51**, 1242–4.
[98] S.G.J. Mathijssen, M. Colle, H. Gomes, E.C.P. Smits, B. de Boer, I. McCulloch, P.A. Bobbert, and D.M. de Leeuw (2007) Dynamics of threshold voltage shifts in organic and amorphous silicon field-effect transistors, *Adv. Mater.* **19**, 2785–9.
[99] N.D. Young and A. Gill (1990) Electron trapping instabilities in polycrystalline silicon thin-film transistors, *Semiconductor Science and Technology* **5**, 72–7.
[100] Y.K. Moon, S. Lee, W.S. Kim, B.W. Kang, C.O. Jeong, D.H. Lee, and J.W. Park (2009) Improvement in the bias stability of amorphous indium gallium zinc oxide thin-film transistors using an O_2 plasma-treated insulator, *Applied Physics Letters* **95**, 013507-1–013507-3.
[101] M.E. Lopes, H.L. Gomes, M.C.R. Medeiros, P. Barquinha, L. Pereira, E. Fortunato, R. Martins, and I. Ferreira (2009) Gate-bias stress in amorphous oxide semiconductors thin-film transistors, *Applied Physics Letters* **95**, 063502-1–063502-3.
[102] A. Suresh and J. F. Muth (2008) Bias stress stability of indium gallium zinc oxide channel based transparent thin film transistors, *Applied Physics Letters* **92**, 033502-1–033502-3.
[103] I.T. Cho, J.M. Lee, J.H. Lee, and H.I. Kwon (2009) Charge trapping and detrapping characteristics in amorphous InGaZnO TFTs under static and dynamic stresses, *Semicond. Sci. Technol.* **24**, 015013-1–015013-6.
[104] J.M. Lee, I.T. Cho, J.H. Lee, and H.I. Kwon (2008) Bias-stress-induced stretched-exponential time dependence of threshold voltage shift in InGaZnO thin film transistors, *Applied Physics Letters* **93**, 093504-1–093504-3.
[105] J.K. Jeong, H.W. Yang, J.H. Jeong, Y.G. Mo, and H.D. Kim (2008) Origin of threshold voltage instability in indium-gallium-zinc oxide thin film transistors, *Applied Physics Letters* **93**, 123508-1–123508-3.
[106] R.B.M. Cross and M.M. De Souza (2006) Investigating the stability of zinc oxide thin film transistors, *Applied Physics Letters* **89**, 263513-1–263513-3.
[107] E. Fortunato, P. Barquinha, G. Goncalo, L. Pereira, and R. Martins (2010) Oxide Semiconductors: From Materials to Devices, in *Transparent Electronics: From Synthesis to Applications*, A. Facchetti and T.J. Marks, eds, Chichester: John Wiley & Sons, Ltd.
[108] D. Hong, G. Yerubandi, H.Q. Chiang, M.C. Spiegelberg, and J.F. Wager (2008) Electrical modeling of thin-film transistors, *Crit. Rev. Solid State Mat. Sci.* **33**, 101–32.
[109] W. Lim, S.H. Kim, Y.L. Wang, J.W. Lee, D.P. Norton, S.J. Pearton, F. Ren, and I.I. Kravchenko (2008) Stable room temperature deposited amorphous $InGaZnO_4$ thin film transistors, *J. Vac. Sci. Technol. B* **26**, 959–62.
[110] C.C. Liu, Y.S. Chen, and J.J. Huang (2006) High-performance ZnO thin-film transistors fabricated at low temperature on glass substrates, *Electron. Lett.* **42**, 824–5.
[111] H. Yabuta, M. Sano, K. Abe, T. Aiba, T. Den, H. Kumomi, K. Nomura, T. Kamiya, and H. Hosono (2006) High-mobility thin-film transistor with amorphous In $GaZnO_4$ channel fabricated by room temperature rf-magnetron sputtering, *Applied Physics Letters* **89**, 112123-1 – 112123-3.

[112] W.B. Jackson, G.S. Herman, R.L. Hoffman, C. Taussig, S. Braymen, F. Jeffery, and J. Hauschildt (2006) Zinc tin oxide transistors on flexible substrates, *J. Non-Cryst. Solids* **352**, 1753–5.

[113] Y.L. Wang, F. Ren, W. Lim, D.P. Norton, S.J. Pearton, I.I. Kravchenko, and J.M. Zavada (2007) Room temperature deposited indium zinc oxide thin film transistors, *Applied Physics Letters* **90**, 232103-1–232103-3.

[114] K. Ip, G.T. Thaler, H. Yang, S. Youn Han, Y. Li, D.P. Norton, S.J. Pearton, S. Jang, and F. Ren (2006) Contacts to ZnO, *Journal of Crystal Growth* **287**, 149–56.

[115] G. Goncalves, V. Grasso, P. Barquinha, L. Pereira, E. Elamurugu, M. Brignone, R. Martins, V. Lambertini, and E. Fortunato (2011) Role of room temperature sputtered high conductive and high transparent indium zinc oxide film contacts in the performance of orange, green, and blue organic light emitting diodes, *Plasma Process. Polym.* **8**, 340–5.

[116] P. Barquinha, A. M. Vila, G. Goncalves, R. Martins, J. R. Morante, E. Fortunato, and L. Pereira (2008) Gallium-indium-zinc-oxide-based thin-film transistors: Influence of the source/drain material, *IEEE Trans. Electron Devices* **55**, 954–60.

[117] P. Stallinga and H.L. Gomes (2006) Modeling electrical characteristics of thin-film field-effect transistors I. Trap-free materials, *Synth. Met.* **156**, 1305–15.

[118] Y. Shimura, K. Nomura, H. Yanagi, T. Kamiya, M. Hirano, and H. Hosono (2008) Specific contact resistances between amorphous oxide semiconductor In-Ga-Zn-O and metallic electrodes, *Thin Solid Films* **516**, 5899–5902.

[119] W.T. Lim, D.P. Norton, J.H. Jang, V. Craciun, S.J. Pearton, and F. Ren (2008) Carrier concentration dependence of Ti/Au specific contact resistance on n-type amorphous indium zinc oxide thin films, *Applied Physics Letters* **92**, 122102-1–122102-3.

[120] H.S. Yang, D.P. Norton, S.J. Pearton, and F. Ren (2005) Ti/Au n-type Ohmic contacts to bulk ZnO substrates, *Applied Physics Letters* **87**, 212106-1–212106-3.

[121] S.W. Luan and G.W. Neudeck (1992) An experimental-study of the source drain parasitic resistance effects in amorphous-silicon thin-film transistors, *Journal of Applied Physics* **72**, 766–72.

[122] J.S. Park, J.K. Jeong, H.J. Chung, Y.G. Mo, and H.D. Kim (2008) Electronic transport properties of amorphous indium-gallium-zinc oxide semiconductor upon exposure to water, *Applied Physics Letters* **92**, 072104-1–072104-3.

[123] H.N. Lee, J. Kyung, M.C. Sung, D.Y. Kim, S.K. Kang, S.J. Kim, C.N. Kim, H.G. Kim, and S.T. Kim (2008) Oxide TFT with multilayer gate insulator for backplane of AMOLED device, J. *Soc. Inf. Disp.* **16**, 265–72.

[124] J.M. Shaw, J.D. Gelorme, N.C. LaBianca, W.E. Conley, and S.J. Holmes (1997) Negative photoresists for optical lithography, *IBM J. Res. Dev.* **41**, 81–94.

[125] A.L. Bogdanov and S.S. Peredkov (2000) Use of SU-8 photoresist for very high aspect ratio x-ray lithography, *Microelectronic Engineering* **53**, 493–6.

[126] A.K. Nallani, S.W. Park, and J.B. Lee (2003) "Characterization of SU-8 as a resist for electron beam lithography," in *Smart Sensors, Actuators, and Mems, Pts 1 and 2*. vol. 5116, J.C. Chiao, V.K. Varadan, and C. Cane, eds, Bellingham: Spie-Int Soc Optical Engineering, 414–23.

[127] A. Olziersky, P. Barquinha, A. Vila, L. Pereira, G. Goncalves, E. Fortunato, R. Martins, and J.R. Morante (2010) Insight on the SU-8 resist as passivation layer for transparent Ga2O3-In2O3-ZnO thin-film transistors, *Journal of Applied Physics* **108**, 064505-1–064505-7.

[128] D. Hong and J.F. Wager (2005) Passivation of zinc-tin-oxide thin-film transistors, *J. Vac. Sci. Technol. B* **23**, L25–L27.

[129] D.H. Zhang (1996) Adsorption and photodesorption of oxygen on the surface and crystallite interfaces of sputtered ZnO films, *Materials Chemistry and Physics* **45**, 248–52.

[130] D.H. Levy, D. Freeman, S.F. Nelson, P.J. Cowdery-Corvan, and L.M. Irving (2008) Stable ZnO thin film transistors by fast open air atomic layer deposition, *Applied Physics Letters* **92**, 192101-1–192101-3.

[131] K. Ellmer (2001) Resistivity of polycrystalline zinc oxide films: current status and physical limit, *J. Phys. D-Appl. Phys.* **34**, 3097–3108.

[132] E.M.C. Fortunato, L.M.N. Pereira, P.M.C. Barquinha, A.M.B. do Rego, G. Goncalves, A. Vila, J.R. Morante, and R.F.P. Martins (2008) High mobility indium free amorphous oxide thin film transistors, *Applied Physics Letters* **92**, 222103-1–222103-3.

[133] K. Okamura, D. Nikolova, N. Mechau, and H. Hahn (2009) Appropriate choice of channel ratio in thin-film transistors for the exact determination of field-effect mobility, *Applied Physics Letters* **94**, 183-503-1–183503-3.

[134] P.K. Nayak, J.V. Pinto, G. Gonçalves, R. Martins, and E. Fortunato, in press.

[135] P. Gorrn, M. Lehnhardt, T. Riedl, and W. Kowalsky (2007) The influence of visible light on transparent zinc tin oxide thin film transistors, *Applied Physics Letters* **91**, 193504-1–193504-3.

[136] P. Barquinha, A. Pimentel, A. Marques, L. Pereira, R. Martins, and E. Fortunato (2006) Effect of UV and visible light radiation on the electrical performances of transparent TFTs based on amorphous indium zinc oxide, *J. Non-Cryst. Solids* **352**, 1756–60.

[137] T.C. Chen, T.C. Chang, T.Y. Hsieh, C.T. Tsai, S.C. Chen, C.S. Lin, M.C. Hung, C.H. Tu, J.J. Chang, and P.L. Chen (2010) Light-induced instability of an InGaZnO thin film transistor with and without SiO(x) passivation layer formed by plasma-enhanced-chemical-vapor-deposition, *Applied Physics Letters* **97**, 192103-1–192103-3.

[138] Y. Jeong, K. Song, D. Kim, C. Y. Koo, and J. Moon (2009) Bias Stress Stability of Solution-Processed Zinc Tin Oxide Thin-Film Transistors, *J. Electrochem. Soc.* **156**, H808–H812.

[139] P.K. Nayak, T. Busani, E. Elamurugu, P. Barquinha, R. Martins, Y. Hong, and E. Fortunato (2010) Zinc concentration dependence study of solution processed amorphous indium gallium zinc oxide thin film transistors using high-k dielectric, *Applied Physics Letters* **97**, 183504-1–183504-3.

[140] E. Fortunato, V. Figueiredo, P. Barquinha, E. Elamurugu, R. Barros, G. Goncalves, S.H.K. Park, C.S. Hwang, and R. Martins (2010) Thin-film transistors based on p-type Cu2O thin films produced at room temperature, *Applied Physics Letters* **96**, 239902-1–239902-3.

[141] E. Fortunato, R. Barros, P. Barquinha, V. Figueiredo, S.H.K. Park, C.S. Hwang, and R. Martins (2010) Transparent p-type SnOx thin film transistors produced by reactive rf magnetron sputtering followed by low temperature annealing, *Applied Physics Letters* **97**, 052105-1–052105-3.

[142] H. Dunwald and C. Wagner (1933) Tests on the appearances of irregularities in copper oxidule and its influence on electrical characteristics, *Z. Phys. Chem. B-Chem. Elem. Aufbau. Mater.* **22**, 212–25.

[143] K. Matsuzaki, K. Nomura, H. Yanagi, T. Kamiya, M. Hirano, and H. Hosono (2008) Epitaxial growth of high mobility Cu2O thin films and application to p-channel thin film transistor, *Applied Physics Letters* **93**, 202107-1–202107-3.

[144] K. Matsuzaki, K. Nomura, H. Yanagi, T. Kamiya, M. Hirano, and H. Hosono (2009) Effects of post-annealing on (110) Cu2O epitaxial films and origin of low mobility in Cu2O thin-film transistor, *Phys Status Solidi A* **206**, 2192–7.

[145] S.Y. Sung, S.Y. Kim, K.M. Jo, J.H. Lee, J.J. Kim, S.G. Kim, K.H. Chai, S.J. Pearton, D.P. Norton, and Y.W. Heo Fabrication of p-channel thin-film transistors using CuO active layers deposited at low temperature, *Applied Physics Letters* **97**, 222109-1–222109-3.

[146] X.A. Zou, G.J. Fang, L.Y. Yuan, M.Y. Li, W.J. Guan, and X.Z. Zhao (2010) Top-Gate Low-Threshold Voltage p-Cu(2)O Thin-Film Transistor Grown on SiO(2)/Si Substrate Using a High-kappa HfON Gate Dielectric, *IEEE Electron Device Lett.* **31**, 827–9.

[147] V. Figueiredo, E. Elangovan, G. Goncalves, P. Barquinha, L. Pereira, N. Franco, E. Alves, R. Martins, and E. Fortunato (2008) Effect of post-annealing on the properties of copper oxide thin films obtained from the oxidation of evaporated metallic copper, *Appl Surf Sci* **254**, 3949–54.

[148] E. Fortunato and R. Martins (2011) Where science fiction meets reality? With oxide semiconductors!, *physica status solidi (RRL) – Rapid Research Letters*, **5**, 336–9.

[149] Y. Ogo, H. Hiramatsu, K. Nomura, H. Yanagi, T. Kamiya, M. Hirano, and H. Hosono (2008) p-channel thin-film transistor using p-type oxide semiconductor, SnO, *Applied Physics Letters* **93**, 032113-1–032113-3.

[150] C.W. Ou, Z.Y. Ho, Y.C. Chuang, S.S. Cheng, M.C. Wu, K.C. Ho, and C.W. Chu (2008) Anomalous p-channel amorphous oxide transistors based on tin oxide and their complementary circuits, *Applied Physics Letters* **92**, 122113-1–122113-3.

[151] Dhananjay, C.W. Chu, C.W. Ou, M.C. Wu, Z.Y. Ho, K.C. Ho, and S.W. Lee (2008) Complementary inverter circuits based on p-SnO(2) and n-In(2)O(3) thin film transistors, *Applied Physics Letters* **92**, 232103-1–232103-3.
[152] H.N. Lee, H.J. Kim, and C K. Kim (2010) p-Channel Tin Monoxide Thin Film Transistor Fabricated by Vacuum Thermal Evaporation, *Japanese Journal of Applied Physics* **49**, 020202.
[153] H. Yabuta, N. Kaji, R. Hayashi, H. Kumomi, K. Nomura, T. Kamiya, M. Hirano, and H. Hosono (2010) Sputtering formation of p-type SnO thin-film transistors on glass toward oxide complimentary circuits, *Applied Physics Letters* **97**, 072111-1–072111-3.
[154] L.Y. Liang, Z.M. Liu, H.T. Cao, Z. Yu, Y.Y. Shi, A.H. Chen, H.Z. Zhang, Y.Q. Fang, and X.L. Sun (2010) Phase and Optical Characterizations of Annealed SnO Thin Films and Their p-Type TFT Application, *Journal of the Electrochemical Society* **157**, H598–H602.
[155] K. Nomura, T. Kamiya, and H. Hosono (2011) Ambipolar Oxide Thin-Film Transistor, *Advanced Materials*, 23, 3431–4.
[156] P.F. Carcia, R.S. McLean, and M.H. Reilly (2006) High-performance ZnO thin-film transistors on gate dielectrics grown by atomic layer deposition, *Applied Physics Letters* **88**, 123509-1–123509-3.
[157] P. Barquinha, L. Pereira, G. Goncalves, R. Martins, D. Kuscer, M. Kosec, and E. Fortunato (2009) Performance and Stability of Low Temperature Transparent Thin-Film Transistors Using Amorphous Multicomponent Dielectrics, *J. Electrochem. Soc.* **156**, H824–H831.
[158] M. Ofuji, K. Abe, H. Shimizu, N. Kaji, R. Hayashi, M. Sano, H. Kumomi, K. Nomura, T. Kamiya, and H. Hosono (2007) Fast thin-film transistor circuits based on amorphous oxide semiconductor, *IEEE Electron Device Lett.* **28**, 273–5.
[159] H. Hartnagel, A. Dawar, A. Jain, and C. Jagadish (1995) *Semiconducting Transparent Thin Films*. Bristol: IOP Publishing.
[160] Z.L. Pei, L. Pereira, G. Goncalves, P. Barquinha, N. Franco, E. Alves, A.M.B. Rego, R. Martins, and E. Fortunato (2009) Room-Temperature Cosputtered HfO_2-Al_2O_3 Multicomponent Gate Dielectrics, *Electrochem. Solid State Lett.* **12**, G65–G68.
[161] G.D. Wilk, R.M. Wallace, and J.M. Anthony (2001) High-kappa gate dielectrics: Current status and materials properties considerations, *Journal of Applied Physics* **89**, 5243–75.
[162] L. Pereira, P. Barquinha, G. Goncalves, A. Vila, A. Olziersky, J. Morante, E. Fortunato, and R. Martins (2009) Sputtered multicomponent amorphous dielectrics for transparent electronics, *Phys. Status Solidi A-Appl. Mat.* **206**, 2149–54.
[163] S. Chang, Y.W. Song, S. Lee, S.Y. Lee, and B.K. Ju (2008) Efficient suppression of charge trapping in ZnO-based transparent thin film transistors with novel Al2O3/HfO2/Al2O3 structure, *Applied Physics Letters* **92**, 192104-1–192104-3.
[164] S.Y. Lee, S. Chang, and J.-S. Lee (2010) Role of high-k gate insulators for oxide thin film transistors, *Thin Solid Films* **518**, 3030–2.

6

Electronics With and On Paper

6.1 Introduction

Today there is a strong interest within the scientific and industrial community concerning the use of biopolymers such as paper for electronic applications, mainly driven by its low-cost ($\approx 10^{-3}$ cent m^{-2}, more than one order of magnitude cheaper than the cheapest based polymer substrate, which is polyethylene terephtalate – PET), lightness, flexibility and ability to be 100% recyclable. Moreover, paper is the Earth's major biopolymer, a well-known technology whose process speed (roll-to-roll – R2R – manufacturing) can exceed 10^2 km h^{-1}, under well controlled and reliable conditions, making it a key factor as far its availability and feasibility is concerned.

Most paper that has been used up to now is intended to achieve the required mechanical and physical properties to be employed as the support for inks of different origins or for packaging applications, independent of the attempts made in the 20th century to use it in electronics applications [1, 2]. These attempts envisaged exploiting paper's potential as a substrate for low-cost flexible electronics [3–6]. Adding to this interest, we must recognize the importance of the aspiration for wireless auto sustained and low energy consumption electronics [7–9]. This can be fulfilled by cellulose paper, the lightest and the cheapest substrate material which is of tremendous global economic importance.

In the future, specific electronic heterogeneous paper sheets could be fabricated designed to create paper fibers with the required bulk and surface functionalities, the proper water/vapor barrier, size and diameter/thickness of the fibrils and the proper control of paper thickness for full integration in electronic devices. This will be the function of the components/devices that are to be incorporated/integrated such as Thin Film Transistors

Transparent Oxide Electronics: From Materials to Devices, First Edition. Pedro Barquinha, Rodrigo Martins, Luis Pereira and Elvira Fortunato.

(TFTs), Complementary Metal Oxide Semiconductor (CMOS) devices; passive electronic components (Resistances Inductors and Capacitors); Memory Transistors; Electrochromics and the Thin Film Paper Batteries. This mixed used of paper as an active component in the building blocks of integrated electronics is extremely attractive if we compare the costs involved with those of conventional silicon technology.

In commercial terms, a transistor made of paper is about 30 000× cheaper (comparing a A4 paper sheet of 80gr. and a silicon wafer 8″ in diameter) than one made of silicon; the materials based on metal oxides such as ZnO that are used with it to fabricate such transistors are about 1000 times less expensive than silicon, in terms of extraction; the technology of the paper is about 8000× more economical because it uses no harmful gases or high process temperatures (at or above 1000°C, as happens with silicon), and does not require such highly sophisticated environments and control tools as those used in conventional microelectronics foundries or electronics fabrications (including integrated circuits). These are the concepts that support the technical and economical importance of combining oxides with paper so as to give rise to the concept of electronics with and on paper, towards a green electronics, that is 100 % recyclable, with a considerable impact on so-called low cost and disposable electronics.

6.2 Paper in Electronics

Cellulose paper has been used in electronics since the beginning of the 20th century, but due to constraints related to its durability, electronic devices do not use it as an active element. The most common applications of cellulose-based materials are as a dielectric for capacitors [10] and supercapacitors [11], permeable membranes in liquid electrolyte batteries [12, 13] or simply the physical support for energy storage devices [14], organic thin film transistor (OTFTs) arrays [15], inkjet-printed sensors and RFID[a] tags [16, 17], screen-printed batteries [18], inorganic powder electroluminescence devices [19], foldable printed circuit boards [20], oxide TFTs [21] and flexible low-voltage electric double-layer TFTs [22] amongst others.

Printing electronics on paper has been tried since the 1960s, where TFTs were deposited on paper substrates on a roll inside a vacuum chamber [23]. The applications in mind at that time were flexible circuits for credit cards, electric sensors, toys and hobby kits but without any industrial or commercial success since it appeared during the boom of the advent of silicon based CMOS devices that fueled a complete revolution in the so-called Information, Communication and Telecommunication fields [24, 25]. Moreover, most of the problems related to paper stability and durability were not properly addressed; mainly hydrophilic paper behavior when the surface is not treated. Even today, paper durability is still a problem and so, is of great importance as far electronics is concerned.

However, a new concept of low cost and disposable electronic devices is now under discussion for applications such as ID tags, pH, medicine indicators and food quality control devices where the recyclability and the environmental impact have to be taken into account. This concept requires power sources and devices, preferably integrating

[a] RFID – **R**adio-**F**requency **ID**entification.

auto-regenerative devices, using environmentally friendly processes. Recent attempts in this direction were the paper batteries proposed by Pushparaj [11] and the systems investigated by the Enfucell (commercialization of the SoftBattery®) and Acreo companies, in printed electronics [16, 20]. The major drawbacks with the battery devices are their incompatibility with all-solid-state electronics and the fact that they are not auto-regenerative (this is not a problem for disposable devices).

Here we have to take into account the fact that paper is a sustainable fiber product whose specific functionalities can be achieved thanks to the proper choice and selection of raw materials (fibers and additives in mass), with coating and calendaring aimed at achieving the expected defined specifications. Indeed, paper is perfectly adapted to recycling but is limited in performance for electronic purposes, due to its typically hydrophilic behavior over which the surface finishing has a strong influence. The only solutions undertaken by industry to make paper hydrophobic is to use an extrusion polymer coating on the paper surface in order to create smoothness and barrier properties, giving it the required dimensional and mechanical stability, with the proper folding ability, enough for the food industry but not enough for microelectronics. Some solutions have already been found by using a biosourced polymer (PLA[b]) [1] or nanofibrils to make the paper fully transparent [26]. Nevertheless, PLA is very sensitive to temperature and can't withstand many processes at temperatures higher than 45°C, limiting its uses in electronics which requires two main grades of paper in the same sheet (heterogeneous paper sheets): one with a very high level of barrier and highly smooth and compact surfaces in order to process highly stable electronics components on its surface or even as a field effect transistor using the paper also as dielectric [2, 27, 28], with electrical performances that are not environmentally dependent; and the other with a nano porous surface, if high storage ions/charges are envisaged, as required for battery or memory applications [2, 7, 8, 27–29]. To achieve the first goal, the selection should involve paper fabricated by the coating of a bioplastic onto the paper surface. To achieve the second goal, a thin layer of nanofibrills of cellulose may be deposited on the surface of the paper and thus create a nanoporous material. This makes the paper surface highly reactive regarding chromatogenic chemistry due to its extremely high content of hydroxyl groups. Thus, the hydrophilic/hydrophobic balance of this material can be tailored for optimized field effect transistors (FET), non-volatile memory TFTs, CMOS and solid state batteries, all of which are printable by using proper electronic inks, or can be processed using R2R vacuum based techniques such as sputtering. That is, it is possible to fabricate complex functionalized paper so as to be integrated in complex electronic systems within the same paper sheet, whose main limitation is the process temperature used. The emerging class of n- and p-type oxide semiconductors is able to overcome most of those restrictions, since they can be prepared as thin films under moderate/low temperature.

Indeed, the recent developments of n- and p-type oxide TFTs [30–54] and in particular the production of n-type paper transistors [55, 56] at room temperature have contributed, as a first step, to the development of disposable, low cost and flexible electronic devices, where paper, besides being a substrate, is also one constituent of them. As we want to process electronic systems that are fully recyclable and environmentally friendly, this involves the selection of green materials connected to abundant sources, as in the case of Ti and Zn based oxides, either for targets or inks formulation, so as to catapult the paper and

[b] PLA – PolyLactic Acid.

electronics industry into a new era of growth, fully open to enabling Information and Communication Technologies in sectors such as *smart tags, labels and interactive display placards*, *disposable bio-sensors*, *interactive journals*, *packaging*, amongst others, besides a variety of novel applications whose limit it is our imagination!

For this type of application, the wireless energy harvesting concept should be also fulfilled in order to make electronics in and with paper fully autonomous and to avoid the problems related to tearing paper under mechanical stress. Moreover, paper electronics should not be confused with "electronic paper" (e-paper) that refers to paper-like electronic displays, such as a tablet, today mostly based on conventional electronics; and it may also be substituted in the future by oxide based integrated electronics, including functionalized paper that we will refer to from now on *paper-e*.

In this chapter we present the properties of different papers used as a substrate (passive) and also as an active component for electronic applications. To do so we start by presenting the typical structure for a set of selected papers with different compositions, paper thickness and mineral contents, followed by the analysis of the electrical properties of the paper, namely their electrical resistivity and capacitance behavior. These properties are relevant when paper is used as an active component in the fabrication of electronic devices.

To evaluate the paper behavior as a substrate, we analyze the electrical characteristics (resistivity and Hall mobility) of high transparent conductive oxides based on indium zinc oxide (IZO) deposited on different papers as well as on a glass substrate, selected for reference. Finally, solid state devices integrating the paper as an active component will be described.

6.3 Paper Properties

6.3.1 Structure, Morphology and Thermal Properties

Paper has a typically crystalline structure, as revealed by cellulose type I, which is evidenced by the lattice plane (101) at $2\theta=15.6°$ and the lattice plane (200) at $2\theta=22.5°$ [57], as depicted by the X-ray diffraction (XRD) pattern shown in Figure 6.1. These results are also in accordance with the available data regarding commercial cellulose [58–60]. The percentage of the degree of crystallinity index (I_c) can also be determined using the empirical method [61, 62]:

$$I_c = \frac{\left(I_{(002)} - I_{(am)}\right)}{I_{(002)}} \times 100, \qquad (6.1)$$

where $I_{(002)}$ is the maximum intensity of the 002 lattice diffraction and $I_{(am)}$ the minimum value between the 002 and 101 lattice diffraction close to 18°, representing the amorphous material in the cellulosic fibers. The result that was obtained shows that $I_c \cong 75\%$. Considering that hemicellulose has a random amorphous structure, it can be concluded that the one evaluated here has a small quantity of hemicellulose.

However, more important than the structure, we have the surface morphology which dictates the type of application to which the paper fits. In Table 6.1 we present a set of five papers with different surface finishing in which the electronic performance of films or devices constructed will be discussed.

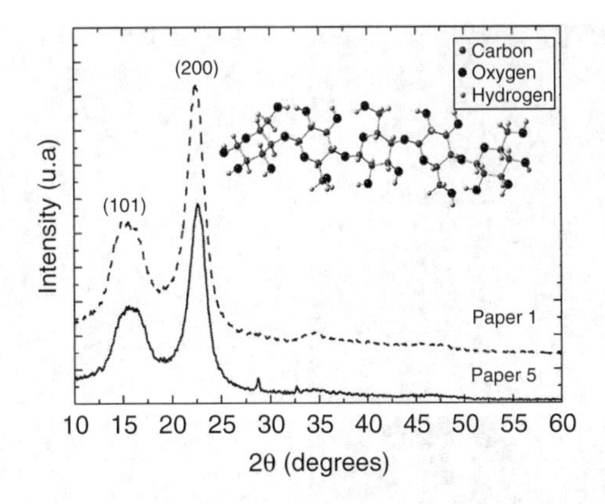

Figure 6.1 *XRD patterns of two types of natural papers (hydrophobic (A) and hydrophilic (B)) used in this study that exhibit different degrees of porosity and compactness. The inset shows the 3D structure of cellulose (adapted from [63]) showing the main atoms that constitute the cellulose. Reprinted with permission from [63] Copyright (2009) Ben Mills.*

Table 6.1 *Characteristics of the papers used in the present work.*

	Thickness (μm)	Roughness (nm)	Weight/ area (g m^{-2})	T_E (°C) (% mass loss)	T_{P1} (°C)	T_{P2} (°C) (% mass loss)
Paper 1	188	3407	190	129 (2.47)	300	329.7 (31.33)
Paper 2	199	9102	200	87.6 (3.33)	267.7	327.1 (39.87)
Paper 3	79	39634	70	76 (4.42)	291.0	344.0 (47.12)
Paper 4	173	8487	160	109.7 (3.23)	287.0	321.9 (38.4)
Paper 5	53	39634	38	139		

Figure 6.2 shows the surface morphology of the set of papers listed in Table 6.1. As far as the papers listed in Table 6.1 are concerned, the data from Figure 6.2 show that in paper-1 the presence of cellulose fiber is visible, despite the fact that the surface is coated with a hydrophilic layer that makes it homogeneous and probably continuous, but still with the roughness of the paper. The paper roughness is moderated (3407 nm). The TGA and DSC data reveal an endothermic peak (T_E = 129°C) in the thermal curve with a weight loss of 2.47 % due to the release of adsorbed water. Above 300°C, two further mass loss steps of 31.33 % occurred, which are due to the pyrolytic decomposition (T_p) of the paper.

The surface morphology of paper-2 is very different from that of paper-1. In this case the hydrophilic layer is thicker than the previous one since we cannot see any cellulose fiber. This paper has a surface roughness of about ~9 μm. For this case T_E = 87.6°C in the thermal curve with a weight loss of 3.33 % due to the release of adsorbed water. Above 267.7°C, two further mass loss steps of 39.87 % occurred, which are due to the pyrolytic decomposition of the paper. That is, this paper is less temperature resistant than paper-1.

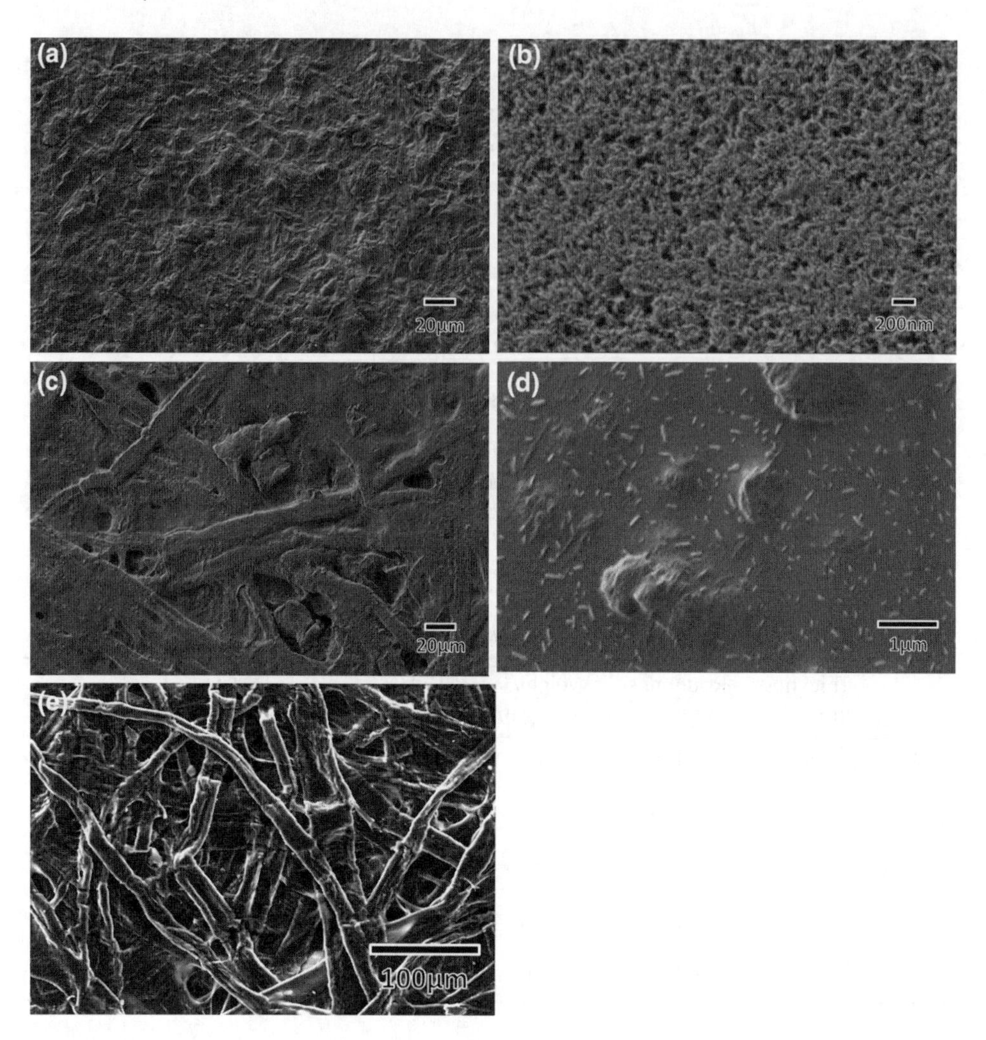

Figure 6.2 SEM image (not with the same magnification, since this depends on the type of surface under analysis) of: a) paper-1; b) paper-2; c) paper-3; d) paper-4; e) paper-5. Reprinted with permission from [65] Copyright (2011) SPIE.

The image for paper-3 reveals the presence of the cellulose fibers with a non smooth surface (roughness of ~40 μm); T_E=87.6°C in the thermal curve with a weight loss of 3.33 % due to the release of adsorbed water. Above 267.7°C, two further mass loss steps of 39.87 % occurred, which are due to the pyrolytic decomposition of the paper. That is, this paper is less temperature resistant than paper-1.

The SEM micrograph for paper-4 reveals a continuous surface, although it presents some agglomerations; T_E=109.7°C in the thermal curve with a weight loss of 3.23 % due to the release of adsorbed water. Above 287°C, two further mass loss steps of 38.4 % occurred, which are due to the pyrolytic decomposition of the paper. That is, this paper is

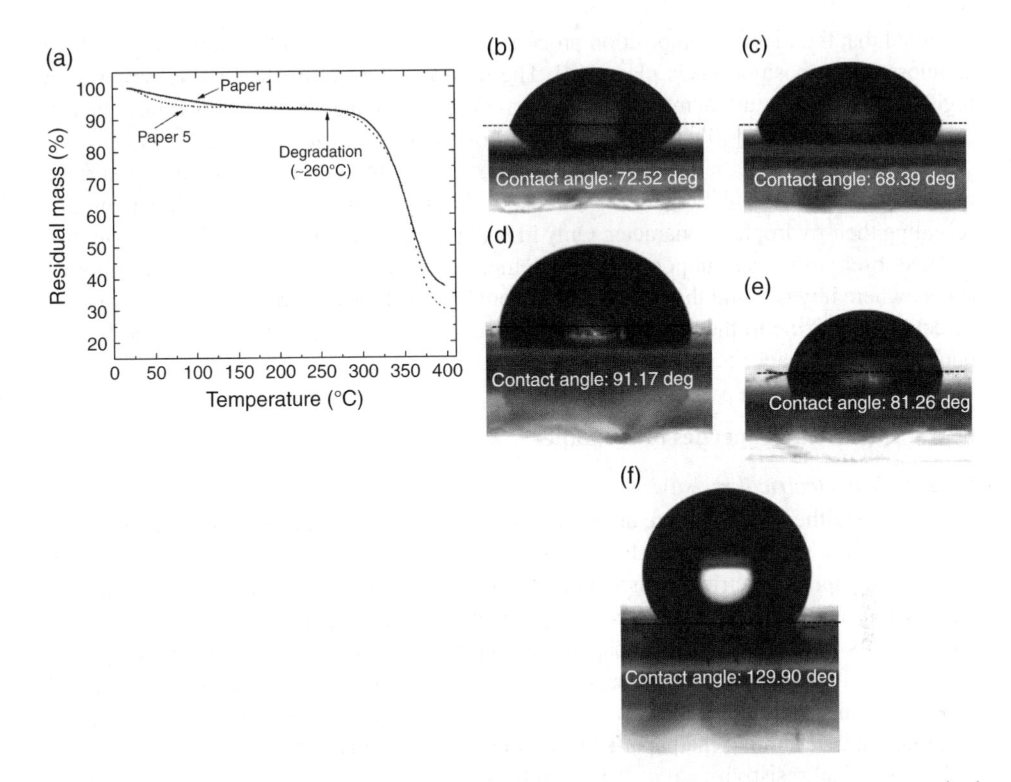

Figure 6.3 *a) Typical thermogravimetry data on two cellulose-based papers studied with different degrees of porosity (B the most porous paper. See Figure 6.1 and also sections 6.5 and 6.6); shape of the drop of water that fell on the surface of: b) paper-1 (hydrophilic); c) paper-2 (hydrophilic); d) paper-3 (transition); e) paper-4 (hydrophilic); f) paper-e (hydrophobic). Reprinted with permission from [65] Copyright (2011) SPIE.*

the one that is more resistant to temperatures as high as 170°C without any appreciable degradation. Finally, the SEM image of paper-5 shows that the paper is constituted by fibers that are randomly distributed with widths in the order of 10–20 μm. The surface roughness measured via profilometry revealed an average value within the range of 3–4 μm.

Figure 6.3a shows a typical dependence of the residual mass weight (wt%) on temperature (thermogravimetry analysis), for two types of papers (designated by A and B, hydrophobic and hydrophilic, respectively, as described in Figure 6.1), representing the two extreme cases of all of the papers studied. This allowed us not only to analyze the paper's behavior as a function of the temperature, but also to define the threshold temperature at which the paper starts decomposing.

Overall, the data reveal two main weight loss regions: the first within the 25–150°C temperature range and the second within the 260–400°C temperature range. The first decomposition region is associated with the moisture and desorption of the low-molecular-weight compounds remaining from the isolation manufacture process. The paper showed moisture losses of 6–8 wt% within the 25–150°C range. For the second weight loss region, it can be observed that the main decomposition process occurred within the 300–400°C range, where cellulose decomposition takes place [64]. The moisture uptake for the first decomposition region, where a faster initial mass reduction was observed, related to the hydrophilic tendency presented by the paper [40], agreeing with most of the data shown in Figures 6.3b to f.

Here, we notice that the water contact angle measured for a drop of water on the surface of the papers depicted in Table 6.1 is in most of the paper samples evaluated less than 90°, revealing their hydrophilic character. Only in paper-5 is the contact angle is higher than 90°.

This observation was supported by an elemental analysis carried out by X-Ray fluorescence, where it was found that paper-5 has about 2 % of fluorine, while no traces have been detected of fluorine in the other papers analyzed in this study, which confirms the hydrophobic nature of paper-5.

6.3.2 Electrical Properties of the Paper

6.3.2.1 The electrical resistivity

To use paper either as a substrate or an electronic component in devices, such as a dielectric or a solid state electrolyte or membrane, it is important to know their electrical performance, namely the paper's electrical resistivity and how the environment influences this behavior, namely the humidity. All of this is highly dependent on the paper surface finishing and mineral bulk composition of the paper, especially that of the ionic and cationic species present. In the following we present electrical resistivity data obtained on non treated paper, with the intention of observing how relevant is the paper content in the presence of humidity, in determining the paper's electrical resistivity behavior.

The electrical resistivity (ρ) measured in bulk (ρ_{volume}, across the paper sheet) and that at the surface ($\rho_{surface}$, gap cell or four point probe configuration) are different (it can be as high as more than two orders of magnitude), due to paper anisotropy and the degree of compactness of the paper's constituents, as revealed by the paper's structure and morphology. Moreover, ρ is highly dependent on the minerals the paper contains and its cationic nature, as well as on the state of relative humidity (*RH*) in which measurements are taken, due to the paper's hydrophilic characteristics. Data adapted from the literature [1,65] show that by changing, for instance, the sodium chloride (NaCl) from a concentration of 0.03Kg t^{-1} to 5.8 Kg t^{-1} the electrical resistivity changes by more than one order of magnitude, which is more pronounced when paper sustains $RH \leq 25\,\%$ (see Figure 6.4).

On the other hand, when *RH* changes from 25 % up to 60 %, ρ can vary by more than three orders of magnitude, being more pronounced at the surface than in the bulk of the paper (see Figure 6.4). Indeed, *RH* will affect the electrical surface and bulky resistivity of the paper in as much as the paper is porous and exhibits a high surface roughness (see, for instance, papers 3 and 5). Under these conditions humidity dependence is very strong. The data depicted in Figure 6.4 shows that either ρ_{volume} or $\rho_{surface}$ dependence on *RH* follows an almost exponential law. This behavior suggests that the carriers' transport mechanism can be by hopping (between the hydroxyl groups), or even by percolation, via protons, or hydronium ions and impurity ions (at high electric fields) [66–69]. Reliable electrical measurements of paper resistivity can be complicated for several reasons. First, the response to a change in the *RH* by adsorption and desorption of water molecules is slow and typically results in a hysteresis of the measured electronic properties of paper. Another concern is the transient behavior when applying a constant voltage [70–74], since the steady state may be

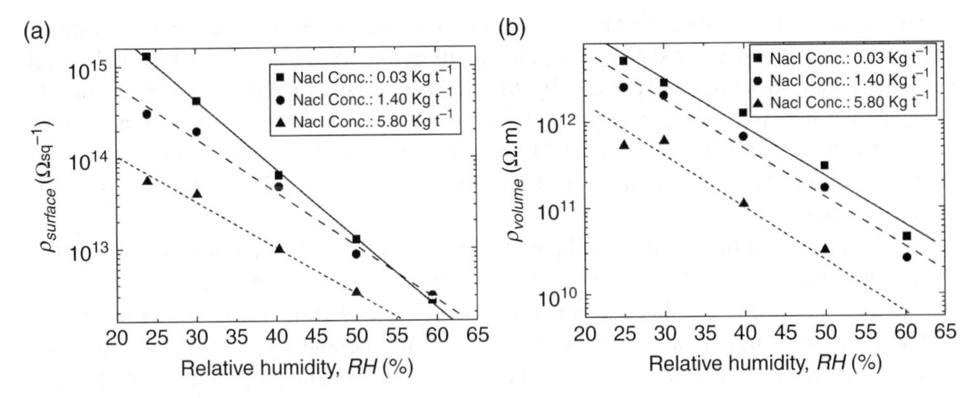

Figure 6.4 The surface (a) and volume (b) resistivity of paper substrates with different concentrations of NaCl, under an electrical bias of 100 V. Reprinted from [66] Copyright (2009) Society for Imaging Science and Technology.

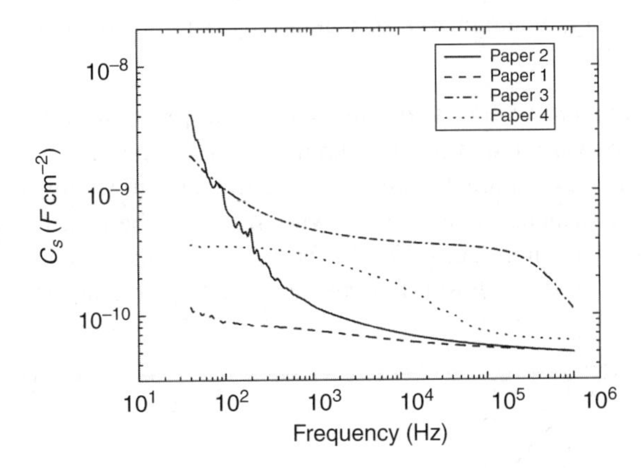

Figure 6.5 Capacitance dependence on frequency for the set of paper samples analysed. Reprinted with permission from [65] Copyright (2011) SPIE.

hard to be achieved. This behavior can be explained by the polarization and the formation of a diffuse double layer of ions at the electrodes, which counteracts the applied electric field.

6.3.2.2 *Electrical capacitance*

The ability of the paper to store charges is highly dependent on the type of fibrils and mineral (neutral and non-neutral species) composition that constitute the bulk of the paper. This is highly relevant when paper is used either as a dielectric in active devices such as transistors or as an electrolyte in solid state batteries. In the following we present data on the capacitance per unit area (C_S) of the set of papers listed in Table 6.1. Figure 6.5 shows the dependence of C_S as a function of frequency (f) for the papers whose characteristics are shown in Table 6.1 and their morphology depicted in Figure 6.2.

For paper-1 we note that the capacitance is almost independent of the frequency, meaning that the capacitance is mostly determined by its geometry. This is consistent with a high compact paper structure, as revealed by the SEM images. For paper-2 we note that the capacitance increases as the frequency decreases. This is consistent with a non-compact paper structure, as revealed by the SEM images. For paper-3 we note a first capacitance increase, followed by a flat region until *f* is as low as about 1 kHz. Below this frequency C_s starts increasing.

The *C(f)* behavior observed is attributed to the role of the different planes that constitute the paper. Thus, we expect a strong role for the paper thickness in the final C_s. For paper-4 we notice that an initial enhancement of C_s as *f* decreases from 1 MHz to 1 kHz. Below 1 kHz the capacitance remains almost constant. From the analysis of the data discussed above we conclude that with a porous and foam-like paper structure, suchas that of paper-2, the static capacitance increases almost exponentially as we approach the steady state condition. Paper-5 was studied with in a larger frequency range than that used for data depicted in Figure 6.6, aiming to better understand the role of the paper's structure in the electrical capacitance of the paper. This information is very important, especially when paper is used either as electrolyte and membrane in all solid state batteries [8, 11, 14] or as a non-volatile memory, where charge storage ability determines the performance of such devices [27, 75, 76].

6.3.2.3 Paper capacitance in less compact and porous paper structures

Figure 6.6 shows the variation of the measured capacitance per unit area (C/A) of paper-5 as a function of the excitation frequency, for different temperatures. The high values obtained for low frequencies are associated with the space charge accumulation along the disperse net of fibers and interfaces [77–80]. Moreover, the increase in loss factor (tan δ in the inset of Figure 6.6) for lower frequencies and higher temperatures suggests that

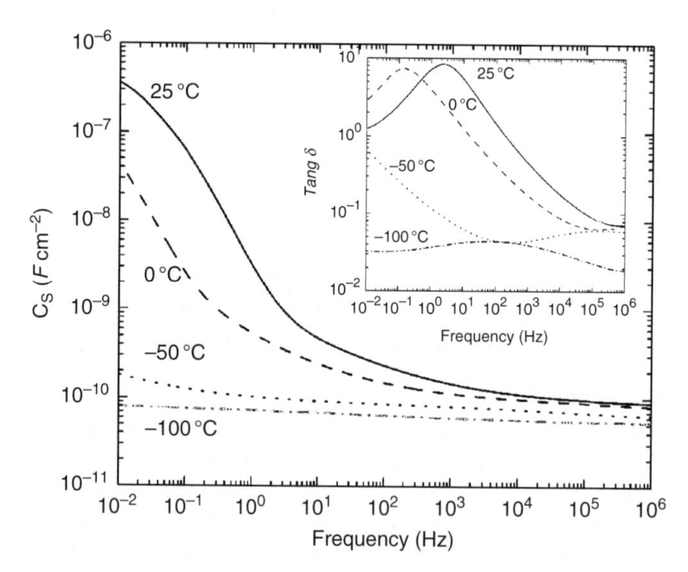

Figure 6.6 *Variation of the capacitance and loss factor tan δ (inset) as a function of frequency and temperature.*

there was charge movement in the bulk of the paper [71]. Three kinds of movement may exist: rotation on fixed sites; short-distance hopping towards the places where the energy barrier for movement is higher, causing charge accumulation; and hopping between sites where the energy barrier is high under AC signals [81]. The first two are associated with the increase in capacitance while the third may contribute to the decrease in resistivity (increase in loss factor; see the behaviour of $tan\ \delta$ in the inset of Figure 6.6), as the data depicted in Figure 6.6 suggests. The paper structure and the defects that exist inside it are indeed barriers to charge movement and potentiate charge accumulation.

To understand the notorious ability of paper to accumulate charge, the contribution of the fibers' surfaces to the enhancement of the overall area where charge accumulation may occur can be considered. That is, it is assumed that the charges are accumulated in the random-distributed fibers dispersed throughout the paper thickness. As such, the capacitance under static conditions depends on how the fibers are distributed within the dielectric volume, which determines the way that the charge will accumulate along them and the corresponding interfaces.

As depicted in Figure 6.2e, the fibers have thicknesses of several micrometers (>4 μm) and a length L', being combined in planes that are mechanically compressed to give the desired final paper thickness. Apart from that, we also notice that between fibers and between planes we can find pores, which may be filled with the ionic resin and/or the glue used.

In this type of foam or sponge like structure we have permanent fixed charges introduced during the paper's fabrication as well as induced balanced charges, when the paper sustains a static or dynamic electric field along its bulk. That is, within the paper we have space charges that we consider to be distributed along the fibers' surface and the different planes that constitute the paper (see Figure 6.7, see color plate section). That is, we may have balanced charges along both surfaces of the same fiber with a thickness d_f or between two close fibers with a gap (filled or not with glue) of d_g or between consecutive planes that contain the fibers, d_p.

Apart from that, we also notice that each fiber is constituted by microfibrils (see Figure 6.7b) with thicknesses below 1 μm where charges can also be accumulated.

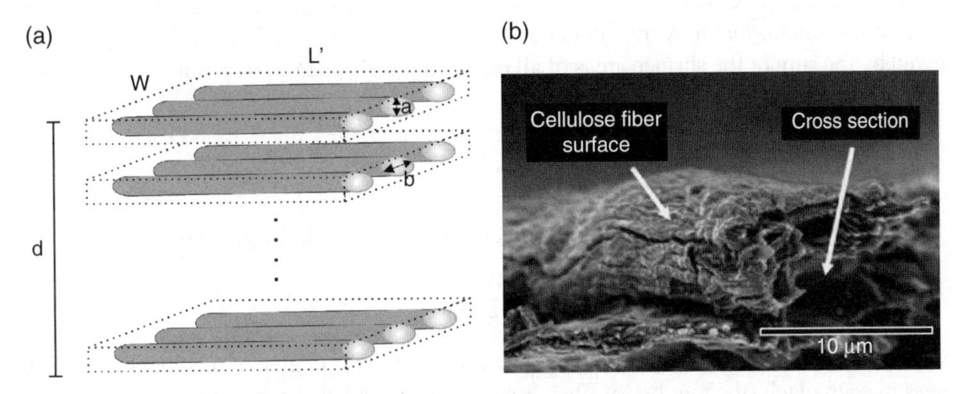

Figure 6.7 (a) Sketch of the distribution of the fibers along the planes, and of how such planes are associated along the dielectric width and its thickness. (b) Cross-section SEM image of a fiber [30]. Reprinted with permission from [30] Copyright (2011) Wiley-VCH.

Under these circumstances and considering the approach to parallel plates with charges uniformly distributed and balanced along both faces we have a charge density given by [29]:

$$Q_s = \sum (Q_{fs} + Q_{is}) = \sum (\varepsilon_f \frac{1}{d_f} + \varepsilon_g \frac{1}{d_g} + \varepsilon_p \frac{1}{d_p}) V(\omega) \tag{6.2}$$

where Q_s is the total charge density per unit area accumulated in the bulk of the paper; Q_{fs} is the total number of fixed charges embedded in the paper during the fabrication process; Q_{is} are the induced density of charges due to the applied filed dynamic field at an angular frequency $\omega = 2\pi f$, $V(\omega)$, with f being the frequency of the applied signal; ε_f is the fiber dielectric constant; ε_g is the glue dielectric constant; ε_f is the dielectric constant related to the plans that contain the fibers.

Under these conditions, we define a capacitance per unit area as being given by:

$$C_s(\omega) = \frac{Q_s(d_i)}{V(\omega)} = \sum (\frac{\varepsilon_f}{d_f} + \frac{\varepsilon_g}{d_g} + \frac{\varepsilon_p}{d_p}) = \sum (C_f(\omega) + C_g(\omega) + C_p(\omega)) \tag{6.3}$$

where $C_f(\omega)$ is the capacitance per unit area related to charges accumulation around the fiber; $C_g(\omega)$ is the capacitance per unit area related to charges accumulated between fiber; $C_p(\omega)$ is the capacitance per unit area related to charges accumulated between planes that contain the fibers.

In general terms, we can say that the capacitance per unit area can be given by the simple equation:

$$\frac{C}{A} = \sum_{i=0}^{3} \frac{\varepsilon_i}{d_i} \tag{6.4}$$

Using a simplified approach, the paper can be regarded as being constituted by a series of planes with W width and L length along the paper thickness, where in each plane there is a number n_f of fibers distributed parallel to one another, as depicted in Figure 6.7a. Along these fibers, the process of charge accumulation occurs irrespective of their nature or the type of movement that they may undertake. In each plane, the surface area (A_{Splane}) will be given by the sum of the surface areas of all of the fibers (A_f), which are assumed to be equal. That is:

$$A_{S\ plane} \equiv \sum_{n_f} A_{S\ f} \equiv n_f A_{S\ f}. \tag{6.5}$$

As an approximation, the number of fibers in each plane will be given by

$$n_f \equiv \frac{W}{b} F_{plane}, \tag{6.6}$$

where b is the width of the fiber and $F_{plane} < 1$ is related to the degree of the fibers' compactness in each plane (the number of fibers that can be placed along the width of each plane). Under these conditions:

$$A_{S\ plane} \equiv A_{S\ f} \frac{W}{b} F_{plane}. \tag{6.7}$$

Based on the foregoing, the total surface area of the dielectric (A_{Spaper}) corresponds to the value determined in equation (6.7), multiplied by the total number of planes (all assumed to be equal) along its thickness d:

$$A_{S\,paper} \equiv \sum_{n_p} A_{Splane} \equiv n_p\, A_{Splane}, \tag{6.8}$$

where n_p, the number of planes that can be placed along thickness d, is given by:

$$n_p \equiv \frac{d}{a} F_d, \tag{6.9}$$

where a is the average height of the fibers incorporated in each plane, and $F_d < 1$ represents the degree of the planes' compactness along thickness d.

Combining equations (6.7) to (6.9) yields the following:

$$A_{S\,paper} \equiv \frac{Wd}{ab} A_{S\,f} F, \tag{6.10}$$

where $F = F_{plane}.F_d$.

If it is considered that the fibers are cylindrical, equation (6.7) can be simplified into:

$$A_{S\,paper} \equiv W \cdot L' \frac{d}{2\pi \cdot r} F, \tag{6.11}$$

From equation (6.11), the potential trapping sites for charge accumulation can be realized, and consequently, the low-frequency capacitance will be highly dependent on the paper structure (compactness, fibers' diameter, and paper thickness), as well as on how the fibers are arranged in it. In Figure 6.7b, it can be seen that the fiber surface area may be even higher as it consists of several microfibrils. Moreover, the inner part of the fibers is not compact, which further enhances the paper's ability to accumulate charges. Of course, the paper composition and preparation methods will determine the type of mobile charges that can be accumulated. This will allow the observed experiment values of the capacitance at low frequencies (see Figure 6.6) to be understood; where an enhancement of more than two orders of magnitude was observed as the static regime was approached.

This means that we cannot use the geometric value of the paper capacitance, a function of its dielectric constant, in order to evaluate the static capacitance of the devices in which paper is employed. Indeed, the static capacitances for the same geometry may change by more than four orders of magnitude leading to strong errors in device dynamic parameters such as mobility which can be overestimated.

6.4 Resistivity Behaviour of Transparent Conductive Oxides Deposited on Paper

To test the ability of paper to be used as a substrate able to be coated with films that are electronically graded, papers with different surface morphologies such as those referred to in Table 6.1 and in Figure 6.2 were used. Those films were coated with transparent conductive oxides (TCO) based on amorphous indium zinc oxide (a-IZO) grown by rf

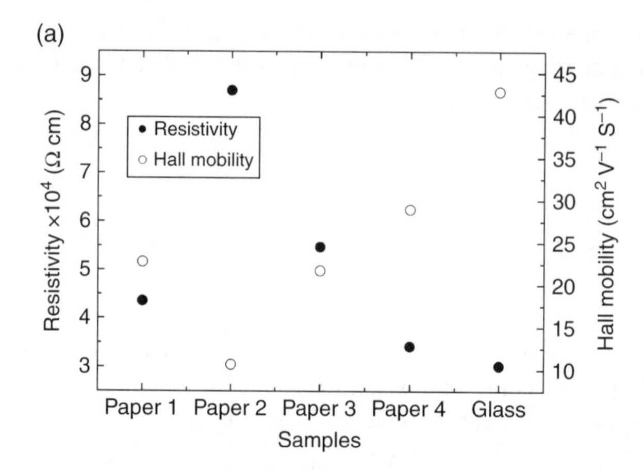

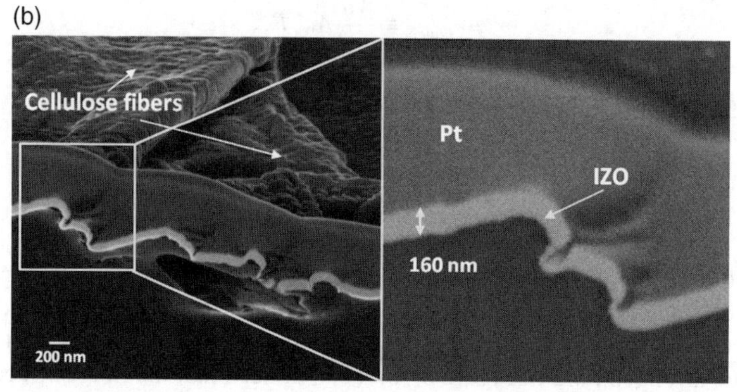

Figure 6.8 *a) Resistivity and Hall mobility of a-IZO films deposited on different sample substrates. b) Scanning electron microscopy cross section of a-IZO deposited on paper, where good step coverage of the oxide over the cellulose fibers was obtained. Reprinted with permission from [65] Copyright (2011) SPIE.*

magnetron sputtering technique [82–85] with a thickness of around 0.12 μm. The aim is to see how the surface finishing and composition of paper influences the electrical performance of the films deposited. The electrical resistivity and mobility of those TCOs were obtained with Hall Effect measurements and compared with those achieved in a glass substrate used as reference (see Figure 6.8). The IZO that exhibits the best electric performance is the one deposited on paper-4 that exhibits an electrical resistivity of about 3.42×10^{-4} (Ω cm), with a Hall mobility of about 28.9 ($cm^2\ V^{-1}\ s^{-1}$). This behavior can be related to the lower value of roughness and degree of compactness than the others papers analyzed in this study. The values achieved are quite close to those obtained in the glass reference substrate, proving its superiority when used as substrate for electronic purposes.

Although paper-3 exhibits the highest roughness, the resistivity and mobility achieved for a-IZO films is only about a factor of two higher than the one achieved on paper-4. This clearly means that we have good surface step coverage by a-IZO (see Figure 6.8b). For paper-2 we note the highest resistivity and lowest mobility exhibited by the a-IZO films.

This behavior could be attributed to the porous like paper structure, in-line with SEM and the capacitance measurements performed.

For paper-1 the values of the electrical resistivity and of the Hall mobility are similar to those achieved for paper-3, in line with a high compact paper structure with a high surface roughness.

The results achieved show that papers as paper-4 can be used as a substrate for electronics applications, where surface energy, roughness and porosity are respectively adequate neutralized, smooth and compact, by a proper paper coating. Indeed, the typical requirement of very smooth and non-absorbing substrates is an important issue for electronic substrates, fully satisfied by paper-4, which is able to sustain a deposition process undertaken under vacuum conditions, as it is the case with magnetron sputtering. Moreover, it means that the coating of the paper surface leads to the required surface smoothness, so as to give the required surface continuity and impermeability. While these paper coatings compromise low cost and recyclability, this might still be acceptable for relatively high-value electronic applications that require relatively expensive materials, many processing steps, and encapsulation [1, 86]. The cost of a plastic-coated paper might, however, become too large (or a too substantial part) in situations when the electronics are truly low cost and robust, i.e., consisting of R2R low-cost materials, and do not require encapsulation. Under these conditions we should use a highly compact paper as is the case with paper-1, even if one loses some electrical performance. In this way, recyclable paper substrates with relatively high smoothness and good barrier properties can be achieved, as is required for electronics.

6.5 Paper Transistors

To build up transistors, such as field-effect transistors (FETs) we need the combination of three types of different materials: high conductive materials, such as metals or transparent conductive oxides, able to guarantee the proper signal extraction or device fed with energy with minimum ohmic losses; semiconductors, the core of the device that allows proper channel conductance modulation through an applied electric field sustained between a gate conducting electrode and the semiconductor, having in between a high insulating layer; the dielectric, whose performance should be such as to limit as much as possible the leakage current through it and to exhibit the correct permittivity or capacitance per unit area that enables the proper charge accumulation in the channel (semiconductor).

The semiconductor and the semiconductor/dielectric interface are often the most critical parts. Inorganic semiconductors usually have a better performance and stability, while organic semiconductors typically have a superior processability and flexibility. However, for most known covalent inorganic semiconductors, the temperature used to process the film is usually incompatible with paper substrates. Examples of semiconducting inorganic materials are Si, oxides of transition metals, and chalcogenides. The crystallinity of Si affects the electronic properties to a large extent leading to materials with electron mobilities higher than $200\,cm^2\,V^{-1}\,s^{-1}$, but they require high process temperatures (≈ 1000°C) [87], incompatible with the maximum temperatures that paper may sustain (typically, below 150°C). Besides crystalline silicon, we have hydrogenated amorphous Si (a-Si:H) that may be processed at temperatures as low as 150°C [88–90], but the corresponding charge carrier mobility is two to three orders of magnitude lower than that of single-crystalline Si. Since flexible paper and plastic substrates are incompatible with such high temperatures, when high electron mobilities are required, a complicated step of transferring the deposited Si layer from a rigid substrate to a paper substrate has been depicted [4, 91, 92].

Another class of inorganic semiconductors is the binary or multicomponent oxides of transition metals, which usually are also transparent [30, 45, 54, 93] and require active thickess in the nano scale range. Films are typically deposited by using RF-magnetron sputtering [33, 51] or pulsed-laser techniques [49] at room temperature, or at temperatures below 250°C by a chemical solution such as sol gel [93–95], spray pirolysis [96–98] or ink-jet printing [100–102]. Those semiconductors may exhibit mobilities as high as 100 $cm^2 V^{-1} s^{-1}$ and have been achieved by using multicomponent ZnO-based oxides, such as IZO [31, 34, 103–105]; Ga_2O_3 -In_2O_3 -ZnO (GIZO) [44–46, 96, 106–109], Gallium Indium Oxide (IGO) [40], Gallium Tin Zinc Oxide (GZTO) [39] amongst others. This makes these inorganic materials the most suitable to be used as semiconductors to be used with and on paper [2].

Besides the semiconductor, to fabricate a transistor we also need electrical insulating materials for preventing shortage of crossing conducting wires. A high capacitance per unit area value is often desirable, not only in capacitors, but also in FETs in order to obtain low-voltage operation [43, 110].This can be achieved by using a thin dielectric layer with a large permittivity. It is, however, also important to prevent leakage currents or electrical shortage through the dielectric, which is common in dielectric layers that are thin, porous, inhomogeneous, and have a low dielectric strength. As shown above, paper, under certain conditions does offer the ability to exhibit a high capacitance per unit area and so, to be used as dielectric in building block process of active electronic devices such as transistors. Here the bottleneck is the required active thickness to make it possible to modulate the charges in the channel layer using low applied voltages and to have the required degree of compactness to decrease leakage current as much as possible.

The possibility of integrating electronic and optoelectronic functions within the production methods of the paper industry is therefore of current interest, in order to enhance and to add new functionalities to conventional cellulose fiber based-paper. To fulfill these demands materials and methods should be developed for cheap and mass production. Fortunato and her group [56] proposed in 2008 to use the paper not only as substrate but also as a dielectric layer. In this new device approach they used the cellulose fiber-based paper as an "interstate" structure (see Figure 6.9, see color plate section) since the device is built on both sides of the cellulose sheet: the gate electrode, based on a transparent conductive oxide, is deposited on one side and on the other side the active semiconductor, to be used as the channel layer, and the highly conductive drain and source regions [44, 82, 111].

Indeed, in order to use paper as an electrically active part of devices, it is first necessary to integrate electronic and optoelectronic functions into conventional cellulose fiber paper. The methods to be developed should be compatible with low processing temperatures, such as those used for the development of semiconductor oxide TFTs, which are processed at room temperature [38, 85]. Taking this into account paper transistor devices were processed using shadow masks at room temperature using the RF-sputtering technique. One side of the paper was coated with RF-sputtered highly conductive a-IZO [83–85, 112] (better than 20 $\Omega\,\square^{-1}$), for use as the gate electrode, using a ceramic target with the composition In_2O_3-ZnO, 5:2 mol% film. The other side of the paper was coated with amorphous a-GIZO [44–46] (with a thickness below 100 nm), for use as the active channel layer, using also a ceramic target with the composition Ga_2O_3-In_2O_3-ZnO, 1:2:1 in molecular weight. Here we

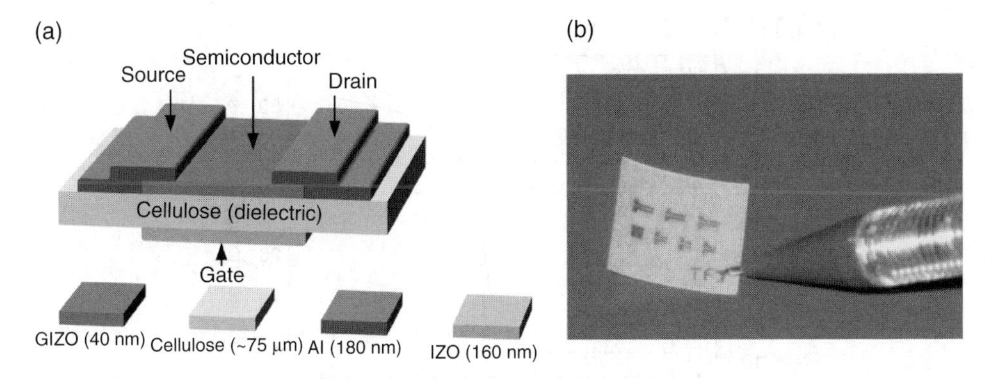

Figure 6.9 *a) Schematic representation of the field effect transistors structure using the cellulose sheet as the gate dielectric, where are also seen the other constituents of the final paper transistor; b) picture of the set of paper transistors.*

should emphasize that the use of amorphous thin films is very important since no extra problems will be raised due grain boundaries and intra grain defects associated with the paper fibers/oxides interfaces. Patterned metal layers were then deposited at the edges of the channel layer to serve as source/drain electrodes and to promote the proper contact with the channel layer of the transistor.

The transistors processed in this way have an enhancement n-type operation mode and exhibit an almost zero threshold voltage, a channel saturation mobility exceeding 30 $cm^2 V^{-1} s^{-1}$, a drain-source current I_{ON}/I_{OFF} modulation ratio above 10^4 and a sub-threshold gate voltage swing of about 0.8 V/ decade. Even two months after processing the device performances were unchanged revealing that they are environmentally stable (stored in air ambient conditions). Reference GIZO thin film transistors (TFT) deposited on Si doped substrate and using as a dielectric thermal grown silicon dioxide (SiO_2) exhibit similar mobilities, but a leakage current of about four orders of magnitude lower than that of the paper transistor, and better subthreshold gate voltage swing (S). That is, the reference TFT had an on/off ratio of about 8×10^8 an $S \approx 0.4$ V dec^{-1}.

The performances of the proposed paper transistors depend greatly on the properties of the paper used. In Figure 6.10 we present the transfer characteristics of transistors processed using different papers, namely paper-4 and paper-5 whose characteristics are listed in Table 6.1 and their surface morphology is shown in Figure 6.2.

The performance of the transistor produced on paper-4 with a paper capacitance per unit area of 6.56 nF cm^{-2} is given by an threshold voltage of 0.24 V, a channel saturation mobility of approximately 3.75 $cm^2 V^{-1} s^{-1}$, a drain-source current on/off modulation ratio above 10^5, and S of about 3.34 V dec^{-1}. Moreover, we note that the drain saturation current (I_D) is achieved only for gate voltages (V_G) above 35 V. This is mainly explained by the high degree of paper compactness (we have an almost continuous film on the paper surface) and of the thickness of the paper (almost all of the paper thickness is contributing to the active electrical layer thickness).

For the TFT processed on paper-5 with a paper capacitance per unit area of 25 nF cm^{-2} we obtain S about a factor of five less than in paper-4, leading to a voltage to get the I_D saturation of about 5 V and an on/off ratio of about 2.9×10^4. Moreover, the channel mobility

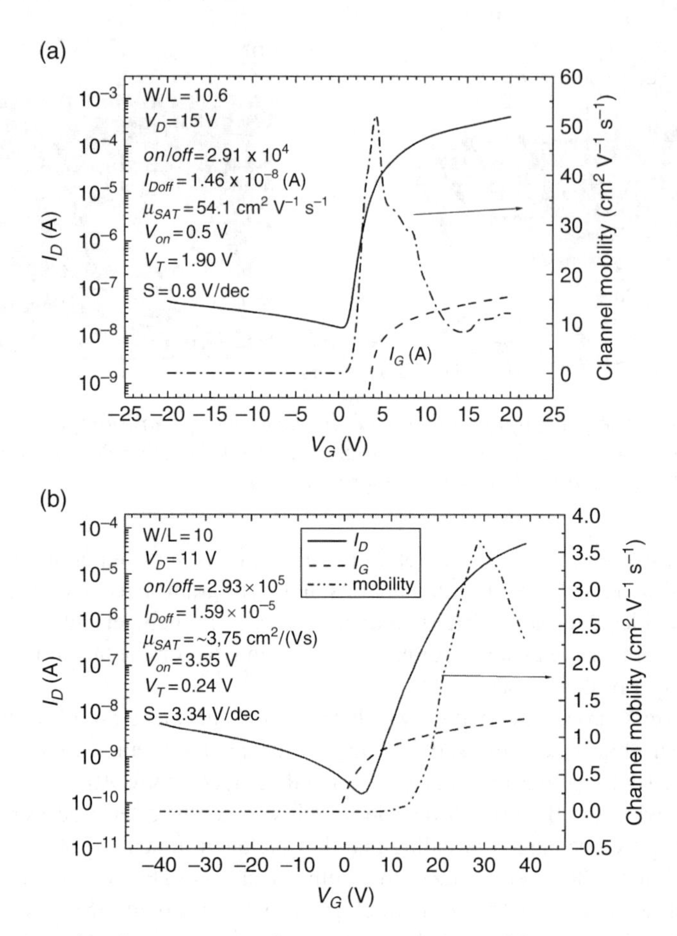

Figure 6.10 *Transfer characteristics of TFT using as channel layer a GIZO film (<100 nm thick) for: a) paper-5; b) paper-4. For mobility the paper capacitance was estimated from data depicted in Figures 6.5 and 6.6. Reprinted with permission from [65] Copyright (2011) SPIE.*

is about one order of magnitude larger than that reported for the TFT produced on and with paper-4. This behavior is ascribed to the fact that the thickness of paper-5 is about a factor of 3.5 less thick than paper-4, besides exhibiting a more sponge like structure, translated by the high capacitance per unit area measured in this paper. To explain this discrepancy, the present authors have proposed a transport model [28, 29, 77] whose main characteristics are described below.

6.5.1 Current Transport in Paper Transistors

To understand the current transport in the proposed interstrate paper field effect transistors, let's consider the structure depicted in Figure 6.11 (see color plate section) where is shown the drain and source regions and the "channel region" composed by the deposited GIZO semiconductor in a non continuous manner, but all interconnected at the channel borders

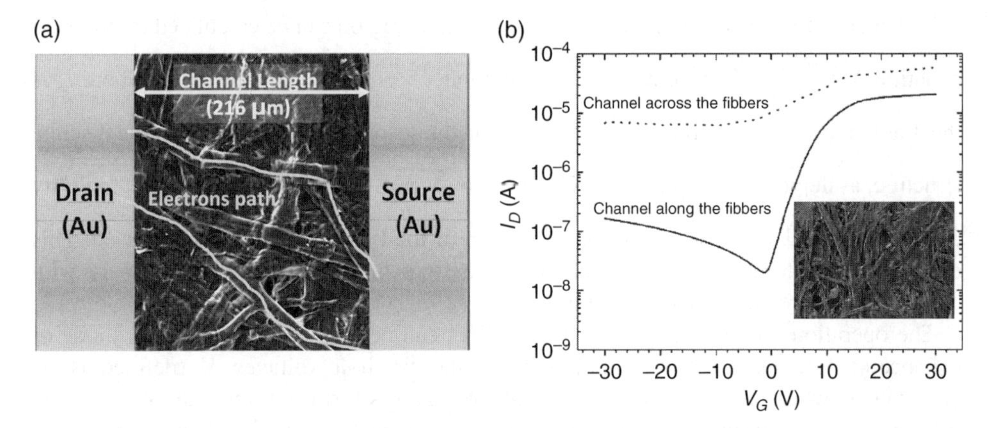

Figure 6.11 *a) A real image of the 2D surface of the paper TFT showing the discrete channel region limited by the continue drain and source metal regions. Within the fibers are sketched two possible carrier pathways between drain and source under a certain V_D, for a fixed V_G applied to the back gate electrode; b) transfer characteristics of paper TFTs where the channel was processed along the fibers and across the fibers, in the paper structure shown in the inset, respectively.*

by the deposited metal layers that constitute respectively the drain and source regions. In the present study the reference/ground terminal is the source.

Using the paper with the aforementioned characteristics, devices were fabricated employing the structure shown in Figure 6.9a. The operation of the device relies on the ability of the paper fibers to retain charges, which depends on their mineral content [13, 55, 72–74, 76, 113]. The process through which the mobile charges are induced into the foam like paper structure can be explained as follows. When a positive gate voltage is applied, charges move to the paper foam structure close to and within the channel interface, and they induce an image charge of equal density and an opposite sign in the channel layer, similar to the case of organic transistors gated by ionic liquids or solid state electrolytes [24]. The major advantage of the foam like paper structure is that the specific capacitance per unit area is exceptionally large, which results in a high current throughput collected via the different possible existing pathway between source and drain, as depicted in Figure 6.11, at low operating voltage, as observed experimentally.

That is, the collected drain to source current, I_D, is given by $I_D \approx \sum_{n=1}^{n} \Delta i_D$, where Δi_D is the current collected via the discrete fibers; n is the number of fibers that define the pathway between the drain and source. As far as the mobility is concerned, this can be taken from the relation:

$$I_D \cong \frac{W}{2L} \times \frac{C}{A} \times \mu (V_G - V_T)^2 \tag{6.12}$$

where W is the channel width; L is the channel length; μ the free charge mobility; V_G is the gate to source applied voltage; V_T is the threshold voltage.

The charge density accumulated within the channel region can be calculated through the equation $N \cong \frac{\Delta V_{sf}}{q} \times \frac{C}{A}$, leading to values of about 5×10^{12}–10^{13} charges cm^{-2}, where I_D is the drain to the source current collected (Δi_{ds}) via the set of pathways defined in the discrete structure, as depicted in Figure 6.11. That is $I_D \approx \sum_{n=1}^{n} \Delta i_D$. By doing so, we reach a mobility above 22 cm^2 V^{-1} s^{-1} or 30 cm^2 V^{-1} s^{-1}, function of the type of paper structure used. This explains the differences recorded on the transistors fabricated using two papers with different structure and surface finishing.

The operation mode of the n-type paper TFTs can be classified as depletion mode or enhancement mode, in terms of the polarity of the threshold voltages, V_T (defined as the gate voltage where a strong charge accumulation layer is formed at the interface between the paper layer (working as the dielectric) and the channel region of the transistor), above which the drain current (I_D) starts increasing rapidly. The output characteristics (I_D as a function of the applied drain voltage, using as a common reference, the source, V_D, for different applied gate voltages, V_G) of these devices have well-defined linear and saturation regimes, while the transfer characteristics (I_D as a function of V_G) reveal the existence of an hysteresis behavior, where V_T ascribed with the applied positive V_G (first charging cycle) is positive. The hysteresis behavior recorded is due to the existing fixed and induced charges, which are accumulated within the foam like paper structure (see Figure 6.2e).

As V_G increases, the transistor channel accumulates electrons leading to a drastic change in its conductance and so to an enhancement of I_D, that saturates at 0.95 mA for V_G above +3.5 V for paper-5 and about 0.01 mA for paper-4, for a V_G above +35 V. These changes in behavior are fully dependent on the type of structure and surface finishing of the paper used. An additional proof of this concept was obtained in the laboratory by the present authors by processing in a paper structure similar to paper-4, two FETs, one with channel deposited along the fibers (full line) and another (dotted line) across the fibers. As expected, we can promote the connection of the current along the fibers via the drain and source contacts, when the channel is oriented with the fibers (as indicated in Figure 6.11a), but the same is not true when the channel is processed across the fibers, where the leakage current is much higher than in the previous case, leading to a substantial reduction in the on/off ratio, besides it not being possible to acheive hard saturation currents.

6.6 Floating Gate Non-volatile Paper Memory Transistor

Ever since the pioneering work of Kahng and Sze [114] in 1967, in which they proposed the first floating gate memory device, major changes have transpired in the microelectronics memory industry involving either the need for high-performance devices within a nanoscale range or the need to turn the available devices into low-cost, flexible, disposable devices. In the first case, the driving force is centred on the development of higher-permittivity dielectrics [115–119], such as ferroelectric crystals [109, 119], or of novel complex oxides [116, 117, 121–124] to replace silicon dioxide in the next generation of CMOS nanoscale devices for low cost electronics applications. Apart from that, novel approaches involving either the use of carbon nanotubes [124–126] or nanowires [127, 128] with high dielectric constant oxide as a first step towards the ultra-high integration density of three-dimensional (3D) ferroelectric random-access memories (RAMs) [129–133] are also being tried. In spite of the excellent electronic performance demonstrated by the aforementioned devices, their fabrication processes require the use of high temperatures. As such, they are not compatible with low-cost flexible substrates.

In the second case, the main motivation is low-end applications where low-cost, flexible, lightweight, and low-temperature processing are the driving forces rather than the pursuit of outstanding electronic performance devices. These demands are mainly fulfilled by organic devices, with several attempts being made to use them as memories. Some examples of such devices are the electrochromic conducting polymers that have been used to write and read information [123, 134–138], molecular electronic random-access memory circuits for storing and retrieving information [10, 139–142], and organic memory transistors, such as those based on the gold nanoparticle pentacene [143–145], aiming to produce non-volatile memories [145–147]. Nevertheless, all of these organic memory devices, even for low-performance electronic applications, still suffer from low carrier mobility (below 3–7 $cm^2 V^{-1} s^{-1}$), which limits the device switching time, besides exhibiting low charge retention times (ranging from 4500s [124] to about one week for polymer ferroelectric FETs [148–153]), which limits their fields of application.

In 2008 Martins and his co-workers [27–29, 77] reported the production of self-sustained hybrid natural-cellulose-fiber inorganic-floating-gate memory field-effect transistors that are able to write-erase and read many times and whose electronic performance surpasses that of the known organic devices, making this inexpensive, lightweight, and flexible memory technology viable for a wide range of disposable devices. This opens up a new field of applications for natural cellulose fibers, the material that is known to be the lightest, cheapest, and easiest to recycle. Shown herein is the possibility of using natural cellulose paper fibers simultaneously as a substrate and dielectric in non-volatile selective n-type memory FETs [55, 56] whose channel and gate electrode layers are based respectively in multicomponent amorphous active and passive oxide semiconductors, such as GIZO [44–46] and IZO [83–85, 112, 154]. This opens up a new era of lighter, efficient, and reliable disposable electronic devices, the so-called “paper-e” [155], the green electronics for the future. The floating gate non-volatile memory transistor was processed in a similar manner to that described for the n-type field effect transistor, but now exploiting the sponge like paper structures (see Figures 6.12, see color plate section, and 6.13).

The operation of the device relies on the ability of the paper fibers to exhibit spontaneous polarization. That is, the paper behaves like electrets [78], with the capability to retain the charge, function of its structure, and its constitution. The process through which the mobile charges induced into the foam like structure can be explained as follows. When a positive gate voltage is applied, charges move to the paper foam structure close and within the channel interface, and they induce an image charge of equal density and opposite sign in the channel layer, similar to the case of organic transistors gated by ionic liquids or solid state electrolytes [156–158]. The major advantage of the foam like structure is that the specific capacitance per unit area is exceptionally large, which results in a high current throughput collected via the different possible existing pathway between source and drain, at low operating voltage, giving rise to the typical large and broad transfer hysteresis characteristics, as depicted in Figure 6.14 for two types of paper, designated as paper A and paper B [27].

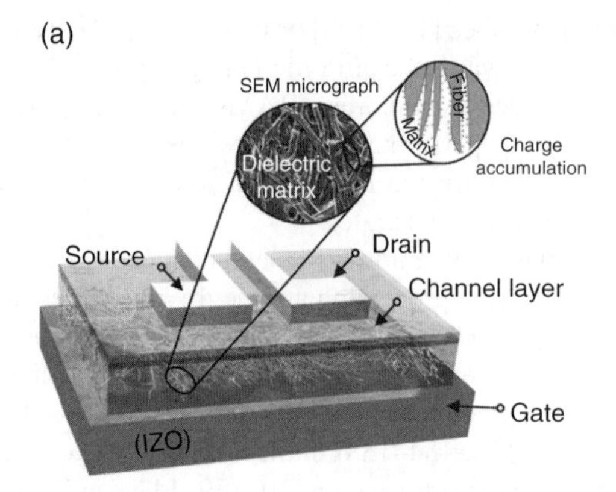

Figure 6.12 *a) Sketch of the distribution of the fibers along the planes, and of how such planes are associated along the dielectric width and its thickness. Reprinted with permission from [27] Copyright (2008) American Institute of Physics.*

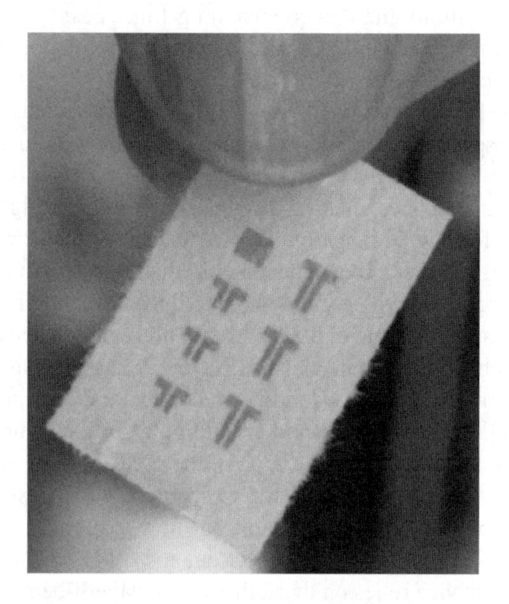

Figure 6.13 *Image of a floating gate non-volatile paper memory transistor* [28]. *Reprinted with permission from [28] Copyright (2009) Wiley-VCH.*

The transistors also exhibit a subthreshold swing (S) of about 0.65 V dec^{-1} and off-currents of the order of 10^{-7} A and 5×10^{-7} A, similar to the values achieved for the gate leakage currents obtained in both types of paper-dielectric used. As V_G increases, the transistor channel accumulates electrons leading to a drastic change in its conductance and so to an enhancement of I_D that saturates at 0.26 mA and 0.95 mA for V_G above +3.5 V, with on/off ratios of 0.5×10^3 and 3×10^4, respectively for papers A and B.

The data depicted in Figure 6.14 also reveal that by decreasing V_G in a double sweep measurement, the shoulder of the saturation I_D plateau is shifted towards negative values, from about +3.5 V to −10 V for both types of devices studied. This means that charges

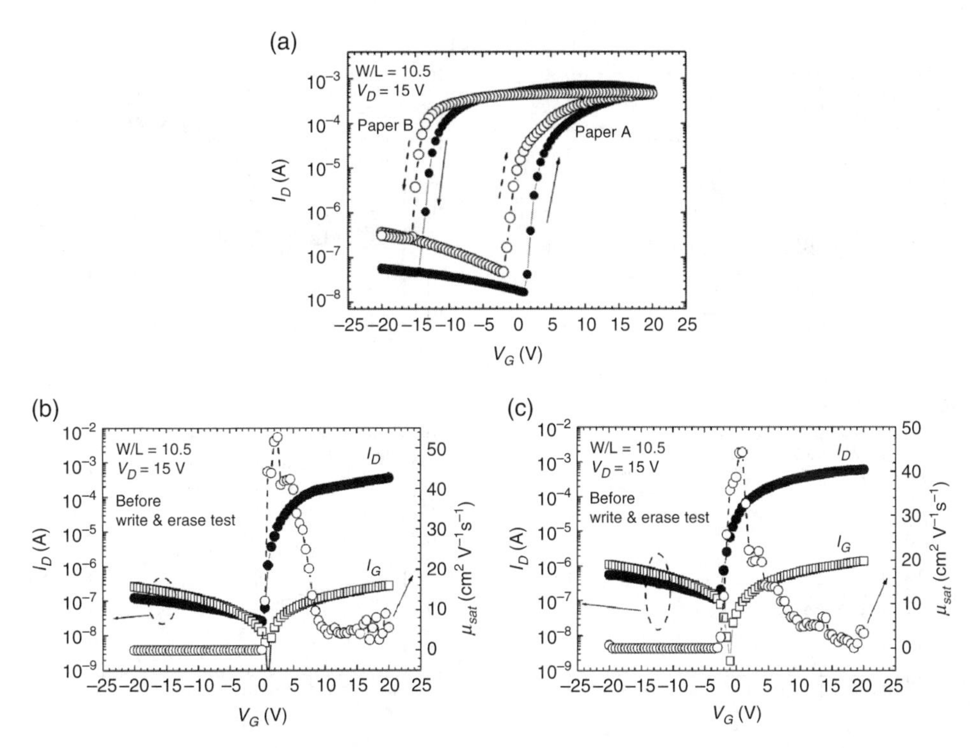

Figure 6.14 Current-Voltage transfer characteristics of the paper memory paper transistor devices: a) double sweep measurements showing the large hysteresis obtained for two different type of papers (A and B); single sweep measurements showing the μ_{sat} versus V_G evolution for papers A b) and B c), respectively, before being under several cycles of write and erase information. In all of the figures is also depicted the behavior of the gate-source leakage current I_G. Reprinted with permission from [27] and [28] Copyright (2008) and (2009) Wiley-VCH and American Institute of Physics.

induced via the cellulose fibers are now trapped within the bulk of the foam like paper structure and so the transistor is kept in the on-state even when no gate bias is applied, allowing the transistor operation to be dominated by the charge memory effect of the device and so, by the interface between the fibers and the channel region. Decreasing further V_G down to −15 V for both devices, I_D is reduced, reaching the same off-current as the one obtained in the initial state. This cycle corresponds to write and to erase the information (charges) stored in the device. Apart from that, we notice that both set of devices analyzed exhibit on/off ratios (the ratio between the off current-below V_T and the on current-when I_D saturates) above 10^3.

6.6.1 Memory Paper Device Feasibility and Stability

The device stability tests [77] performed are depicted in Figure 6.15 for both sets of paper memory transistors studied. After applying to the gate electrode a symmetric square wave pulse (peak to peak voltage (V_{pp}) of 20 V, with offset of 0 V) at 1 Hz for 1 hour to cycle the device between the on and off states (i.e., write and erase operations), the μ_{sat} estimated following the procedure described above for both devices decreases. For device A, based

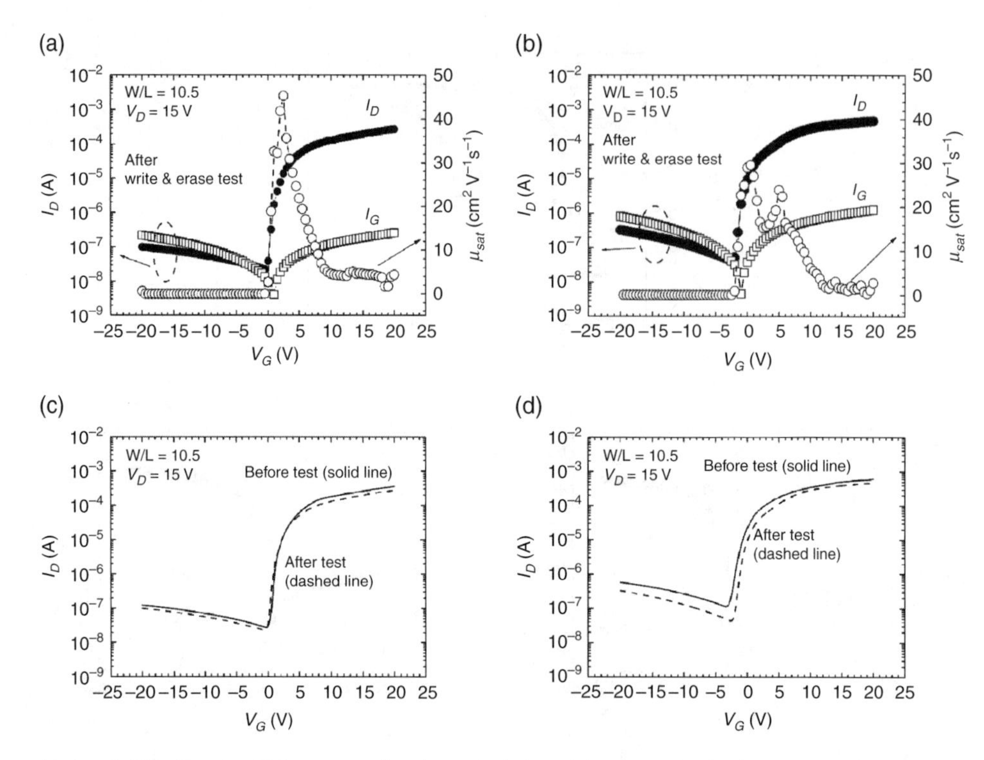

Figure 6.15 *Current-Voltage transfer characteristics of the devices: single sweep measurements obtained after the On-Off stress test (square wave pulse, peak to peak voltage of 20 V, offset of 0 V, frequency of 1 Hz, for 1 hr) showing the μ_{sat} as a function of V_{GS} evolution for papers A a) and B b) (compare with data depicted in Figure 14b and c, respectively); comparison of single sweep measurements before and after the On-Off stress test for papers A c) and B d). Reprinted with permission from ref [28] Copyright (2009) Wiley-VCH.*

on more compact cellulose fibers as dielectric and having the highest thickness, the μ_{sat} peak estimated decreases slightly, more pronounced for the less compact paper. This behavior is attributed to the dielectric architecture of the latter devices, related to how fibers are randomly distributed along the entire volume of the gate dielectric that overshadows the value of μ_{sat}. Nevertheless, besides the described μ_{sat} variation, the memory paper transistor is highly stable after sustaining more than 10^3 on-off cycles, as proved by the comparison of the transfer characteristics measured before and after the stress test (see Figures 15c and d). Even after repeating the same test several times, the devices' stability persisted, meaning that information can be written and erased many times without any apparent loss of performance.

6.6.2 Memory Selective and Charge Retention Time Behaviors

For both types of paper, after sustaining a set of several write and erase cycles, if the absolute value of the off-state V_G (erase) is smaller than the on-state V_G (write), the electrons trapped/stored along the fibers are not fully removed and so they repel the electrons

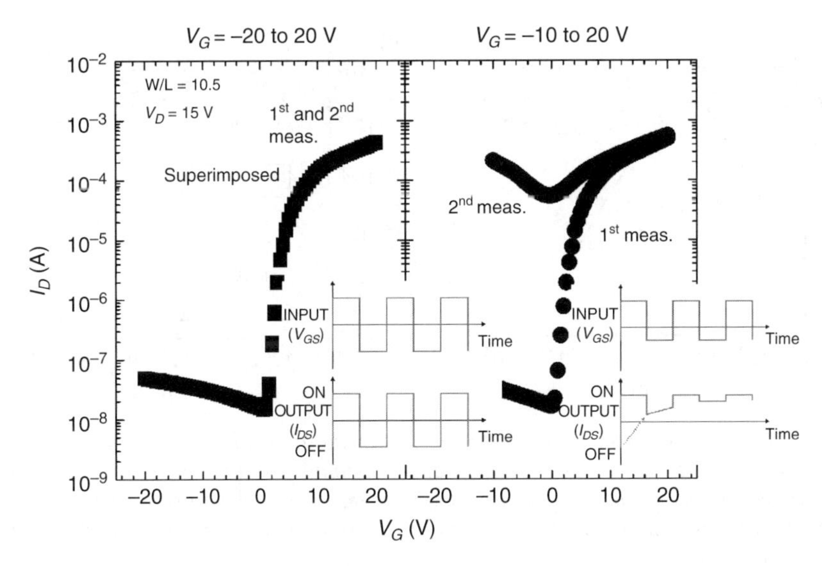

Figure 6.16 Successive single sweep current-voltage transfer characteristics of the WERMT-FET device based on paper A for different V_G ranges: −20 to 20 V and −10 to 20 V. The insets show the schematics for input (V_G) and output (I_D) signals for each V_G range. Reprinted with permission from [27] Copyright (2008) American Institute of Physics.

accumulated in the channel region, thus avoiding the pull-down of I_D towards its off-state value, leading to a small difference between on and off states, i.e. to a low channel conductivity modulation and consequently to on/off ratio close to 1 [28, 77]. That is, the on-state cannot be erased unless a symmetric V_G to the one used for writing/store the information is used (see Figure 6.16).

To demonstrate the memory capacity of the developed floating gate memory transistor devices, after changing the devices from off to the on-state (write operation, accomplished by a single sweep transfer characteristic measurement), the gate electrode was opened and I_D evolution with the time monitored for more than 15 hours, keeping the drain voltage (V_D) constant (15 V), as shown in Figure 6.17. The set of data obtained reveals an exponential dependence of the current as a function of the time, as expected [80], where the devices exhibit different kinds of behaviors, highly dependent on the type, shape and distribution of fibers, including the fibers' compactness.

For the paper memory based on paper A (the most compact one), the data depicted in Figure 6.17 reveal that after opening the circuit on the gate electrode, I_D increases above the initial value recorded after the single sweep transfer characteristic measurement, by a factor of about three. This peak is reached 30 minutes after opening the gate electrode. After this, I_D starts decreasing, taking about 30 h to reach almost the same initial value, showing good retention characteristics up to 2×10^6 seconds, that is, around 540 hours. On other hand, paper B shows better charge retention. On/off state ratio above 10^3 is obtained for 2×10^7 seconds, which corresponds to more than 5550 h, that is, around 230 days. For paper B, even after 6×10^7 seconds (almost 16 000 h, more than 1.5 years), the current ratio is still >10 between the two states, which is adequate for modern sense-amplifiers, depending on the architecture of the memory [55].

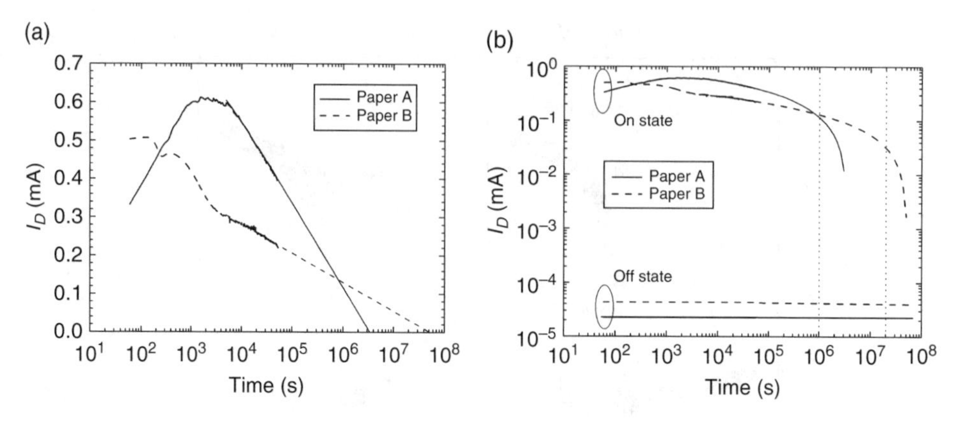

Figure 6.17 *Charge retention characteristics, showing the evolution of I_D with time, after a single sweep transfer characteristic and after opening the gate electrode and keeping a constant V_D= 15 V, for both kind of papers: a) semi-log plot of the on-state current, b) log-log plot of current versus time. Adapted with permission from ref [27] Copyright (2008) American Institute of Physics.*

The initial behavior is attributed to charge de-trapping, from point defects or shallow states and its accumulation along the fibers, leading to a charge build up effect in the channel region, responsible for the observed I_D enhancement. This is related to the fact that the fibers are randomly distributed along the channel area and throughout the paper thickness, leading to the formation of stratified stacked plans of fibers distributed along the paper thickness (see Figures 6.7a and 6.12). This leads to an overall enhancement of *C/A* by more than two orders of magnitude, as shown by the model proposed, when compared to the apparent geometric capacitance. For the less compact structure (paper B) the same charge crowding effect as that observed in the previous device is not seen. This behavior can be attributed to a less critical interface gap between the fibers and the active semiconductor region (due to the fact that the fraction of the paper volume occupied by air in paper B is larger than in paper A and thus the charges are well distributed along the fibers from the beginning) or to fibers with fewer defects along their surfaces. The discharge behavior is also different, it being possible, in the time interval where the measurements were taken to observe two distinct discharge mechanisms that lead to an overall estimated charge retention time of about 16 000 hours (more than 1.5 years!). This enhancement on the retention charge time is attributed to the sponge like structure of this dielectric and to the cross section shape of the fibers that allows a larger number of charges to be stored when compared with the previous devices. After 12 months of the paper memory devices being fabricated their electronic performances are reproducible within an interval range of ±2%, revealing that these flexible and disposable memories are highly environmentally stable.

For the purposes of comparison, in Table 6.2, the performances of similar nonvolatile memories based on inorganic, organic, and hybrid devices are presented [78]. The data clearly show the high performance of the memory paper transistor even at the early and non-optimized stage, compared with all of the organic memory devices [123, 124, 142, 152]. Indeed, the data presented show that the memory paper floating gate transistor has a superior performance compared to those listed as organic polymers [78, 159] as well as

Table 6.2 *Characteristics of different memory devices.*

Performance	Flash Memory Nor [159]	Flash Nand [159]	Inorganic NFGM [159]	Inorganic FeFET [159]	Polymer FeFET [148]	This Work [29]
Feature size, F	130 nm	130 nm	90 nm	12 μm	–	216 μm
Cell area	10 F^2	5 F^2	16 F^2	8 F^2	8 F^2	5 F^2
Write voltage	12V	15V	±6V	±6V	15V	>5V
Read voltage	2.5V	2.5V	2.5V	2.5V	−1V	>5V
Erase voltage	–	–	–	–	−10V	−5V
Retention time	>10 years	>10 years	200 days	30 days	> week	>1.5 years
Cycling endurance	>10^5	>10^5	>10^4	>10^{12}	10^3	>10^7
Writing time*	1 μs	1 ms	1-10 μs	500 ns	0.3 ms	<0.2 ms
Erasing time*	10 ms	0.1 ms	10-100 ms	500 ns	0.5 ms	≥0.1 ms

*Response time to a low-frequency (1KHZ) applied pulse. Reprinted with permission from [28] Copyright (2009) Wiley-VCH.

inorganic NFGM [159] and organic FeFET [148, 160], which are the only existing clear-bitted inorganic flash memory devices [60].

Overall, we note that the compatibility of these low cost devices with large-scale/large-area conventional fabrication deposition techniques, together with their excellent electronic performances, such as stability, very low operating bias, high mobility and high charge retention times, delineates a promising approach of using natural cellulose fiber paper on low cost high-performance flexible and disposable electronics such as paper displays, smart labels, smart packaging, point-of-care systems for self analysis in bio-applications, RFID amongst others.

6.7 Complementary Metal Oxide Semiconductor Circuits With and On Paper – Paper CMOS

The challenge for paper electronics is to be able to produce low power energy consumption devices that allow for densely packed integrated circuits, for a plethora of applications such as computer memory chips, digital logic circuits, microprocessors, and analog (linear) circuits, thus fueling the microelectronics revolution in the so-called technologies of information and communication [24, 87]. For instance, consider a temporary register [161]. In a static circuit the contents of the register remain fixed until new information arrives to be stored (unless the power goes out or the computer is turned off, then all information is lost). In a dynamic circuit, the contents of the register leak away and must be refreshed periodically. The advantage of dynamic circuits is that they do not draw current between refreshing; the disadvantage is that refreshing requires additional circuitry including clocks to synchronize the updating with the operation of the register and thereby makes the designer's job harder [24, 161]. As done in the past for silicon, CMOS are required [24, 161], which naturally do not draw power as occurs with any type of transistor combinations of the same type, and which are very fast and can easily be implemented as static circuits without the need for clocking circuitry, whose chief example is a DC inverter. Indeed,

CMOS technology has driven the revolution in microelectronics thanks to its low power consumption and high-density integration of electronic circuits, which inherently lends itself to simplicity in design.

As we are dealing with paper, integrated electronics demand the proper management of heat loss. To allow high integration and proper control of heat loss it is almost mandatory for future large electronic integration to use CMOS devices in substrates where heat dissipation is a problem, since they do not sustain high temperatures continuously. This combination of low power circuitry, low temperature process and recyclable substrate is a significant step forward in green electronics, as our society moves towards a sustainable environment. Moreover, it opens up a variety of new applications for paper, ranging from smart labels and sensors on clothing and packaging to electronic displays printed on paper pages for use in newspapers, magazines, books, signs and advertising billboards. As the CMOS inverter circuit reported here is the fundamental building block for digital logic circuits, this development creates the potential to have computers seamlessly layered onto paper. Thus, building CMOS circuits on and with paper opens the door for a range of novel applications such as electronic displays that are attached to printed media, clothes and packaging.

Traditionally, CMOS architecture has required a high temperature process and the use of crystalline silicon (c-Si) substrates [24, 25, 87, 161]. The high temperatures required for conventional CMOS technology means that it cannot be used to meet the growing demand for light, flexible [2, 162, 163] and cheap devices. This has resulted in a move towards newly emerging thin-film semiconductor materials such as organics, [162–168] hybrids [169–174] and inorganic semiconductors such as amorphous/nanocrystalline silicon [88, 175, 176] and metal oxides [33, 45, 49, 177–181]. Although these materials cannot compete with c-Si CMOS in terms of the performance required for high-speed computation and digital signal processing applications, they are an alternative for a range of newly emerging applications such as disposable electronics and, in particular, for human-machine interfacing where the need for speed is intrinsically limited by biology.

CMOS architecture is based on connecting two transistors of opposite switching polarities (the so-called n- and p-type transistors), leading to complementary operation, as depicted in Figure 6.18 (see color plate section).

The circuit uses both n-channel and p-channel FETs connected in series, with a common gate electrode. The input signal is applied simultaneously to the gate of both transistors having a common dielectric (paper) that also acts as the substrate. The output signal is taken from the node between the serially connected drains of the two transistors. Previous attempts involving low temperature processes to implement CMOS circuits using thin-film organic or amorphous silicon FETs [163, 171, 174, 175] have met with limited success due to the poor n-type conduction in organics [162, 164] and the poor p-type conduction in amorphous silicon [175]. Recently we have demonstrated n-channel oxide FETs [34, 44–46, 64] with mobilities of at least one order of magnitude higher than the organic p-channel or n-channel amorphous silicon FET, even when processed at low temperatures and implemented on paper [2, 55, 56]. This together with the deposition of p-type oxide FET at low temperatures with mobilities exceeding $1.2\,cm^2V^{-1}s^{-1}$ [179] creates the opportunity for layering CMOS circuits using paper as a substrate and dielectric in which both p-channel and n-channel are seamlessly integrated.

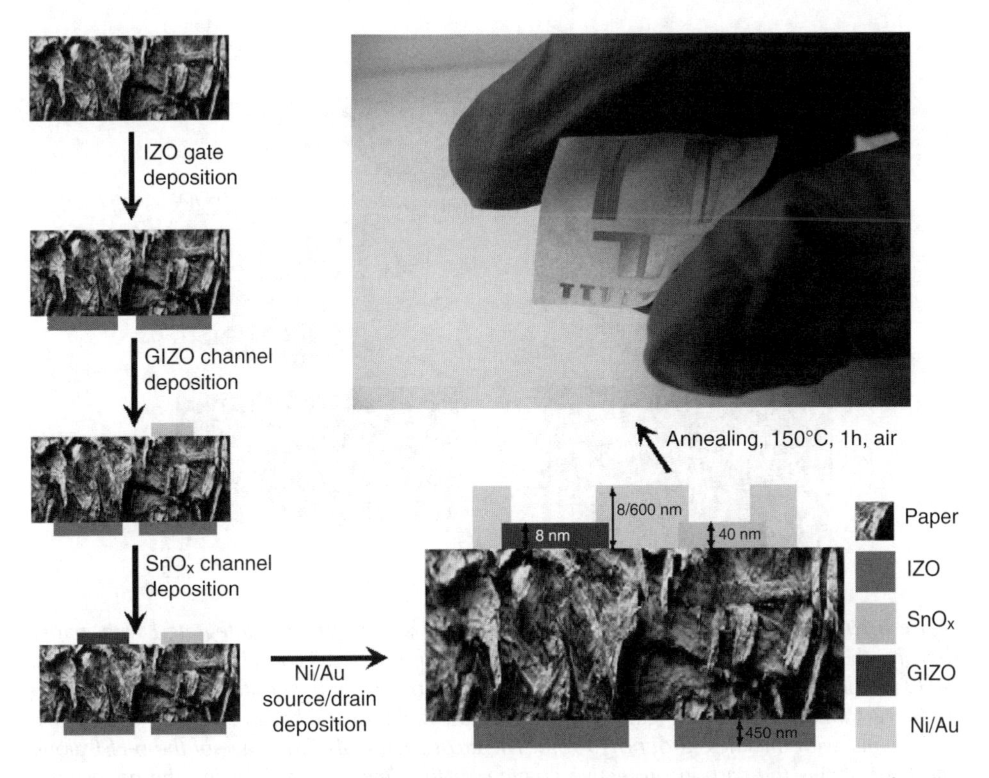

Figure 6.18 Cross sectional schematic of fabrication sequence of the paper-CMOS showing all layers that constitute the final device and how they are interconnect as well as image of the real device. Reprinted with permission from [182] Copyright (2011) Wiley-VCH.

Table 6.3 Summary of FET channel dimensions used in the CMOS inverter circuit parameters and its effect on the drive to load ratio (β_R). Note that circuits 1 and 2 have identical geometry but different supply voltages.

	V_{DD} (V)	W_p (μm) / L_p (μm) = $(W/L)_p$	W_n (μm) / L_n (μm) = $(W/L)_n$	$(W/L)_p$ / $(W/L)_n$	$-\beta_R$ (Experimental)	$-\beta_R$ (Theoretical)
1	17	12500/658 ≈ 19.0	2000/200 ≈ 10	1.90	4.6±0.2	5.4±0.3
2	15	12500/658 ≈ 19.0	2000/200 ≈ 10	1.90	4.6±0.2	5.4±0.3
3	17	6500/1200 ≈ 5.4	1000/330 ≈ 3.0	1.80	4.86±0.2	5.2±0.3

The n- and p-type TFT of the CMOS inverter on paper referred to here use as a channel layer respectively GIZO and non-stoichiometric tin oxide (SnO_x, x<2), 40 nm and 8 nm thick; while the gate electrode is based on a-IZO, 450 nm thick and the drain/source contacts on Ni/Au (see Figure 6.19, see color plate section). Shadow masks were used to define the gate electrodes, and n- and p-channel FET islands, as well as to define the metallization regions. Finally an electron-beam evaporator was used to deposit a Ni/Au (8/120 nm) double-layer, which acts as the drain and a source contacts as well as circuit interconnectors. Table 6.3 summarizes the channel width and length of FETs used in the inverter circuit discussed here.

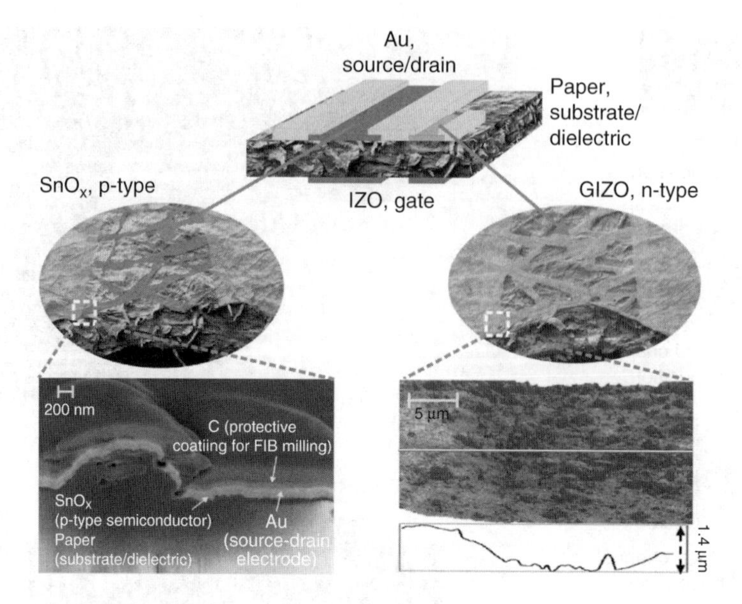

Figure 6.19 *Schematic of the paper-CMOS inverter layered on, and integrated with, paper showing the top and back view of the n-and p-channel FETs whose channels are based on GIZO (40 nm) and SnO_x (8 nm), respectively. The drain and source contacts are based on Ni/Au (8/120 nm thick) films and gate electrodes on the backside are based on highly conductive a-IZO films, with thickness around 450 nm. The cross sectional SEM image of the p-FET along one fiber, shows the carbon protective coating on the device before milling the paper by a focus ion beam, along with the Au on the drain and the SnO_x film. The AFM image on the lower right shows the GIZO surface on the paper substrate.*

6.7.1 Capacitance-Voltage and Current-Voltage Characteristics of N/P-Type Paper Transistors

Electrical characterization of the field effect transistors that constitute the inverter circuits of the CMOS device analyzed here was performed in dark at the room temperature with relative humidity of 35–40% and ambient atmosphere conditions. Capacitance-voltage (C-V) measurements were performed on n- and p-channel FETs using the configuration shown in Figure 6.20 (see color plate section) in a frequency range from 100 Hz to 1 MHz.

Figure 6.21a and b illustrate the C-V behavior of the n- and p-channel devices, respectively. The data reveal the strong frequency dependence of the capacitance on the frequency. At very low frequencies, discrete paper fibers respond to the measurement signal according to the main carriers involved (holes or electrons).

The C–V curves exhibit distinct regions of charge depletion and accumulation for both FETs. In the accumulation regime, a thin conductive sheet of charge is formed in the channel close to the interface of the channel/paper interface, and therefore the capacitance measured in this regime is effectively that of the paper dielectric. In the depletion regime, the drop in the total capacitance is a result of the combined channel and paper dielectric capacitances.

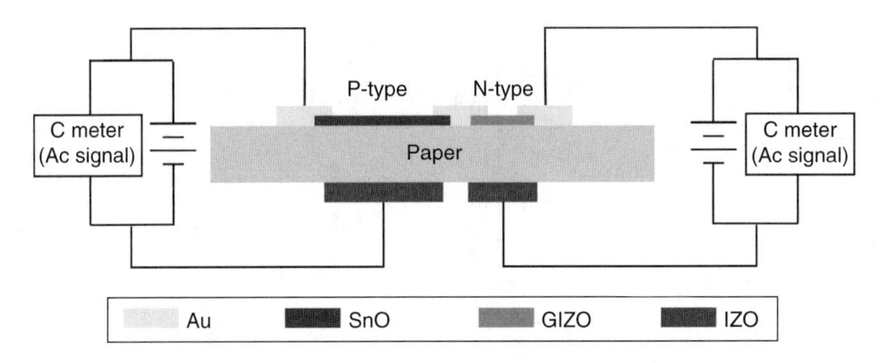

Figure 6.20 *Schematic of the capacitance-voltage measurement.*

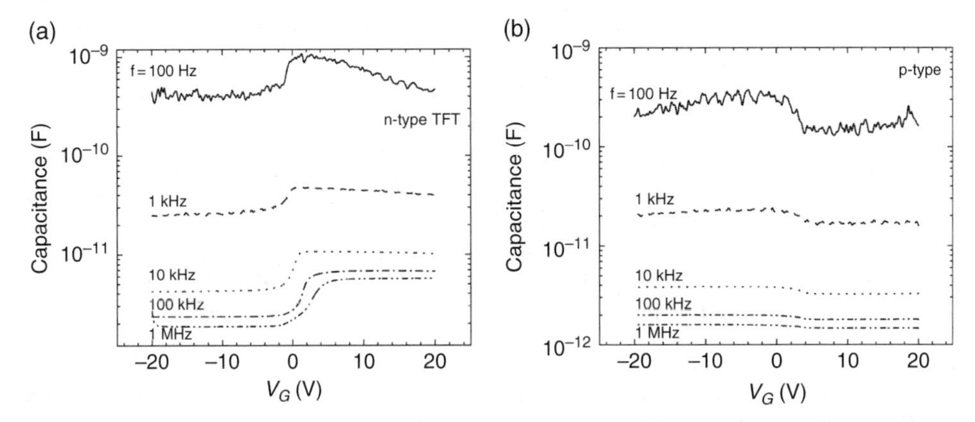

Figure 6.21 *Frequency dependence of capacitance-voltage characteristics of (a) n-channel and (b) p-channel FETs using paper gate dielectric. Reprinted with permission from [182] Copyright (2011) Wiley-VCH.*

The C–V curve shapes accord with that which was expected for transport due to electrons and holes, respectively. Besides, the C–V data show strong frequency dependence, as expected. The C–V curves exhibit distinct regions of charge depletion and accumulation for both FETs. In the accumulation regime, a thin conductive sheet of charge is formed in the channel close to the interface of the channel/paper interface, and therefore the capacitance measured in this regime is effectively that of the paper dielectric. In the depletion regime, the drop in the total capacitance is a result of the combined channel and paper dielectric capacitances.

The transfer and output characteristics of the n- and p-channel FETs are depicted in Figure 6.22. Both devices exhibit high gate leakage currents (I_G). This can be attributed to the foam like nature of the paper dielectric and ion bombardment of the paper during the growth process, coupled with the use of large gate electrodes. The gate leakage current density can be decreased by using a more compact paper structure, with nanofibrils instead of microfibrils, along with full charge compensation of existing ions within the bulk of the paper. Furthermore, a lower gate leakage current can be achieved by scaling down the device size and thus the gate electrode area. The anti-clockwise hysteresis behavior in the transfer curves of both devices is believed to stem from a combination of ion migration and charge

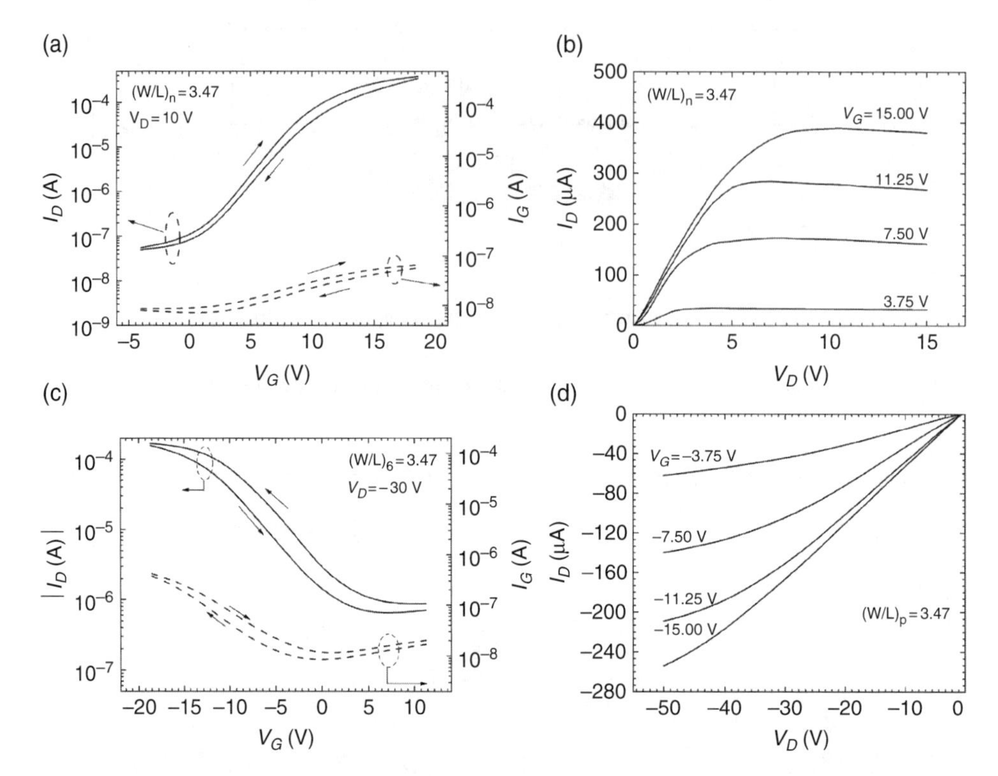

Figure 6.22 *a), b) Transfer and output characteristics, respectively, of the n-channel FET for W/L ratio of 3.47. The TFT has a threshold voltage of 2.1±0.5 V, mobility of 23±2.0 cm² V⁻¹ s⁻¹, sub-threshold slope of 3.1 V/decade and on/off ratio of 10⁴. c), d) Transfer and output characteristics, respectively, of p-channel FET for W/L ratio of 3.47. The TFT has a threshold voltage of 1.4±0.5 V, mobility of 1.0±0.2 cm² V⁻¹ s⁻¹, sub-threshold slope of 6.9 V dec⁻¹ and on/off ratio of over 10².*

trapping effects [11]. It is possible to decrease the hysteresis and leakage currents by minimizing the number of charge carrier traps, increasing of the smoothness of the paper, and using low ion bombardment deposition processes.

The p-channel FET performance is sufficient for use in CMOS architecture, proving the feasibility of layering CMOS devices on, and integrating with, paper. This is despite the fact that the p-channel FET works in depletion mode with a gate leakage current about one order of magnitude higher than that of the n-channel FET. This behavior can be partly attributed to the high radio-frequency power density used during the deposition of SnO_x film and the subsequent damage to the paper dielectric. Furthermore, the poly-crystaline nature of SnO_x yields a film with a rougher surface and less compact structure than its GIZO [45, 46] counterpart. This leads to an enhancement of defects at fiber interfaces, thus contributing to the observed decrease in on/off ratio, sub-threshold slope and field effect mobility, as compared to the n-channel FETs. The p-channel FET's output characteristics do not show hard saturation behavior, which is in-line with previous arguments concerning leakage current and operation mode. As discussed later, this behavior would have an influence on the input-output characteristic of the CMOS inverter circuit presented here (Figure 6.22).

The mobilities of the n- and p-channel FETs were estimated in the saturation regime, by using the derivative of the square-root of the drain-source current $I_D^{1/2}$ with respect to the gate voltage (V_G) based on the ubiquitous square law terminal characteristic, $I_D=(\mu C_{paper}W/L)[(V_G-V_T)^2/2]$ [183]. Here, V_T is the threshold voltage, μ the device mobility, L and W the channel length and width, respectively, and C_{paper} the gate capacitance per unit area. As paper acts as the gate dielectric with a high capacitance at low frequencies, this can lead to an enhancement of FET properties and, in particular, a high transconductance (g_m) which favors a low turn-on gate voltage [184] and, facilitates the drift of charged species though cation exchange at the negatively charged carboxyl and phenolic hydroxyl sites in the paper matrix, and anion migration due to the negative zeta potential of paper [110]. This leads to the formation of electric double layers along the fibers beneath the channel region and on the gate electrode. Moreover, the effective dielectric thickness is much smaller than the geometric paper thickness as a consequence of its fiber network.

In conventional FETs, the transconductance is limited by the geometric attributes of the device and the carrier mobility, since the device geometry is implicitly a function of gate dielectric thickness and dielectric constant, and therefore the gate dielectric capacitance. Since the paper fibers are more than two orders of magnitude thicker than the active channel layers, non-continuous mesh-like channel layers are formed, covering mainly the paper fibers with large gaps between fibers (see Figure 6.19).

From a circuit design standpoint, a high g_m is advantageous as it allows the desired drain-source current to be achieved at lower operating voltages; and this is what the use of paper as a dielectric makes possible in this circuit. The foam like structure of paper allows us to obtain a quasi-static gate capacitance of 40 nF cm^{-2} corresponding to an apparent active dielectric thickness of 0.5 μm, which is significantly smaller than the paper's actual geometric thickness.

Both n and p-type field effect transistors that constitute the CMOS device analyzed here exhibit electron and hole mobilities greater than 21 $cm^2 V^{-1} s^{-1}$ and 1.0 $cm^2 V^{-1} s^{-1}$, respectively. As the paper can act as both substrate and dielectric, this means that complexity of transferring a stand-alone electronic circuit onto a paper page is reduced.

6.7.2 N- and P-channel Paper FET Operation

Figures 6.23 and 6.24 show the simplified band diagram for the n- and p-channel FET devices respectively, where part a refers to the FET's operation at the flat-band condition and part b represents the device operated in the above-threshold regime.

The band structure of the paper fiber is yet to be fully investigated and is therefore estimated as shown to depict energy levels consistent with that of an insulating material. However, included in Figures 6.23 (see color plate section) and 6.24 (see color plate section) are charged species whose migration leads to the enhancement of gate capacitance. As shown in Figure 6.23, the n-channel FET is an enhancement mode device with a positive flat-band voltage. A further increase in the positive bias leads to accumulation of electrons at the interface of GIZO and paper dielectric where there is a downward bending of the energy bands. Figure 6.24 represents the depletion mode p-channel FETs. In this mode of operation, the flat-band voltage is positive, and therefore at zero gate bias the device is switched on. Further reduction in the gate bias leads to increase in the accumulation of holes in the channel due to upward bending of the energy bands.

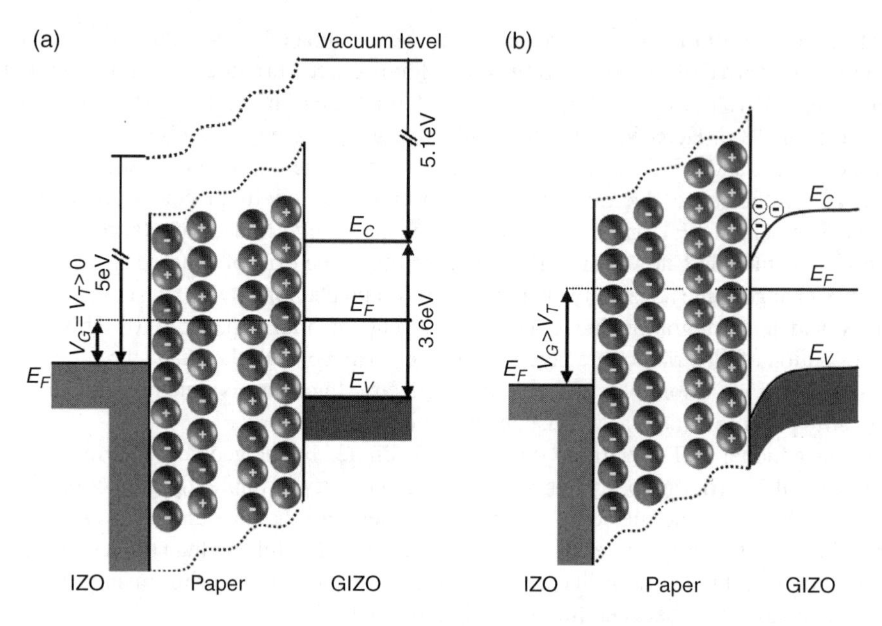

Figure 6.23 *Simplified band diagram of an n-channel FET for (a) $V_G = V_T > 0$ and (b) $V_G > V_T$. Reprinted with permission from [182] Copyright (2011) Wiley-VCH.*

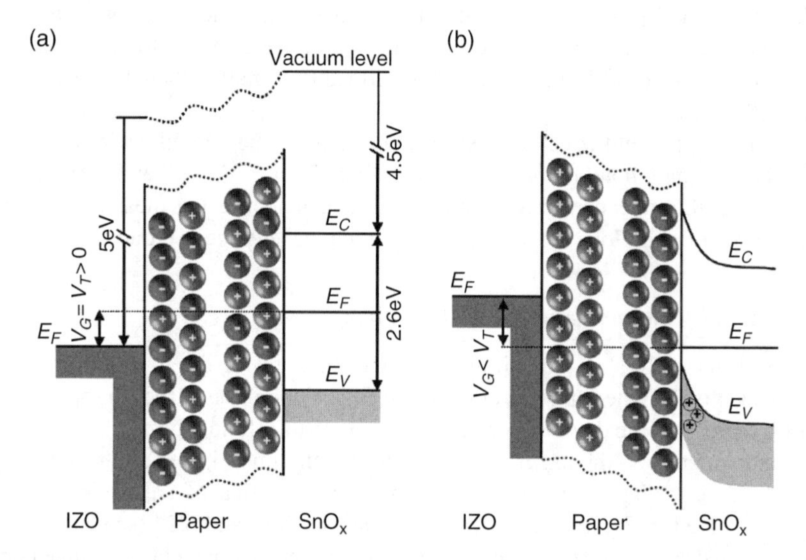

Figure 6.24 *Simplified band diagram of an p-channel FET for (a) $V_G = V_T > 0$ (b) and $|V_G| > V_T$. Reprinted with permission from [182] Copyright (2011) Wiley-VCH.*

6.7.3 CMOS Inverter Working Principles

The operation of a CMOS inverter can be divided in five basic regimes (see Figure 6.25). In regime I, the p-channel FET is switched on (linear) and the n-channel FET is off. Here, the output voltage (V_{OUT}) is the supply voltage (V_{DD}) less the voltage drop across the p-channel FET (i.e. $V_{OUT} = V_{Dn} = V_{DD} - V_{Dp}$).

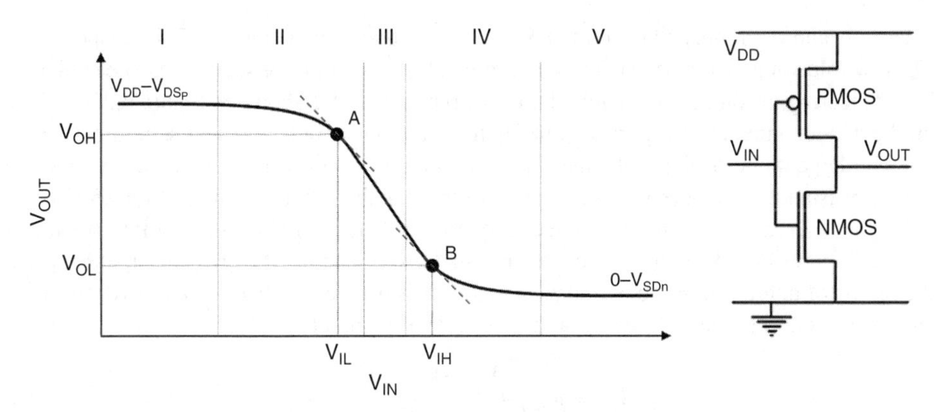

Figure 6.25 *CMOS inverter transfer characteristics indicting the various operating regimes and corresponding circuit schematic; the n- and p-channel FETs are denoted as NMOS and PMOS, respectively; A and B depict points of unity gain,* $\Delta V_{OUT}=\Delta V_{IN}=-1$. *Reprinted with permission from [182] Copyright (2011) Wiley-VCH.*

Further increase in the input voltage (V_{IN}) leads to the n-channel FET switching on (saturation) and p-device remaining in the linear region. In Figure 6.25 this is identified as regime II. The operation of the n-channel FET leads to a drop in the output voltage V_{OUT} as a function of V_{IN}. Since the current though both FETs is the same, equations 6.13 to 6.17 below relate V_{OUT} with V_{IN} and other circuit parameters [161, 183].

$$I_{DSn} = g_{m-n}\frac{\left(V_{IN}-V_{Tn}\right)^2}{2} \tag{6.13}$$

$$I_{DSp} = g_{m-p}\left[\left(V_{IN}-V_{DD}-V_{Tp}\right)\left(V_{OUT}-V_{DD}\right)-\frac{\left(V_{OUT}-V_{DD}\right)^2}{2}\right] \tag{6.14}$$

$$I_{DSp} = I_{DSn} \tag{6.15}$$

$$V_{OUT} = \left(V_{IN}-V_{Tp}\right)+\left(\left(V_{IN}-V_{DD}-V_{Tp}\right)^2-\frac{g_{m-n}}{g_{m-p}}\left(V_{IN}-V_{Tn}\right)^2\right)^{1/2} \tag{6.16}$$

Here, V_{Tn} and V_{Tp} are the threshold voltages of the n- and p-channel FETs, respectively, and g_{m-n} and g_{m-p} the associated transconductances.

The conduction through the n-channel FET leads to a current path from the supply rail, V_{DD} to ground via the n- and p-channel FET chain. As V_{IN} is further increased, this current also increases, leading to a higher voltage drop across the p-channel FET, which eventually starts to operate in saturation. This is indicated as regime III in Figure 6.25. Here, the static current consumption peaks, owing to the fact that both FETs are in saturation, and the slope of the voltage transfer curve (VTC) is the highest. This range also qualifies the use of the inverter as an analogue amplifier with its gain defined as:

$$\text{Gain} = -\frac{V_{OUT}}{V_{IN}} = (g_{m-n}+g_{m-p})(r_{DSn}\,\|\,r_{DSp}) \tag{6.17}$$

where r_{DSn} and r_{DSp} represent the output resistances of the n and p-channel FETs, respectively. The low value of gain reported here (see Figure 6.26, see color plate section) can be explained by considering the output characteristics of the p-channel FET shown in Figure 6.22d, which unlike its n-channel counterpart seen in Figure 6.22b does not show clear saturation behavior. The large slope of the p-channel FET's output curves even at large V_D leads to a low output resistance r_{DSp}, thereby lowering the inverter's gain as defined in equation (6.17).

Further increase in the input voltage leads to the n-channel FET moving into the linear region (regime IV). The analysis of this regime is similar to that of regime II, except that the region of operation of the n- and p-channel FETs is reversed. Equations (6.18) to (6.21) can be written to relate V_{OUT} to V_{IN} and other circuit parameters:

$$I_{Dp} = g_{m-p}\frac{\left(V_{IN} - V_{DD} - V_{Tp}\right)^2}{2} \tag{6.18}$$

$$I_{Dn} = g_{m-n}\left[\left(V_{IN} - V_{Tn}\right)V_{OUT} - \frac{V_{OUT}^{\ 2}}{2}\right] \tag{6.19}$$

$$I_{Dn} = I_{Dp} \tag{6.20}$$

$$V_{OUT} = \left(V_{IN} - V_{Tn}\right) + \left(\left(V_{IN} - V_{Tn}\right)^2 - \frac{g_{m-p}}{g_{m-n}}\left(V_{IN} - V_{DD} - V_{Tp}\right)^2\right)^{1/2} \tag{6.21}$$

Eventually when V_{IN} is raised beyond the V_{TP}, the inverter enters regime *V*, with the p-channel FET switched off and n-channel device operating in the linear region. While the output voltage should be ideally zero, the voltage drop across the n-channel FET leads to $V_{OUT} > 0\,V$. Points A and B in Figure 6.25 indicate where the slope of the transfer curve is −1 (i.e. $\Delta V_{OUT} = -\Delta V_{IN}$), and are used for calculation of inverter's noise margins:

$$\text{Low Noise Margin: } NM_L = \left|V_{inA} - V_{outB}\right| = \left|V_{IL} - V_{OL}\right| \tag{6.22}$$

$$\text{High Noise Margin: } NM_H = \left|V_{outA} - V_{inB}\right| = \left|V_{OH} - V_{IH}\right| \tag{6.23}$$

Table 6.3 summarizes the theoretical and experimental values of driver to load ratio (β_R). The theoretical value of β_R was extracted using $\beta_R = \beta_D / \beta_L \approx g_{m-n} / g_{m-p} \approx 1/2 \times \mu_n / \mu_p \times \left[(W/L)_n / (W/L)_p\right]$ and its experimental value obtained from the square root of the slope of the VTC in the steep transition region. It is noteworthy that the VTC curves show asymmetry due to asymmetry in the inverter circuit geometry used.

6.7.4 Paper CMOS Performance

The CMOS inverters using the n- and p-channel FET devices described above were fabricated on the same paper substrate with different geometric aspect ratios $(W/L)_p/(W/L_n)$ (see Figure 6.26a) and were tested under different supply voltages. Figure 6.26b shows the associated voltage transfer characteristics (VTC) of the inverters along with their gain and leakage current in Figure 6.26c. A summary of the electrical performance under the different testing conditions is given in Table 6.4. The values shown represent a statistical

Table 6.4 *Effect of FET geometry $[(W/L)_p/(W/L)_n]$ and supply voltage (V_{DD}) on the DC performance of the paper CMOS inverter.*

Ref. Sample	$(W/L)_p/(W/L)_n$	V_{DD} (V)	NM_H (V) [a]	$\lvert NM_L \rvert$ (V) [b]	TW (V) [c]	V_l (V) [d]
1	1.90	17	9.8±0.8	1.0±0.1	4.7±0.3	13.5±0.7
2	1.90	15	9.8±0.7	1.0±0.1	4.3±0.3	12.4±0.6
3	1.80	17	6.5±0.6	0.6±0.1	3.5±0.2	10.5±0.5

[a] $NM_H=V_{OH}-V_{IH}$ high noise margin; [b] $NM_L=V_{IL}-V_{OL}$ low noise margin; [c] the transition width ($V_{IH}-V_{IL}$); [d] the logic swing ($V_{OH}-V_{OL}$). Here, V_{OH}, V_{IH}, V_{IL} and V_{OL}, are associated with the knee points ($\partial V_{OUT}/\partial V_{IN}=-1$) of the VTC curves (see Figure 6.25).

average of measurements performed for various devices, taking into account measurement errors. Within the limits of experimental error, the results shown are consistent and reproducible. The absence of hard saturation in the p-channel device manifests itself in a relatively low gain. Regardless, the CMOS inverter does show large voltage discrimination between the high (V_{OH}) and low (V_{OL}) states, where V_{OH} and V_{OL} defined the onset of the steep transitions in the voltage transfer characteristic (see Figure 6.25) [161, 183]. The inverter exhibits a logic voltage swing (V_l) [183] that is over 85% of V_{OH} as required for most static CMOS applications (see Table 6.4).

As discussed, a CMOS inverter circuit is configured such that the two gates of the n- and p-channel FETs are connected to form an input node, with $V_{IN}=V_{Gn}=V_{DD}-V_{Gp}$, while the source of p-channel FET is connected to the drain of n-channel FET to provide an output node, with $V_{OUT}=V_{Dn}=V_{DD}-V_{Dp}$. Analysis of the VTC curves shown in Figure 6.26b shows that $V_{OH}\neq V_{DD}$ and that $V_{OL}\neq 0V$. This means that $V_{Dp}\neq 0$ when input is low (n-channel FET in saturation and p-channel FET is linear regions), and $V_{Dn}\neq 0V$ when input is high (n-channel FET in linear and p-FET in saturation regions).

Increasing the p-channel FET's channel length (L_p) used in the inverter circuit from 658 μm to 1200 μm while maintaining $V_{DD}=17V$ (curves 1 and 3, Figure 6.26b), leads to a reduction in V_{OH} from 15.6±0.8 V to 12.1±0.6 V, corresponding to a increase in V_{Dp} from 4.75±0.12 V to 4.45±0.32 V. The effect of L_p on V_{OH} can be explained by considering the effect of L_p on the p-channel FET's output resistance, r_{SDp}. The voltage drop across the drain and source contacts of the p-channel FET devices can be related to r_{SDp} and in turn to the channel length and width based on $V_{Dp3}/V_{Dp1}\approx r_{DSp3}/r_{DSp1}\approx[L_{p3}/L_{p1}]\times[W_{p1}/W_{p3}]$. This suggests that $V_{Dp3}/V_{Dp1}=(4.7\pm0.32)/(1.45\pm0.12)$, should be approximately equal to $[L_{p3}/L_{p1}]\times[W_{p1}/W_{p3}]\approx 0.96$ within the range of experimental error (below 5%) and accounting for contact resistance.

A similar analysis for V_{OL} shows a reduction from 2.10±0.15 V to 1.55±0.11 V as L_p is increased from 658 μm to 1200 μm, which again can be explained based on the change in r_{SDp}. At the low output state, V_{OUT} fails to reach 0 V due to the voltage drop across the n-channel FET (V_{Dn}) which can be attributed to the output resistance, r_{DSn}. Increasing L_n from 200 nm to 330 nm, while maintaining a constant V_{DD} (curves 1 and 3, Figure 6.26b) leads to a similar trend as the one observed for V_{Dp}, with a reduction in V_{DS}. Contrary to the effects of channel length (L_p and L_n) on V_{OH} and V_{OL}, decreasing V_{DD} from 17 V to 15 V for a fixed circuit geometry (curves 1 and 2, Figure 6.26b), does not lead to any change in the value of V_{Dp}: 4.45±0.12 V. This is expected and is in line with the analysis and reproducibility of the devices.

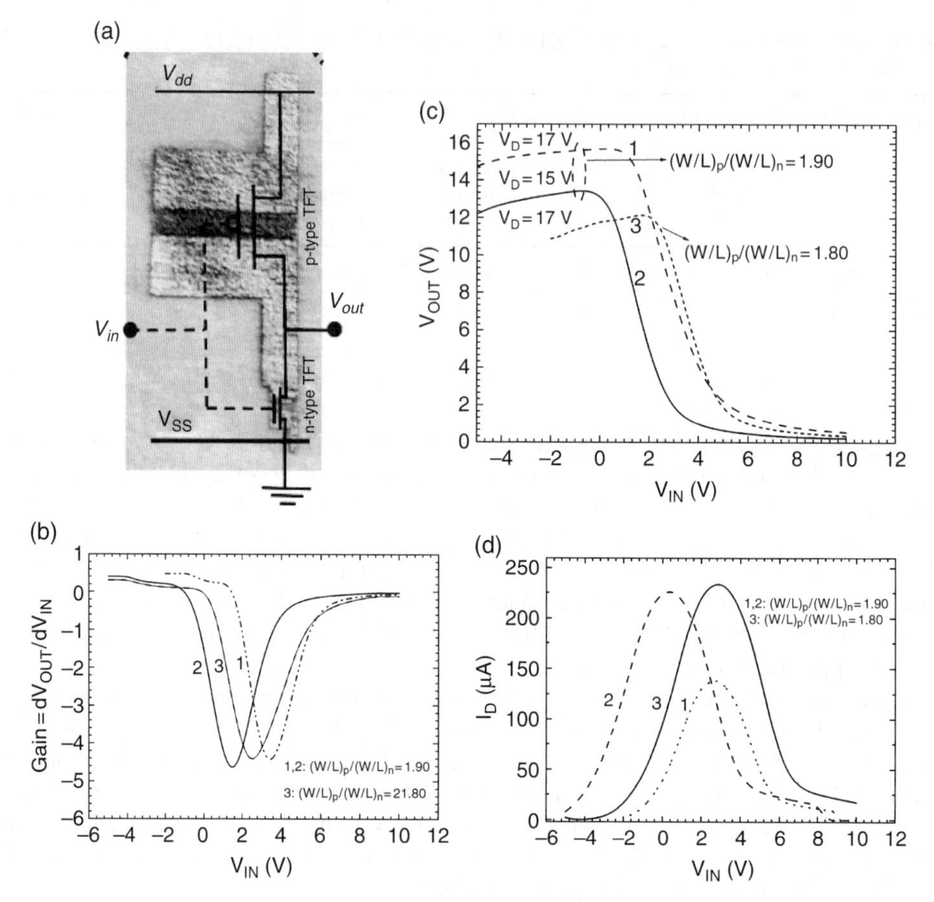

Figure 6.26 *(a) Image of the CMOS on paper where the large $(W/L)_p \approx 19.0$ and small $(W/L)_n \approx 10.0$ correspond to the p-FET and n-FET, respectively; (b) VTC of the CMOS inverter for the different configurations as numbered in Table 6.4. This is used to extract the high and low states, associated with the input and output voltages (V_{IL}, V_{OH} and V_{IH}, V_{OL}). (c) Gain and circuit leakage current, I_D, for different configurations as indicated in Table 6.4. Reprinted with permission from [182] Copyright (2011) Wiley-VCH.*

The paper-CMOS static power dissipation broadly compares favorably with modern silicon CMOS inverters [185] and their low power thin-film counterparts [164]. In all cases, the static leakage current of the circuit indicated as I_{DD} in Figure 6.26c at the high or low states is less than 1.9 pA, which corresponds to a maximum static power dissipation of 32 pW per inverter. The static leakage current stems from a combined contribution of sub-threshold and gate leakage [185, 186], both of which can be minimized by optimizing the geometric attributes of the circuit, such as channel width and area. For example, the sub-threshold leakage can be reduced by minimizing the channel width [187]. The origin of gate leakage current can be attributed to ionic conduction along the cellulose fibers [188, 189], as highlighted in Figures 6.21 and 6.22. Thus, the gate leakage scales with the total number of ions available for conduction, and subsequently the gate electrode area [190]. Given the inherent scope for scalability, the size of the circuit presented here can be significantly reduced leading to a substantial reduction in the power consumed per inverter. Indeed, the static power dissipation per channel width and per channel area is respectively around 16.1 fW μm^{-1} and 4.1 aW μm^{-2} respectively. The increase in the leakage current during the brief period of switching between states, shown in Figure 6.26c, it is characteristic of the CMOS architecture. This is in contrast to a unipolar architecture such as nMOS, in which current flows as long as the input is active leading to higher power consumption.

Although in this first attempt of fabricating a CMOS device with and on paper a high leakage current is observed, it is not surprising given the advance from a rigid substrate to paper. Additionally the conventional FET gate dielectric is now replaced with paper. The use of paper dielectrics and layering of p- and n-channel FETs is highly promising. The performance of the circuit presented here does not inhibit the implementation of CMOS on paper, thus creating an opportunity for light weight, low cost and fully recyclable complementary circuits – green electronics. This is expected to create applications in disposable and recyclable electronics that range from smart labels, tags, sensors and memories to TFT driven electro-chromic paper displays and integrated systems.

6.8 Solid State Paper Batteries

To fulfil the demand for paper electronics wireless, it is necessary to prove the concept of self powered devices. In the case of paper electronics, this implies demonstrating the idea of self regenerated thin film paper batteries and its integration with other electronic components. This concept was proved by Ferrcira and her co-workers [7, 8] that paper is able to to work simultaneously as substrate, electrolyte and membrane in self regenerating paper batteries. The paper battery consists of using a porous paper sheet with the required bulk contents, composed of metal cationic species leading to higher current density, hydroxyl groups and hydronium ions that when covered on both faces with thin layers of opposite electrochemical materials, a voltage appears between both electrodes – a paper battery, which is also self-regenerating. Moreover, paper has a high concentration of ions such as Al^{3+}, Na^{+}, Cl^{-}, Ca^{2+}, and others, as a function of the paper's mineral content, whose presence is responsible for the density of the current in the battery.

The value of the potential depends upon the materials used for anode and cathode. An open circuit voltage of 0.5V and a short-circuit current density of 1 μA cm^{-2} were obtained in the simplest structure produced (Cu(≅100 nm)/paper/Al(≅100 nm)). For actuating the gate of a paper transistor, seven paper batteries were integrated in the same substrate in series, supplying a voltage of 3.4 V. This enables proper on/off control of the paper transistor. Apart from that, transparent conductive oxides can be also used as cathode/anode materials allowing for the production of thin film batteries with transparent electrodes compatible with flexible, invisible, self powered and wireless electronics.

The porosity of the paper is an advantage for the absorbtion of agents from atmosphere, especially water vapor. The presence of this agent in non encapsulated regions of the paper promotes chemical reactions inside the cellulose fiber matrix (paper) with a consequent emergence of electrons in the electrodes of the regenerative battery – for instance, the use of a WO_3 interlayer between the paper electrolyte and the cathode contribute to enhancing the open circuit voltage (V_{oc}) of the battery cell to 0.68 V. Replacing the metal of the cathode by a transparent conductive oxide layer based on Gallium Zinc Oxide (GZO) we are able to produce an invisible cathode while the V_{oc} of battery is improved to 0.75 V (see Table 6.5).

Table 6.5 Different open circuit voltages recorded on design paper (DP) and copy paper (CP) using different materials for anode and cathode.

Matrix	Anode	Cathode	V_{oc} (V)
DP	Al	ZnO:Al/Cu	0.78
DP	Al	Y_2O_5/Cu	0.73
DP	Al/$LiAlF_4$	$LiAlF_4$/WO_3/Cu	0.71
DP	Al/$LiAlF_4$	$LiAlF_4$/Cu	0.68
CP	Al	ZnO:Al/Cu	0.68
DP	Al/V_2O_5:WO_3	Cu	0.63
DP, CP	Al	Cu	0.50

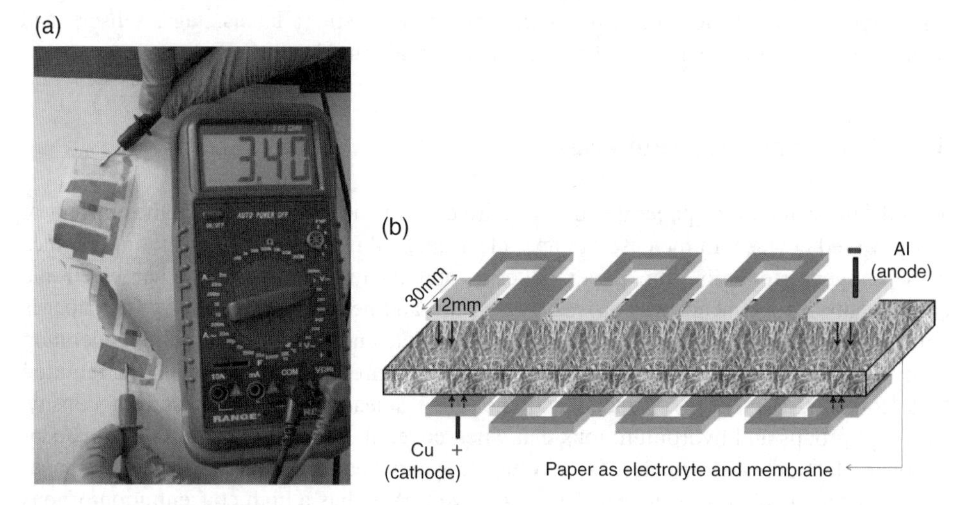

Figure 6.27 Paper battery (sketch and structure): a) photograph of the paper battery, with seven elements integrated in series, during the measurement of open circuit voltage with a multimeter; b) Schematic of the integrated (in series) paper battery, revealing its constituents, as depicted in the sketch.

Discharge tests performed on this cell element show that open circuit potential drops in the first minutes by more than 0.2 V, and then stabilizes at around 0.5 V for more than 18 hours under standard environment conditions. Moreover, the battery when exposed to a relative humidity above 45%, recovers the charge capacity. This behavior leads us to assume that the performance of the battery is atmospherically dependent. This result is related to the adsorption of atmospheric elements, as if the batteries breathe. This is related to the OH terminations of the cellulose network. Besides that, the introduction of some additives into the cellulose matrix during the papermaking process may play an important role in the electrochemical process involved in this type of battery. The photograph and the scheme of the battery proposed by Ferreira and her co-workers are depicted [7, 8] in Figure 6.27 (see color plate section).

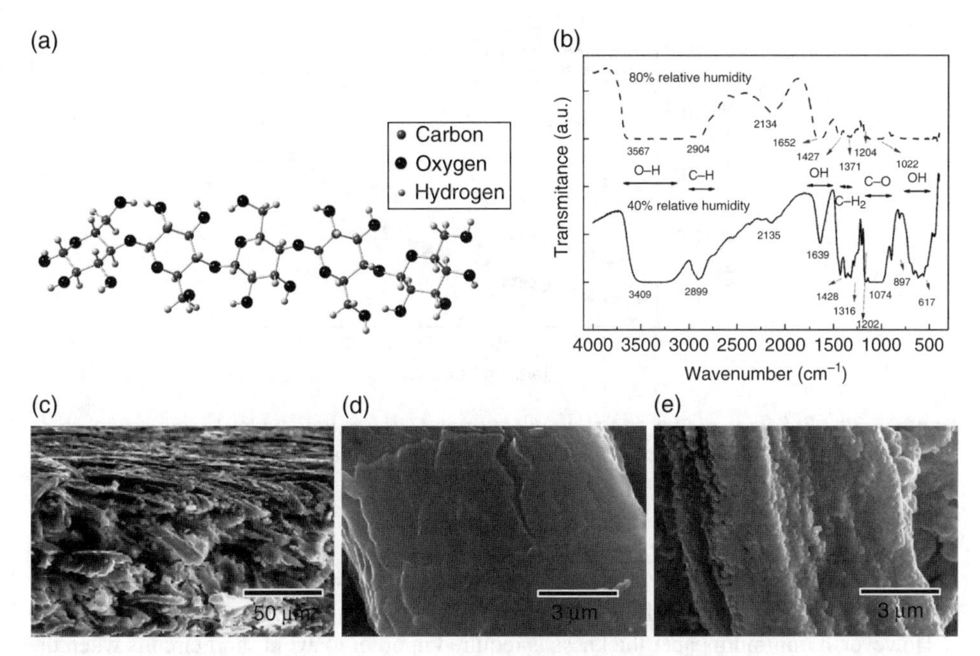

Figure 6.28 *Cellulose paper matrix: a) Schematic representation of a cellulose fiber repetitive unit in its simplest form and several hydrogen bridges between water vapor molecules and hydroxyl groups in the cellulose structure (dashed line). Oxygen atoms in black; Hydrogen in white and small; Carbon in gray; b) FTIR spectra of paper under different relative humidity: 40% (a) and 80% (b). The most important characteristic bonds of cellulose are indicated; c) SEM cross section image of the paper used; d) SEM image of the paper fibers without any coating; e)SEM image of the paper fibers cover with thin film layers with a total thickness of 500nm.*

In Figure 6.28 we show a cellulose fiber repetitive unit in its simplest form, where the increased moisture content favors ion transport (diffusion) between the two electrodes [190–192]. FTIR spectra obtained in a 100 µm paper are also shown in Figure 6.28b where is presented the data achieved under a relative humidity (*RH*) of 40 % and 80 %, respectively. The data show that the peaks related to OH vibrations modes suffer a dramatic decrease of transmission, as RH increases, leading to an over saturation of the spectrum in several wave number regions, especially the OH stretching vibrations due to hydrogen bridges in the range of 3550cm^{-1} to 3000cm^{-1}. This OH absorption also has an effect on the current of the devices. The short circuit current density of the devices with 80 % of humidity is enhanced by two orders of magnitude when compared with 60 % of humidity, respectively from 1 µA cm^{-2} to 100µA cm^{-2} (see Figure 6.29). Two possible mechanisms can be held responsible for the registered increase in current: electronic charge transfer through nearby hydrogen bridges and/or the swelling/tunneling of the cellulose structure due to increased number of hydrogen bridges between the hydroxyl groups in cellulose and in the water vapor. Thus, pathways within the paper, through which ions can flow, become enlarged. Both mechanisms are favored if the matrix is porous and has continuous pathways/channels through which charge transfer occurs. Being constituted by matrixes of

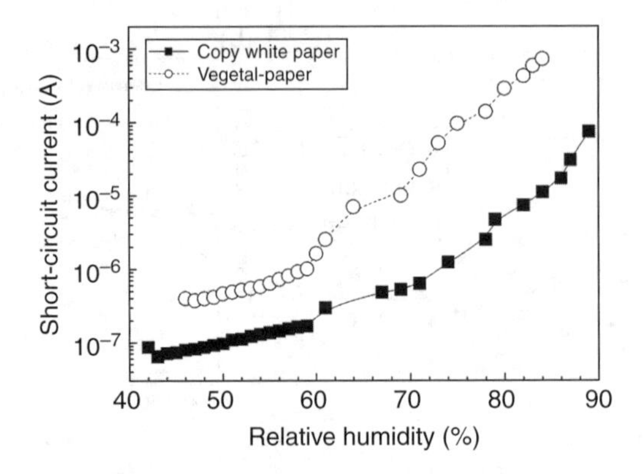

Figure 6.29 *Behavior of the short circuit current of the paper battery when exposed to different percentages of humidity, for two papers with different type of porosity. The most porous one leads to higher short circuit current as RH increases.*

mashed fibers, the papers employed in these devices are ideal for this process. Due to the fact that the charges (ion and/or electrons) have to travel along the paper thickness to the electrodes, the performance of the devices can be limited by the thickness of the paper.

However, a minimum paper thickness is required in order to avoid short circuits when the electrodes are deposited on the both faces. The paper battery only works properly if both electrodes are formed by continuous films covering the paper fibers, enabling the full interconnection of the fibers, around which the charges are accumulated. Indeed, the geometric paper thickness does not determine the current density or the voltage of the devices.

This is conditioned more by the final active paper thickness which depends mostly on the possible interconnections among the different plans in which the paper is constituted (see Figure 28c) and the fiber step coverage for the anode/cathode deposited films (see Figure 28e) and how they interconnect the different fibers, in a similar manner as that described for the paper memory [27–29, 77].

6.9 Electrochromic Paper Transistors

Paper has been also used as a component (mainly as substrate) in electrochromic devices based chiefly on organic components [194, 195]. This pioneer work is due mainly to companies like Philips, and Sony Inc, amongst others, and by technology providers such as E-ink as well as labs of which we have to highlight the work of the Magnus Berggren group [196, 197] aiming to bring costs down by using simpler processes such as printing based and using organic EC materials that can merge with the well established technology of paper substrates. The manufacturing process involves mainly three steps: patterning of the EC, electrolyte deposition and device encapsulation. For this work the Berggren group used as EC the system poly(3,4-ethylenedioxythiophene) doped with poly(styrenesulfonate) (PEDOT:PSS) [198] due to its high conductivity and chemical stability. Moreover, this material can be reversibly controlled, resulting in switching of the bulk conductivity as well as color via the reduced and oxidized states of the films (see Figure 6.30). They observe that by printing the devices on paper it inhibits full reduction of the polymer to its

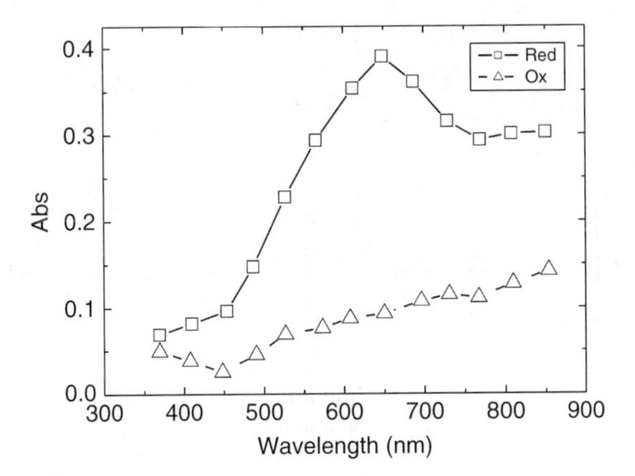

Figure 6.30 *Absorption spectra of reduced and oxidized states of PEDOT.PSS* [174, 197]. *Reprinted from [197] Copyright (2009).*

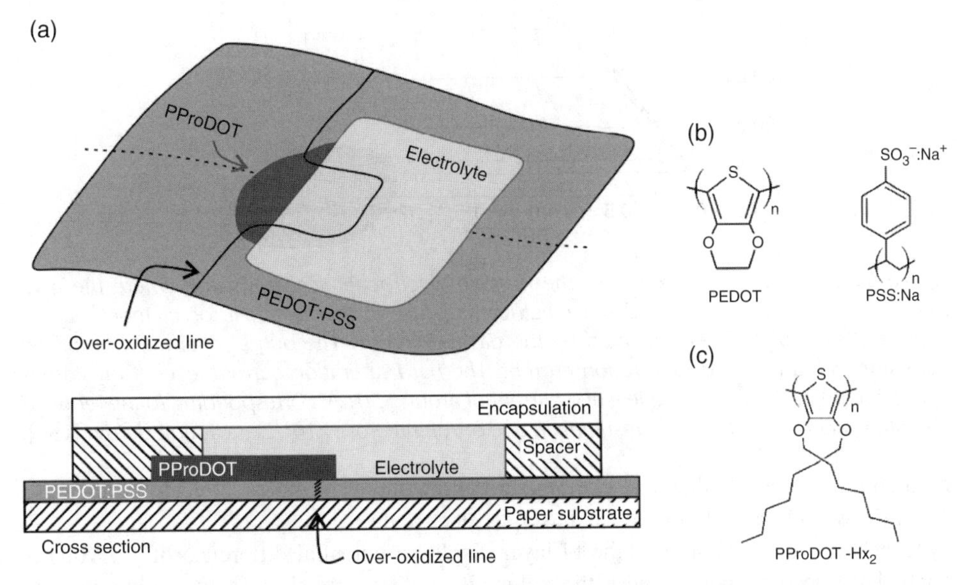

Figure 6.31 *a) The device architecture of the bi-layer device paper display; b) the repeated unit structure; c) PProDOT-Hx2. Reprinted with permission from* [196] *Copyright (2007) Wiley-VCH.*

electrically insulating state. This effect is attributed to polaronic charge carriers [197]. Moreover, the electrical conductivity can be controlled over five orders of magnitude and the optical absorption can be modulated in the visible spectrum range, mainly towards the red spectrum region.

Exploiting this behavior they develop a bi-layer display that includes a PProDOT-Hx2 deposited on the top of a PEDOT:PSS coated paper substrates, patterned by using a rotary screen printer with a conducting squeegee [196] (see Figure 6.31, see color plate section).

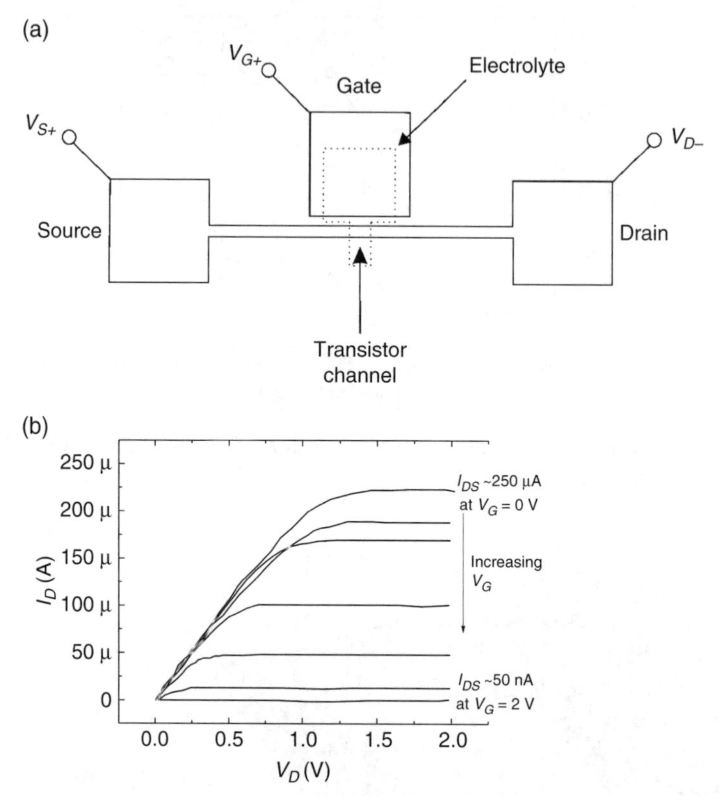

Figure 6.32 *a) Top view sketch of the organic electrochemical transistor, where the gate electrode is ionically connected but electrically insulated to the transistor channel via the deposited electrolyte layer indicated by the dashed line; b) The output characteristic of the transistor with a nice hard saturation current. The hard saturation current level at on state is about 250* μA *at* $V_G = 0\,V$*, with a leakage current of around 50 nA, corresponding to the reduced OFF state, achieved for* $V_G = 2\,V$*, resulting in one on/off ratio of* 5×10^3*. Data adapted from* [196].

As a counter electrode they use platinized titanium and as a reference electrode they use Ag/AgCl with 0.1 M solution of $LiCF_3SO_3$ as electrolyte.

To achieve the switching of the bi-layer display a potential difference of 1.5 Volts is applied to the electrodes, where the color contrast was maximized by optimizing the thickness of the PProDOT-Hx2 film and leaving unchanged the thickness of the bottom PEDOT:PSS film. From the alphanumeric display fabricated by the Berggren group [197] the overall switching time (defined as the time required to achieve 90% color saturation of a full switch) to go from *on* to *off* state and vice versa was 6.7 seconds and 4.5 seconds respectively. This asymmetric response time behavior is attributed by the authors to the relatively poor ionic conductivity of the PProDOT-Hx2in comparison to the PEDOT:PSS film, as well as to the possible diffusion contribution of the electrolyte.

To build up active matrix displays on paper, it has been also proposed to combine electrochemical three terminal organic PEDOT:PSS based transistors [194] with display cells [196, 197]. The typical structure and output characteristics of these transistors are depicted in Figure 6.32. By applying a positive gate voltage the conductivity of the transistor channel decreases via reduction of the PEDOT to its neutral state, according to the equation [196]:

$$\mathrm{PEDOT^{+}PSS^{-^{\circ}} + M^{+} + e^{-} \rightarrow PEDOT^{0} + M^{+}PSS^{-}} \qquad (6.24)$$

Reversible, by grounding the gate electrode, the PEDOT returns to its high conducting state via-re-oxidation. The matrix display, to address the same display cells as those described above, typically uses low voltages (up to 2 or 3 V), but suffers from the fact that it has to stay in *on* state without any voltage applied, requiring the application of the voltage to be on *off* state, which is a disadvantage when it comes to inorganic based active matrices [45]. Moreover, the organic transistors used usually present very low hole mobilities, leading to low speed switching operations (switching times of several seconds as against tenths of milliseconds), which is not compatible with most of the required industry performances for displays, namely the dynamic ones. Nevertheless, they offer great potential for static display applications where switching time is not so crucial, opening up the door for future hybrid based display applications where transistors are based in inorganic materials, such as oxide, as described in the previous sections, benefiting from the fact that the normal state of the transistor is the OFF state; the paper is not only the substrate, but also the dielectric and all devices can have printing compatibility and be processed at low temperatures, as happens with the proposed organic based devices.

In conclusion, we demonstrate the ability to process full autonomous devices with and on paper, going from transparent conductive oxides with performance close to that exhibited in standard glass substrates; thin film transistors, memory devices, CMOS circuits, all-solid-state battery and electrochromic devices. As far as thin film transistors are concerned we note that the device performances are highly sensitive to paper thickness and structure and that the transport behavior is controlled by the discrete nature of the fibers that constitute the paper. Moreover, the battery is auto- regenerated and therefore can be used for more than one application. This is certainly the beginning of tremendously challenging innovations as far as auto-sustained low cost, flexibility and when necessary transparent electronics on paper, "paper-e", are concerned.

6.10 Paper UV Light Sensors

Besides the above mentioned applications, paper has also recently been exploited as a light sensor, making all processes easy and cheap [199], when compared with conventional UV sensors, based on ZnO structures [200].

The paper sensor proposed using paper as a porous matrix to hold ZnO crystals on its surface and whose electrodes are just pencil-drawn graphite lines [199], as depicted in Figure 6.33 (see color plate section).

The typical resistivity of the pencil lines is of the order of 10^4 Ω.cm [201] which is more than two to three orders of magnitude smaller than that exhibited by ZnO [200]. Moreover, exploiting the strong sensibility of the ZnO films to UV light, which is highly dependent on the size of its nanostructure and the state of the grain boundaries [200], it is possible to produce highly cheap and efficient UV sensors. In this case, the reaction that takes place in the ZnO surface involves the adsorption of molecular oxygen from the atmosphere that may

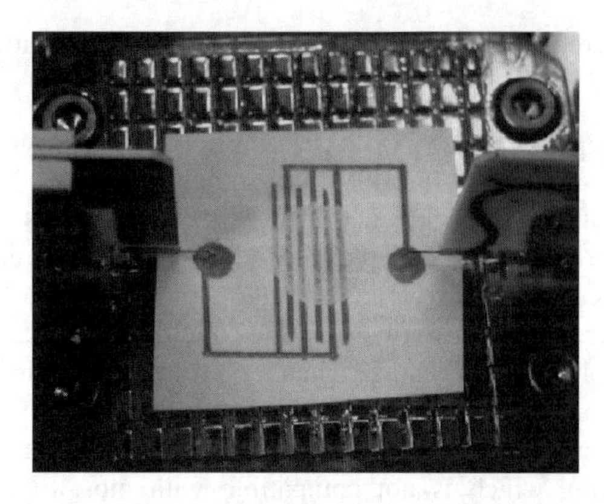

Figure 6.33 Interdigitized sensor in a paper substrate connected to a probe station [199]. *Reprinted with permission from [198] Copyright (2011) American Chemical Society.*

even lead to the ozone formation when UV light is exposed [202, 203], leading to a tremendous decrease (orders of magnitude) of the ZnO resistivity. In the cheapest and fasted configuration, Gimenez and co-workers [199] prepared a 10 mL water suspension with 0.1 gr. of ZnO powder with sizes below 5 μm, which was dispersed over a porous paper sheet using an ultrasonic bath for 5 min to avoid precipitation and clustering and then properly rinsed. Above this "deposited" film the electrodes were drawn using a pencil. The set of data achieved reveal that the leakage current obtained on the paper is of the order of nA, about four orders of magnitude higher than the data achieved on glass substrates, This data agrees well with the leakage current observed for the active devices studied in this chapter and are attributed to the porous paper structure. Moreover, the data achieved by these sensors reveal that they are highly sensitive to UV light, exhibiting a UV photosensitivity in the range of two orders of magnitude, when using a 395 nm LED lamp as a UV light source placed 10 cm above the sensor, with response times (from dark to UV light illumination) below 5 seconds.

References

[1] D. Tobjork and R. Osterbacka (2011) Paper Electronics, *Advanced Materials* **23**, 1935–61.
[2] R. Martins, I. Ferreira, and E. Fortunato (2011) Electronics with and on paper, *Phys. Status Solidi RRL* **1**, 332–5.
[3] A. Manekkathodi, M.Y. Lu, C.W. Wang, and L.J. Chen (2010) Direct Growth of Aligned Zinc Oxide Nanorods on Paper Substrates for Low-Cost Flexible Electronics, *Advanced Materials* **22**, 4059–63.
[4] D.H. Kim, Y.S. Kim, J. Wu, Z.J. Liu, J.Z. Song, H.S. Kim, Y.G.Y. Huang, K.C. Hwang (2009) and J. A. Rogers, Ultrathin Silicon Circuits With Strain-isolation Layers and Mesh Layouts for High-Performance Electronics on Fabric, Vinyl, Leather, and Paper, *Advanced Materials* **21**, 3703–7.
[5] M. Berggren, D. Nilsson, and N.D. Robinson (2007) Organic materials for printed electronics, *Nature Materials* **6**, 3–5.
[6] Y.G. Sun and J.A. Rogers (2007) Inorganic semiconductors for flexible electronics, *Advanced Materials* **19**, 1897–1916.
[7] I. Ferreira, B. Bras, N. Correia, P. Barquinha, E. Fortunato, and R. Martins (2010) Self-Rechargeable Paper Thin-Film Batteries: Performance and Applications, *Journal of Display Technology* **6**, 332–5.

[8] I. Ferreira, B. Bras, J. I. Martins, N. Correia, P. Barquinha, E. Fortunato, and R. Martins (2011) Solid-state paper batteries for controlling paper transistors, *Electrochimica Acta* **56**, 1099–1105.
[9] A.C. Baptista, J.I. Martins, E. Fortunato, R. Martins, J.P. Borges, and I. Ferreira (2011) Thin and flexible bio-batteries made of electrospun cellulose-based membranes, *Biosensors & Bioelectronics* **26**, 2742–5.
[10] Y. Tanaka, N. Ishii, J. Okuma, and R. Hara (1999) "Electric doublelayer capacitor having a separator made from a cellulose fiber." U.S., patent number 5963419.
[11] V.L. Pushparaj, M.M. Shaijumon, A. Kumar, S. Murugesan, L. Ci, R. Vajtai, R.J. Linhardt, O. Nalamasu, and P.M. Ajayan (2007) Flexible energy storage devices based on nanocomposite paper, *Proceedings of the National Academy of Sciences of the United States of America* **104**, 13574–7.
[12] K.B. Lee (2006) Two-step activation of paper batteries for high power generation: design and fabrication of biofluid- and water-activated paper batteries, *Journal of Micromechanics and Microengineering* **16**, 2312–17.
[13] G. Nystrom, A. Razaq, M. Stromme, L. Nyholm, and A. Mihranyan (2009) Ultrafast All-Polymer Paper-Based Batteries, *Nano Letters* **9**, 3635–9.
[14] J.J. Kucherovsky and *et al.* (2002) "Method of making a thin film battery." U.S., patent number 6379835.
[15] Y.H. Kim, D.G. Moon, and J.I. Han (2004) Organic TFT array on a paper substrate, *Ieee Electron Device Letters* **25**, 702–4.
[16] M.M. Tentzeris, L. Yang, A. Rida, A. Traille, R. Vyas, and T. Wu (2007) Inkjet-printed RFID tags on paper-based substrates for UHF "cognitive intelligence" applications, in *2007 Ieee 18th International Symposium on Personal, Indoor and Mobile Radio Communications, Vols 1–9*, 1185–8.
[17] L. Yang, A. Rida, R. Vyas, and M.M. Tentzeris (2007) RFID tag and RF structures on a paper substrate using inkjet-printing technology, *Ieee Transactions on Microwave Theory and Techniques* **55**, 2894–2901.
[18] M. Hilder, B. Winther-Jensen, and N.B. Clark (2009) Paper-based, printed zinc-air battery, *Journal of Power Sources* **194**, 1135–41.
[19] J.Y. Kim, S.H. Park, T. Jeong, M. J. Bae, S. Song, J. Lee, I.T. Han, D. Jung, and S. Yu (2010) Paper as a substrate for inorganic powder electroluminescence devices, *Ieee Transactions on Electron Devices* **57**, 1470–4.
[20] A.C. Siegel, S.T. Phillips, M.D. Dickey, N. S. Lu, Z.G. Suo, and G.M. Whitesides (2010) Foldable printed circuit boards on paper substrates, *Advanced Functional Materials* **20**, 28–35.
[21] W. Lim, E.A. Douglas, D.P. Norton, S.J. Pearton, F. Ren, Y.W. Heo, S.Y. Son, and J.H. Yuh (2010) Low-voltage indium gallium zinc oxide thin film transistors on paper substrates, *Applied Physics Letters* **96**, 053510.
[22] A.X. Lu, M.Z. Dai, J. Sun, J. Jiang, and Q. Wan (2011) Flexible low-voltage electric-double-layer TFTs self-assembled on paper substrates, *Ieee Electron Device Letters* **32**, 518–20.
[23] T.P. Brody (1984) The thin-film transistor – a late flowering bloom, *Ieee Transactions on Electron Devices* **31**, 1614–28.
[24] A.L. Robinson (1984) CMOS future for microelectronic circuits, *Science* **224**, 705–7.
[25] S.A. White (1969) A complementary MOS (CMOS) NAND/NOR gate, *Ieee Journal of Solid-State Circuits* **SC 4**, 50.
[26] M. Nogi, S. Iwamoto, A.N. Nakagaito, and H. Yano (2009) Optically transparent nanofiber paper, *Advanced Materials* **21**, 1595–8.
[27] R. Martins, P. Barquinha, L. Pereira, N. Correia, G. Goncalves, I. Ferreira, and E. Fortunato (2008) Write-erase and read paper memory transistor, *Applied Physics Letters* **93**, 203501.
[28] R. Martins, P. Barquinha, L. Pereira, N. Correia, G. Goncalves, I. Ferreira, and E. Fortunato (2009) Selective floating gate non-volatile paper memory transistor, *Physica Status Solidi-Rapid Research Letters* **3**, 308–10.
[29] R. Martins, L. Pereira, P. Barquinha, N. Correia, G. Goncalves, I. Ferreira, C. Dias, and E. Fortunato (2010) "Floating gate memory paper transistor," in *Oxide-Based Materials and Devices*. vol. 7603, F.H. Teherani, D.C. Look, C.W. Litton, and D.J. Rogers, eds, pp. 760314–11.

[30] E. Fortunato and R. Martins (2011) Where science fiction meets reality? With oxide semiconductors!, *Phys. Status Solidi RRL,* **1**, 336–9.
[31] E. Fortunato, P. Barquinha, G. Goncalves, L. Pereira, and R. Martins (2008) High mobility and low threshold voltage transparent thin film transistors based on amorphous indium zinc oxide semiconductors, *Solid-State Electronics* **52**, 443–8.
[32] E. Fortunato, P. Barquinha, L. Pereira, G. Goncalves, and R. Martins (2007) Advanced materials for the next generation of thin film transistors, in *Idmc'07: Proceedings of the International Display Manufacturing Conference 2007*, C.H. Chen and Y.S. Tsai, eds, *Taipei: Society Information Display*, Taipei Chapter, 371–3.
[33] E. Fortunato, P. Barquinha, A. Pimentel, A. Goncalves, A. Marques, L. Pereira, and R. Martins (2005) Recent advances in ZnO transparent thin film transistors, *Thin Solid Films* **487**, 205–11.
[34] E. Fortunato, P. Barquinha, A. Pimentel, L. Pereira, G. Goncalves, and R. Martins (2007) Amorphous IZO TTFTs with saturation mobilities exceeding 100 cm^2/Vs, *Physica Status Solidi-Rapid Research Letters* **1**, R34–R36.
[35] V. Figueiredo, E. Elangovan, G. Goncalves, P. Barquinha, L. Pereira, N. Franco, E. Alves, R. Martins, and E. Fortunato (2008) Effect of post-annealing on the properties of copper oxide thin films obtained from the oxidation of evaporated metallic copper, *Appl. Surf. Sci.* **254**, 3949–54.
[36] E. Fortunato, L. Pereira, P. Barquinha, I. Ferreira, R. Prabakaran, G. Goncalves, A. Goncalves, and R. Martins (2009) Oxide semiconductors: Order within the disorder, *Philosophical Magazine* **89**, 2741–58.
[37] E.M.C. Fortunato, P.M.C. Barquinha, A. Pimentel, A.M.F. Goncalves, A.J.S. Marques, R.F. P. Martins, and L.M.N. Pereira (2004) Wide-bandgap high-mobility ZnO thin-film transistors produced at room temperature, *Applied Physics Letters* **85**, 2541–3.
[38] E.M.C. Fortunato, P.M.C. Barquinha, A. Pimentel, A.M.F. Goncalves, A.J.S. Marques, L.M.N. Pereira, and R.F.P. Martins (2005) Fully transparent ZnO thin-film transistor produced at room temperature, *Advanced Materials* **17**, 590–4.
[39] E.M.C. Fortunato, L.M.N. Pereira, P.M.C. Barquinha, A.M.B. do Rego, G. Goncalves, A. Vila, J.R. Morante, and R.F.P. Martins (2008) High mobility indium free amorphous oxide thin film transistors, *Applied Physics Letters* **92**, 222103.
[40] G. Goncalves, P. Barquinha, L. Pereira, N. Franco, E. Alves, R. Martins, and E. Fortunato (2009) High Mobility a-IGO Films Produced at Room Temperature and Their Application in TFTs, *Electrochemical and Solid State Letters* **13**, II20–II22.
[41] R. Martins, P. Barquinha, L. Pereira, I. Ferreira, and E. Fortunato (2007) Role of order and disorder in covalent semiconductors and ionic oxides used to produce thin film transistors, *Applied Physics a-Materials Science & Processing* **89**, 37–42.
[42] P. Barquinha, G. Goncalves, L. Pereira, R. Martins, and E. Fortunato (2007) Effect of annealing temperature on the properties of IZO films and IZO based transparent TFTs, *Thin Solid Films* **515**, 8450–4.
[43] P. Barquinha, L. Pereira, G. Goncalves, D. Kuscer, M. Kosec, A. Vila, A. Olziersky, J.R. Morante, R. Martins, and E. Fortunato (2010) Low-temperature sputtered mixtures of high-kappa and high bandgap dielectrics for GIZO TFTs, *Journal of the Society for Information Display* **18**, 762–72.
[44] P. Barquinha, L. Pereira, G. Goncalves, R. Martins, and E. Fortunato (2008) The effect of deposition conditions and annealing on the performance of high-mobility GIZO TFTs, *Electrochemical and Solid State Letters* **11**, H248–H251.
[45] P. Barquinha, L. Pereira, G. Goncalves, R. Martins, and E. Fortunato (2009) Toward high-performance amorphous GIZO TFTs, *Journal of the Electrochemical Society* **156**, H161–H168.
[46] P. Barquinha, L. Pereira, G. Goncalves, R. Martins, D. Kuscer, M. Kosec, and E. Fortunato (2009) Performance and stability of low temperature transparent thin-film transistors using amorphous multicomponent dielectrics, *Journal of the Electrochemical Society* **156**, H824–H831.
[47] H.A. Klasens and H. Koelmans (19640 A tin oxide field-effect transistor, *Solid-State Electronics* **7**, 701–2.
[48] G.F. Boesen and J.E. Jacobs, ZnO field-effect transistor, *Proceedings of the Institute of Electrical and Electronics Engineers* **56**, 2094–5.

[49] K. Nomura, H. Ohta, A. Takagi, T. Kamiya, M. Hirano, and H. Hosono (2004) Room-temperature fabrication of transparent flexible thin-film transistors using amorphous oxide semiconductors, *Nature* **432**, 488–92.
[50] R.L. Hoffman, B.J. Norris, and J.F. Wager (2003) ZnO-based transparent thin-film transistors, *Applied Physics Letters* **82**, 733–5.
[51] P.F. Carcia, R.S. McLean, M.H. Reilly, I. Malajovich, K.G. Sharp, S. Agrawal, and G. Nunes (2003) "ZnO thin film transistors for flexible electronics," in *Flexible Electronics-Materials and Device Technology*. vol. 769, N. Fruehauf, B.R. Chalamala, B.E. Gnade, and J. Jang, eds, 233–8.
[52] P.F. Carcia, R.S. McLean, M.H. Reilly, and G. Nunes (2003) Transparent ZnO thin-film transistor fabricated by rf magnetron sputtering, *Applied Physics Letters* **82**, 1117–19.
[53] L.J. Shao, K. Nomura, T. Kamiya, and H. Hosono (2011) Operation characteristics of thin-film transistors using very thin amorphous in-Ga-Zn-O channels, *Electrochemical and Solid State Letters* **14**, H197–H200.
[54] M. Lorenz, A. Lajn, H. Frenzel, H. V. Wenckstern, M. Grundmann, P. Barquinha, R. Martins, and E. Fortunato (2010) Low-temperature processed Schottky-gated field-effect transistors based on amorphous gallium-indium-zinc-oxide thin films, *Applied Physics Letters* **97**, 243506.
[55] E. Fortunato, N. Correia, P. Barquinha, C. Costa, L. Pereira, G. Goncalves, and R. Martins (2009) Paper field effect transistor, in *Proceedings of SPIE*, SPIE, 2009 72170K–72170K-11.
[56] J.Z. Wang, E. Elamurugu, V. Sallet, F. Jomard, A. Lusson, A.M.B. do Rego, P. Barquinha, G. Goncalves, R. Martins, and E. Fortunato (2008) Effect of annealing on the properties of N-doped ZnO films deposited by RF magnetron sputtering, *Appl. Surf. Sci.* **254**, 7178–82.
[57] L.N. Zhang, D. Ruan, and S.J. Gao (2002) Dissolution and regeneration of cellulose in NaOH/thiourea aqueous solution, *Journal of Polymer Science Part B-Polymer Physics* **40**, 1521–9.
[58] E.L. Hult, M. Iotti, and M. Lenes (2010) Efficient approach to high barrier packaging using microfibrillar cellulose and shellac, *Cellulose* **17**, 575–86.
[59] E.L. Hult, T. Iversen, and J. Sugiyama (2003) Characterization of the supermolecular structure of cellulose in wood pulp fibres, *Cellulose* **10**, 103–10.
[60] Z.T. Liu, Y.N. Yang, L.L. Zhang, Z.W. Liu, and H.P. Xiong (2007) Study on the cationic modification and dyeing of ramie fiber, *Cellulose* **14**, 337–45.
[61] J.J. Creely, L. Segal, and L. Loeb (1959) An X-ray study of new cellulose complexes with diamines containing 3, 5, 6, 7, AND 8 carbon atoms, *Journal of Polymer Science* **36**, 205–14.
[62] L. Segal (1954) On the preparation of cellulose-III with ethylamine, *Textile Research Journal* **24**, 861–2.
[63] http://en.wikipedia.org/wiki/Cellulose.
[64] P. Barquinha, A.M. Vila, G. Goncalves, R. Martins, J. R. Morante, E. Fortunato, and L. Pereira (2008) Gallium-indium-zinc-oxide-based thin-film transistors: Influence of the source/drain material, *Ieee Transactions on Electron Devices* **55**, 954–60.
[65] R. Martins, B. Brás, I. Ferreira, L. Pereira, P. Barquinha, N. Correia, R. Costa, T. Busani, A. Gonçalves, A. Pimentel, E. Fortunato (2011) Away from silicon era: the paper electronics, *Proceedings SPIE* 7940, 79400P.
[66] P. Sirvio, J. Sidaravicius, R. Maldzius, K. Backfolk, and E. Montrimas (2009) Effect of NaCl and Moisture Content on Electrical and Dielectric Properties of Paper, *Journal of Imaging Science and Technology* **53**, 020501.
[67] M. Nilsson and M. Stromme (2005) Electrodynamic investigations of conduction processes in humid microcrystalline cellulose tablets, *Journal of Physical Chemistry B* **109**, 5450–5.
[68] J.H. Christie, S.H. Krenek, and I.M. Woodhead (2009) The electrical properties of hygroscopic solids, *Biosystems Engineering* **102**, 143–52.
[69] S. Simula and K. Niskanen (1999) Electrical properties of viscose-kraft fibre mixtures, *Nordic Pulp & Paper Research Journal* **14**, 243–6.
[70] I. Brodie, Dahlquis.Ja, and A. Sher (1968) Measurement of charge transfer in electrographic processes, *Journal of Applied Physics* **39**, 1618.
[71] K. Backfolk, J. Sidaravicius, P. Sirvio, R. Maldzius, T. Lozovski, and J.L.B. Rosenholm (2010) Effect of base paper grammage and electrolyte content on electrical and dielectric properties of coated papers, *Nordic Pulp & Paper Research Journal* **25**, 319–27.

[72] R. Maldzius, P. Sirvio, J. Sidaravicius, T. Lozovski, K. Backfolk, and J.B. Rosenholm (2010) Temperature-dependence of electrical and dielectric properties of papers for electrophotography, *Journal of Applied Physics* **107**, 114904.

[73] S. Samula (2002) "Electrical Properties: II. Applications and Measurement methods," in *Handbook of Physical Testing of Paper*, 2nd ed. vol. 2, J. Borch, M. Lyne, R. Mark, and C.C. J. Habeger, eds, New York: Marcel Dekker, 361–87.

[74] G. Baum (2002) Properties: I. Theory in *Handbook of Physical Testing of Paper*, 2nd ed. vol. 2, J.J. Borch, M. Lyne, R. Mark, and C. J. Habeger, Eds.: Marcel Dekker, 333–59.

[75] W. Lim, E.A. Douglas, S.H. Kim, D.P. Norton, S.J. Pearton, F. Ren, H. Shen, and W.H. Chang (2009) High mobility $InGaZnO_4$ thin-film transistors on paper, *Applied Physics Letters* **94**, 072103.

[76] S. Simula, S. Ikalainen, K. Niskanen, T. Varpula, H. Seppa, and A. Paukku (1999) Measurement of the dielectric properties of paper, *Journal of Imaging Science and Technology* **43**, 472–7.

[77] R. Martins, L. Pereira, P. Barquinha, N. Correia, G. Gonçalves, I. Ferreira, C. Dias, N. Correia, M. Dionísio, M.M. Silva, and E. Fortunato (2009) Self sustained n-type memory transistor devices based on natural cellulose paper fibers, *Journal of Information Display* **10**, 80–8.

[78] Q.D. Ling, D.J. Liaw, C.X. Zhu, D.S.H. Chan, E.T. Kang, and K.G. Neoh (2008) Polymer electronic memories: Materials, devices and mechanisms, *Progress in Polymer Science* **33**, 917–78.

[79] Y.Q. Zhao, G.Q. Zhang, and H.L. Li (2006) Electrochemical characterization on layered lithium ruthenate for electrochemical supercapacitors, *Solid State Ionics* **177**, 1335–9.

[80] G.Q. Zhang and X.G. Zhang (2003) A novel alkaline Zn/MnO2 cell with alkaline solid polymer electrolyte, *Solid State Ionics* **160**, 155–9.

[81] A. Sheoran, S. Sanghi, S. Rani, A. Agarwal, and V.P. Seth (2009) Impedance spectroscopy and dielectric relaxation in alkali tungsten borate glasses, *Journal of Alloys and Compounds* **475**, 804–9.

[82] R. Martins, P. Almeida, P. Barquinha, L. Pereira, A. Pimentel, I. Ferreira, and E. Fortunato (2006) Electron transport and optical characteristics in amorphous indium zinc oxide films, *Journal of Non-Crystalline Solids* **352**, 1471–4.

[83] R. Martins, P. Barquinha, A. Pimentel, L. Pereira, and E. Fortunato (2005) Transport in high mobility amorphous wide band gap indium zinc oxide films, *Physica Status Solidi a-Applications and Materials Science* **202**, R95–R97.

[84] G. Goncalves, E. Elangovan, P. Barquinha, L. Pereira, R. Martins, and E. Fortunato (2007) Influence of post-annealing temperature on the properties exhibited by ITO, IZO and GZO thin films, *Thin Solid Films* **515**, 8562–6.

[85] G. Goncalves, V. Grasso, P. Barquinha, L. Pereira, E. Elamurugu, M. Brignone, R. Martins, V. Lambertini, and E. Fortunato (2011) Role of Room Temperature Sputtered High Conductive and High Transparent Indium Zinc Oxide Film Contacts on the Performance of Orange, Green, and Blue Organic Light Emitting Diodes, *Plasma Processes and Polymers* **8**, 340–5.

[86] J.R. Sheats, D. Biesty, J. Noel, and G.N. Taylor (2010) Printing technology for ubiquitous electronics, *Circuit World* **36**, 40–7.

[87] J.R. Tillman (1966) Silicon semiconductor technology, *Nature* **210**, 559.

[88] R. Martins, L. Raniero, L. Pereira, D. Costa, H. Aguas, S. Pereira, L. Silva, A. Goncalves, I. Ferreira, and E. Fortunato (2009) Nanostructured silicon and its application to solar cells, position sensors and thin film transistors, *Philosophical Magazine* **89**, 2699–2721.

[89] M.B. Schubert and R. Merz (2009) Flexible solar cells and modules, *Philosophical Magazine* **89**, 2623–44.

[90] J. Ni, J.J. Zhang, Y. Cao, X.B. Wang, X. L. Chen, X.H. Geng, and Y. Zhao (2011) Low temperature deposition of high open-circuit voltage (>1.0 V) p-i-n type amorphous silicon solar cells, *Solar Energy Materials and Solar Cells* **95**, 1922–6.

[91] J.A. Rogers and Y.G. Huang (2009) A curvy, stretchy future for electronics, *Proceedings of the National Academy of Sciences of the United States of America* **106**, 10875–6.

[92] J. Song, Y. Huang, J. Xiao, S. Wang, K.C. Hwang, H.C. Ko, D.H. Kim, M.P. Stoykovich (2009) and J.A. Rogers, Mechanics of noncoplanar mesh design for stretchable electronic circuits, *Journal of Applied Physics* **105**, 123516.

[93] L. Pereira, P. Barquinha, E. Fortunato, and R. Martins (2007) Low temperature high-k dielectric on poly-Si TFTs, in *22nd International Conference on Amorphous and Nanocrystalline Semiconductors*, Breckenridge, CO, pp. 2534–7.

[94] P.K. Nayak, T. Busani, E. Elamurugu, P. Barquinha, R. Martins, Y. Hong, and E. Fortunato (2010) Zinc concentration dependence study of solution processed amorphous indium gallium zinc oxide thin film transistors using high-k dielectric, *Applied Physics Letters* **97**, 183504.

[95] S. Jeong, J.Y. Lee, S.S. Lee, Y. Choi, and B.H. Ryu (2011) Impact of metal salt precursor on low-temperature annealed solution-derived Ga-doped In_2O_3 semiconductor for thin-film transistors, *Journal of Physical Chemistry C* **115**, 11773–80.

[96] W.F. Chung, T.C. Chang, H.W. Li, S.C. Chen, Y.C. Chen, T.Y. Tseng, and Y.H. Tai (2011) Environment-dependent thermal instability of sol-gel derived amorphous indium-gallium-zinc-oxide thin film transistors, *Applied Physics Letters* **98**, 152109.

[97] M.G. Kim, M.G. Kanatzidis, A. Facchetti, and T.J. Marks (2011) Low-temperature fabrication of high-performance metal oxide thin-film electronics via combustion processing, *Nature Materials* **10**, 382–8.

[98] K.K. Banger, Y. Yamashita, K. Mori, R.L. Peterson, T. Leedham, J. Rickard, and H. Sirringhaus (2011) Low-temperature, high-performance solution-processed metal oxide thin-film transistors formed by a 'sol-gel on chip' process, *Nature Materials* **10**, 45–50.

[99] G. Adamopoulos, S. Thomas, P.H. Wobkenberg, D.D.C. Bradley, M.A. McLachlan, and T. D. Anthopoulos (2011) High-Mobility Low-Voltage ZnO and Li-Doped ZnO Transistors Based on ZrO2 High-k Dielectric Grown by Spray Pyrolysis in Ambient Air, *Advanced Materials* **23** (2011), 1894–8.

[100] D.H. Lee, Y.J. Chang, G.S. Herman, and C.H. Chang (2007) A general route to printable high-mobility transparent amorphous oxide semiconductors, *Advanced Materials* **19**, 843–7.

[101] Z.C. Qu, G.X. Chen, B.L. Tang, and S.S. Wen (20110 Effect of paper surface characteristics on dot gain in ink-jet printing, in *Printing and Packaging Study*. vol. 174, O.Y. Yun, X. Min, and Y. Li, eds, 227–30.

[102] R.N. Das, H.T. Lin, J.M. Lauffer, and V.R. Markovich (2011) Printable electronics: towards materials development and device fabrication, *Circuit World* **37**, 38–45.

[103] S. Lee, H. Park, and D.C. Paine (2011) A study of the specific contact resistance and channel resistivity of amorphous IZO thin film transistors with IZO source-drain metallization, *Journal of Applied Physics* **109**, 063702.

[104] Y. Wu, E. Girgis, V. Strom, W. Voit, L. Belova, and K. V. Rao (2011) Ultraviolet light sensitive In-doped ZnO thin film field effect transistor printed by inkjet technique, *Physica Status Solidi a-Applications and Materials Science* **208**, 206–9.

[105] J. Sun, J. Jiang, A.X. Lu, B. Zhou, and Q. Wan (2011) Low-Voltage Transparent Indium-Zinc-Oxide Coplanar Homojunction TFTs Self-Assembled on Inorganic Proton Conductors, *Ieee Transactions on Electron Devices* **58**, 764–8.

[106] C.W. Chien, C.H. Wu, Y.T. Tsai, Y.C. Kung, C.Y. Lin, P.C. Hsu, H.H. Hsieh, C.C. Wu, Y.H. Yeh, C.M. Leu, and T.M. Lee (2011) High-Performance Flexible a-IGZO TFTs Adopting Stacked Electrodes and Transparent Polyimide-Based Nanocomposite Substrates, *Ieee Transactions on Electron Devices* **58**, 1440–46.

[107] D. Geng, D.H. Kang, and J. Jang (2011) High-performance amorphous indium-gallium-zinc-oxide thin-film transistor with a self-aligned etch stopper patterned by back-side UV exposure, *Ieee Electron Device Letters* **32**, 758–60.

[108] M. Mativenga, M.H. Choi, D.H. Kang, and J. Jang (2011) High-performance drain-offset a-IGZO thin-film transistors, *Ieee Electron Device Letters* **32**, 644–6.

[109] R.I. Kondratyuk, K. Im, D. Stryakhilev, C.G. Choi, M.G. Kim, H. Yang, H. Park, Y.G. Mo, H.D. Kim, and S.S. Kim (2011) A Study of Parasitic Series Resistance Components in In-Ga-Zn-Oxide (a-IGZO) Thin-Film Transistors, *Ieee Electron Device Letters* **32**, 503–5.

[110] A. Facchetti, M.H. Yoon, and T.J. Marks (2005) Gate dielectrics for organic field-effect transistors: New opportunities for organic electronics, *Advanced Materials* **17**, 1705–25.

[111] P. Barquinha, A. Vila, G. Goncalves, L. Pereira, R. Martins, J. Morante, and E. Fortunato (2008) The role of source and drain material in the performance of GIZO based thin-film transistors, *Physica Status Solidi a-Applications and Materials Science* **205**, 1905–9.

[112] G. Goncalves, P. Barquinha, L. Raniero, R. Martins, and E. Fortunato (2008) Crystallization of amorphous indium zinc oxide thin films produced by radio-frequency magnetron sputtering, *Thin Solid Films* **516**, 1374–6.

[113] K. Niskanen and S. Simula (1999) Thermal diffusivity of paper, *Nordic Pulp & Paper Research Journal* **14**, 236–42.

[114] D. Kahng and S. Sze (1967) A floating-gate and its application to memory devices, *Bell System Technical Journal* **46**, 1288–95.

[115] J. McGinnes, P. Corry, and P. Proctor (1974) Amorphous-semiconductor switching in melanins, *Science* **183**, 853–5.

[116] M.G. Masud, B.K. Chaudhuri, and H.D. Yang (2011) High dielectric permittivity and room temperature magneto-dielectric response of charge disproportionate $La_{0.5}Ba_{0.5}FeO_3$ perovskite, *Journal of Physics D-Applied Physics* **44**, 255403.

[117] Z.M. Dang, T. Zhou, S.H. Yao, J.K. Yuan, J.W. Zha, H.T. Song, J.Y. Li, Q. Chen, W.T. Yang, and J. Bai (2009) Advanced Calcium Copper Titanate/Polyimide Functional Hybrid Films with High Dielectric Permittivity, *Advanced Materials* **21**, 2077–82.

[118] S.E. Ahn, M.J. Lee, Y. Park, B.S. Kang, C.B. Lee, K.H. Kim, S. Seo, D. S. Suh, D.C. Kim, J. Hur, W. Xianyu, G. Stefanovich, H.A. Yin, I.K. Yoo, A.H. Lee, J.B. Park, I.G. Baek, and B.H. Park (2008) Write current reduction in transition metal oxide based resistance-change memory, *Advanced Materials* **20**, 924–8.

[119] S. Guillemet-Fritsch, Z. Valdez-Nava, C. Tenailleau, T. Lebey, B. Durand, and J. Y. Chane-Ching (2008) Colossal permittivity in ultrafine grain size $BaTiO_{3-x}$ and $Ba_{0.95}La_{0.05}TiO_{3-x}$ materials, *Advanced Materials* **20**, 551–5.

[120] M. Mushrush, A. Facchetti, M. Lefenfeld, H.E. Katz, and T.J. Marks (2003) Easily processable phenylene-thiophene-based organic field-effect transistors and solution-fabricated nonvolatile transistor memory elements, *Journal of the American Chemical Society* **125**, 9414–23.

[121] L. Pereira, P. Barquinha, E. Fortunato, R. Martins, D. Kang, C.J. Kim, H. Lim, I. Song, and Y. Park (2008) High k dielectrics for low temperature electronics, *Thin Solid Films* **516**, 1544–8.

[122] L. Pereira, P. Barquinha, G. Goncalves, A. Vila, A. Olziersky, J. Morante, E. Fortunato, and R. Martins (2009) Sputtered multicomponent amorphous dielectrics for transparent electronics, *Physica Status Solidi a-Applications and Materials Science* **206**, 2149–54.

[123] S. Moller, C. Perlov, W. Jackson, C. Taussig, and S.R. Forrest (2003) A polymer/semiconductor write-once read-many-times memory, *Nature* **426**, 166–9.

[124] J.E. Green, J.W. Choi, A. Boukai, Y. Bunimovich, E. Johnston-Halperin, E. DeIonno, Y. Luo, B.A. Sheriff, K. Xu, Y. S. Shin, H.R. Tseng, J.F. Stoddart, and J.R. Heath (2007) A 160-kilobit molecular electronic memory patterned at 10(11) bits per square centimetre, *Nature* **445**, 414–17.

[125] J. Shen, C.Y. Zhang, and Q. Chen (2011) Resistive switching of crossbar memories with carbon nanotube electrodes, *Physica Status Solidi-Rapid Research Letters* **5**, 205–7.

[126] Y. Jung, R. Agarwal, C.Y. Yang, and R. Agarwal (2011) Chalcogenide phase-change memory nanotubes for lower writing current operation, *Nanotechnology* **22**, 254012.

[127] O.A. Tretiakov, Y. Liu, and A. Abanov (2011) Power optimization for domain wall motion in ferromagnetic nanowires, *Journal of Applied Physics* **109**, 07D505.

[128] X.X. Zhu, Q.L. Li, D.E. Ioannou, D.F. Gu, J.E. Bonevich, H. Baumgart, J.S. Suehle, and C.A. Richter (2011) Fabrication, characterization and simulation of high performance Si nanowire-based non-volatile memory cells, *Nanotechnology* **22**, 254020.

[129] C.C. Chang, Z.W. Pei, and Y.J. Chan (2008) Artificial electrical dipole in polymer multilayers for nonvolatile thin film transistor memory, *Applied Physics Letters* **93**, 143302.

[130] X. Cao, X.M. Li, X.D. Gao, X.J. Liu, C. Yang, R. Yang, and P. Jin (2011) All-ZnO-based transparent resistance random access memory device fully fabricated at room temperature, *Journal of Physics D-Applied Physics* **44**, 255104.

[131] T. Kawahara (2011) Challenges toward gigabit-scale spin-transfer torque random access memory and beyond for normally off, green information technology infrastructure (invited), *Journal of Applied Physics* **109**, 07D325.

[132] D. Lencer, M. Salinga, and M. Wuttig (2011) Design Rules for Phase-Change Materials in Data Storage Applications, *Advanced Materials* **23**, 2030–58.

[133] M.A. Dehkordi, A.S. Shamsabadi, B.S. Ghahfarokhi, and A. Vafaei (2011) Novel RAM cell designs based on inherent capabilities of quantum-dot cellular automata, *Microelectronics Journal* **42**, 701–8.

[134] X.Y. Deng and K.Y. Wong (2011) Polymeric memory device with dual electrical and optical reading modes, *Optical Engineering* **50**, 044003.
[135] Y. Zhang, S.H. Lee, A. Mascarenhas, and S.K. Deb (2008) An UV photochromic memory effect in proton-based WO3 electrochromic devices, *Applied Physics Letters* **93**, 203508.
[136] Y. Sanehira, S. Uchida, T. Kubo, and H. Segawa (2008) A distinguished retentive memory using polyethylene glycol electrolyte solvent for viologen modified titania electrochromic device, *Electrochemistry* **76**, 150–3.
[137] Y.C. Nah, K.S. Ahn, and Y.E. Sung (2003) Effects of tantalum oxide films on stability and optical memory in electrochromic tungsten oxide films, *Solid State Ionics* **165**, 229–33.
[138] S. Moller, S.R. Forrest, C. Perlov, W. Jackson, and C. Taussig (2003) Electrochromic conductive polymer fuses for hybrid organic/inorganic semiconductor memories, *Journal of Applied Physics* **94**, 7811–19.
[139] B. Fabre (2010) Ferrocene-Terminated Monolayers Covalently Bound to Hydrogen-Terminated Silicon Surfaces. Toward the Development of Charge Storage and Communication Devices, *Accounts of Chemical Research* **43**, 1509–18.
[140] B.C. Das and A.J. Pal (2009) Enhancement of electrical bistability through semiconducting nanoparticles for organic memory applications, *Philosophical Transactions of the Royal Society a-Mathematical Physical and Engineering Sciences* **367**, 4181–90.
[141] S. Barman, F.J. Deng, and R.L. McCreery (2008) Conducting polymer memory devices based on dynamic doping, *Journal of the American Chemical Society* **130**, 11073–81.
[142] R.L. McCreery (2004) Molecular electronic junctions, *Chemistry of Materials* **16**, 4477–96.
[143] A.K. Rath and A.J. Pal (2007) Conductance switching in an organic material: From bulk to monolayer, *Langmuir* **23**, 9831–5.
[144] Y. Yang, J. Ouyang, L.P. Ma, R.J.H. Tseng, and C.W. Chu (2006) Electrical switching and bistability in organic/polymeric thin films and memory devices, *Advanced Functional Materials* **16**, 1001–14.
[145] C. Novembre, D. Guerin, K. Lmimouni, C. Gamrat, and D. Vuillaume (2008) Gold nanoparticle-pentacene memory transistors, *Applied Physics Letters* **92**, 103314.
[146] C.H. Cheng, F.S. Yeh, and A. Chin (2011) Low-Power High-Performance Non-Volatile Memory on a Flexible Substrate with Excellent Endurance, *Advanced Materials* **23**, 902–5.
[147] B.H. Park, B.S. Kang, S.D. Bu, T.W. Noh, J. Lee, and W. Jo (1999) Lanthanum-substituted bismuth titanate for use in non-volatile memories, *Nature* **401**, 682–4.
[148] R.C.G. Naber, B. de Boer, P.W.M. Blom, and D.M. de Leeuw (2005) Low-voltage polymer field-effect transistors for nonvolatile memories, *Applied Physics Letters* **87**, 203509.
[149] Q.M. Zhang, H.F. Li, M. Poh, F. Xia, Z.Y. Cheng, H.S. Xu, and C. Huang (2002) An all-organic composite actuator material with a high dielectric constant, *Nature* **419**, 284–7.
[150] W. Lehmann, H. Skupin, C. Tolksdorf, E. Gebhard, R. Zentel, P. Kruger, M. Losche, and F. Kremer (2001) Giant lateral electrostriction in ferroelectric liquid-crystalline elastomers, *Nature* **410**, 447–50.
[151] K.H. Lee, G. Lee, K. Lee, M.S. Oh, S. Im, and S.M. Yoon (2009) High-mobility nonvolatile memory thin-film transistors with a ferroelectric polymer interfacing ZnO and pentacene channels, *Advanced Materials* **21**, 4287–91.
[152] R.C.G. Naber, P.W.M. Blom, G.H. Gelinck, A.W. Marsman, and D.M. de Leeuw (2005) An organic field-effect transistor with programmable polarity, *Advanced Materials* **17**, 2692–5.
[153] Z.M. Dang, Y.H. Lin, and C.W. Nan (2003) Novel ferroelectric polymer composites with high dielectric constants, *Advanced Materials* **15**, 1625–9.
[154] R. Martins, P. Barquinha, A. Pimentel, L. Pereira, E. Fortunato, D. Kang, I. Song, C. Kim, J. Park, and Y. Park (2008) Electron transport in single and multicomponent n-type oxide semiconductors, *Thin Solid Films* **516**, 1322–5.
[155] E. Fortunato, R. Martins, L. Pereira, P. Barquinha, N. Correia, "Use of cellulose and bio-organic based fibbers simultaneously as physical support and charged memory component in field effect electronic devices": EU (PPE 42416/10); (PAT 40050/09); (PTI-MX 42418/1); (PTI-BR 42419/1); (PTI-US 42420/1); (PTI-CA 42421/1); Russia (PTI-RU 42422/1); (PTI-IN 42423/1); (PTI-CN 42424/1); (PTI-JP 42425/1); (PTI-KR 42426/1); (PTI-AU 42427/1).

[156] J. Sun, Q. Wan, A.X. Lu, and J. Jiang (2009) Low-voltage electric-double-layer paper transistors gated by microporous SiO2 processed at room temperature, *Applied Physics Letters* **95**, 222108.
[157] A.X. Lu, J. Sun, J. Jiang, and Q. Wan (2009) Microporous SiO_2 with huge electric-double-layer capacitance for low-voltage indium tin oxide thin-film transistors, *Applied Physics Letters* **95**, 222905.
[158] J. Jiang, Q. Wan, J. Sun, and A.X. Lu (2009) Ultralow-voltage transparent electric-double-layer thin-film transistors processed at room-temperature, *Applied Physics Letters* **95**, 152114.
[159] Semiconductor Industry Association, Austin (2005), in *International technology roadmap for semiconductors (ITRS), Emerging research devices*.
[160] G.H. Gelinck, A.W. Marsman, F.J. Touwslager, S. Setayesh, D.M. de Leeuw, R.C.G. Naber, and P.W.M. Blom (2005) All-polymer ferroelectric transistors, *Applied Physics Letters* **87**, 092903.
[161] N. Weste and K. Eshraghian (1993) *Principles of CMOS VLSI Desing: A Systems Prespective*, 2nd edn, New York: Addison-Wesley.
[162] H. Klauk, U. Zschieschang, J. Pflaum, and M. Halik (2007) Ultralow-power organic complementary circuits, *Nature* **445**, 745–8.
[163] D.R. Hines, V.W. Ballarotto, E.D. Williams, Y. Shao, and S.A. Solin (2007) Transfer printing methods for the fabrication of flexible organic electronics, *Journal of Applied Physics* **101**, 024503.
[164] A.L. Briseno, S.C.B. Mannsfeld, M.M. Ling, S.H. Liu, R.J. Tseng, C. Reese, M.E. Roberts, Y. Yang, F. Wudl, and Z.N. Bao (2006) Patterning organic single-crystal transistor arrays, *Nature* **444**, 913–17.
[165] S.P. Cummings, J. Savchenko, and T. Ren (2011) Functionalization of flat Si surfaces with inorganic compounds-Towards molecular CMOS hybrid devices, *Coordination Chemistry Reviews* **255**, 1587–1602.
[166] M.J. An, H.S. Seo, Y. Zhang, J.D. Oh, and J.H. Choi (2011) Air-stable, hysteresis-free organic complementary inverters produced by the neutral cluster beam deposition method, *Journal of Physical Chemistry C* **115**, 11763–7.
[167] W.S.C. Roelofs, S.G.J. Mathijssen, J.C. Bijleveld, D. Raiteri, T.C.T. Geuns, M. Kemerink, E. Cantatore, R.A.J. Janssen, and D.M. de Leeuw (2011) Fast ambipolar integrated circuits with poly(diketopyrrolopyrrole-terthiophene), *Applied Physics Letters* **98**, 203301.
[168] A.G. Ismail and I.G. Hill (2011) Stability of n-channel organic thin-film transistors using oxide, SAM-modified oxide and polymeric gate dielectrics, *Organic Electronics* **12**, 1033–42.
[169] L.L. Chua, J. Zaumseil, J.F. Chang, E.C.W. Ou, P.K.H. Ho, H. Sirringhaus, and R.H. Friend (2005) General observation of n-type field-effect behaviour in organic semiconductors, *Nature* **434**, 194–9.
[170] S. Gowrisanker, M.A. Quevedo-Lopez, H.N. Alshareef, B.E. Gnade, S. Venugopal, R. Krishna, K. Kaftanoglu, and D.R. Allee (2009) A novel low temperature integration of hybrid CMOS devices on flexible substrates, *Organic Electronics* **10**, 1217–22.
[171] C. Melzer and H. von Seggern (2010) Organic field-effect transistors for CMOS devices, in *Organic Electronics* **223**, 213–57.
[172] J.C. Ribierre, T. Fujihara, T. Muto, and T. Aoyama (2010) Patterning by laser annealing of complementary inverters based on a solution-processible ambipolar quinoidal oligothiophene, *Organic Electronics* **11**, 1469–75.
[173] W.Y. Chou, B.L. Yeh, H.L. Cheng, B.Y. Sun, Y.C. Cheng, Y.S. Lin, S.J. Liu, F.C. Tang, and C.C. Chang (2009) Organic complementary inverters with polyimide films as the surface modification of dielectrics, *Organic Electronics* **10**, 1001–5.
[174] L. Wang, M.H. Yoon, G. Lu, Y. Yang, A. Facchetti, and T.J. Marks (2006) High-performance transparent inorganic-organic hybrid thin-film n-type transistors, *Nature Materials* **5**, 893–900.
[175] I.C. Cheng and S. Wagner (2002) Hole and electron field-effect mobilities in nanocrystalline silicon deposited at 150 degrees C, *Applied Physics Letters* **80**, 440–2.
[176] C.H. Lee, A. Sazonov, A. Nathan, and J. Robertson (2006) Directly deposited nanocrystalline silicon thin-film transistors with ultra high mobilities, *Applied Physics Letters* **89**, 252101.

[177] H. Hosono, K. Nomura, Y. Ogo, T. Uruga, and T. Kamiya (2008) Factors controlling electron transport properties in transparent amorphous oxide semiconductors, *Journal of Non-Crystalline Solids* **354**, 2796–2800.

[178] Y. Ogo, H. Hiramatsu, K. Nomura, H. Yanagi, T. Kamiya, M. Hirano, and H. Hosono (2008) p-channel thin-film transistor using p-type oxide semiconductor, SnO, *Applied Physics Letters* **93**, 032113.

[179] E. Fortunato, R. Barros, P. Barquinha, V. Figueiredo, S.H.K. Park, C.S. Hwang, and R. Martins (2010) Transparent p-type SnOx thin film transistors produced by reactive rf magnetron sputtering followed by low temperature annealing, *Applied Physics Letters* **97**, 052105.

[180] E. Fortunato, V. Figueiredo, P. Barquinha, E. Elamurugu, R. Barros, G. Goncalves, S.H.K. Park, C.S. Hwang, and R. Martins (2010) Thin-film transistors based on p-type Cu_2O thin films produced at room temperature, *Applied Physics Letters* **96**, 192102.

[181] K. Nomura, T. Kamiya, H. Hosono, and D. (2011) Article first published online: 1 JUL 2011, Ambipolar Oxide Thin-Film Transistor, *Advanced Materials* **23**, 3431–4.

[182] R. Martins, E. Fortunato, R. Barros, L. Pereira, P.P. Barquinha, N. Correia, R. Costa, A. Ahnood, I. Ferreira, and A. Nathan (2011) Complementary Metal Oxide Semiconductor Technology with and on Paper, *Advanced Materials* **23**, 4491–6.

[183] J. Uyemura (1988) *Fundamentals of MOS Digital Integrated Circuits*. New York: Addison-Wesley Publishing company.

[184] P.J. Simons, M. Spiro, and J.F. Levy (1998) Electrical properties of wood – Determination of ionic transference numbers and electroosmotic water flow in Pinus sylvestris L. (Scots pine), *Journal of the Chemical Society-Faraday Transactions* **94**, 223–6.

[185] W.K. Henson, N. Yang, S. Kubicek, E.M. Vogel, J.J. Wortman, K. De Meyer, and A. Naem (2000) Analysis of leakage currents and impact on off-state power consumption for CMOS technology in the 100-nm regime, *Ieee Transactions on Electron Devices* **47**, 1393–1400.

[186] N.S. Kim, T. Austin, D. Blaauw, T. Mudge, F. Krisztian, J.S. Hu, M.J. Irwin, M. Kandemir, and V. Narayanan (2003) Leakage current: Moore's law meets static power, *Computer* **36**, 68–75.

[187] P. Servati and A. Nathan (2002) Modeling of the static and dynamic behavior of hydrogenated amorphous silicon thin-film transistors, *Journal of Vacuum Science & Technology a-Vacuum Surfaces and Films* **20**, 1038–42.

[188] J.H. Christie, I.M. Woodhead, S. Krenek, and J.R. Sedcole (2002) A new model of DC conductivity of hygroscopic solids - Part II: Wool and silk, *Textile Research Journal* **72**, 303–8.

[189] J.H. Christie and I.M. Woodhead (2002) A new model of DC conductivity of hygroscopic solids - Part 1: Cellulosic materials, *Textile Research Journal* **72**, 273–8.

[190] E.H. Snow, A.S. Grove, B.E. Deal, and C.T. Sah (1965) Ion Transport Phenomena in insulating films, *Journal of Applied Physics* **36**, 1664.

[191] K. Shimizu, H. Habazaki, P. Skeldon, G.E. Thompson, and G.C. Wood (2000) Migration of sulphate ions in anodic alumina, *Electrochimica Acta* **45**, 1805–9.

[192] J.O. Bockris and L.V. Minevski (1993) On the mechanism of the passivity of aluminium and aluminium-alloys, *Journal of Electroanalytical Chemistry* **349**, 375–414.

[193] L.F. Lin, C.Y. Chao, and D.D. Macdonald (1981) A point-defect model for anodic passive films. 2. Chemical breakdown and pit initiation, *Journal of the Electrochemical Society* **128**, 1194–8.

[194] D. Nilsson, R. Forchheimer, M. Berggren, and N. Robinson (2005) "The electrochemical transistor and circuit design considerations," in *Proceedings of the 2005 European Conference on Circuit Theory and Design, Vol 3*, F. Oregan and C. Wegemer, eds, New York: Ieee, 349–52.

[195] M.X. Chen (2005) Printed electrochemical devices using conducting polymers as active materials on flexible substrates, *Proceedings of the Ieee* **93**, 1339–47.

[196] P. Andersson, R. Forchheimer, P. Tehrani, and M. Berggren (2007) Printable all-organic electrochromic active-matrix displays, *Advanced Functional Materials* **17**, 3074–82.

[197] P. Tehrani, L.O. Hennerdal, A.L. Dyer, J.R. Reynolds, and M. Berggren (2009) Improving the contrast of all-printed electrochromic polymer on paper displays, *J. Mater. Chem.* **19**, 1799–1802.

[198] B.L. Groenendaal, F. Jonas, D. Freitag, H. Pielartzik, and J.R. Reynolds (2000) Poly(3,4-ethylenedioxythiophene) and its derivatives: past, present, and future, *Advanced Materials* **12**, 481–94.

[199] A.J. Gimenez, J.M. Yanez-Limon, and J.M. Seminario (2011) ZnO-Paper Based Photoconductive UV Sensor, *Journal of Physical Chemistry C* **115,** 282–7.
[200] R. Martins, E. Fortunato, P. Nunes, I. Ferreira, A. Marques, M. Bender, N. Katsarakis, V. Cimalla, and G. Kiriakidis (2004) Zinc oxide as an ozone sensor, *Journal of Applied Physics* **96**, 1398–1408.
[201] L.D. Woolf and H.H. Streckert (1996) Graphite pencil line for exploring resistance, *Phys. Teach.* **34**, 440–1.
[202] G. Goncalves, A. Pimentel, E. Fortunato, R. Martins, E.L. Queiroz, R.F. Bianchi, and R.M. Faria (2006) UV and ozone influence on the conductivity of ZnO thin films, *Journal of Non-Crystalline Solids* **352**, 1444–7.
[203] A. Pimentel, A. Goncalves, A. Marques, R. Martins, and E. Fortunato (2006) Role of the thickness on the electrical and optical performances of undoped polycrystalline zinc oxide films used as UV detectors, *Journal of Non-Crystalline Solids* **352**, 1448–52.

7

A Glance at Current and Upcoming Applications

7.1 Introduction: Emerging Areas for (Non-)transparent Electronics Based on Oxide Semiconductors

The study and optimization of oxide conductors, semiconductors and dielectrics and their integration in oxide TFTs is only the first step in the field of transparent electronics. In order to show the definitive added-value of a certain material technology for electronics, one that is useful either to replace current materials/devices or to create novel application areas, electronic circuits or other products of interest need to be demonstrated.

Transparency is naturally one of the key properties differentiating oxides from other semiconductor materials, being one of the motivations for using oxides. Nevertheless, oxide semiconductors exhibit many other great advantages, such as low cost, low temperature processing and remarkable electrical properties, making them also desirable for a large range of applications where transparency is not strictly required.

The aim of this chapter is to present the potential of oxides in a variety of current and novel applications, where oxides rival or exceed the functionality, reliability and production costs of the materials currently used in the semiconductor industry, based mainly on silicon technology. In addition, one must consider the great relevance that complex transparent circuits will have in areas such as security and control, where "invisibility" is an issue. For instance, if one aims to control the interconnectivity of a power network circuit so as to reduce losses due to malfunctioning, this could be done using invisible electronics embedded in front-protective glasses, with little impact in terms of costs on the systems' assembly. In this case we are looking at future power electronics, where transparent diodes will also be required.

Transparent Oxide Electronics: From Materials to Devices, First Edition. Pedro Barquinha, Rodrigo Martins, Luis Pereira and Elvira Fortunato.

This chapter illustrates briefly some of the most relevant current and upcoming applications of oxide semiconductors and oxide TFTs, tackling market needs and potential growth, namely displays, electronic circuits, oxide heterojunctions and biosensors, presenting the prototypes fabricated at CENIMAT. Many other applications are left to the imagination of our readers.

7.2 Active Matrices for Displays

7.2.1 Display Market Overview and Future Trends

Displays constitute one of the most interesting markets in the electronics area, both technologically and economically. Nowadays, flat panel displays (FPDs) dominate the market, with FPD shipments representing more than 90 000 million US dollars (MUSD), with 125 000 MUSD expected by 2015 [1]. Liquid crystal displays (LCDs) are currently the most relevant FPDs. These offer immediate advantages over the old cathode ray tubes (CRTs) in terms of power consumption, weight and space, allowing to one explore two parallel markets: on one hand, large area TV sets, and on the other integration of displays in a variety of portable applications, such as cell phones and notebooks, something that was unthinkable with the bulky and heavy CRT technology. Recently, in 2007–2008, LCD shipments began to supplant CRTs [2] and LCDs are increasingly becoming the dominant FPD technology even for large size screens (more than 40″), a market which until 2006 was dominated by another FPD technology, plasma displays (PDPs) [3]. The continuous development in the area of displays does not show any signs of slowing down with a new technology emerging at the beginning of the millennium, which is expected to surpass LCDs in terms of market share over the coming decades: organic light emitting devices (OLED) displays (Figure 7.1, see color plate section).

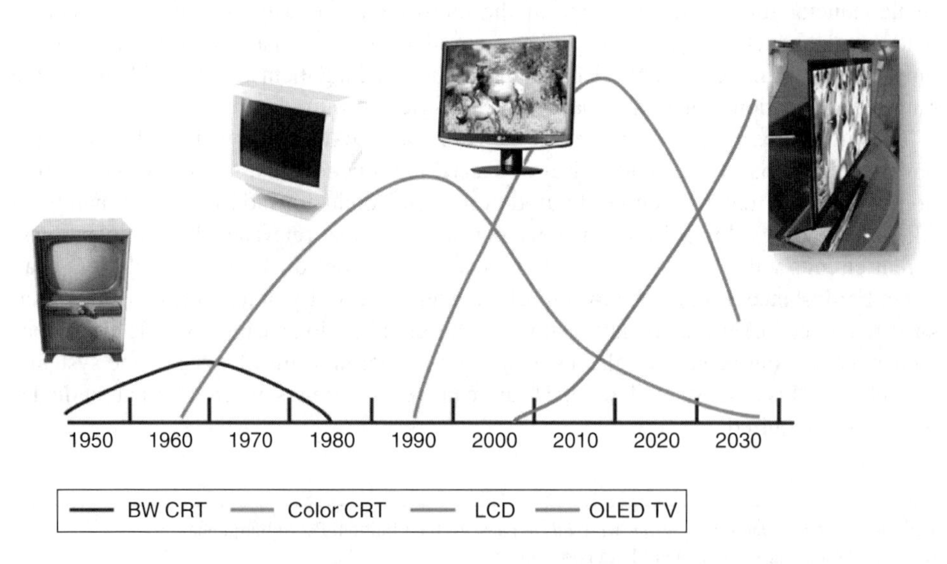

Figure 7.1 Life cycle forecast of the most relevant display technologies (adapted from Display Bank). Adapted with permission from [7] Copyright (2011) DisplayBank.

Two new trends are emerging for future displays, enabled by frontplane technologies such as OLEDs and electrophoretic displays [4], together with low-temperature backplane technologies such as organic and oxide TFTs:

- Flexible displays, which are predicted to expand by a factor of 35 from 2007 to 2013. It is foreseen that by that time revenues of 2800 MUSD will be achieved [5], increasing to 12 000 MUSD by 2017 [6];
- Transparent (and flexible) displays, an area closely related to oxide TFTs that is now making its debut, with forecasted revenues of $87.2 billion by 2025 (see Chapter 1, Figure 1.2) [7].

7.2.2 Driving Schemes and TFT Requirements for LCD and OLED Displays

Given that LCDs and OLEDs are the dominant frontplane technologies, this section provides a generic overview of the schemes and transistor requirements for these displays.

A LCD is an example of a non-emissive display, because liquid crystals do not emit light, but rather work as independent light switches in each pixel. Liquid crystals were discovered more than a century ago, but their useful electro-optic effects and stability were only developed during the 1960s and 70s [8]. Regarding their usage in displays, liquid crystals can be seen as light modulators, since their function is to allow or block light transmission. This is achieved by the reorientation of the liquid crystal director by an applied external voltage [8].

Two driving techniques can be distinguished for LCDs: passive and active matrices (PM and AM, respectively). PMs consist of simple x–y electrodes in a stripe configuration, with those on top of the liquid crystal aligned perpendicularly to those on the bottom and each cross section constituting a pixel. This is a simple addressing scheme and was used in the first LCDs. However, PMs have severe limitations on the maximum display size and resolution achievable and suffer from large crosstalk (i.e., the electrical signal applied to a line has an undesirable effect on neighboring lines, which is especially critical in displays with large pixel densities and LCDs in particular due to the capacitor characteristics of the liquid crystal layer) [8, 9]. AMs solve these problems, by employing an array of TFTs in a backplane (i.e., below the liquid crystal), with each pixel being controlled independently by a TFT,[a] while an unpatterned transparent electrode is placed on the top of the liquid crystal. AMs are nowadays used in all the high resolution LCDs, with both small and large areas. Figure 7.2a shows a schematic of the electronic circuit of an AM LCD pixel. The presence of the storage capacitor improves the retention characteristics, allowing one to bias the liquid crystal pixel even if the corresponding scan line is not selected [8].

In AM LCDs, the TFTs act as switches, their main requirements being large *on/off* ratios and close to 0 threshold voltage (V_T), in order to allow for high image contrast, to simplify circuit design and to minimize power dissipation. a-Si:H, poly-Si, organic and oxide TFT technologies can fulfill these requirements, at least for the current displays with small-to-moderate sizes and resolutions. However, for future LCDs, which will have even larger sizes, resolutions and frame rates, field-effect mobility (μ_{FE}) starts to become an important parameter to consider and can hinder the usage of a-Si:H or organic TFTs. In the case of an ultra definition (UD, or 4k × 2k pixels) AM LCD, a-Si:H TFTs enable operation of a 82″

[a] In a color LCD, each pixel has at least three subpixels, each one with a TFT, in order to obtain red, green and blue colors and their combinations.

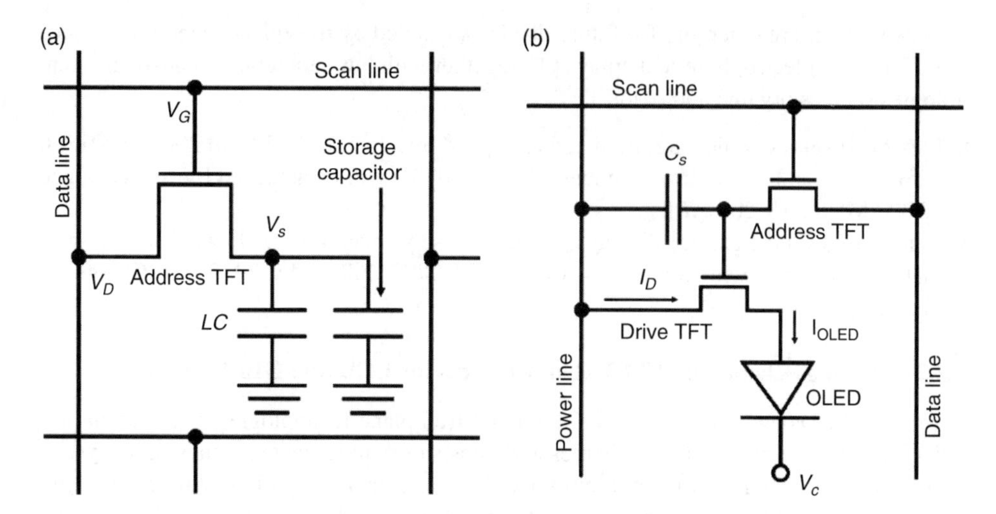

Figure 7.2 *Schematics of the pixel circuit for a) LCD and b) OLED displays.*

display at only 60Hz. To achieve higher frequencies (120 Hz or above) μ_{FE}> 3 cm^2 V^{-1} s^{-1} is required, because of the need to charge the storage capacitor with shorter pulse time [10].

In contrast to LCDs, OLED displays are emissive, because OLEDs emit light. This process is achieved by the recombination of holes and electrons, injected from an anode and a cathode, with the bandgap of the organic semiconductors determining the emission wavelength [8]. Electroluminescence in organic materials was initially observed in the 1950s by Bernanose and co-workers, but the first OLED was only invented in the late 1980s by Tang and Vanslyke, for Kodak [11, 12]. Since OLED displays are emissive they do not need any backlight, allowing these displays to exhibit much higher contrast ratios than LCDs, because in the *off*-state the pixel is completely dark. Besides this, the structure of the display is considerably simpler, which will contribute so as to reduce the production costs once this technology is widely implemented. In brief, the main advantages of OLED displays as compared to LCDs are higher contrast ratios, lower response times, wider viewing angle, lower display thickness and lower power consumption [13].

As with LCDs, PMs and AMs can be used as driving techniques in OLED displays. PMs suffer from the same typical problems as in LCDs, such as crosstalk and limitations on screen size and resolution. Given this, AMs are also preferred as driving schemes for OLEDs. However, the electronic circuit of an AM OLED pixel is more complicated than an AM LCD (Figure 7.2b). Here, at least two TFTs need to be used, one as a switching element (address TFT), the other as a current driver to switch on the OLED (drive TFT). For the drive TFT a high current density is required. Although some a-Si:H TFT architectures can provide this, very large W/L and high supply voltages would have to be used, which results in low aperture ratios and high power consumption [13, 14]. A considerably better solution, employed in some of the current AM OLEDs, is to use poly-Si TFTs: since μ_{FE} is much higher than in a-Si:H TFTs, the required current densities can be achieved using smaller devices and lower operating voltages. However, the uniformity problems in large areas can be a problem for the success of poly-Si TFTs in future OLED displays. These problems, as well as those related to stability, can be attenuated by using pixel structures with more than two TFTs (compensation circuits), but this increases production costs and decreases the aperture ratio. Due to all of this, oxide TFTs appear to be potential candidates for future AM OLED displays: besides having considerably larger μ_{FE} than a-Si:H TFTs, oxide TFTs can be made larger without compromising the aperture ratio, due to their transparency. Furthermore, the fact that they can be produced at very low temperatures fits, together with the OLED frontplane, the needs for flexible electronics.

7.2.3 Displays with Oxide-Based Backplanes

Given the close relation between TFTs and displays, it is not strange that the most significant application of oxide TFTs is in displays, mostly LCDs and OLEDs. From 2007, the Society for Information Display (SID) and International Meeting on Information Display (IMID) exhibitions have been the stage for the presentation of several prototypes employing oxide TFT technology.

At CENIMAT, oxides have been used both for PM and AM backplanes.[b] The PM backplane was produced using room-temperature sputtered IZO [15], patterned using conventional photolithographic processes, on 5×3 cm glass and polymeric substrates. Transparent alphanumeric displays with seven segments, usable as head-up displays (HUDs) in the automotive industry, were obtained by integration with chipLED frontplane technology (Figure 7.3, see color plate section).

AM backplanes for 2.8″ LCDs were fabricated on glass, consisting of an array of 128×128 pixels, with each pixel having an approximate area of 350×350 μm (Figure 7.4a and b, see color plate section) [16]. For demonstration purposes, the architecture was the simplest possible, with a single TFT per pixel. Backplane production consisted of a standard five mask process, using either Ti/Au or IZO source-drain and gate electrodes (for opaque and transparent AMs, respectively), GIZO channel layer, TSiO dielectric and SU-8 passivation. The process flow was similar to the one presented in Figure 5.35, with the final backplanes being annealed at 150°C. The AM backplanes were then successfully integrated with reflective LCD frontplanes (Figure 7.4c). However, evidence of some shorted gates were observed and attributed to the lack of reliability and reproducibility of the TSiO dielectric, as mentioned in section 5.2.6.1. Improved yield should be obtained by using multilayer dielectrics.

Several companies are currently working on oxide TFTs for demonstrating flexible and/or transparent displays. Some examples, all driven by GIZO TFTs, are those from Samsung SDI, with a 4.1″ 176×220 resolution full-color OLED display, having a transmittance higher than 20% [17]; from LG Electronics Inc., with a 3.5″ 176×220 resolution OLED display on a 0.1 mm thick stainless steel plate [18]; from Toppan Printing Inc., with 4″ 320×240 resolution electrophoretic display on polyethylene naphthalate (PEN) plastic substrate (Figure 7.5a, see color plate section) [19]. With atomic layer deposited ZnO TFTs, Park *et al.* also reported a 2.5″ 220×176 resolution OLED transparent display on a glass substrate [20].

[b] The integration of the PM and AM backplanes presented here with frontplane technologies was done at Centro Ricerche Fiat (CRF-Italy) and Hewlett Packard (HP-Ireland), respectively, under the framework of Multiflexioxides project.

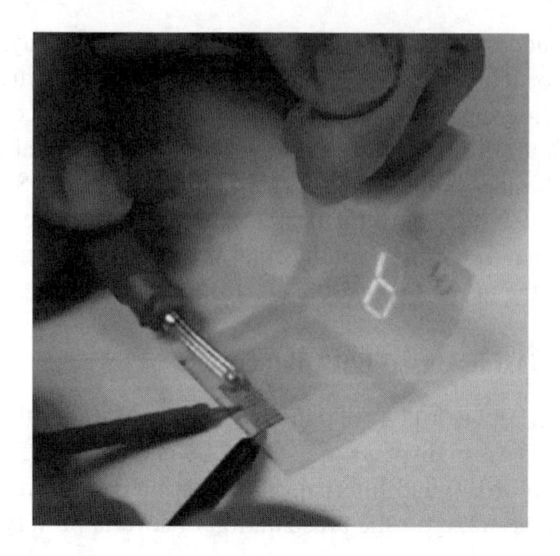

Figure 7.3 *Transparent PM chipLED display using IZO electrodes.*

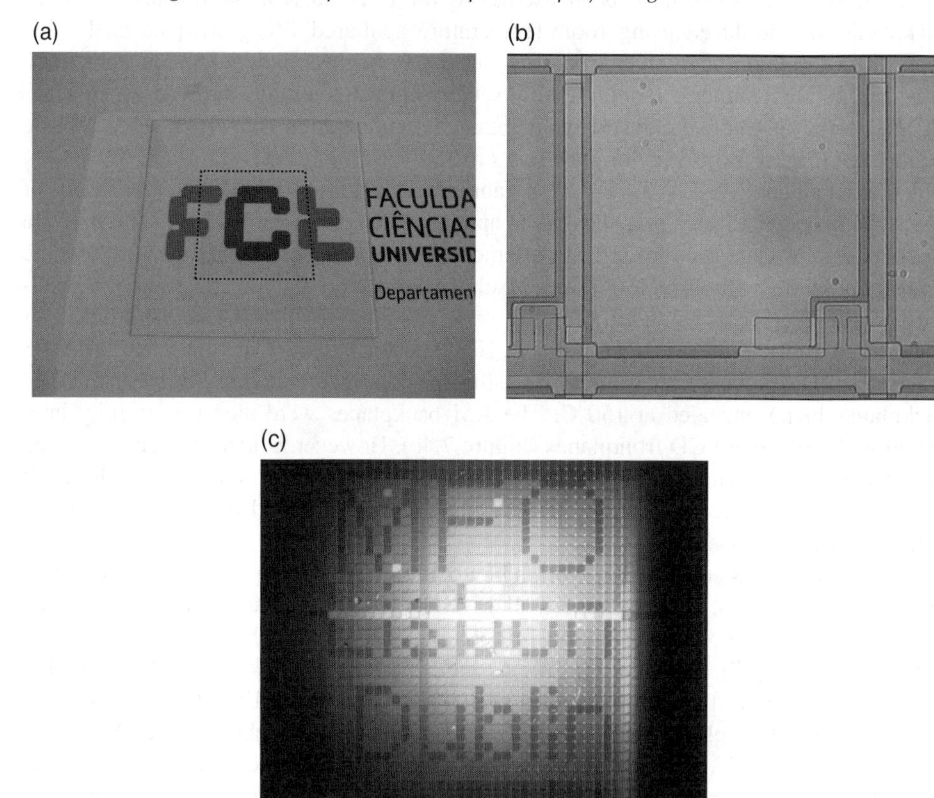

Figure 7.4 *AM backplane with GIZO TFTs produced at CENIMAT for a LCD display: a) photograph of a transparent AM on glass; b) micrograph of pixel area (350×350 μm); c) prototype after integration with LCD frontplane. Reprinted with permission from [16] Copyright (2009) Electrochemical Society.*

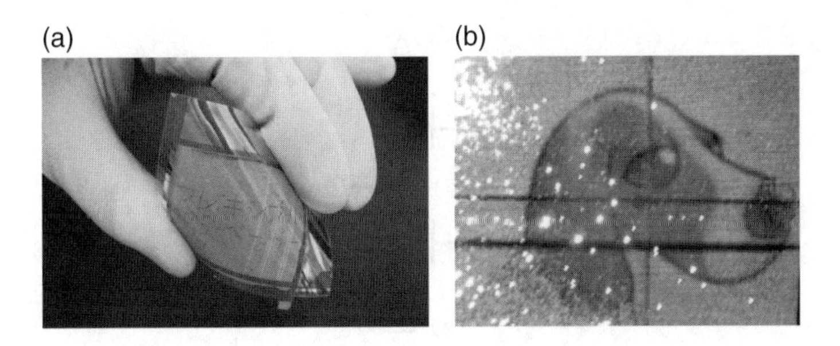

Figure 7.5 Prototypes of displays with oxide TFT backplanes: a) flexible electrophoretic display. Reprinted with permission from [19] Copyright (2008) Wiley-VCH; b) OLED display driven by solution-processed IZO TFTs [23]. Reprinted from [23] Copyright (2009) Society for Information Display.

However, even when transparency of the overall display is not persuaded, oxide TFTs are of great interest. Their high μ_{FE} and large area uniformity make it possible to have high-end displays, such as the 70″ UD LCD 3DTV with a high scanning frequency (240 Hz) demonstrated by Samsung in 2010 [21].

In most of the reported displays employing oxide TFTs, the channel layer is produced by sputtering. However, despite the early stages of research on solution-processed oxide TFTs, LCD and OLED displays are already being fabricated using this lower cost processing technology. In 2008, Chiang *et al.* reported sol-gel $Zn_{0.97}Zr_{0.03}O$ TFTs driving a 4.1″ LCD display [22]. The first demonstration with OLEDs is attributed to Samsung Electronics, with a 2.2″ monochromatic display with 128×160 resolution (Figure 7.5b), being the spin-coated IZO TFTs processed at a maximum temperature of 350°C [23]. Other works followed, with increased panel sizes and resolution, such as the 4.1″ LCD display with 320×240 resolution driven by spin-coated GIZO TFTs, as reported by the Taiwan TFT LCD Association (TTLA)/Inpria Corp./Oregon State University [24].

The stage is now set for a variety of displays using oxide TFT technology, with small area flexible OLEDs and large area rigid high-end LCD and OLED displays being the probable main targets for commercial products in the near future. The numerous prototypes already shown by most of the great display companies are certainly an indication that products within the "transparent electronics" concept, or better, employing oxide semiconductors but not necessarily resulting in transparent displays, are not far from entering in our daily lives. In fact, as pointed out in Chapter 1, mass-production of the first transparent 22″ LCD panels was announced by Samsung in 2011. These displays exhibit a 1680×1050 resolution, a contrast ratio of 500:1 and a transparency of 15 %. Instead of a backlight, ambient light is used, resulting in 90 % less power consumption than conventional LCDs [25].

7.3 Transparent Circuits

7.3.1 Inverters and Ring Oscillators

For any semiconductor technology, the fabrication of inverters and ring oscillators (ROs) is a basic requirement for producing a complete range of logic circuits. A simple inverter is

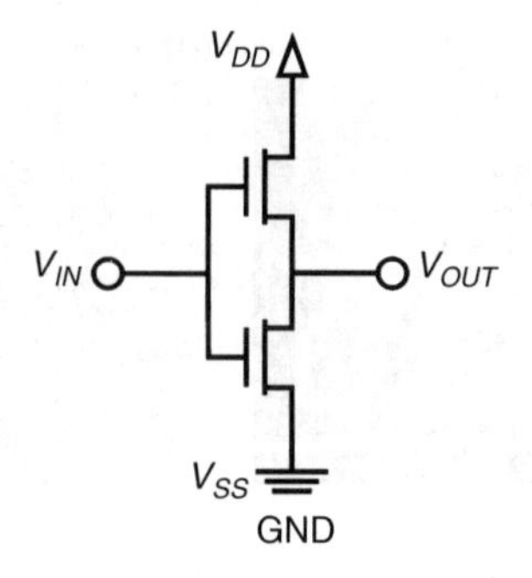

Figure 7.6 *Schematic of a nMOS inverter.*

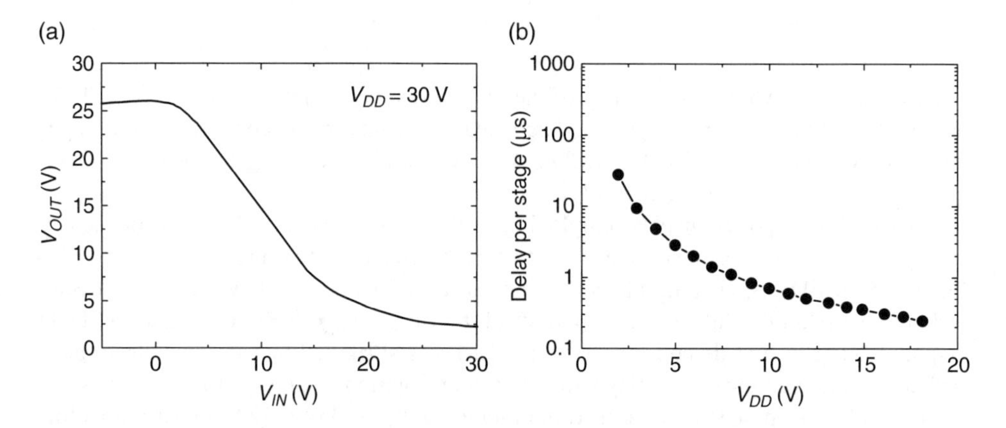

Figure 7.7 *Electrical properties of initial oxide TFT based circuits: a) VTC for a transparent inverter using IGO TFTs. Adapted with permission from [27] Copyright (2006) Elsevier Ltd; b) Propagation delay in a RO based on GIZO TFTs (adapted from [26]).*

composed by two transistors (control and load), as depicted in Figure 7.6. When the input voltage (V_{IN}) is low the control transistor is in the *off*-state and the load transistor pulls the output (V_{OUT}) to a "high" level. When V_{IN} is high, current flows through the control transistor, causing the output to be at a "low" level. A RO consists of any odd number of inverters connected in series. The evaluation of the performance of these simple circuits can provide a valuable tool used to define the range of applications of a TFT technology. For instance, the propagation delay of a RO is a widely accepted benchmark regarding the speed at which a TFT can operate [26].

To the authors' knowledge, the first transparent circuits with oxide semiconductors were reported in 2006 by Presley *et al.* [27]. The authors reported sputtered IGO TFT based inverters and five-stage ROs on glass. Figure 7.7a shows the voltage transfer characteristic (VTC) of the inverters, which present a peak gain magnitude (dV_{OUT}/dV_{IN}) of ≈1.5. A maximum oscillation frequency of 9.5 kHz and a propagation delay of 11 μs/stage are achieved for the RO. These values are significantly affected by the large parasitic capacitance that arises due to the large source/gate and drain/gate overlap, used here to facilitate device fabrication.

In 2007 finer design rules and improved TFT performance allowed Ofuji *et al.* [26] to report on five-stage ROs based on GIZO TFTs with considerably improved performance, such as an oscillation frequency of 410kHz and a propagation delay of 0.24μs/stage (Figure 7.7b). The authors report that the propagation delay achieved is half and almost one third of that achieved using a-Si:H and organic TFTs, respectively.

In 2008 the fastest oxide circuits on glass were reported by Sun *et al.* [28]. ALD deposited ZnO TFTs processed at 200°C were used to fabricate a seven-stage RO that operates at 2.3MHz, corresponding to a propagation delay of only 31ns/stage.

In 2011, eleven-stage ROs were also fabricated in flexible substrates by Mativenga *et al.* [29], requiring maximum processing temperatures of 220°C [29]. These operate at a frequency of 94.8kHz with a propagation delay of 0.48μs/stage. The authors also reported a flexible two-clock shift register (gate driver) with AC operation. Based on gate bias stress measurements, an estimated lifetime of more than 10 years was announced for the gate driver, which is highly promising for flexible display applications.

Besides TFTs, ZnO metal-semiconductor field-effect transistors (MESFETs) were used to fabricate integrated inverters, using Ag_xO Schottky diodes as level shifters, as reported by Frenzel *et al.* [30] in 2010. For this work, ZnO was grown by pulsed laser deposition (PLD) on sapphire substrates at 675°C. Very low operation voltage is achieved with MESFET technology, with the inverters showing high gain values up to 197 at 3V. The additional diode as second input allows implementing the logic NOR-function in the circuits. Oxide MESFETs seem to be quite promising for fast and low power consumption transparent integrated circuits, and GIZO was used to fabricate MESFETs with excellent electrical properties ($S = 69\,mV\,dec^{-1}$ and $\mu_{FE} = 15\,cm^2\,V^{-1}\,s^{-1}$) with a maximum temperature of 150°C [31], although GIZO MESFET inverters were not demonstrated.

Based on these initial results, a similar scenario to that depicted for displays appears to arise regarding oxide-based circuits: transparency can be an added-value property for specific applications, but the main focus of research is on exploring the great electrical properties and low processing temperatures of oxide TFTs, that enable low cost and high performance circuit fabrication, even in flexible substrates.

7.3.2 The Introduction of Oxide CMOS

All the circuit applications reported in section 7.3.1 make use of only n-type oxide TFTs, i.e. are based on NMOS logic. However, both n- and p-type oxide TFTs are required if the fabrication of complementary metal oxide semiconductor (CMOS) circuits is envisaged. CMOS technology offers great advantages over NMOS, especially with regard to power dissipation and higher density of logic functions on a chip.

Initial results for transparent CMOS inverters were obtained at CENIMAT,[c] by using the sputtered GIZO TFTs and SnO TFTs presented in Chapter 5, with multilayer hafnium-based dielectric and IZO electrodes. At this initial stage, a passivation layer was not used. After production, the CMOS inverters were annealed at 200°C for 30min in air. Figure 7.8 (see color plate section) shows a 2.5×2.5cm glass substrate with several transparent CMOS inverters.

[c] This work was done in collaboration with Dr. Sang-Hee Ko Park and Dr. Chi-Sun Hwang from ETRI (Korea).

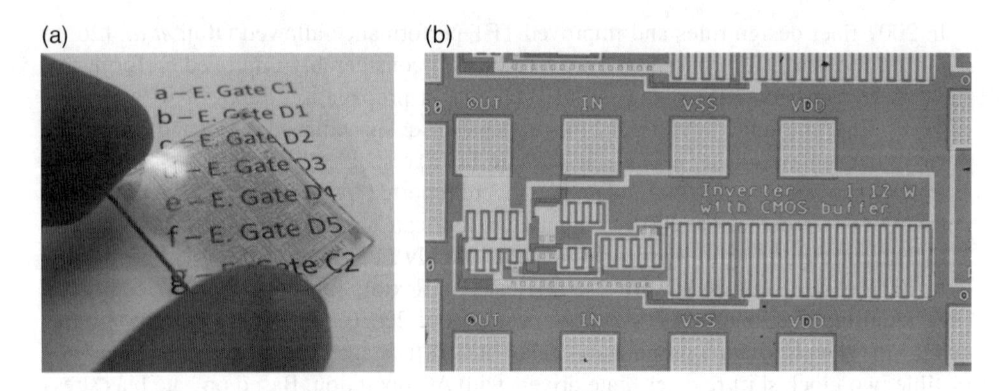

Figure 7.8 *a) Transparent CMOS inverters on glass; b) optical micrograph of an oxide CMOS inverter.*

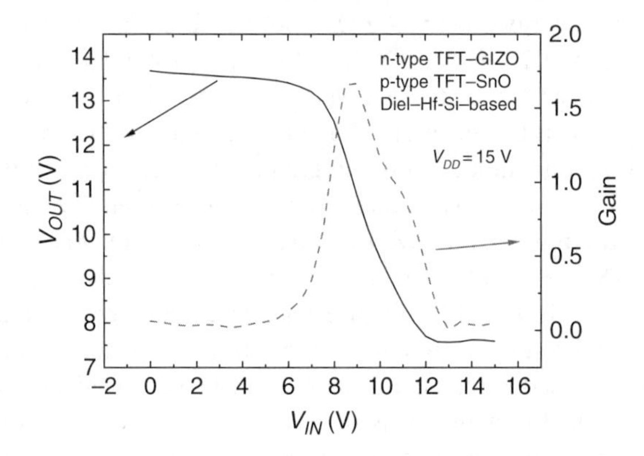

Figure 7.9 *VTC and gain for a transparent CMOS inverter with n-type GIZO and p-type SnO TFTs fabricated at CENIMAT.*

The VTC for a transparent CMOS inverter is presented in Figure 7.9. The peak gain is ≈1.7, being that a gain magnitude above 1 is required in order to sustain signal propagation in integrated circuits using inverters, such as ROs. However, even for low V_{IN} the p-type TFT has some non-negligible current flowing between source and drain, causing $V_{OUT}<V_{DD}$. The fact that the p-type TFT cannot be entirely switched off also contributes to the fact that for larger V_{IN} we have $V_{OUT}>0$ V. Greater detailed knowledge regarding how to control the subgap density of states in p-type oxide semiconductors will certainly allow one to improve device and circuit performance, but even at this initial stage these results already prove that transparent CMOS circuit design is achievable using transparent oxide semiconductors with maximum (post-)processing temperatures of 200°C.

Reports on oxide semiconductor based CMOS are fairly recent. Given that p-type oxide TFTs are still at the initial stage of research, the first CMOS inverters were based on hybrid solutions, comprising n-type oxide TFTs and p-type organic TFTs. Dhanajay *et al.* [32] reported in 2008 on a hybrid CMOS inverter in Si substrates based on ambipolar TFTs comprising a stack of In_2O_3 and pentacene as the channel layer. Ambipolar behavior

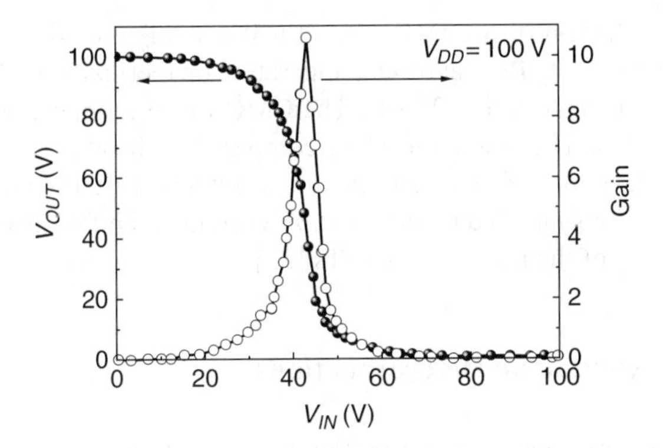

Figure 7.10 *VTC and gain for a CMOS inverter with p-type* SnO_x *and n-type* In_2O_3 *TFTs. Reprinted with permission from [33] Copyright (2008) American Institute of Physics.*

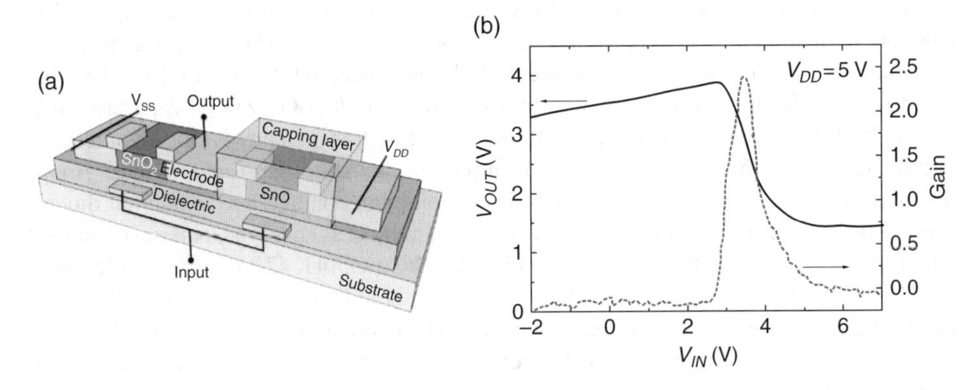

Figure 7.11 *Schematic of a conceptual CMOS inverter using n-type* SnO_2 *and p-type SnO TFTs (adapted from [34]); b)VTC for an ambipolar SnO CMOS inverter. Adapted from [35] Copyright (2011) Wiley-VCH.*

depends critically on the thickness and morphology of the In_2O_3 layer. Two identical ambipolar TFTs, fabricated with a maximum temperature of 750°C, were used to build a CMOS inverter that exhibits a peak gain of 9. However, in 2008, the same authors [33] reported for the first time fully oxide based CMOS inverters, by combining p-type SnO_x and n-type In_2O_3 TFTs, being the channel layers produced by reactive evaporation. The inverter operates at fairly high voltages and exhibits a peak gain of ≈11 (Figure 7.10).

In 2010, Yabuta *et al.* [34] reported the formation of both n-type SnO_2 TFTs and p-type SnO TFTs in the same substrate, by using a SiO_x capping layer on top of sputtered SnO films, which allows one to fine-tune the oxygen content of the SnO film during an annealing process (300°C), which is crucial to obtaining p-type behaviour in SnO. A conceptual design of a CMOS inverter based on this production method was presented (Figure 7.11a, see color plate section), but no actual CMOS devices were fabricated.

The ambipolar effect observed in SnO films was successfully applied to a CMOS inverter by Nomura *et al.* [35] in 2011. But in this case, rather than using organic and oxide semiconductors as in [32], only SnO was used to achieve the ambipolar behaviour, making this the first demonstration of a CMOS circuit using a single oxide semiconductor. The SnO films were deposited by PLD at room temperature on thermally oxidized Si substrates, the final devices being annealed at 250°C. The CMOS inverter operates with a peak gain of ≈2.5 (Figure 7.11b). The low gain when compared with other ambipolar transistors containing channel layer of different materials such as microcrystalline silicon and organics (voltage gains of 5–20) is justified by the large *off*-current and imbalance of the SnO TFTs, caused by the low μ_{FE} of the n-type SnO TFTs [35].

7.4 Oxide Semiconductor Heterojunctions

7.4.1 Oxide Semiconductor Heterojunctions in the Literature

Throughout this book we have centered our discussion on oxide materials for TFTs application. Nevertheless, transparent circuits are not made only of transistors and other active or passive components such as thin film capacitors, resistors or diodes are needed. The latter are particularly interesting as they can also work as photodiodes or photoemitters.

As far as the development of transparent diodes is concerned, Sato *et al.* [36] were the first to report the formation of a semi-transparent p-NiO/i-NiO/i-ZnO/n-ZnO structure, with only 20% visible transparency. Although this low transparency is not favorable for "transparent electronics" it represented an important breakthrough. Later, several groups reported the fabrication of p–i–n as well as p–n homojunction and heterojunction diodes. Zinc oxide has been the most studied material on this subject as an n-type material combined with p-type materials such as delafossites [37–40], $SrCu_2O_2$ [41, 42] or nickel oxide (NiO) [43–46]. However, the processing of these devices is still a challenge, as high quality material and interfaces are required, namely when photoemitters are aimed. So, those diodes that are based entirely on oxides normally use high temperature processed layers, typically above 250°C.

If thin film oxide diodes are aimed at transparent electronic circuits one key parameter is the forward/reverse current ratio. If we focus ourselves on heterojunctions produced in

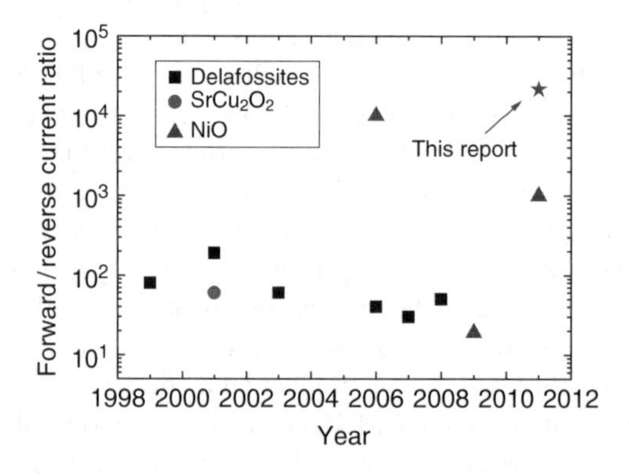

Figure 7.12 *Evolution of the forward/reverse current ratio on thin film all-oxide heterojuntions produced on glass substrates* [37–41, 44–48].

glass substrates, the values found in the literature hardly exceed 100 (Figure 7.12, see color plate section). Only two papers were found reporting values above 1000, by Vygranenko *et al.* [46], and more recently, by Mistry *et al.* [44], both on ZnO/NiO heterojunctions.

As emphasized throughout this book, the reduction of the processing temperature is one important goal in order to allow for low-cost substrates and move towards flexible electronic circuits. Among the data plotted in Figure 7.12, only the heterojunction reported by Vygranenko *et al.* was done without involving intentional substrate heating. We have already seen in Chapter 5 that GIZO sputtered at low temperatures is a suitable semiconductor for TFTs. However, this material will be much more relevant for "transparent electronics" if other devices are possible, such as diodes.

7.4.2 GIZO Heterojunctions Fabricated at CENIMAT

Sputtering was used to produce GIZO/NiO heterojunctions in glass substrates with the structure presented in Figure 7.13.[d] All the layers were deposited without any intentional substrate heating. The bottom electrode is IZO while the top one is palladium (Pd) in order to ensure a small difference in work function with NiO, that has a value around 5.3 eV [49]. The selection of the top electrode is very important in order to avoid a rectifying electrical contact in p-type material. Gold (Au) or platinum (Pt) are also suitable options, but of course transparent electrodes are desirable for completely transparent devices.

The current density-voltage (J–V) characteristics of these heterojunctions are presented in Figure 7.14. In order to improve the rectifying behavior of the devices, the surface of GIZO must be plasma treated. The relevance of this process is evident as good rectifying characteristics can be obtained with a forward/reverse current ratio around 3×10^4 and a threshold voltage of 0.67 V [48]. To our knowledge this is the highest value found in heterojunctions of Zn-based oxide/NiO. If the plasma treatment is not used the current ratio hardly exceeds 10.

As shown in Figure 7.15 (see color plate section) the top electrode of these heterojunctions is not totally transparent. In the future this must be replaced by a transparent one with good work function matching with NiO. Nevertheless, all the other layers are transparent with a total average transmittance in the visible range above 60 %.

This proves that not only TFTs but also other devices can be produced using oxide semiconductors produced at low temperature. Another remarkable point is that these junctions are able to support high currents, as required for invisible transparent rectifiers. This opens the way for something that is no longer a dream: environmentally friendly materials and processes for transparent devices, transparent electronics and a sustainable transparent future for mankind.

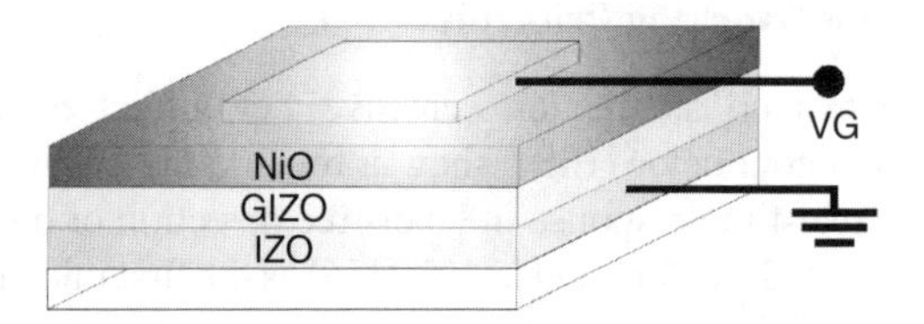

Figure 7.13 Structure of the GIZO/NiO heterojunction on glass substrates.

[d] This work was done within the framework of a research project with Saint-Gobain Recherche (France).

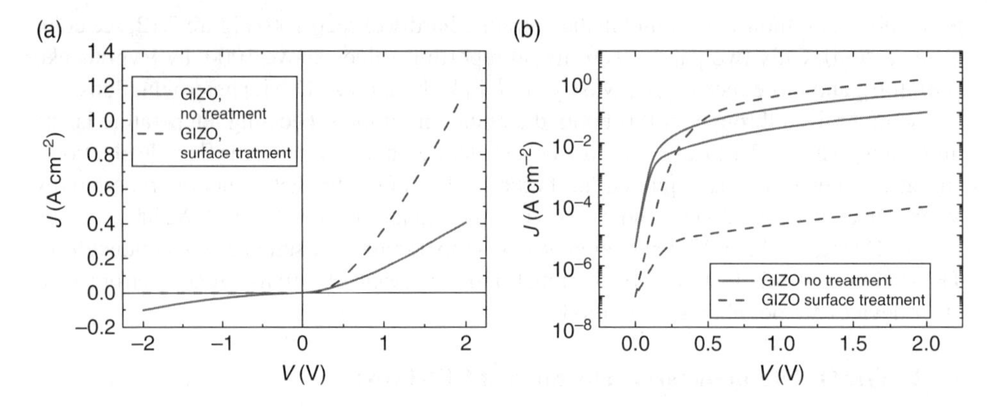

Figure 7.14 J–V characteristics of GIZO/NiO heterojunction highlighting the importance of GIZO surface treatment: a) normal plot and b) J–V characteristics reduced to the first quadrant where is possible to confirm the forward/reverse ratio above 10^4 if plasma treatment is used [internal report].

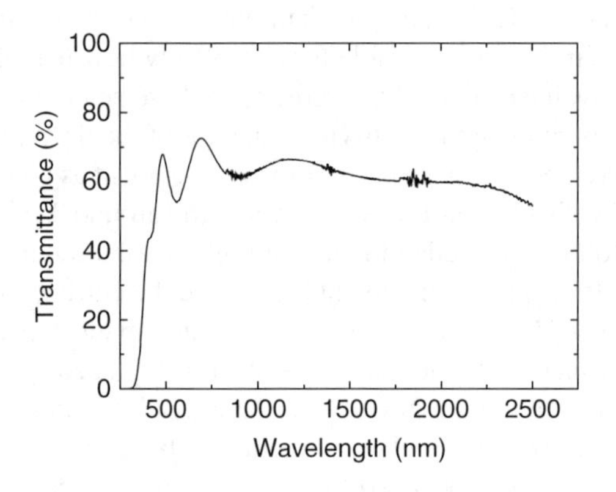

Figure 7.15 Total transmittance of the stack glass/IZO/GIZO/NiO.

7.5 Field Effect Biosensors

7.5.1 Device Types and Working Principles

Field effect biosensors, such as the ion sensitive field-effect transistor (ISFET) and electrolyte-insulator-semiconductor (EIS), show a remarkable capability for miniaturization and versatility for real-time response in label-free detection of biomolecules [50–54].

The ISFET was introduced by Bergveld in 1970 [55] as the first miniaturized silicon-based chemical sensor. The ISFET's structure and operating principles are similar to a MOSFET (metal oxide semiconductor field-effect transistor) where an electrolyte and a reference electrode substitute for the gate contact. EIS devices mimic the gate structure of ISFETs and are similar to MOS capacitors, nonetheless the sensing mechanism is identical for both field-effect based sensors. The gate dielectric is in direct contact with the solution and hence the threshold voltage (ISFET) or flat-band potential (EIS) modulation is dependent on electrolyte composition. According to the site-binding model [56], the oxide's surface is assumed to be amphoteric, meaning that the ionisable surface hydroxyl groups can take up or release protons working as a hydrogen ion buffer. The surface's protonation state and hence the surface potential vary with the pH of the solution and so the threshold or flat band voltage modulation is due to the potential oxide/solution interface. The relationship between this interfacial potential and the solution's pH is determined by the buffer capacity of the oxide surface. In principle, all oxides commonly used as gate dielectrics are pH sensitive, nevertheless Ta_2O_5 has a higher surface buffer capacity which is ideal for achieving maximum sensitivity [50].

The simplest ISFET is pH sensitive, nevertheless bio-sensitivity and specificity can be achieved with functionalization of the gate insulator with a biological layer.

Enzyme-modified sensors (EnFETs) can be constructed by coating the gate insulator with an enzyme layer. The detection mechanism is pH-based, thus it is required that the reaction of the enzyme with the analyte induces a local change of pH, either by consuming or generating protons. When the enzymatic reaction takes place, the underlying ISFET measures the local change of pH, which can be directly correlated with the concentration of analyte in solution. A wide range of EnFETs have been constructed and the most commonly used enzymes are penicillinase for penicillin detection, glucose oxidase for glucose and urease for urea, amongst others [57].

In DNA-modified sensors (DNAFETs) the detection of a target DNA molecule occurs via hybridization to single stranded DNA (ssDNA) probes functionalized on the sensors' surface. In this case the sensing mechanism is based on the intrinsic molecular charge of DNA that arises from its phosphate backbone. Upon hybridization an increase of negative surface charge induces a voltage shift directly correlated with the concentration of complementary DNA strands in solution.

The diversity of biological functionalization allows for the production of sensors for ions of biological interest (other than protons), enzyme-modified sensors and DNA sensors. The area of application for this type of sensors is therefore limitless, embracing health care, environmental, food and quality control, process monitoring and even bioterrorism.

Given that most of these biosensors rely on silicon technology, the migration to oxide semiconductor based field effect devices brings the added benefits of low production temperature, low cost materials and substrates and lower environmental impact in addition to the versatility of this type of sensors.

7.5.2 Oxide-Based Biosensors Fabricated at CENIMAT

Sputtered GIZO ISFETs have been produced using similar process flows and materials to those described in Chapter 5. The major modification to the conventional TFT structure is the use of an extended-gate electrode (Figure 7.16) to ensure that only the sensitive area is in contact with the electrolyte, having the added advantage of protecting the device from exposure to liquid solutions that can affect its performance [58].

The pH sensitivity of ISFETs is assessed by applying a V_G with respect to the reference electrode (V_{REF}) while maintaining a constant V_D. The transfer (I–V) characteristics of these sensors are thus obtained and sensitivity is assessed by the threshold voltage shift (ΔV_T) induced by pH variations. ISFETs based on GIZO have yielded promising results for pH sensing, showing a 40 mV/pH sensitivity in a pH range between 4 and 10 (Figure 7.17).

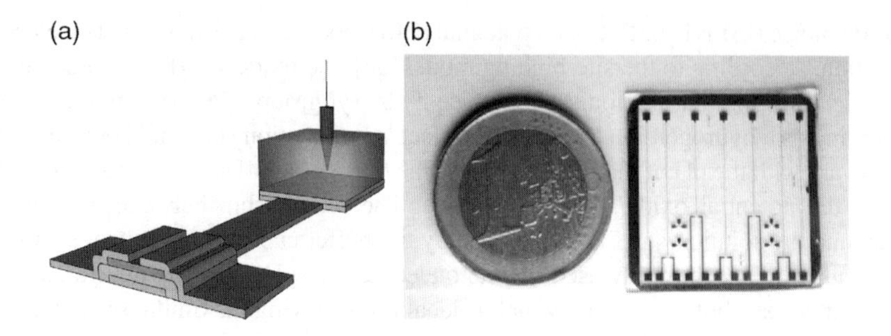

Figure 7.16 a) Schematic of an extended-gate GIZO ISFET; b) Photograph of a 2.5×2.5 cm substrate with several ISFETs produced at CENIMAT.

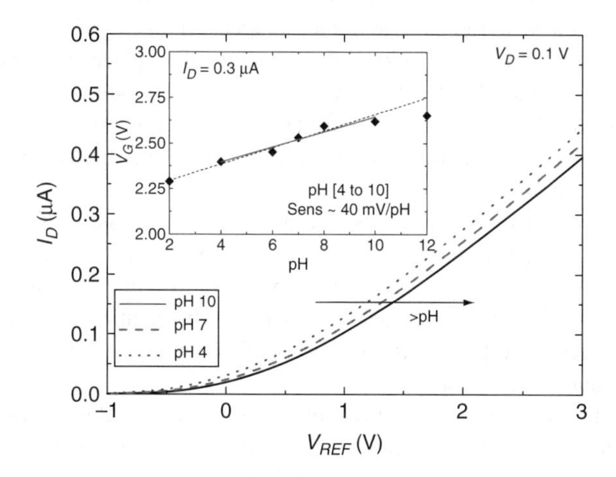

Figure 7.17 Transfer characteristics of GIZO ISFETs at V_D=0.1 V for different pH solution. The inset shows the sensitivity to pH, defined as V_G at I_D=0.3 μA.

EIS oxide based sensors have also been produced at CENIMAT, using Si wafers and the tantalum-based sputtered dielectrics discussed in Chapter 4. These oxide EIS show interesting results for pH, penicillin and DNA detection. The sensitivity of EIS-based structures is assessed by electrochemical impedance measurements. A bias voltage with a superimposed ac voltage of fixed frequency is applied with respect to the reference electrode. The capacitance–voltage (C–V) characteristics of these sensors are thus obtained and sensitivity is assessed by the flat band voltage shift induced by electrolyte variations.

A high pH sensitivity of 59 mV pH^{-1} in a 2 to 12 pH range was obtained for EIS sensors with sputtered Ta_2O_5 sensitive layer annealed at 200°C. This value is almost the maximum expected voltage shift induced by one pH unit variation (59.5 mV at 25°C), as calculated according to the fundamental electrochemistry Nernst equation [55]. Functionalization of the sensor with a physically adsorbed penicillinase layer yields penicillin EnFET. When the enzymatic reaction occurs, penicillin is decomposed by penicillinase and protons are produced, which results in a pH decrease that is directly proportional to penicillin concentration in solution. A sensitivity of 29 mV mM^{-1} was obtained in a penicillin concentration range up to 5 mM (Figure 7.18).

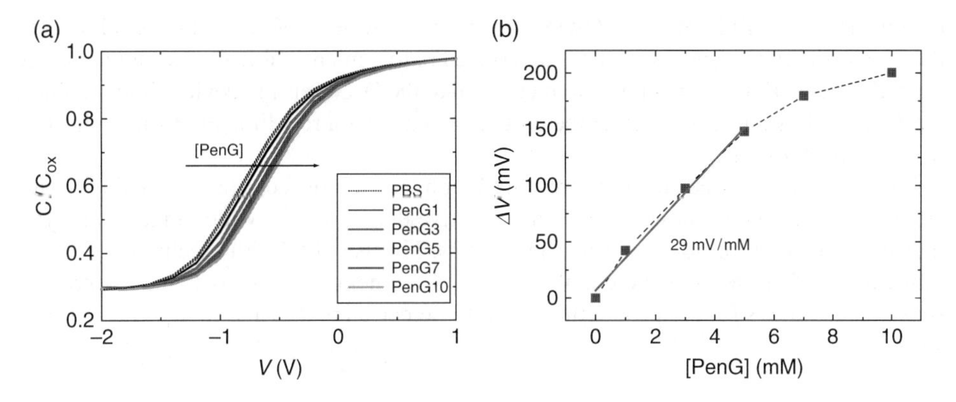

Figure 7.18 *a) C-V response and b) sensitivity of an EIS sensor. Penicillinase was immobilized directly on the Ta_2O_5 films and the curves were obtained for different penicillin G concentrations (1 mM<[PenG]<10 nM) in a 50mM phosphate buffer pH 7 (PBS).*

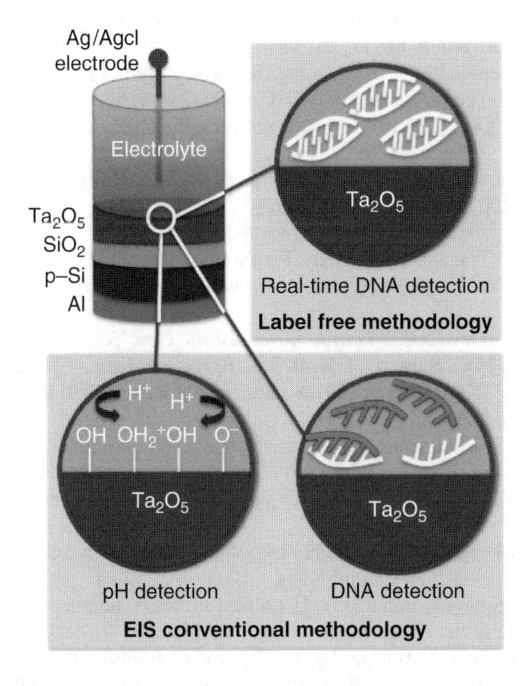

Figure 7.19 *Representation of the Ta_2O_5 EIS sensor structure and comparison of conventional and label free detection methodologies. Adapted from [59] Copyright (2011) Elsevier Ltd.*

Recently, we also reported the fabrication of an intrinsic charge based EIS sensor with these sputtered Ta_2O_5 films, for label free quantitation of DNA (Figure 7.19) [59]. The increase of DNA concentration induces a net charge increase that is detected by the Ta_2O_5 EIS sensor. The use of the developed sensor coupled to Polymerase Chain Reaction (PCR) for real-time quantification of DNA was demonstrated without the need for additional labeling and/or reporter molecules. Amplification of c-MYC, a proto-oncogene that is amplified in most human cancers [60], could be followed in real-time by monitorization of DNA amplification at each PCR cycle. Owing to the small size, ease of fabrication and low-cost, the developed Ta_2O_5 sensor may be incorporated onto a microfluidic device and then used for real-time PCR. Optimization and integration of the DNA sensing devices into a suitable platform could significantly lower the costs associated with real-time PCR for application in cancer diagnostics.

Based on the initial results presented above, field-effect devices using oxide technology have been shown to provide sensitive detection of a wide range of analytes that are of interest for point-of-care diagnostics. The use of thin film technology for the fabrication of these sensors enables the development of low-cost and disposable sensors and allows for the production of multi detection platforms, such as lab-on-chip systems.

References

[1] http://www.displaysearch.com/cps/rde/xchg/displaysearch/hs.xsl/090219_oled_display_revenues_expected_to_reach_6b_in_2015.asp.
[2] http://www.displaysearch.com/cps/rde/xchg/displaysearch/hs.xsl/6138.asp.
[3] http://techon.nikkeibp.co.jp/article/HONSHI/20070424/131608/.
[4] D. Graham-Rowe (2007) Electronic paper rewrites the rulebook for displays, *Nat Photon* **1**, 248–51.
[5] http://www.ciol.com/Semicon/Biz-Watch/News-Reports/Flexible-displays-to-expand-by-factor-of-35-from-2007-13/10608106900/0/.
[6] http://www.displaybank.com/eng/info/show.php?c_id=2312.
[7] DisplayBank (2011) Transparent Display Technology and Market Forecast.
[8] J.-H. Lee, D.N. Liu, and S.-T. Wu (2008) *Introduction to Flat Panel Displays*. Chichester: John Wiley & Sons, Ltd.
[9] C.R. Kagan and P. Andry (2003) *Thin-film transistors*. New York: Marcel Dekker, Inc.
[10] H.N. Lee, J. Kyung, M.C. Sung, D.Y. Kim, S.K. Kang, S.J. Kim, C.N. Kim, H.G. Kim, and S.T. Kim (2008) Oxide TFT with multilayer gate insulator for backplane of AMOLED device, *J. Soc. Inf. Disp.* **16**, 265–72.
[11] A. Bernanose (1955) Electroluminescence of organic compounds, *British Journal of Applied Physics*, S54–S56.
[12] C.W. Tang and S.A. Van Slyke (1987) Organic electroluminescent diodes, *Applied Physics Letters* **51**, 913–15.
[13] F. Zhu (2009) OLED Activity and Technology Development, in *Symposium on Sustainability Driven Innovative Technologies* Hong Kong.
[14] D. Redinger and V. Subramanian (2007) High-performance chemical-bath-deposited zinc oxide thin-film transistors, *IEEE Trans. Electron Devices* **54**, 1301–7.
[15] G. Goncalves, V. Grasso, P. Barquinha, L. Pereira, E. Elamurugu, M. Brignone, R. Martins, V. Lambertini and E. Fortunato (2011) Role of Room Temperature Sputtered High Conductive and High Transparent Indium Zinc Oxide Film Contacts on the Performance of Orange, Green, and Blue Organic Light Emitting Diodes, *Plasma Process. Polym.* **8**, 340–5.
[16] P. Barquinha, L. Pereira, G. Goncalves, R. Martins, D. Kuscer, M. Kosec and E. Fortunato (2009) Performance and Stability of Low Temperature Transparent Thin-Film Transistors Using Amorphous Multicomponent Dielectrics, *J. Electrochem. Soc.* **156**, H824–H831.
[17] J.S. Park, W.-J. Maeng, H.-S. Kim, and J.-S. Park (2011) Review of recent developments in amorphous oxide semiconductor thin-film transistor devices, *Thin Solid Films*, in press.
[18] H.-N. Lee, J. Kyung, M.-C. Sung, D.Y. Kim, S. K. Kang, S.-J. Kim, C.N. Kim, H.-G. Kim and S.-T. Kim (2008) Oxide TFT with multilayer gate insulator for backplane of AMOLED device, *J. Soc. Inf. Disp.* **16**, 265–72.

[19] M. Ito, M. Kon, C. Miyazaki, N. Ikeda, M. Ishizaki, R. Matsubara, Y. Ugajin, and N. Sekine (2008) Amorphous oxide TFT and their applications in electrophoretic displays, *physica status solidi (a)* **205**, 1885–94.
[20] S.-H. K. Park, C.-S. Hwang, M. Ryu, S. Yang, C. Byun, J. Shin, J.-I. Lee, K. Lee, M. S. Oh, and S. Im (2009) Transparent and Photo-stable ZnO Thin-film Transistors to Drive an Active Matrix Organic-Light- Emitting-Diode Display Panel, *Advanced Materials* **21**, 678–82.
[21] http://www.zdnet.com/blog/gadgetreviews/samsung-unveils-70-inch-ultra-definition-240hz-3dtv/19779.
[22] S.C. Chiang, C.C. Yu, F.W. Chang, S.W. Liang, C.H. Tsai, B.C. Chuang, and C.Y. Tsay (2008) $Zn_{0.97}Zr_{0.03}O$-TFT fabricated by sol-gel method and its application for active matrix LCDs, in *2008 Sid International Symposium, Digest of Technical Papers, Vol Xxxix, Books I-Iii*. vol. 39 Playa Del Rey: Soc Information Display, 1188–91.
[23] M.K. Ryu, K. Park, J.B. Seon, J. Park, I. Kee, Y. Lee, and S.Y. Lee (2009) "AMOLED Driven by Solution-Processed Oxide Semiconductor TFT," in *2009 Sid International Symposium Digest of Technical Papers, Vol Xl, Books I–Iii* Campbell: Soc Information Display, 188–90.
[24] D.A. Keszler (2009) Solution processing for transparent electronics, presented at E-MRS Spring Meeting 2009, Symposium F – Advances in transparent electronics: from materials to devices III.
[25] http://www.gizmag.com/samsungs-transparent-lcd-display/18283/picture/132495/.
[26] M. Ofuji, K. Abe, H. Shimizu, N. Kaji, R. Hayashi, M. Sano, H. Kumomi, K. Nomura, T. Kamiya, and H. Hosono (2007) Fast thin-film transistor circuits based on amorphous oxide semiconductor, *IEEE Electron Device Lett.* **28**, 273–5.
[27] R.E. Presley, D. Hong, H.Q. Chiang, C.M. Hung, R.L. Hoffman, and J.F. Wager (2006) Transparent ring oscillator based on indium gallium oxide thin-film transistors, *Solid-State Electron.* **50**, 500–3.
[28] J. Sun, D.A. Mourey, D.L. Zhao, S.K. Park, S.F. Nelson, D.H. Levy, D. Freeman, P. Cowdery-Corvan, L. Tutt, and T.N. Jackson (2008) ZnO thin-film transistor ring oscillators with 31-ns propagation delay, *IEEE Electron Device Lett.* **29**, 721–3.
[29] M. Mativenga, M. H. Choi, J. W. Choi, and J. Jang (2011) Transparent Flexible Circuits Based on Amorphous-Indium-Gallium-Zinc-Oxide Thin-Film Transistors, *IEEE Electron Device Lett.* **32**, 170–2.
[30] H. Frenzel, A. Lajn, M. Brandt, H.V. Wenckstern, G. Biehne, H. Hochmuth, M. Lorenz, and M. Grundmann (2008) ZnO metal-semiconductor field-effect transistors with Ag-Schottky gates, *Applied Physics Letters* **92**, 192108-1 - 192108-3.
[31] M. Lorenz, A. Lajn, H. Frenzel, H. V. Wenckstern, M. Grundmann, P. Barquinha, R. Martins, and E. Fortunato (2010) Low-temperature processed Schottky-gated field-effect transistors based on amorphous gallium-indium-zinc-oxide thin films, *Applied Physics Letters* **97**, 243506-1–243506-3.
[32] Dhananjay, C.W. Ou, C.Y. Yang, M.C. Wu, and C.W. Chu (2008) Ambipolar transport behavior in $In_{(2)}O_{(3)}$/pentacene hybrid heterostructure and their complementary circuits, *Applied Physics Letters* **93**, 033306-1–033306-3.
[33] Dhananjay, C.W. Chu, C.W. Ou, M.C. Wu, Z.Y. Ho, K.C. Ho, and S.W. Lee (2008) Complementary inverter circuits based on p-$SnO_{(2)}$ and n-$In_{(2)}O_{(3)}$ thin film transistors, *Applied Physics Letters* **92**, 232103-1–232103-3.
[34] H. Yabuta, N. Kaji, R. Hayashi, H. Kumomi, K. Nomura, T. Kamiya, M. Hirano, and H. Hosono (2010) Sputtering formation of p-type SnO thin-film transistors on glass toward oxide complimentary circuits, *Applied Physics Letters* **97**, 072111-1–072111-3.
[35] K. Nomura, T. Kamiya, and H. Hosono (2011) Ambipolar Oxide Thin-Film Transistor, *Advanced Materials*, **23**, 3431–4.
[36] H. Sato, T. Minami, S. Takata, and T. Yamada (1993) Transparent conducting p-type NiO thin films prepared by magnetron sputtering, *Thin Solid Films* **236**, 27–31.
[37] A.N. Banerjee, S. Nandy, C.K. Ghosh, and K.K. Chattopadhyay (2007) Fabrication and characterization of all-oxide heterojunction p-$CuAlO_{2+x}$/n-$Zn_{1-x}Al_xO$ transparent diode for potential application in "invisible electronics", *Thin Solid Films* **515**, 7324–30.
[38] R.L. Hoffman, J.F. Wager, M.K. Jayaraj, and J. Tate (2001) Electrical characterization of transparent p-i-n heterojunction diodes, *Journal of Applied Physics* **90**, 5763–7.

[39] D.-S. Kim, T.-J. Park, D.-H. Kim, and S.-Y. Choi (2006) Fabrication of a transparent p–n heterojunction thin film diode composed of p-$CuAlO_2$/n-ZnO, *physica status solidi (a)* **203**, R51–R53.
[40] A. Banerjee and K. Chattopadhyay (2008) Electro-optical properties of all-oxide p-$CuAlO_2$/n-ZnO: Al transparent heterojunction thin film diode fabricated on glass substrate, *Central European Journal of Physics* **6**, 57–63.
[41] A. Kudo, H. Yanagi, K. Ueda, H. Hosono, H. Kawazoe, and Y. Yano (1999) Fabrication of transparent p-n heterojunction thin film diodes based entirely on oxide semiconductors, *Applied Physics Letters* **75**, 2851–3.
[42] H. Ohta, M. Orita, M. Hirano, and H. Hosono (2001) Fabrication and characterization of ultraviolet-emitting diodes composed of transparent p-n heterojunction, p-$SrCu_2O_2$ and n-ZnO, *Journal of Applied Physics* **89**, 5720–5.
[43] X. Chen, K. Ruan, G. Wu, and D. Bao (2008) Tuning electrical properties of transparent p-NiO/n-MgZnO heterojunctions with band gap engineering of MgZnO, *Applied Physics Letters* **93**, 112112-1 - 112112-3.
[44] B.V. Mistry, P. Bhatt, K.H. Bhavsar, S.J. Trivedi, U.N. Trivedi, and U.S. Joshi (2011) Growth and properties of transparent p-NiO/n-ITO (In_2O_3:Sn) p-n junction thin film diode, *Thin Solid Films* **519**, 3840–3.
[45] R.K. Gupta, K. Ghosh, and P.K. Kahol (2009) Fabrication and characterization of NiO/ZnO p-n junctions by pulsed laser deposition, *Physica E: Low-dimensional Systems and Nanostructures* **41**, 617–20.
[46] Y. Vygranenko, K. Wang, and A. Nathan (2006) Low leakage p-NiO/i-ZnO/n-ITO heterostructure ultraviolet sensor, *Applied Physics Letters* **89**, 172105-1 - 172105-3.
[47] K. Tonooka, H. Bando, and Y. Aiura (2003) Photovoltaic effect observed in transparent p-n heterojunctions based on oxide semiconductors, *Thin Solid Films* **445**, 327–331.
[48] L. Pereira (2011) *Internal Report*.
[49] T. Dutta, P. Gupta, A. Gupta, and J. Narayan (2010) High work function (p-type NiO_{1+x})/ $Zn_{0.95}Ga_{0.05}O$ heterostructures for transparent conducting oxides, *Journal of Physics D: Applied Physics* **43**, 105301-1 - 105301-7.
[50] P. Bergveld (2003) Thirty years of ISFETOLOGY: What happened in the past 30 years and what may happen in the next 30 years, *Sensors and Actuators B: Chemical* **88**, 1–20.
[51] J. Fritz, E.B. Cooper, S. Gaudet, P.K. Sorger, and S.R. Manalis (2002) Electronic detection of DNA by its intrinsic molecular charge, *Proc. Natl. Acad. Sci. U. S. A.* **99**, 14142–6.
[52] T.-W. Lin, D. Kekuda, and C.-W. Chu (2010) Label-free detection of DNA using novel organic-based electrolyte-insulator-semiconductor, *Biosensors and Bioelectronics* **25**, 2706–10.
[53] A. Poghossian, M.H. Abouzar, M. Sakkari, T. Kassab, Y. Han, S. Ingebrandt, A. Offenhäusser, and M.J. Schöning (2006) Field-effect sensors for monitoring the layer-by-layer adsorption of charged macromolecules, *Sensors and Actuators B: Chemical* **118**, 163–70.
[54] A. Poghossian, M.H. Abouzar, F. Amberger, D. Mayer, Y. Han, S. Ingebrandt, A. Offenhausser, and M.J. Schoning (2007) Field-effect sensors with charged macromolecules: Characterisation by capacitance-voltage, constant-capacitance, impedance spectroscopy and atomic-force microscopy methods, *Biosens. Bioelectron.* **22**, 2100–7.
[55] P. Bergveld (1970) Development of an Ion-Sensitive Solid-State Device for Neurophysiological Measurements, *Biomedical Engineering, IEEE Transactions on* **BME-17**, 70–1.
[56] D.E. Yates, S. Levine, and T.W. Healy (1974) Site-binding model of the electrical double layer at the oxide/water interface, *Journal of the Chemical Society, Faraday Transactions 1: Physical Chemistry in Condensed Phases* **70**, 1807–18.
[57] M.J. Schoning and A. Poghossian (2002) Recent advances in biologically sensitive field-effect transistors (BioFETs), *Analyst* **127**, 1137–51.
[58] F. Yan, P. Estrela, Y. Mo, P. Migliorato, H. Maeda, S. Inoue, and T. Shimoda (2005) Polycrystalline silicon ion sensitive field effect transistors, *Applied Physics Letters* **86**, 053901-1 - 053901-3.
[59] R. Branquinho, B. Veigas, J.V. Pinto, R. Martins, E. Fortunato, and P.V. Baptista (2011) Real-time monitoring of PCR amplification of proto-oncogene c-MYC using a Ta_2O_5 electrolyte-insulator-semiconductor sensor, *Biosensors and Bioelectronics*, **28**, 44–9.
[60] C.V. Dang, K.A. O'Donnell, K.I. Zeller, T. Nguyen, R.C. Osthus, and F. Li (2006) The c-Myc target gene network, *Semin. Cancer Biol.* **16**, 253–64.

Index

Transparent Oxide Electronics: From Materials to Devices, First Edition. Pedro Barquinha, Rodrigo Martins, Luis Pereira and Elvira Fortunato.